Applications of Chalcogenides: S, Se, and Te

Gurinder Kaur Ahluwalia
Editor

Applications of Chalcogenides: S, Se, and Te

Springer

Editor
Gurinder Kaur Ahluwalia
Materials and Nanotechnology
Research Laboratory
Department of Physics
College of The North Atlantic
Labrador City, NL, Canada

ISBN 978-3-319-41188-0 ISBN 978-3-319-41190-3 (eBook)
DOI 10.1007/978-3-319-41190-3

Library of Congress Control Number: 2016952256

Printed on acid-free paper

This Springer imprint is published by Springer Nature
The registered company is Springer International Publishing AG
The registered company address is: Gewerbestrasse 11, 6330 Cham, Switzerland

To My Son Rajatdeep

"Somewhere, something incredible is waiting to be known."

— Carl Sagan

Preface

The changing landscape of research from pure scientific ventures towards industrial applications in recent years has propelled the writing of this book on materials covering a wide spectrum of topics at the forefront of research such as optical fibers, electrochemical sensors, imaging and detection, infrared biochemical sensing, drug delivery and biomedical applications, data storage, photovoltaics, and infrared detectors. The industry roadmap is ready to lay out a research and development plan that is not centered around Moore's law, but will follow the More than Moore strategy. Rather than making the chips better and letting applications follow, it will start with applications—from smartphones and supercomputers to data centers in the cloud. Looking at the current topics of interest to the industry, government institutions, and society on the whole, I realized that looking beyond silicon, chalcogens (i.e., Group VI elements of the periodic table) form an integral part of materials required for all of these applications. The peculiar properties of chalcogens arise due to the lone pair orbitals forming the valence band as a result of their electronic configuration. At the same time dangling bonds play an important role to modify the electrical behavior of chalcogens in amorphous forms. The book focuses on the three main chalcogens Sulfur (S), Selenium (Se), and Tellurium (Te) and their compounds which find applications in a variety of devices used in the electronics and optoelectronics industry. At some places ZnO is discussed due to its importance in the field of application. Wherever applicable, a comparison with competing materials and technologies is given. A chapter on techniques for characterization presents some of the latest theoretical and experimental methods for materials characterization using synchrotron and optical methods. All the chapters include extensive referencing of the contents in order to facilitate further learning. The book is intended to be a useful addition to the learning experience for a wide spectrum of audience to include researchers at all stages of their careers as well as nonresearchers to satisfy their curiosity and quest for knowledge. Starting with fundamentals, basic principles, and properties of materials which make them suitable for various applications, the book reviews and presents several new and unpublished results by outstanding researchers in their respective fields, at the same time presenting issues to be addressed in the different fields which, in particular, young researchers may find interesting for further investigations. In view of current trends in materials towards nanostructured forms, chapters providing guidelines to go about these for medical and nonmedical applications have been included.

I extend my sincere gratitude to all the authors, who are experts in their respective fields, for sparing their valuable time and making almighty efforts to put up the best quality contributions in the book in a timely manner. My sincere thanks to several senior members of the scientific community who directly or indirectly provided input and participated in the making of this book. I acknowledge with thanks the valuable input from organizers and participants of various International and National conferences where the work was presented.

Thanks to all my coworkers at the College of the North Atlantic, for their enthusiastic support of my research activities, particularly Hailey in the library for getting the literature and Holly for assisting with typing parts of some of the chapters. I would like to thank all my friends, teachers, and family members, especially my parents and Mandeep and Rajat for their consistent patience and encouragement throughout the preparation of the book.

I extend my sincere thanks to the publisher Springer for their consistent and sincere efforts in bringing the book to its form.

Labrador City, NL, Canada
15 February 2016

Gurinder Kaur Ahluwalia

Contents

Contributors

Gurinder Kaur Ahluwalia, Ph.D. Materials and Nanotechnology Research Laboratory, Department of Physics, College of The North Atlantic, Labrador City, NL, Canada

Mandeep Singh Bakshi, Ph.D. Department of Natural and Applied Sciences, University of Wisconsin - Green Bay, Green Bay, WI, USA

Bruno Bureau, Ph.D. Glass and Ceramics, ISCR UMR-CNRS 6226, University of Rennes 1, Rennes, France

Pierre Lucas, Ph.D. University of Arizona, Tucson, AZ, USA

James J. Noël, Ph.D. Department of Chemistry, The University of Western Ontario, London, ON, Canada

Ranjan Patro, Ph.D. Physics Department, College of the North Atlantic, Happy Valley Goose Bay, NL, Canada

P.K. Shishodia, Ph.D. Department of Physics & Electronics, Zakir Husain Delhi College, University of Delhi, Delhi, India

Y.M. Yiu, Ph.D. Chemistry Department, University of Western Ontario, London, ON, Canada

Part I
Introduction

Chapter 1
Fundamentals of Chalcogenides in Crystalline, Amorphous, and Nanocrystalline Forms

Gurinder Kaur Ahluwalia

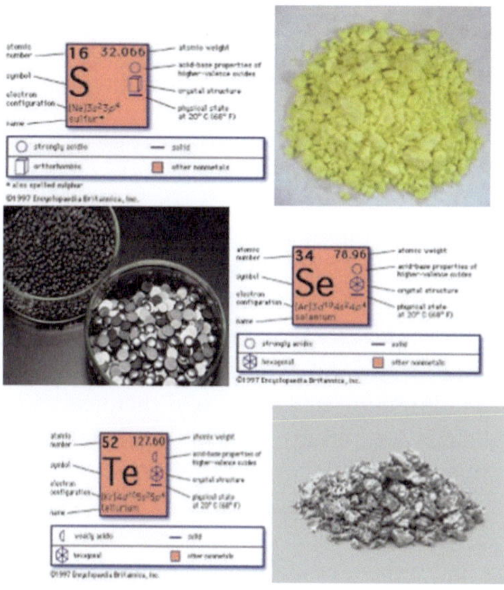

Chalcogens are Group 16 elements of the periodic table finding applications in a range of devices such as computer memories, telecommunications, chemical sensors, photo-detectors, and ionic sensors (Fig. 1.1).

G.K. Ahluwalia, Ph.D. (✉)
Materials and Nanotechnology Research Laboratory, Department of Physics College of The North Atlantic, Labrador West Campus, 1600 Nichols Adam Highway, Labrador City, A2V 0B8, NL, Canada
e-mail: gurinder.kaur@cna.nl.ca

G.K. Ahluwalia (ed.), *Applications of Chalcogenides: S, Se, and Te*,
DOI 10.1007/978-3-319-41190-3_1

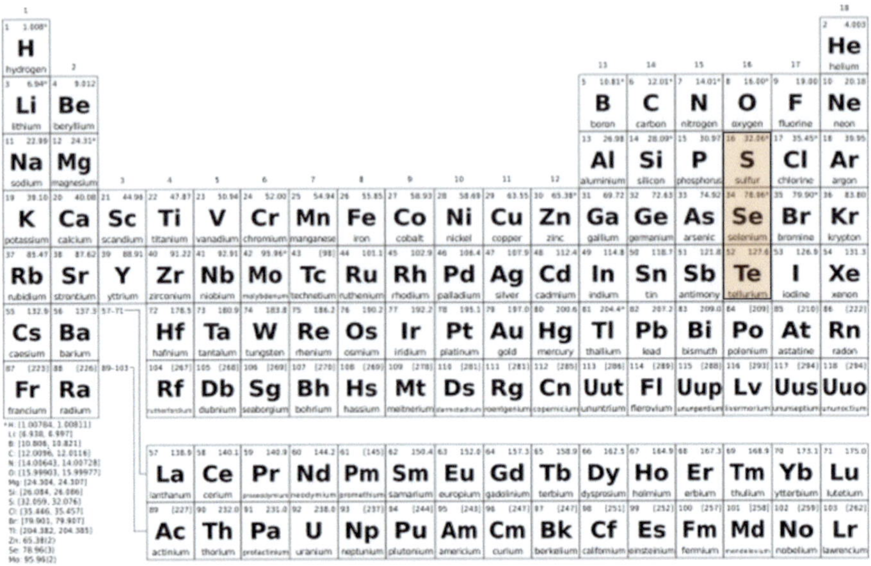

Fig. 1.1 Chalcogens are Group VI elements of the Periodic Table. This book deals with the three most important ones: sulfur, S; selenium, Se; and tellurium, Te

In particular sulfur (S), selenium (Se), and tellurium (Te) find applications in a variety of devices used in the electronics and optoelectronics industry. Sulfur based materials exhibit interesting properties such as a high refractive index, large Kerr nonlinearities, ability to be directly patterned by exposure to near band gap light, good transmission in the IR beyond 1.5 μm. In addition many chalcogenide glass compositions are non-hygroscopic and have good chemical resistance which makes them desirable candidates for various practical applications such as fiber Bragg gratings, fiber-optic communication, and evanescent wave fiber sensors. Selenium based materials find wide applications in rectifiers, solar cells, photographic exposure meters, xerography, and anticancer agents. It is also used in the glass industry to eliminate bubbles and remove undesirable tints produced by iron. In addition, selenium also has a high reactivity towards a wealth of chemicals that can be potentially exploited to convert selenium into other functional materials such as CdSe, ZnSe, Ag_2Se, and so on. Selenium based materials in amorphous form are used for imaging and biomedical applications. In crystalline form, their combination with other materials such as lead (Pb), Copper (Cu). and Indium (In) is widely used for photovoltaic and photo-detection applications. Se and Te are lone pair, polymeric divalent materials with chain structures. They can be designed to bond with cross-linking elements of different bond strengths such as Ge, Sb, and As. The cross-linked Te/Se alloys have a huge number of non-bonded lone pair electrons which could easily be excited by optical and electrical fields. When the amorphous phase cannot contain the excitation energy, a phase change to a crystalline phase occurs. Such polymeric lone pair structures are vibronic in nature wherein electronic transitions are made possible by vibrational motion of the chains. These simultaneous vibrational and electronic transitions form the basis of optical and electrical phase change memories. Tellurium based materials are widely used for data storage devices based on phase change characteristics exhibited by them. State-of-the-art microprocessors are now having transistors that are just 14 nanometers wide, and with Moore's law finally approaching the ultimate physical limits, nanostructuring of materials is at the forefront for all industrial applications.

1.1 Chalcogenide Materials and Their Classification

The word "chalcogen" means "ore forming" materials and is derived from a combination of the Greek word *khalkós* principally meaning copper (*ore or coin*), and the Latinized Greek word *genēs*, meaning *born* or *produced*. Chalcogenides are chemical compounds consisting of at least one chalcogen anion and at least one more electropositive element. Chalcogenides are sometimes referred to as the elements of the oxygen family and includes sulfides, selenides, and tellurides, and the radioactive element polonium (Po). Often, oxygen is treated separately from the other chalcogens, sometimes even excluded from the scope of the term "chalcogen" altogether, due to its very different chemical behavior from sulfur, selenium, tellurium, and polonium. Here, we focus on the applications of compounds formed by the three important chalcogens—sulfur (S), selenium (Se), and tellurium (Te). In particular, S-based materials find applications in optoelectronic devices. Se-based materials are more commonly used for imaging and biomedical applications. They also find applications for ion-selective sensors and modern day solar cells. Te-based materials are extensively used in memory devices based on phase change and Infrared detectors.

Chalcogen derivatives are known for all of the main group elements except the noble gases. In most cases their stoichiometries follow the classical valence trends, e.g., SiS_2, B_2S_3, and Sb_2S_3. Many exceptions exist however, e.g., P_4S_3, an ingredient in strike-anywhere matches. The structures of many of the main group materials are dictated, not by close packing, but by directional covalent bonding. The chalcogen is assigned positive oxidation states for halides, nitrides, and oxides. These compounds, e.g., SCl_2, are not usually classified as chalcogenides. Some specific categories are discussed below.

1.1.1 Alkali Metal and Alkaline Earth Chalcogenides

Alkali metal and alkaline earth mono-chalcogenides are salt like, colorless compounds often soluble in water. The sulfides tend to undergo hydrolysis to form derivatives containing disulfide anions. The alkali metal chalcogenides often crystallize with the antifluorite structure and the alkaline earth salts in the sodium chloride motif. Alkali heavy metal chalcogenides such as CsHgS and CsCdTe are promising candidates for nuclear radiation detectors in comparison to the conventionally investigated semiconductors such as Ge, HgI_2, CdTe, and $Cd_{1-x}Zn_xTe$ [1, 2] which suffer from limitations such as requirement for cryogenic cooling, lack of sufficient sensitivity to gamma rays despite impressive performance because of its low density and low atomic number [3]. Furthermore, while $Cd_{1-x}Zn_xTe$ is currently a leading material for room temperature radiation detectors, growth of large homogeneous crystals remains a challenge. Peters et al. [4] have found the band gap energies of the $Cs_2Hg_6S_7$ and $Cs_2Cd_3Te_4$ compounds to be 1.63 eV and 2.45 eV respectively. The mobility-lifetime products for charge carriers are of the order of ~10^{-3} cm^2/V for electrons and ~10^{-4} cm^2/V for holes. Detectors fabricated from the ternary compounds $Cs_2Hg_6S_7$ shows well resolved spectroscopic features at room temperature in response to γ-rays at 122 keV from ^{57} Co source, indicating its potential as a radiation detector.

1.1.2 Transition Metal Chalcogenides (TMCs)

Most widely investigated TMCs are zinc (ZnS, ZnSe ZnTe,), cadmium (CdS, CdSe, CdTe), MoS_2, and Mercury (HgCdTe) chalcogenides. Transition metal chalcogenides occur with many stoichiometries and many structures. The most important being simple stoichiometries, such as 1:1 and 1:2. TMCs exhibit highly covalent, non-ionic behavior indicated by their zinc-blende or wurtzite structure. In the first case, the sulfide atoms pack in cubic symmetry and the Zn^{2+} ions occupy half of the tetrahedral holes resulting in a diamondoid framework. In the second case, viz., the wurtzite structure, the main atom connectivities are similar (tetrahedral), but the crystal symmetry is hexagonal. A third motif for TMCs is the nickel arsenide lattice, where the metal and chalcogenides each have octahedral and trigonal prismatic coordination, respectively. This motif is commonly subject to non-stoichiometry. Among the monosulfides are some of the minerals and ores such as chalcopyrites (FeS), and pigments such as cadmium sulfide. These can further be categorized as *mono- and di-chalcogenides* on the basis of the bonding configurations with the transition metals. Metal *mono-chalcogenides* (MCs) have the formula MC, where "M" is a transition metal and "C" is one of the chalcogenides—S, Se, or Te.

Metal *Dichalcogenides* (MDCs) have the formula MC_2, where "M" is a transition metal and C=S, Se, or Te. Most common MDCs are insoluble diamagnetic sulfides, exhibiting semiconducting properties. The important ones of these include titanium disulfide investigated in prototype cathodes for secondary batteries, exploiting its ability to reversibly undergo intercalation by lithium. Molybdenum disulfide (also called Molybdenite is the main ore of Molybdenum) has been the subject of much investigations, is used as a solid lubricant and catalyst for hydro sulfurization. The corresponding ditellurides and diselenides are known and need more investigations for applications.

MDCs typically adopt either octahedral structures similar to CdI_2 or trigonal prismatic structures as that of MoS_2. The strong bonding between the metal and chalcogenide ligands, contrasts with the weak chalcogenide–chalcogenide bonding between the layers. These contrasting bond strengths enable these materials to engage in intercalation by alkali metals. The intercalation process is accompanied by charge transfer, reducing the M(IV) centers to M(III). In contrast to classical metal dichalcogenides, iron pyrite (FeS_2), a common mineral, is usually described as consisting of Fe^{2+} and the persulfido anion S_2^{2-}. The sulfur atoms within the disulfide dianion are bound together via a short S–S bond. Late transition metal disulfides such as Mn, Fe, Co, and Ni almost always adopt the pyrite or the related marcasite motif, in contrast to early metals such as V, Ti, Mo, and W, which adopt 4+ oxidation state with two chalcogenide dianions.

Several metals, mainly the early metals such as Ti, V, Cr, and Mn groups also form trichalcogenides. These materials are usually described as M_4^+ (C_2^{2-}) (C_2^-), where C=S, S, Te. A well-known example is niobium triselenide. Amorphous MoS_3 is produced by treatment of tetrathiomolybdate with acid:

$$MoS_4^{2-} + 2H^+ \rightarrow MoS_3 + H_2S$$

1.2 Nanostructured Chalcogenides

Nanocrystals (NCs) or quantum dots (QDs) of the IV–VI semiconductors PbS, PbSe, and PbTe provide unique properties for investigating the effects of strong confinement on electrons and phonons. Bulk PbSe and PbS materials have a cubic (rock salt) crystal structure and a narrow direct band gap (0.28–0.41 eV at 300 K), fourfold degenerated at the L-point of the Brillouin zone. Their high dielectric constant ($\varepsilon = 18.0$–24.0) and small electron and hole effective masses (<0.1 m*; m*: free electron mass) create excitons with a relatively large effective Bohr radii ($a_{B(PbSe)} = 46$ nm, eight times larger than that of CdSe). Interband optical studies of colloidal PbSe NCs reveal well-defined band-edge excitonic transitions tuning between 0.3 and 1.1 eV, a small Stoke's shift and sub-nanosecond and nanosecond lifetimes for the intraband and interband transitions, respectively [5–10]. An amplified spontaneous emission was demonstrated [10] from PbSe NCs with gain parameters similar to those observed in CdSe NCs [11].

In PbSe, the electron, hole, and exciton Bohr radii are 23, 23, and 46 nm, respectively. These large radii allow strong confinement to be achieved in relatively large structures. Thus, QDs of IV–VI materials may have properties reflecting all of the benefits of strong quantum confinement, with reduced influence from surface effects; for the same level of confinement as that of QDs of II–VI or III–V materials, the surface-to-volume ratio can be quite low in IV–VI materials. Furthermore, these large Bohr radii also allow for materials to be fabricated with much stronger quantum confinement than is possible with II–VI or III–V materials [12]. Indeed, studies of extremely confined IV–VI QDs have revealed that these materials have unique vibrational modes [13], can exhibit extremely weak electron–phonon coupling [14], have negligible exchange and Coulomb energies [15], and can have a temperature-independent band gap [16]. Lead-salt quantum dots are among the few materials that can provide size-quantized electronic transitions at technologically important infrared wavelengths. These structures may find use in optoelectronic applications as well as in biophysical applications such as fluorescence microscopy. Monodisperse PbS [17] and PbSe [18] QDs have been produced in glass hosts. However, some applications will benefit from the processability of colloidal solutions. The degree of confinement of charge carriers can be many times stronger than in most II–VI and III–V semiconductors, and lead salt nanostructures may be the only materials in which the electronic energies are determined primarily by quantum confinement.

1.3 Chalcogenide Glasses (ChGs)

Semiconducting glasses containing one or more of the chalcogens are called chalcogenide glasses (ChGs). They behave as semiconductors or more precisely, they exhibit amorphous semiconductor behavior with band gap energies from 1 to 3 eV. They are also referred to as lone-pair semiconductors, since in these; the valence band is derived from the lone pair electrons of the chalcogen atom and the conduction band from the antibonding states. These can further be divided into two groups:

(a) *Elements and compounds*: These materials possess considerable range of order extending locally in one or two dimensions. They consist of molecular complexes of larger size and are held together by weak van der Waal's forces. Some examples are Se-Te-S, As_2Se_3, and As_2S_3.
(b) *Cross-linked networks*: These materials are characterized by three dimensional cross-linked network structures and approximate more closely the ideal structural disorder. Their properties are quite readily affected by minute amounts of impurities. Some examples of this class are: Ge-Sb-Se, Ge-As-Se, Si-Ge-As-Te, and As-Se-Te.

ChGs can be easily prepared by (1) condensation from vapor and (2) cooling from the melt which renders them suitable for applications such as xerography and imaging, where large area films are required.

ChGs possess properties intermediate between those of organic polymers and oxide glasses. The atomic bonding structure is, in general more rigid than that of organic polymers and more flexible than that of oxide glasses. They exhibit excellent transmission in the near and infrared (IR) spectral region. They are high refractive index materials with a nonlinear refractive index typically 100 times that of silica, with a low maximum phonon energy.

The glasses have good solubility towards rare earth elements ions (REI) such as lanthanum. Their low phonon energy and high refractive index provide them with a low probability of multiphonon relaxation and high radiative transition probability of rare earth ions. In this way the materials have increased luminescence efficiency. In addition they have high chemical and thermal stability which allows the creation of easy to prepare optical devices. The high solubility of REIs in the chalcogenide host makes it possible to realize thin film technologies with high concentration of REIs in active media.

In contrast to amorphous silicon and the other group IV tetrahedrally bonded semiconductors, amorphous chalcogens and chalcogenides do not exhibit a dangling bond spin resonance signal in their clean, as prepared state. Pervasive diamagnetism rather than paramagnetism is the general rule. Chalcogens with their twofold coordination and lone pair orbitals offer the network the dual channels of structural and electronic flexibility for removal of dangling bonds. As a consequence, a typical chalcogenide has a relatively sharp optical absorption edge, single activation energy, and efficient photo-excited conductivity and luminescence. All these properties are characteristic of a well-defined and clean forbidden gap. Nevertheless, field-effect and doping experiments indicate that the Fermi level of a chalcogenide glass is nearly pinned. Luminescent emission occurs at sub band gap energies. Paramagnetism can be photo excited at low temperatures. The latter effects are normally characteristic of states in the gap and are explained in view of negative correlation energies of gap states, thereby leading to pairing of electrons with opposite spins. Distortions of the host network are possible so that bonded; paired electrons are favored over uncorrelated, gap filling, single electron states.

Furthermore, ChGs have high photosensitivity which allows preparation of writing channel waveguides. The writing channel is achieved by laser illumination of the glass at a wavelength close to near the band gap of the glass. As the glasses are photosensitive, it is possible to modify the refractive index of the glass in a controlled manner. This technique is promising because it bypasses the standard photolithographic processes needed to obtain channel waveguides. A very important advantage of using ChGs in integrated optics is their availability as thin films.

1.4 Structure, Bonding, and Band Structure

Band gap: Gray Se—1.74 eV; Te—0.33 eV; a-S_8—2.6 eV.

There are two basic arrangements in crystalline Se. The hexagonal form is stable at room temperature; the atoms in this structure are arranged in long parallel chains with threefold spiral about the *c-axis*-axis, with the three atoms in each unit of the spiral exhibiting D2 molecular symmetry (Fig. 1.2 a). The bonding between atoms in a chain is covalent, and the bonding between chains is of van der Waals type. The α and β monoclinic forms of Selenium have the same molecular unit, which consists of an eight atom puckered ring with D_{4d} molecular symmetry. Alternate atoms in the ring lie in two planes separated by about 1.3 Å, with the four atoms in each plane at the corners of a square. Bonding within the rings is covalent, and between the rings, van der Waals. The vitreous form is readily obtained on cooling the liquid, with the glass transition about 40 °C.

Fig. 1.2 (**a**) Structure of Selenium. (**b**) Structure of sulfur

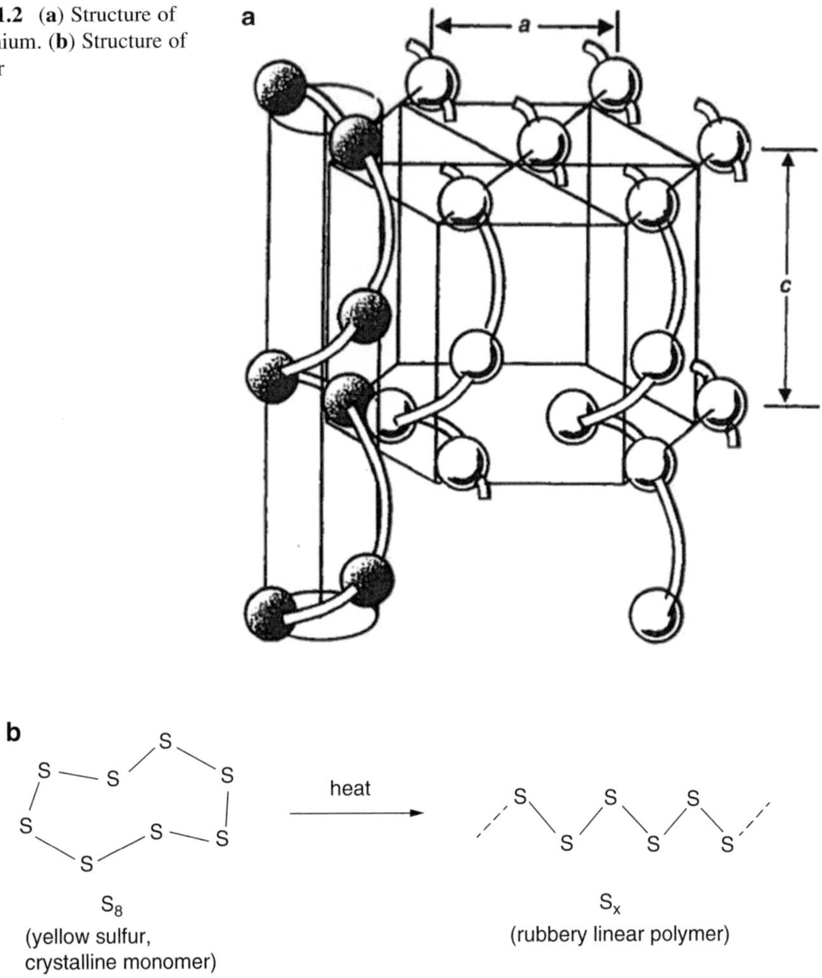

X-Ray studies indicate that amorphous sulfur may have a helical structure with eight sulfur atoms per spiral Fig. 1.2b. Crystalline sulfur seems to be made of rings, each containing eight sulfur atoms, which fit together to give a normal X-ray pattern.

Chalcogenide glasses obey the 8-N coordination rule, whatever the composition. In contrast to crystalline materials, the bandgap of ChGs contains a spectrum of localized states tailing into the gap from the valence and conduction bands introduced by the interaction of lone pairs with each other and with their local environment. Because the lone pair orbitals are filled at equilibrium, these tail states are paired and thus diamagnetic. Excitation creates empty orbitals (holes) which are compensated by electronic or structural means [19, 20].

Chalcogenides have an s^2p^4 valence shell configuration as shown below (Figs. 1.3 and 1.4).

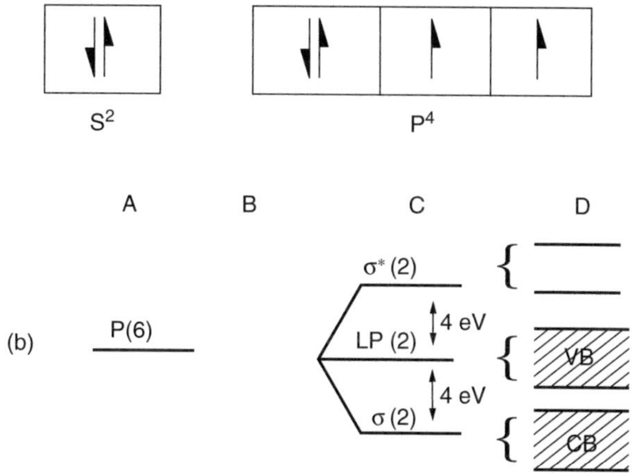

Fig. 1.3 In chalcogenides the valence band (VB) is derived from the lone pair electrons of the chalcogen atom and the conduction band (CB) from the antibonding states [21]

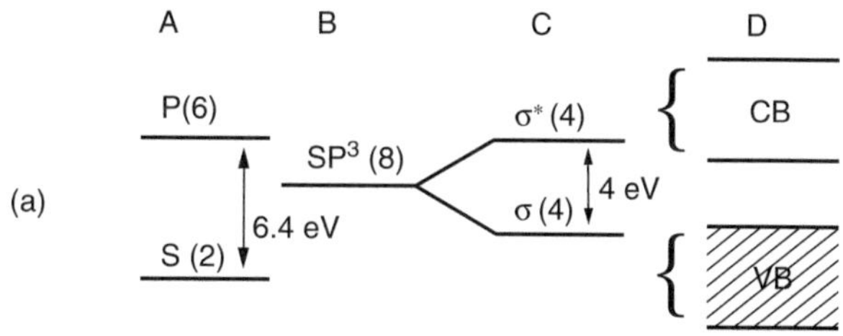

Fig. 1.4 In tetrahedral semiconductors, valence band is obtained from bonding orbitals and conduction band from antibonding orbitals [21]

On the other hand, in non-compensated, tetrahedral, amorphous materials such as germanium and silicon, an equilibrium density of dangling bonds should exist, and these should be reflected in paramagnetically detectable, single occupied states.

Both selenium (Se) and tellurium (Te) exist in helical chains in crystalline form. Se exhibits photovoltaic action by converting light into electricity through photoconductivity due to its reasonably small band gap in the crystalline form. This allows Se to be used in versatile photocells and solar cells [22, 23]. Tellurium also exhibits photoconductivity to a lesser extent. Both Se and Te are p-type semiconductors and hence used in electronics and solid-state applications. Independently, both predominantly form 1D nanostructures [24–29] due to a preferential growth along the c-axis of trigonal hexagonal geometry and exist in the spiral chains. One-dimensional geometries have potential ramifications in electronic device formation [30]. Facile production of such morphologies under economical favorable conditions such as in aqueous phase and at relatively low temperature (<100 °C) is always required. Recently, Se-Te alloys have been found to demonstrate interesting temperature dependence of electrical resistivity [31]. In addition, fine homogeneous nearly monodisperse Se-Te nanorods have also been synthesized simply by following hydrazine reduction [32]. In our recent work 12 related to the synthesis and characterization of PbSe and CdSe nanocrystals (NCs). We have also observed the formation of Se nano- and micro rods in both cases. It shows that Se is more readily reduced in comparison to Pb and Cd, which is understood from its nonmetallic nature as well as its location in the periodic table. The same can be said about Te. Se-Te alloy formation [31, 33] is emerging as a potential field with applications in nanoelectronics; therefore, a comprehensive investigation of their simultaneous reduction requires attention of both reduction chemistry as well as crystallization kinetic perspectives.

1.5 Band Structures and Band Gaps of Selected Chalcogenide Alloys (Theoretical Calculations)

The band structures of selected chalcogenide alloys have been developed theoretically (Y.M. Yiu and Gurinder Kaur, unpublished) based on the transformation of the unit cell with symmetry in the real space to the reciprocal **k** momentum space. The band gap is developed for crystalline semiconductor alloys, while metals encounter zero band gap.

For a crystal with periodic lattice and space symmetry, the reciprocal space is described by the Brillouin zone [34]. A Brillouin zone is defined as a Wigner–Seitz primitive cell in the reciprocal lattice. Band structure calculations are based on the Bloch wave expansion. The single-electron Schrödinger equation is solved for an electron in a lattice-periodic potential, giving Bloch waves as solutions:

$$\psi_{n\mathbf{k}}(\mathbf{r}) = e^{i\mathbf{k}-\mathbf{r}} u_{n\mathbf{k}}(\mathbf{r}) \tag{1.1}$$

where **k** is the wave-vector for the periodic potential of the crystal lattice $u_{n\mathbf{k}}(\mathbf{r})$. $u_{n\mathbf{k}}(\mathbf{r})$ is invariant under a crystal lattice translation. For each value of **k**, there are multiple solutions to the Schrödinger equation labeled by n, the band index, which is the number of the energy bands. There are regions of energy for which no Bloch solution of the wave equation exist. The existence of forbidden regions of energy is a prerequisite to insulators and semiconductors. The width of this forbidden region defines the band gap. The Fermi energy is the energy with filled electrons and usually marks as the zero point energy.

1.5.1 CdSSe Nanostructures

Simulated crystal forms of chalcogenides are demonstrated using the following examples: CdS, CdSSe, ZnS, and ZnSe. CdS, CdSe, and CdSSe system has been modeled as either B3 or B4 states, which possess the cubic or hexagonal symmetry, respectively [35]. The cubic B3 symmetry follows the Zinc-blende system while the hexagonal B4 symmetry possesses the wurtzite symmetry. Their properties have been calculated with the computer code of WIEN2k.

The band structures of CdS, CdSSe, and CdSe have been calculated by WIEN2k. The Brillouin zones of the CdS in B3 and B4 are illustrated in Figs. 1.5 and 1.7 respectively. The corresponding band structures are shown in the graphs, Figs. 1.6 and 1.8. The Brillouin zones are plotted in k-space, the momentum space. The symmetry points in this Brillouin zone are dictated by the crystal symmetry. The calculated band gaps of the simulated E-k curves show that CdS in B3 and B4 are

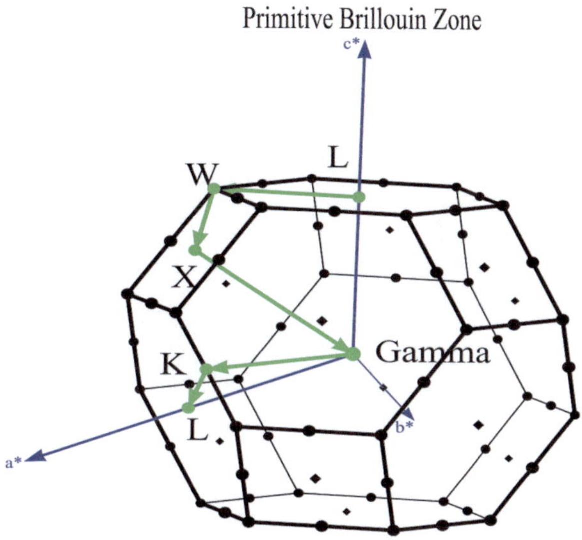

Fig. 1.5 Brillouin zone of CdS (B3)

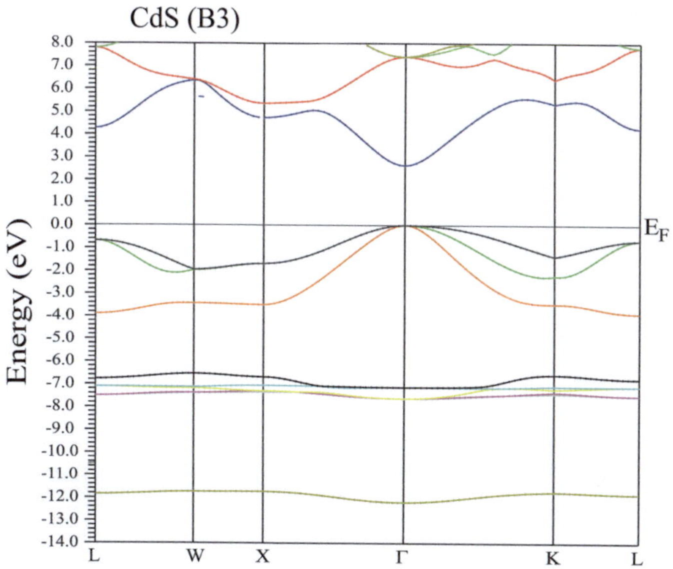

Fig. 1.6 Band structure of CdS in (B3) form

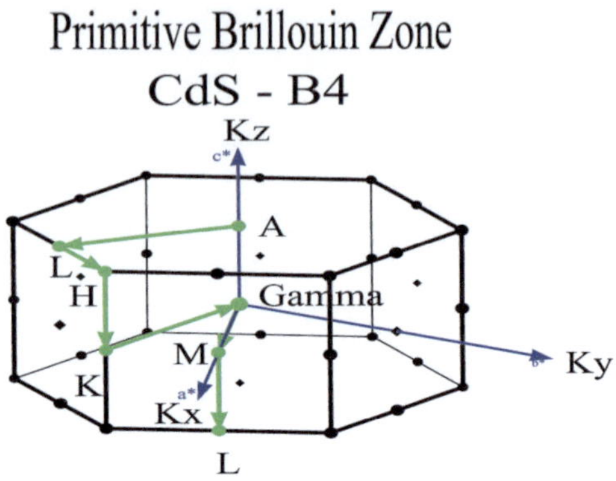

Fig. 1.7 Brillouin zone of CdS in B4 form

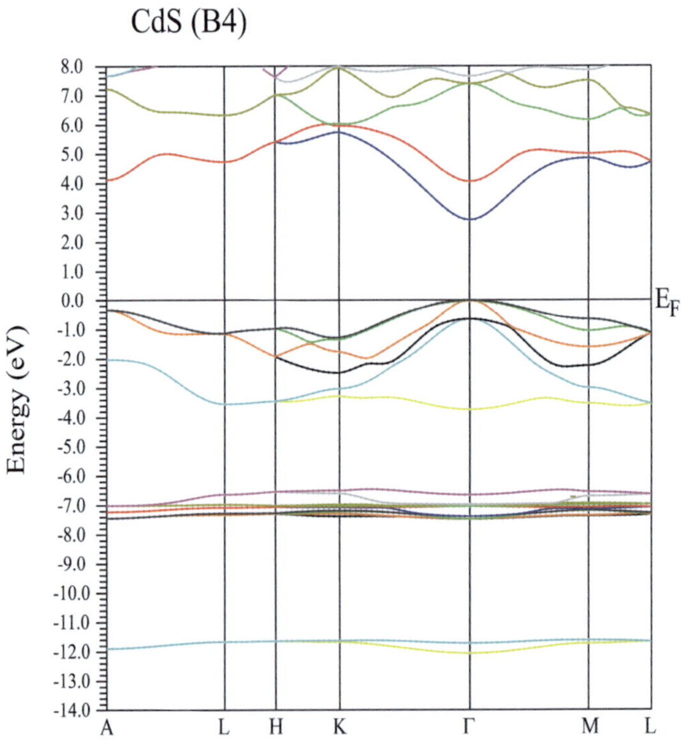

Fig. 1.8 Band structure of CdS in B4 structure

semiconductors with direct band gaps. The band gaps are centered at the Gamma point, which is the central point of the Brillouin zone. The corresponding symmetric points of interest have been evaluated through the band structure *E*-*k* plots. The band structures of CdS of B3 and B4 structures possess direct band gaps, and the smallest openings of the *E*-*k* plots are located at the Gamma point. However, the band gap energies are underestimated using GGA (Generalized Gradient

Table 1.1 The experimental band gap energies of CdS, CdSSe, and CdSe

Nanomaterials	Experimental band gap (XEOL) (eV)
CdS (nanostructure)	2.43
CdSSe (nanostructure)	1.985
CdSe (nanostructure)	1.74

Table 1.2 The calculated band gap energies of CdS, CdSSe, and CdSe in B3 and B4 forms

Materials	Band gap (GGA) (eV)	Band gap (mBJ) (eV)
CdS (B3)	0.963	2.63
CdS (B4)	1.057	2.70
CdSSe (B4)	0.614	2.32
CdSe (B3)	0.464	1.91
CdSe (B4)	0.527	1.96

Approximation). The details of WIEN program and correction of the exchange-correlated mBJ (Modified Becke–Johnson) potential [36] as described in Chap. 2 is evaluated. This band gap energy can then be fitted to the desirable experimental value by the adjustment of A and B parameters of the mBJ potential. However, the global fit of the parameters of this exchange-correlation potential has to follow the majority of semiconductors. The calculated band gaps with the specific parameter A and B are compared to the experimental band gaps of these semiconductors including either wide or narrow band gap materials. It takes long computer time to calculate complicated system even with the highly developed program.

The optical band-gap emission of these CdS_xSe_{1-x} nano-ribbons can be tuned to the range between that of pure CdS (2.43 eV) and CdSe (1.74 eV) by changing the S and Se ratio. The experimental band gaps as measured by XEOL (X-ray excited optical luminescence) are listed in Table 1.1. The results of XEOL (X-ray excited optical luminescence) of CdS_xSe_{1-x} are described in detail in Chap. 2.

The direct band gaps calculated using DFT (Density Functional theory are listed in Table 1.2. The band gaps calculated by DFT are smaller than the experimental band gaps due to the limitation of DFT which assumes relatively uniform electron distributions. Even with GGA correction, the band gap energies are underestimated. Thus the DFT band gaps serve as lower limits for the actual band gaps. To correct the underestimated band gap, the mBJ potential are applied to the converged self-consistent calculation. The computer program WIEN then enters into a second stage of self-consistent iteration. The band gaps calculated by mBJ potential are listed in Table 1.2.

The mBJ potential as calculated by WIEN program is over corrected in comparison to the GGA calculation. To further adjust the input parameters A and B of the mBJ potential, one can fit the experimental band gaps to a closer extent. However, the adjustment of the parameters of the mBJ energy is subject to a global consideration by comparing other wide band gap materials.

1.5.2 $CuInSe_2$

The widely used solar cell material is $CuInSe_2$. The computer model to probe the band structure is developed using the following crystalline model. The crystal structure of $CuInSe_2$ has the group symmetry of I-42d (#122). The crystal constants are $a = 6.1944$ Å and $c = 12.4157$ Å. Cu is at (0, 0, 0), In is at (0, 0, 1/2), and Se is at (0.235, 1/4, 1/8). The unit cell is shown in Fig. 1.9.

The Band gap of $CuInSe_2$ is 0.0633 $R_y = 0.861$ eV, as calculated by the mBJ potential. Using photoluminescence (PL), photoluminescence excitation (PLE) and optical absorption (OA) techniques, the measure band gap of $CuInSe_2$ is 1.0444 eV [37]. The calculated band gap of $CuInSe_2$ is still underestimated.

In comparison, copper indium gallium selenide (CIGS) is a direct band gap material. The photovoltaic cell has the highest efficiency (~20 %) among thin film materials.

The primitive Brillouin zone of $CuInSe_2$ is shown in Fig. 1.10 and the corresponding band structure is shown in Fig. 1.11.

1.5.3 $CuInTe_2$

The model crystal structure of $CuInTe_2$ has the group symmetry of I-42d (#122). The crystal constants are $a = 5.7810$ Å and $c = 11.6422$ Å. Cu is at (0, 0, 0), In is at (0,0,1/2), and Te is at (0.227,1/4,1/8) position. The unit cell, Brillouin zone, and band structure of $CuInTe_2$ are shown in Figs. 1.12, 1.13, and 1.14, respectively.

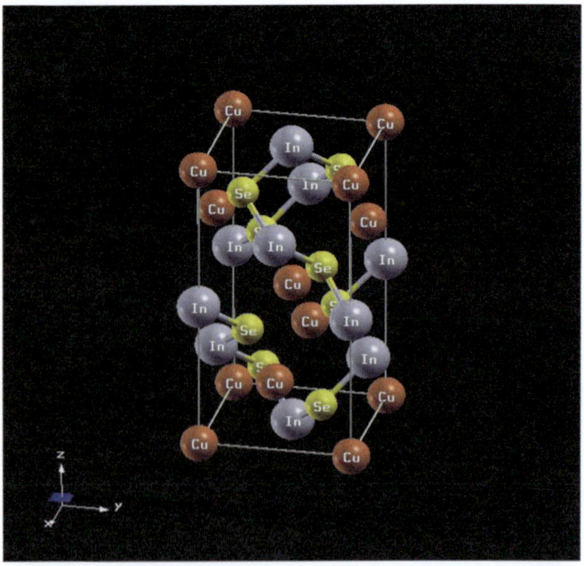

Fig. 1.9 Unit cell of $CuInSe_2$

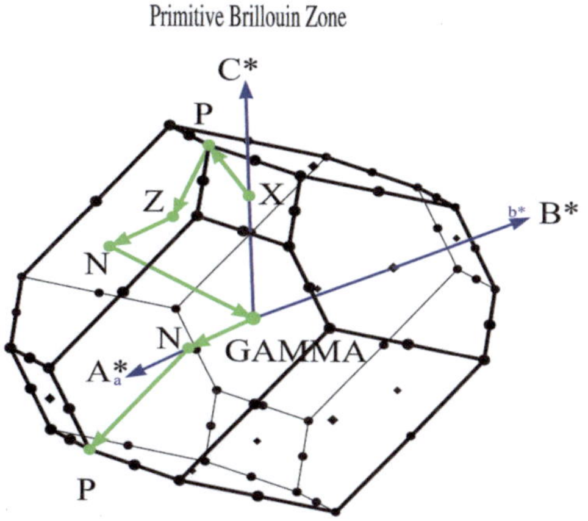

Fig. 1.10 Primitive Brillouin zone of $CuInSe_2$

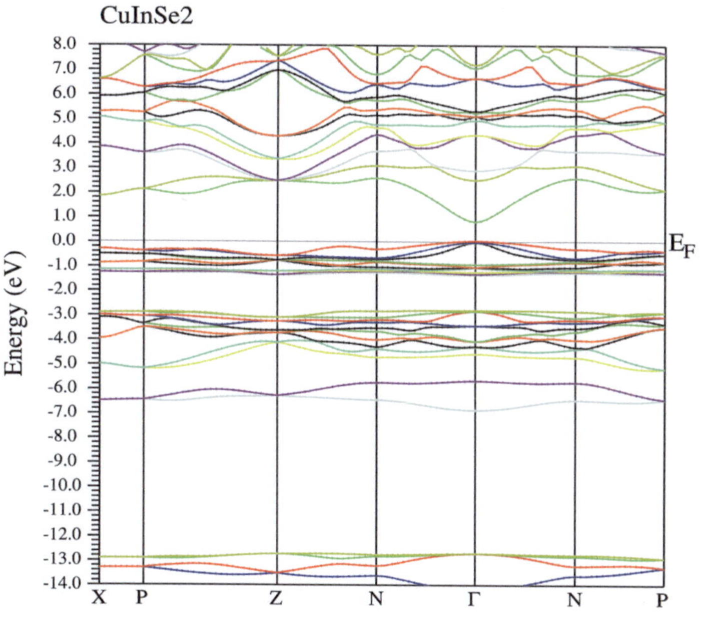

Fig. 1.11 Band structure of $CuInSe_2$

1.5.4 ZnS and ZnSe

The ZnS and ZnSe models possess crystal structures of either cubic or hexagonal nature [38]. The ZnS and ZnSe with cubic symmetry have crystal symmetry of space group, F-43m (#216). The ZnS and ZnSe with hexagonal symmetry have space symmetry of $P6_3mc$ (#186). For the cubic structures, Zn is at the 4a position of (000), S or Se are at the 4c position of (1/4, 1/4, 1/4). The lattice parameters of

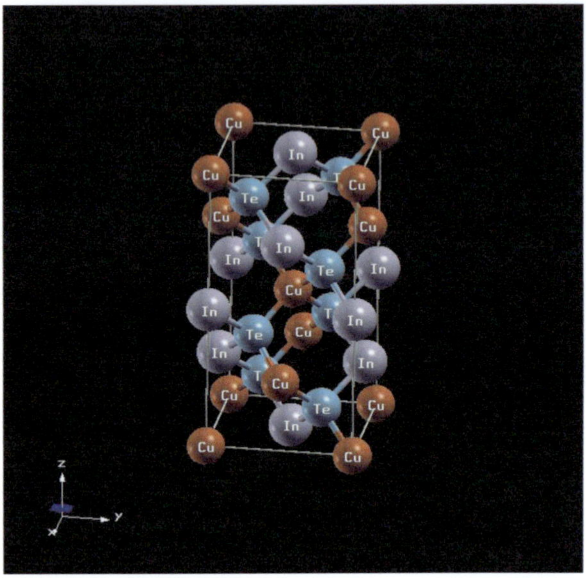

Fig. 1.12 Unit cell of $CuInTe_2$

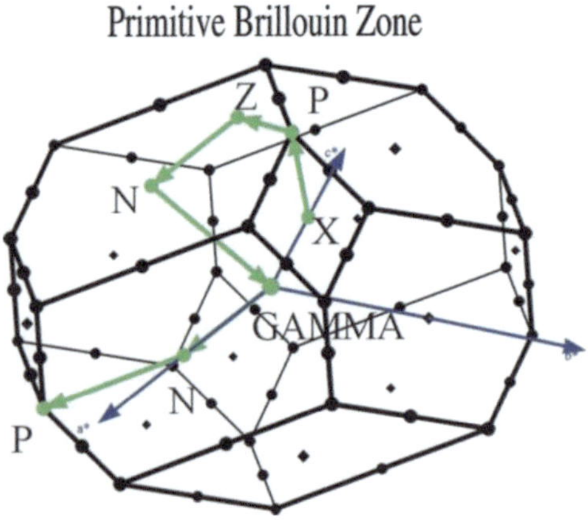

Fig. 1.13 Primitive Brillouin zone of $CuInTe_2$

ZnS and ZnSe in cubic symmetry are 5.4093 Å and 5.6676 Å, respectively. For the hexagonal B4 structures, Zn, S or Se are at the 2b positions of (1/3, 2/3, z) and (2/3, 1/3, $1/2+z$). The lattice constants of ZnS and ZnSe in hexagonal symmetry are $a=3.811$ Å, $c=6.234$ Å and $a=3.98$ Å, c $=6.53$ Å, respectively. The Primitive lattice cells of (a) cubic ZnS and (b) hexagonal ZnS are shown in Fig. 1.15.

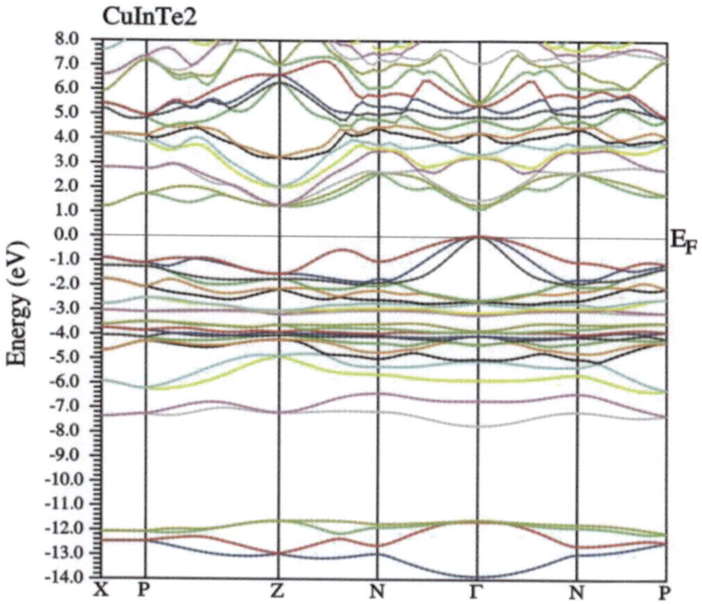

Fig. 1.14 Band structure of $CuInTe_2$

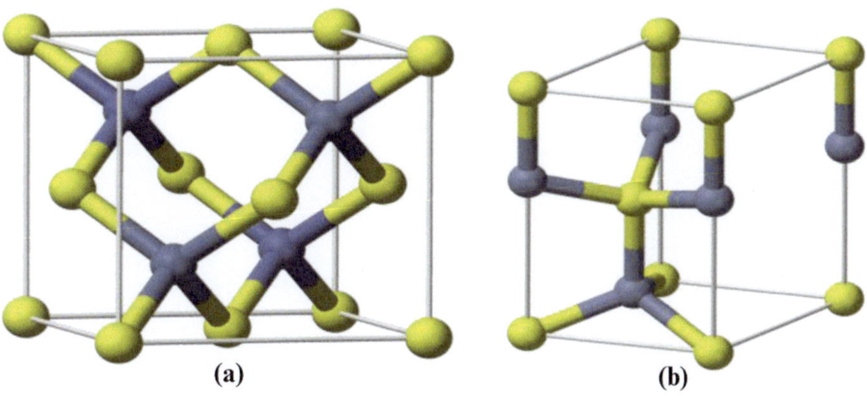

Fig. 1.15 Primitive lattice cells of (**a**) cubic ZnS and (**b**) hexagonal ZnS

The Brillouin zone of ZnS in cubic symmetry is shown in Fig. 1.16.
The band structure of ZnS in cubic symmetry is illustrated in Fig. 1.17.
The Brillouin zone of the hexagonal ZnSe is shown in Fig. 1.18.
The band structure of hexagonal ZnSe is shown in Fig. 1.19.

Fig. 1.16 Primitive Brillouin zone of cubic ZnS

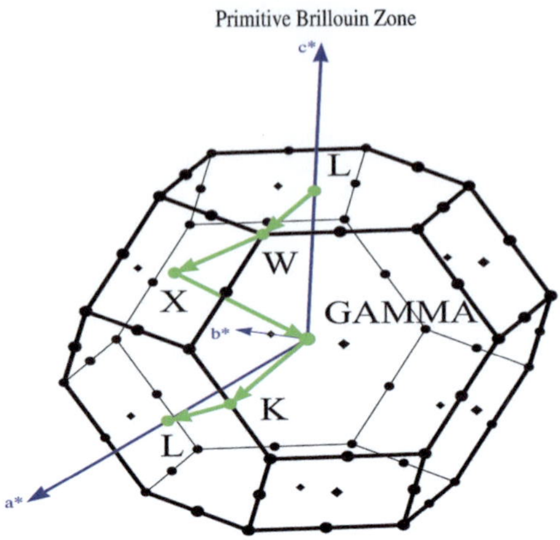

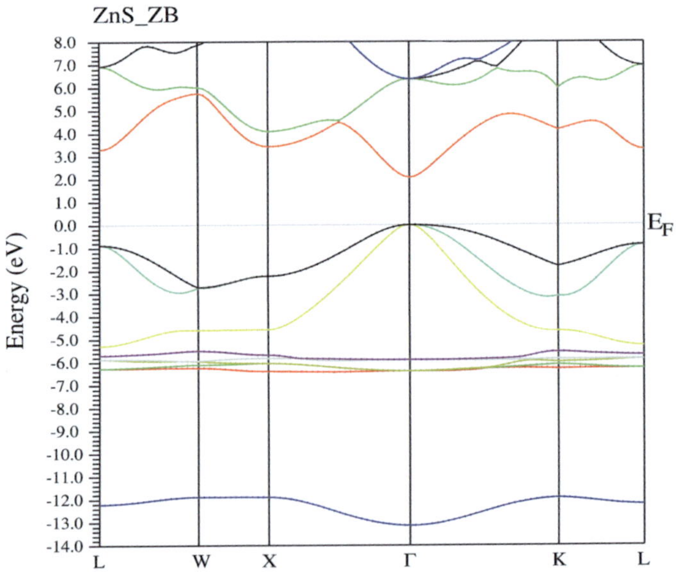

Fig. 1.17 Band structure of cubic ZnS

Fig. 1.18 Primitive Brillouin zone of hexagonal ZnSe

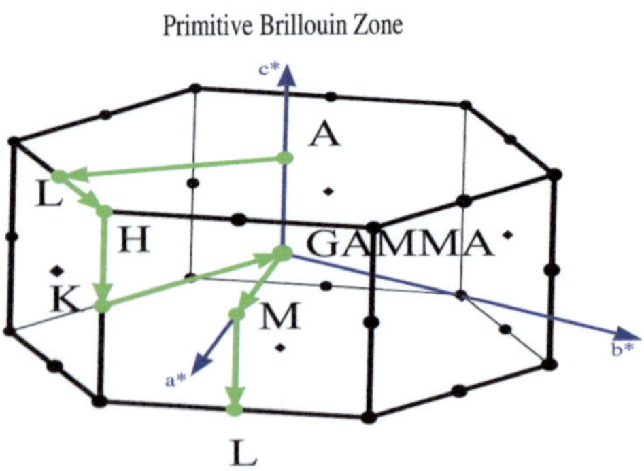

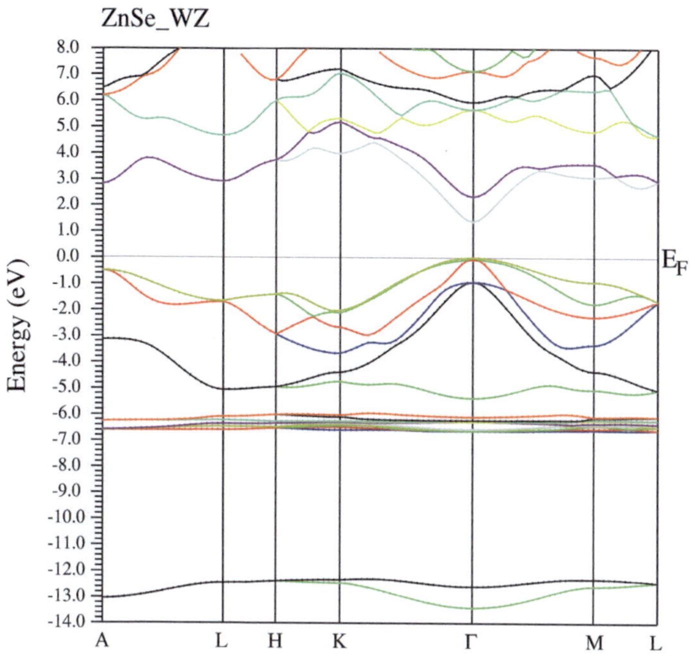

Fig. 1.19 Band structure of hexagonal ZnSe

Table 1.3 The calculated and experimental band gaps of ZnS and ZnSe

Materials	Band gap (GGA) (eV)	Band gap (mBJ) (eV)	Band gap (experimental) (eV)
ZnS-cubic	2.075	3.623	3.68
ZnS-hexagonal	2.210	3.762	3.91
ZnSe-cubic	1.257	2.681	2.82
ZnSe-hexagonal	1.380	2.843	2.795

The experimental [29–32] and theoretical band gaps of ZnS and ZnSe are listed in Table 1.3. The theoretical band gaps are calculated by GGA and mBJ potential.

1.5.5 CdTe

The model crystal structure of CdTe has group symmetry of F-43m. The lattice constant is $a = 6.4827$ Å. Cd is at (0, 0, 0) and Te is at (1/4, 1/4, 1/4) position. The unit cell, Brillouin zone, and band structure of CdTe are shown in Figs. 1.20, 1.21, and 1.22, respectively.

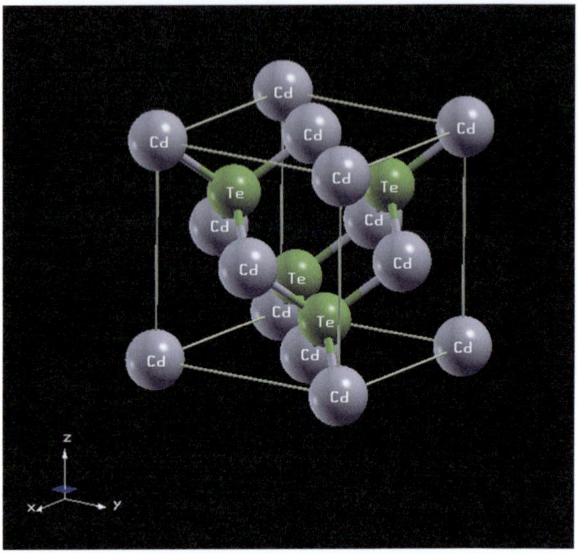

Fig. 1.20 Unit cell of CdTe

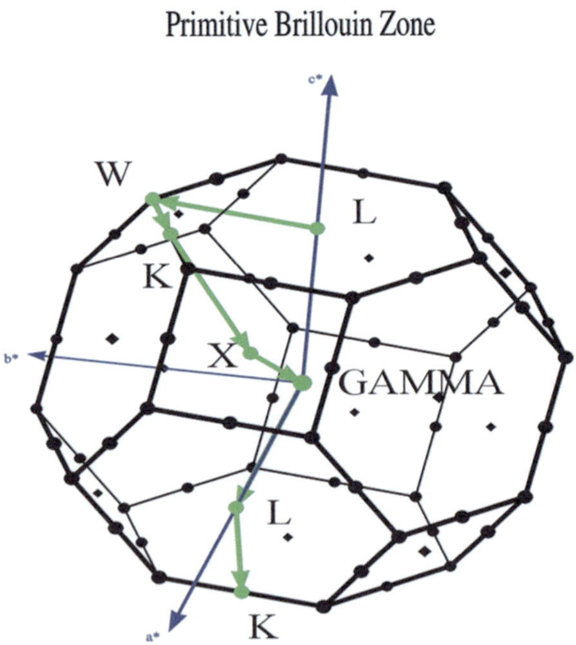

Fig. 1.21 Primitive Brillouin zone of CdTe

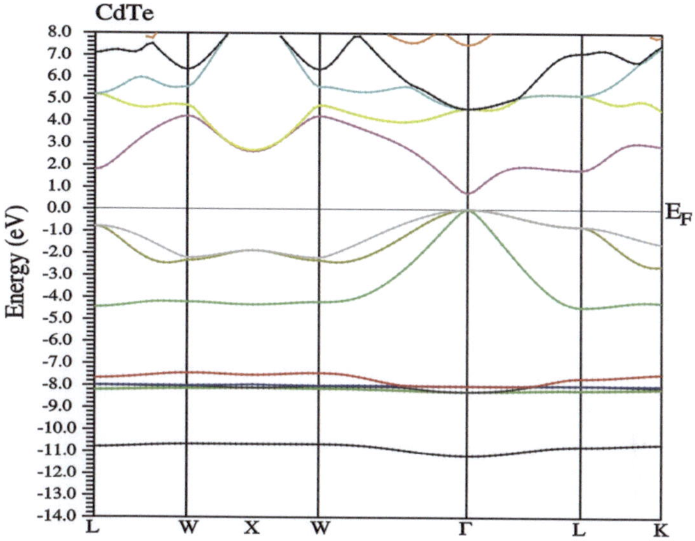

Fig. 1.22 Band structure of CdTe

1.5.6 GeSbTe (GST)

There are two crystalline phases in GeTe-Sb_2Te_3 system. One is a stable hexagonal structure and the other is meta-stable face centered cubic (fcc) structure The crystal structure of the meta-stable phase of GeSbTe is fcc Fm-3 m type. Te occupies the 4a (000) position. Ge and Sb share the 4b (1/2, 1/2, 1/2) positions. The unit cell, Brillouin zone and band Structure of cubic and hexagonal structure of GeSbTe are shown in Figs. 1.23, 1.24, 1.25, 1.26, 1.27, and 1.28.

The crystal structure of the stable hexagonal state is plotted as the following graph, Fig. 1.26. The Brillouin zone and band structure are illustrated in Figs. 1.27 and 1.28, respectively.

1.5.7 HgCdTe

The model crystal structure of HgCdTe has crystal symmetry of F-43m. Te occupies the (000) position while Cd and Hg possess the (1/4, 1/4, 1/4) and (3/4, ¾, 3/4) positions, respectively. The unit cell, Brillouin zone, and band structure of HgCdTe are illustrated in Figs. 1.29, 1.30, and 1.31, respectively.

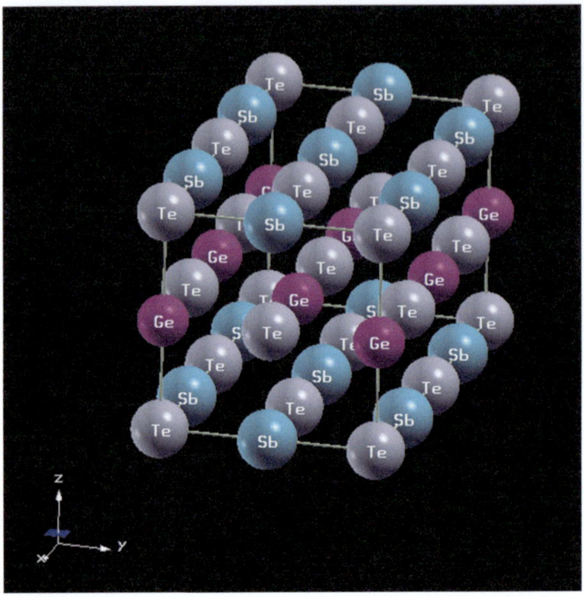

Fig. 1.23 Unit cell of GeSbTe

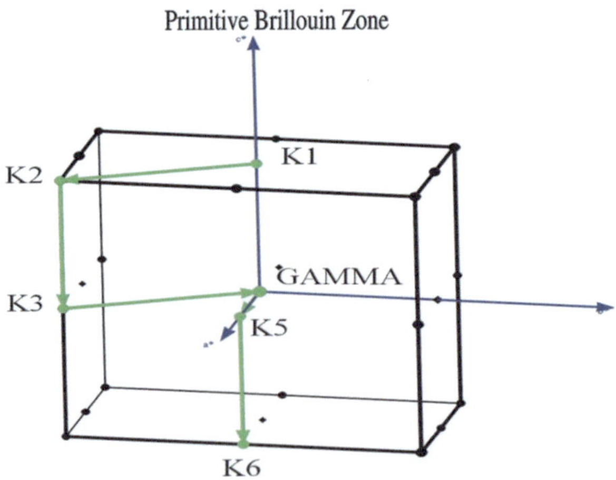

Fig. 1.24 Primitive Brillouin zone of GeSbTe

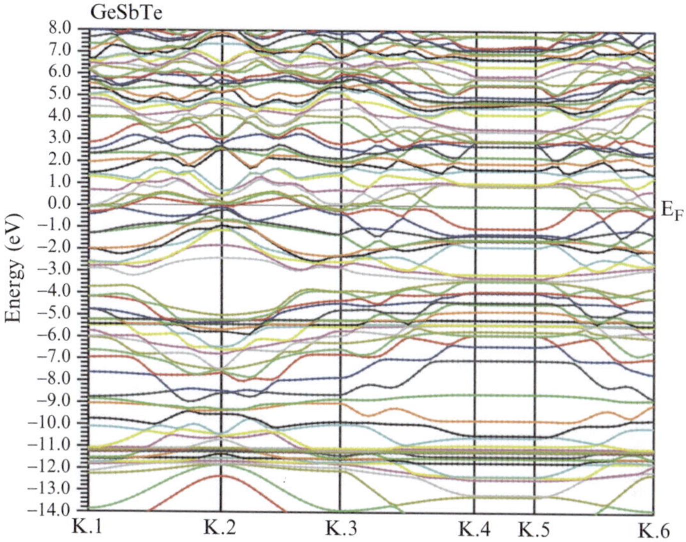

Fig. 1.25 Band structure of GeSbTe

Fig. 1.26 Unit cell of hex-GeSbTe

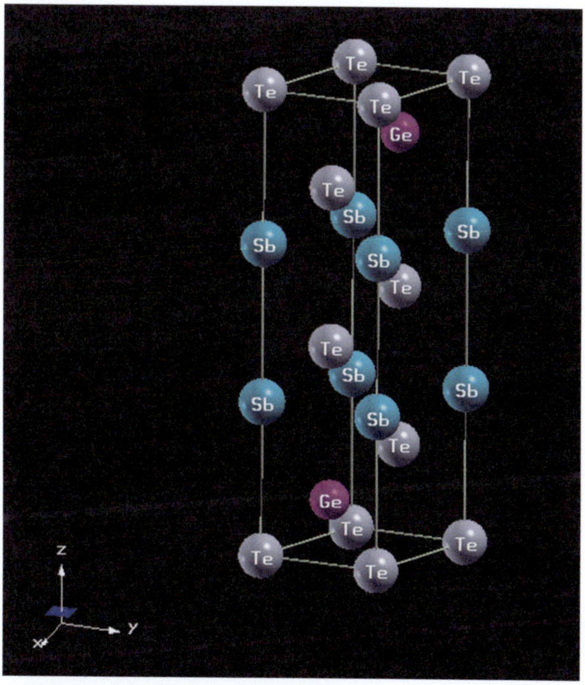

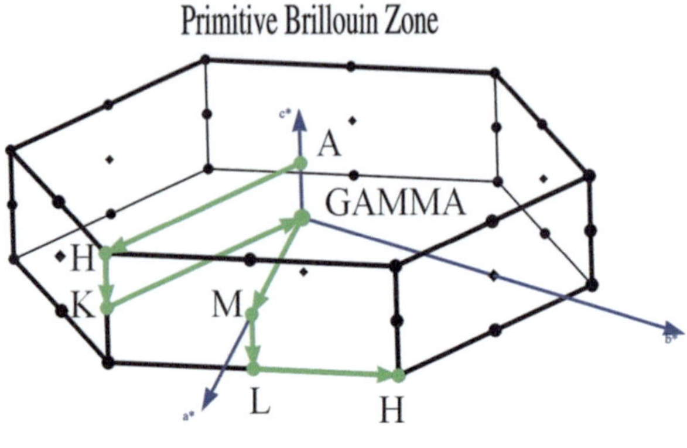

Fig. 1.27 Primitive Brillouin zone of hex-GeSbTe

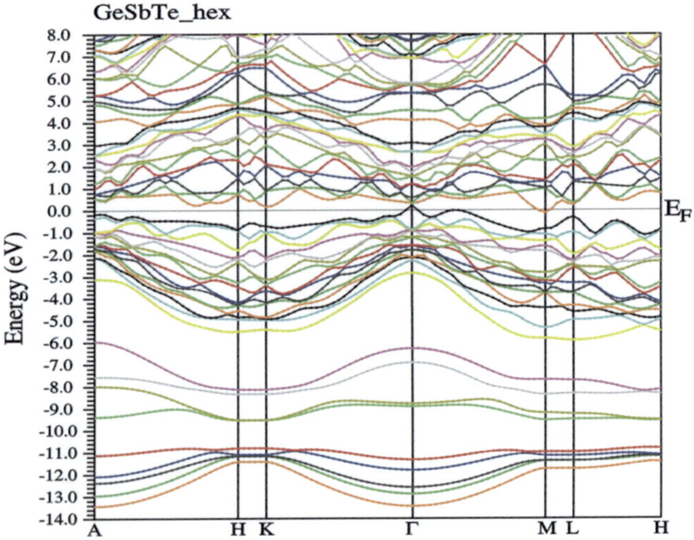

Fig. 1.28 Band structure of hex-GeSbTe

1.5.8 PbSe

Model PbSe crystal has NaCl structure and group symmetry of Fm3m. The lattice constants are $a = 6.117$ Å. Pb occupies the 4a (0, 0, 0) position and Se occupies the 4b (1/2, 1/2, 1/2) position. PbSe has unit cell, Brillouin zone, and band structure as plotted in Figs. 1.32, 1.33, and 1.34.

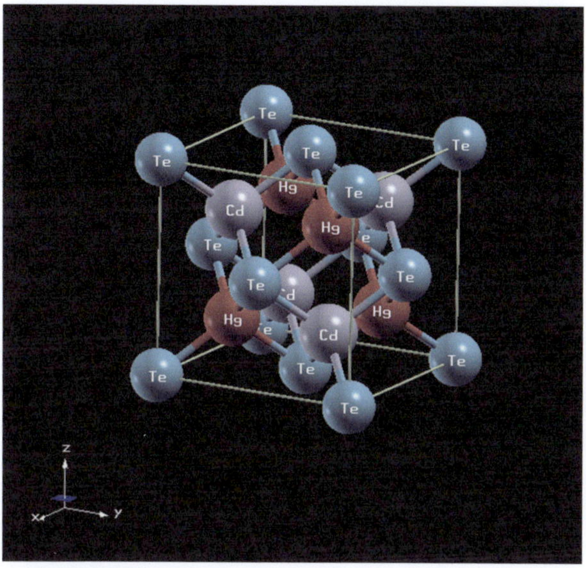

Fig. 1.29 Unit cell of HgCdTe

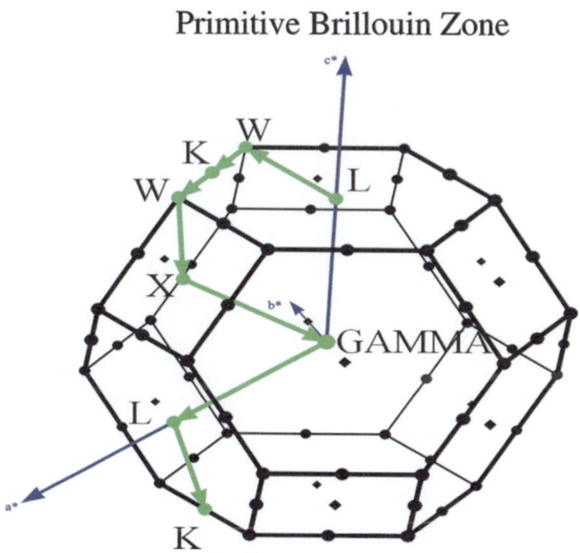

Fig. 1.30 Primitive Brillouin zone of HgCdTe

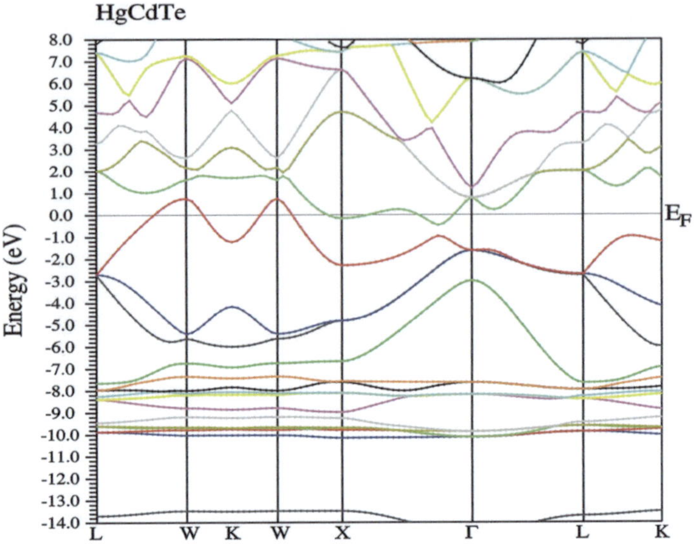

Fig. 1.31 Band structure of HgCdTe

Fig. 1.32 Unit cell of PbSe

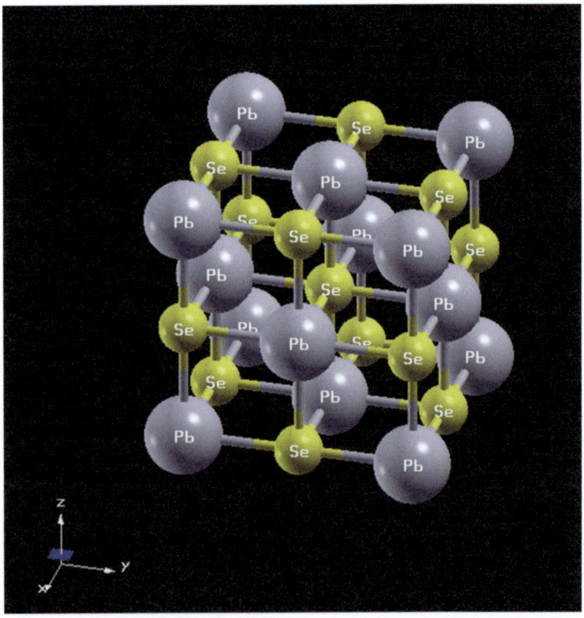

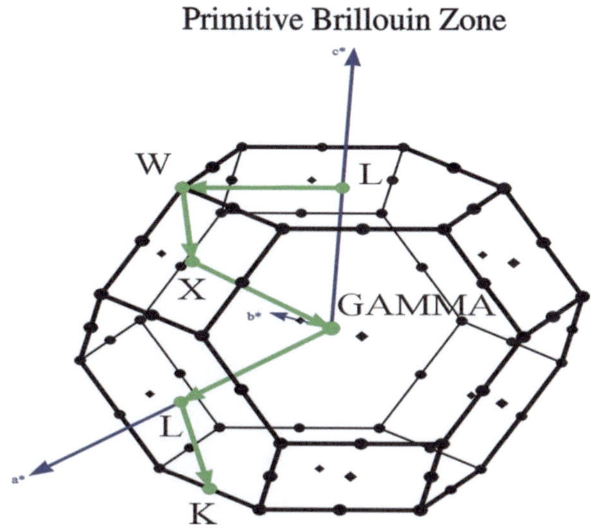

Fig. 1.33 Primitive Brillouin zone of PbSe

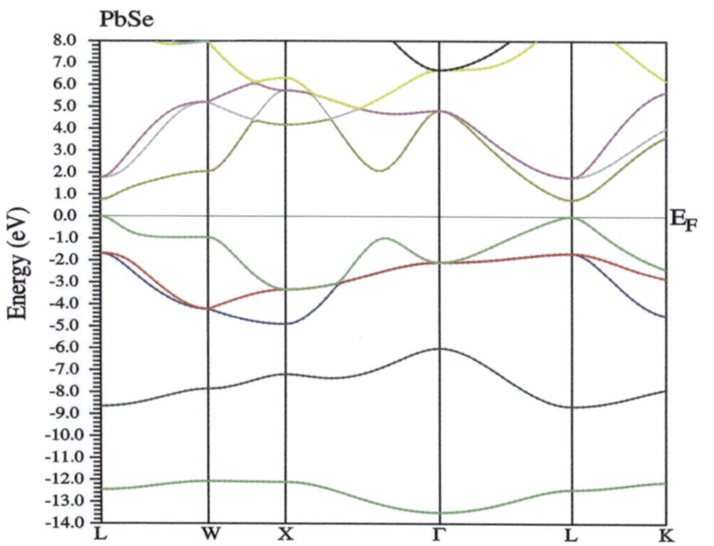

Fig. 1.34 Band structure of PbSe

1.5.9 Se and As and Cl Doped Se

The band structure of trigonal Se have been investigated using the program WIEN2k. The WIEN2k program is discussed in Chap. 2 in detail. We have used the mBJ potential to calculate the band gap energy of the model trigonal Se. The mBJ potential is also presented in Chap. 2. The band gap energy of trigonal Se is calculated as 1.495 eV in comparison to the experimental band gap energy of 1.6 eV [39].

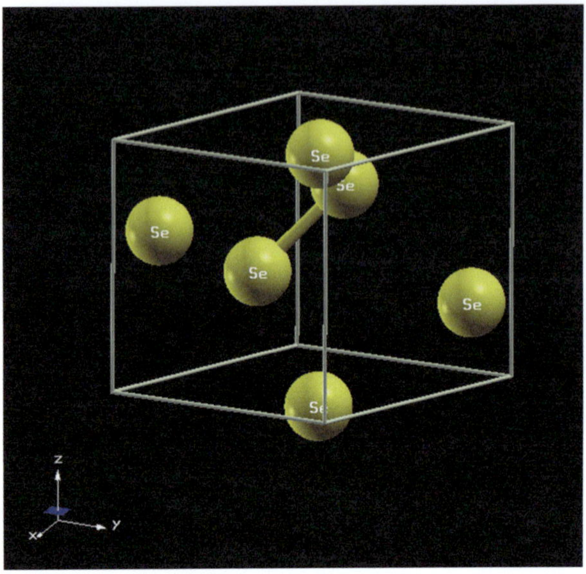

Fig. 1.35 Crystal structure of trigonal Se

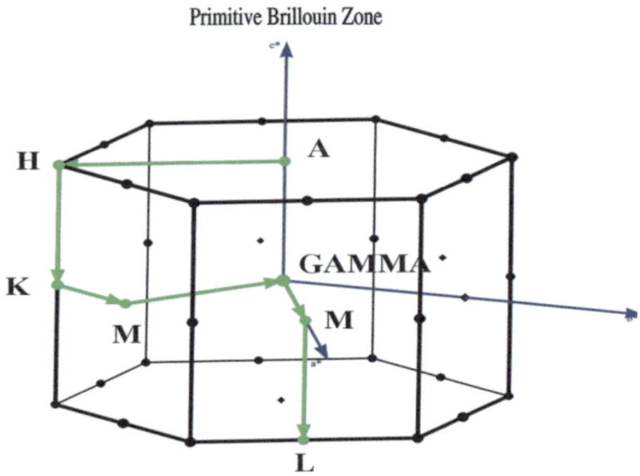

Fig. 1.36 Brillouin zone of trigonal Se

1.5.9.1 Trigonal Se

I. Crystal Structure of Trigonal Se

The trigonal Se has been modeled from the following crystal structure for the density functional calculation, WIEN2k. The crystal structure of the model trigonal Se has lattice constants of $a = 4.3662$ Å, $c = 4.9536$ Å. The space group of the trigonal Se is $P3_121$. The Se atoms occupy a symmetric position at 3 (a) at (x, 0, 1/3) where $x = 0.217$. Its crystal structure is shown in Fig. 1.35.

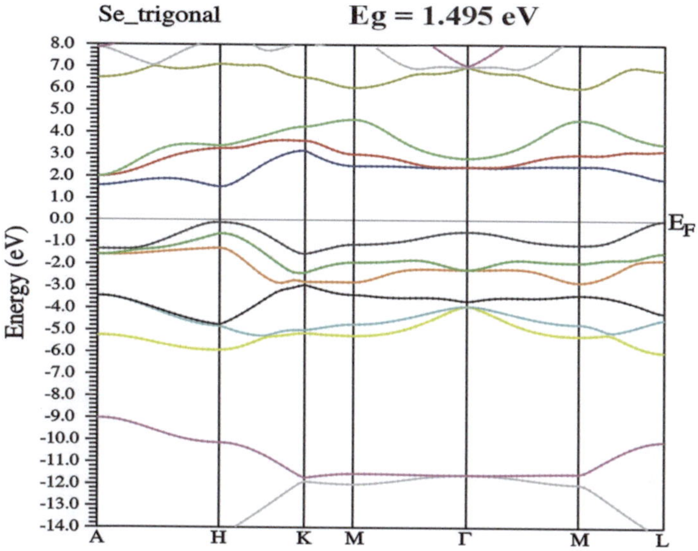

Fig. 1.37 Band structure of trigonal Se

II. Band Structure of Trigonal Se

The band structure of the model trigonal Se has been calculated self-consistently by WIEN2k using the mBJ potential. The Brillouin zone of trigonal Se is illustrated in Fig. 1.36. The center of the Brillouin zone is at the Gamma point of the zone. The band structure of trigonal Se is shown in Fig. 1.37. Density functional calculations with mBJ potential widen the band gap in comparison to the Density functional theory with Generalized Gradient Approximation, as discussed in Chap. 2. The calculated band gap of Se is located at the H point of its Brillouin zone and its band gap energy is 1.495 eV.

III. Density of States of Trigonal Se

The s, p, and d orbital partial densities of states of the trigonal selenium are plotted in Fig. 1.38.

1.5.10 Se Doped with As and Cl

The band structure and density of states of the trigonal Se doped with As and Cl have also been calculated using WIEN2k. The model doped Se alloys is illustrated as the following.

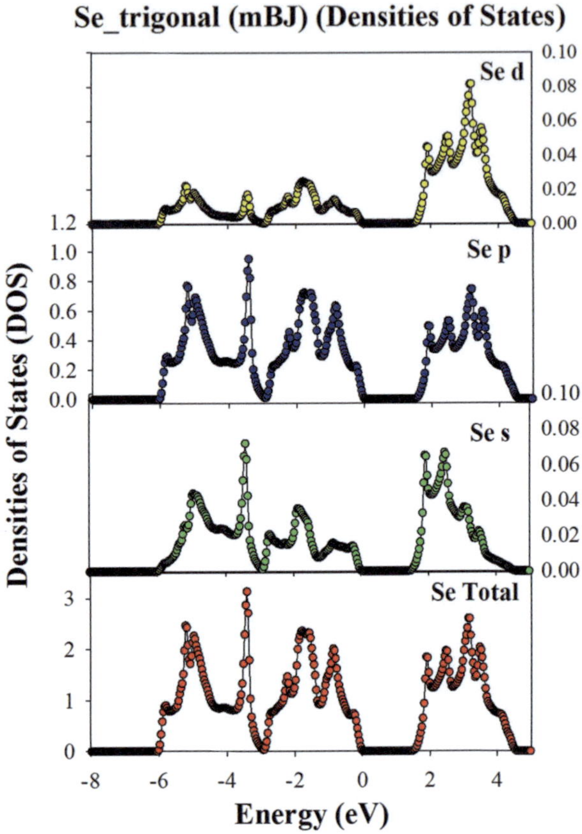

Fig. 1.38 Densities of states of s, p, and d orbitals of trigonal Se

1.5.10.1 Case 1: As Doped Se

The model structure of As doped Se is shown in Fig. 1.39. The ratio of Se to As is 8 to 1. The percentages of Se and As in the crystal are 88.9 at.% and 11.1 at.%, respectively. The band structure is shown in Fig. 1.40. The dopant of As into trigonal Se introduces valence band tail states in the band gap [40], as seen in the partial density of states of As in Fig. 1.41. The conduction band tail state peaks at 1.5 eV above the Fermi energy.

1.5.10.2 Case 2: Cl Doped Se

The model Cl doped Se structure is shown in Fig. 1.42. The ratio of Cl to As is 8 to 1. The percentages of Se and Cl in the crystal are 88.9 at.% and 11.1 at.%, respectively. The band structure of this Cl doped Se have been calculated and plotted in Fig. 1.43. The introduction of Cl into the trigonal Se creates additional

Fig. 1.39 Crystal structure of As doped Se

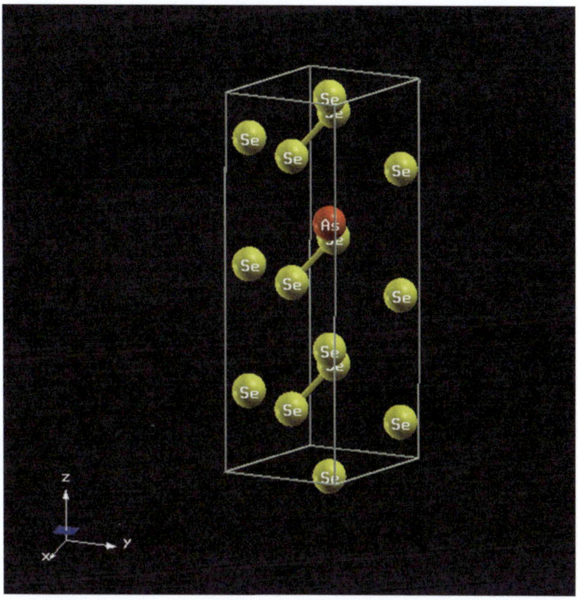

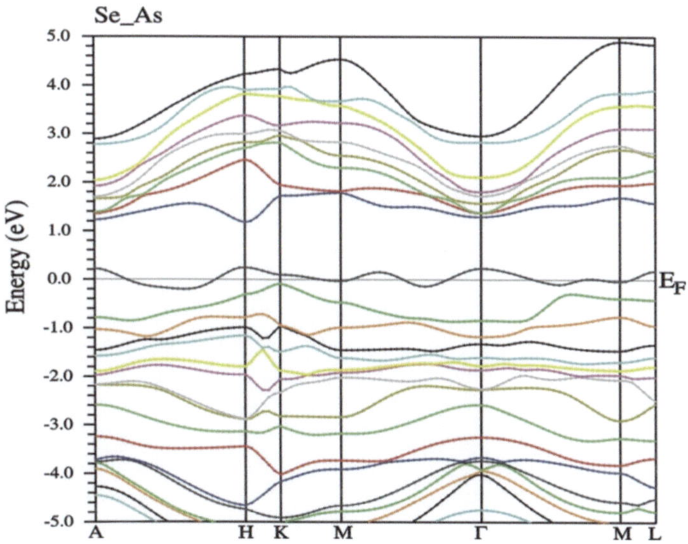

Fig. 1.40 Band structure of As doped Se

levels in the band gap, in particular, it forms the additional acceptor levels in the band gap of trigonal Se. The valence band tail is extended to about 0.33 eV above the valence band. The partial density of states of Cl is presented in Fig. 1.44. The additional defect states are located at about 0.86 eV above the Fermi energy.

Fig. 1.41 Partial density of states of As in As doped Se

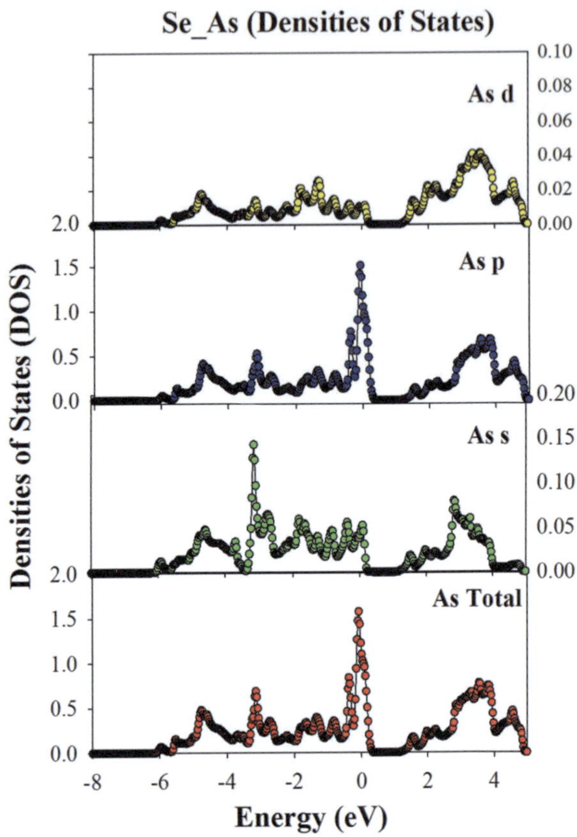

Fig. 1.42 Crystal structure of Cl doped Se

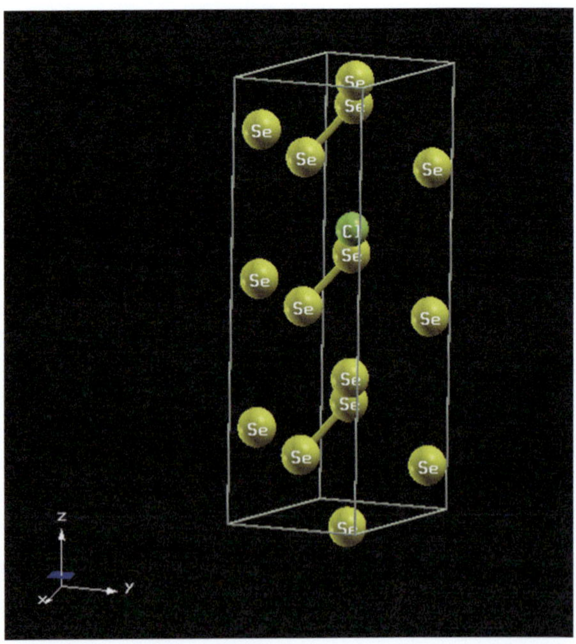

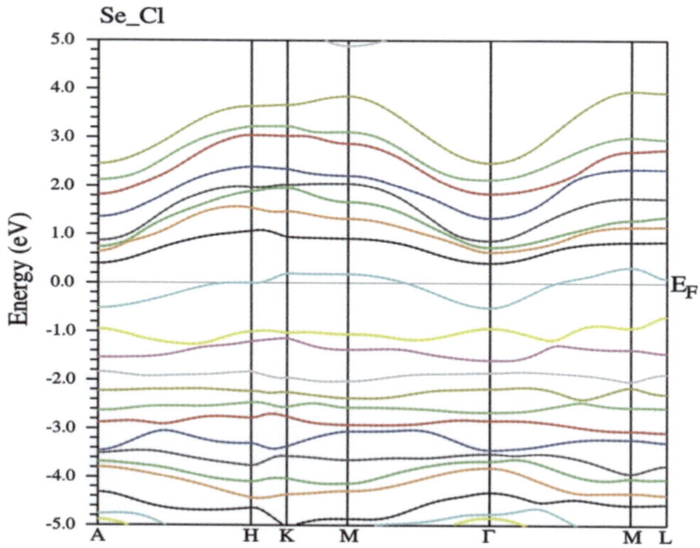

Fig. 1.43 Band structure of Cl doped Se

Fig. 1.44 Partial density of states of Cl in Cl doped Se

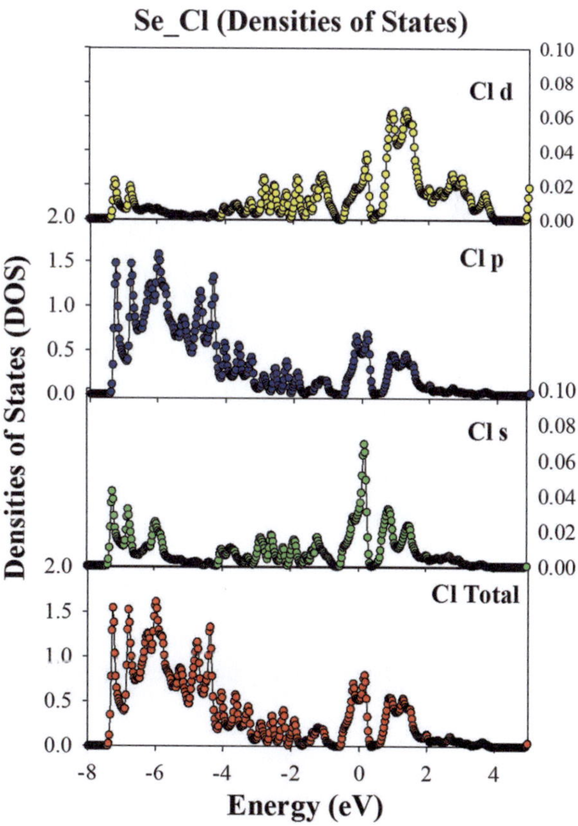

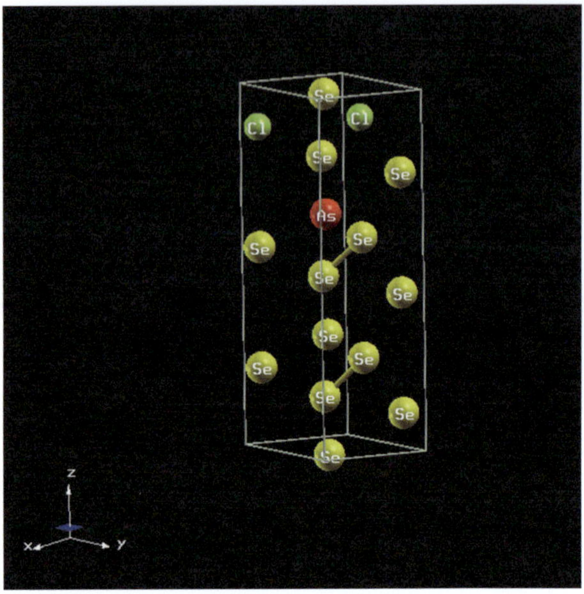

Fig. 1.45 Crystal structure of trigonal Se doped with 1 As and 1 Cl

1.5.10.3 Case 3: As and Cl Doped Se

The As and Cl co-doped Se has been modeled from the super cell of trigonal Se with the composition ratio of Se:As:Cl = 7:1:1. The percentages of Se, As, and Cl in this model alloy are 77.8 at.%, 11.1 at.%, and 11.1 at.%, respectively. The crystal structure is shown in Fig. 1.45 and the band structure As and Cl doped Se is plotted in Fig. 1.46. The partial densities of states of Se, As, and Cl of the trigonal Se doped with As and Cl are plotted in Figs. 1.47, 1.48, and 1.49, respectively. As shown from Fig. 1.48, the As partial densities of states indicate that the presence of As state located deeper in the band gap. The As defect states distribute at 0.57 and 1.4 eV above the Fermi energy. The Cl defect states locate at 0.16, 0.46, and 1.03 eV above the Fermi level. Cl serves as an acceptor and it can accept electrons from the valence band as its distribution occupies near the valence band in the band gap of trigonal Se as shown in Fig. 1.49.

Amorphous selenium (a-Se) is a photoconductive chalcogenide, and it has the inherent ability to generate charge carriers under exposure to X-rays. a-Se spontaneously crystallizes into trigonal t-Se since its glass transition temperature is 40 °C. Adding arsenic to amorphous Se (a-Se) greatly retards the rate of re-crystallization, but introduces charge traps which reduce the photo-conductivity of a-Se. Doping with Cl helps restore the photo-conducting ability of the arsenic-doped selenium alloy [41, 42], and significantly improves the electrical properties of a-Se.

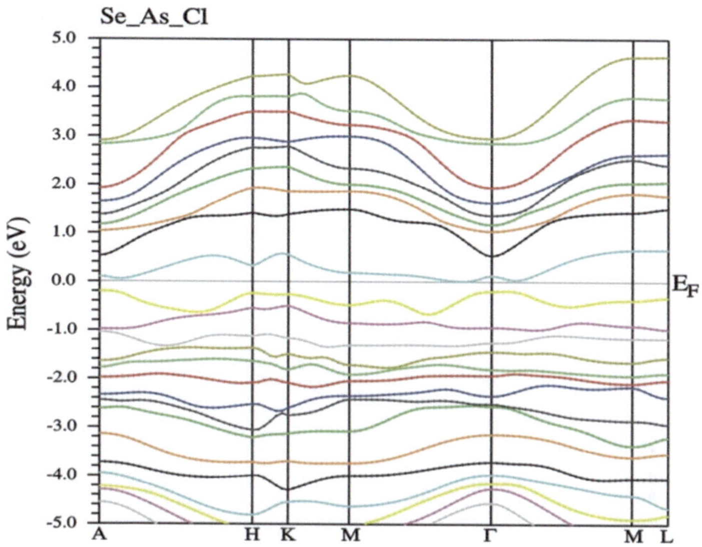

Fig. 1.46 Band structure of trigonal Se doped with 1 As and 1 Cl

Fig. 1.47 Se partial density of states of trigonal Se doped with 1 As and 1 Cl

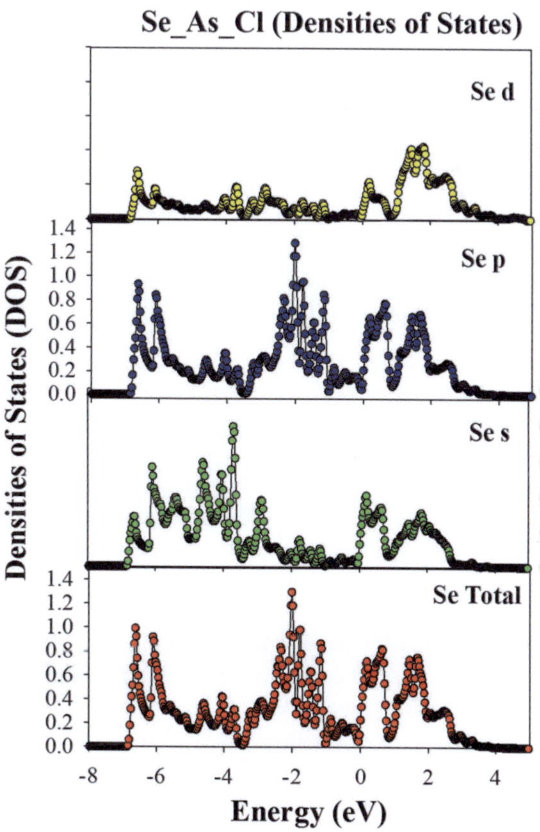

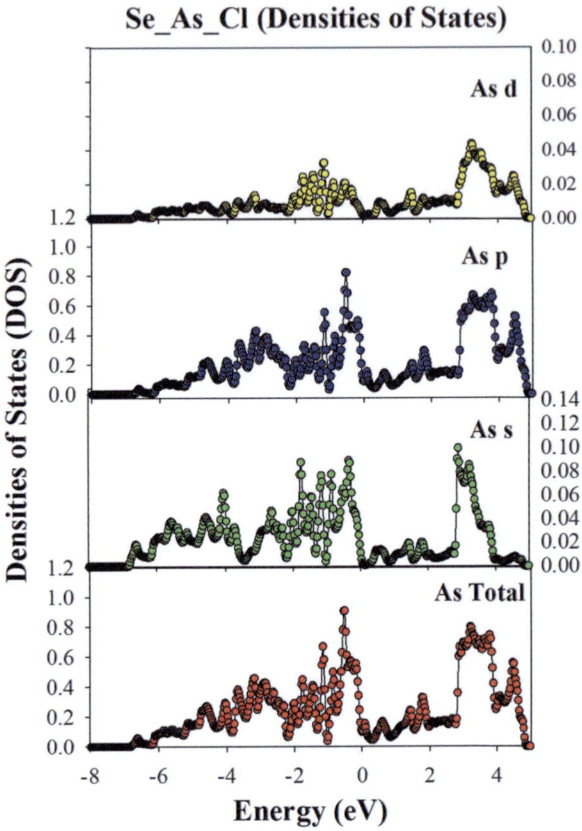

Fig. 1.48 As partial density of states of trigonal Se doped with 1 As and 1 Cl

1.6 Amorphous, Crystalline, and Nanocrystalline Chalcogenides

An important characteristic of chalcogen based materials is their ability to exist in different structural configurations, viz., amorphous (non-crystalline), crystalline, and nanocrystalline forms.

The process of glass transition leads to the freezing of a supercooled liquid at the glass transition temperature, T_g into a non-crystalline solid (glass). As a glass is heated, abrupt changes in thermal expansion, compressibility, specific heat and other second order thermodynamic properties occur at the glass transition temperature.

At T_g, the glass becomes liquid. The additional motional degrees of freedom available in the liquid cause the sudden increase in C at T_g. Above T_g the structure of the supercooled liquid resembles that of the glass. As the temperature is further raised the structure of the liquid may change in a complicated manner.

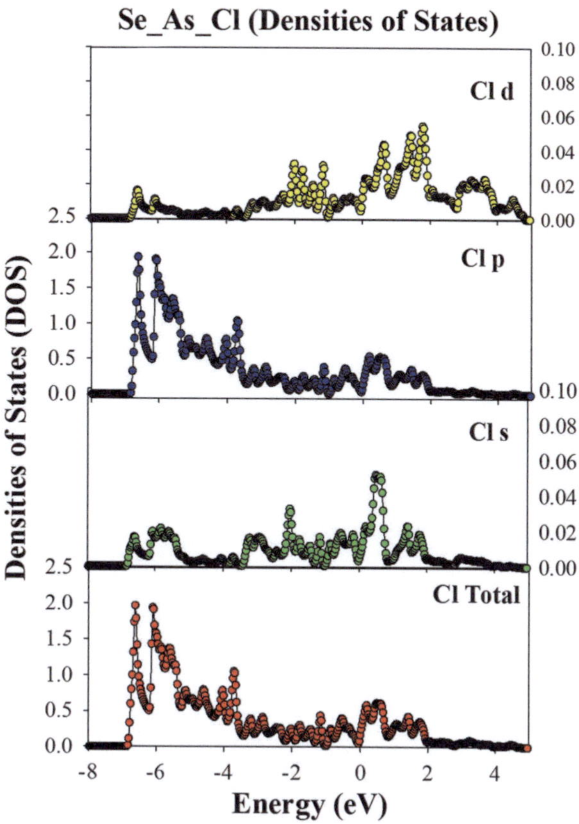

Fig. 1.49 Cl partial density of states of trigonal Se doped with 1 As and 1 Cl

The temperature dependence of the volume of a material which can either form a glass or crystallizes is shown in Fig. 1.50. The crystallization process is manifested by an abrupt change in volume at T_m (melting point), whereas glass formation is characterized by a gradual break in slope. The region over which the change of slope occurs is termed the glass transition temperature. Other thermodynamic quantities such as entropy and enthalpy exhibit a similar behavior.

The basic structural difference between an amorphous solid and its crystalline counterpart is the absence of long range order (LRO) in amorphous materials. They are disordered materials exhibiting disorder such as topological, substitutional, and vibrational. However, a high degree of short range order (SRO) (~10 Å) persists similar to crystalline materials. The absence of LRO in amorphous materials modifies the band structure of these materials and introduces localized states into the forbidden gap.

The concept of localization introduced by Anderson in 1958 [43] in his paper titled "Absence of diffusion in certain random lattices" has had a profound effect on our understanding of the behavior of electrons in non-crystalline materials. Anderson considered a crystalline array of potential wells as illustrated in Fig. 1.51 and

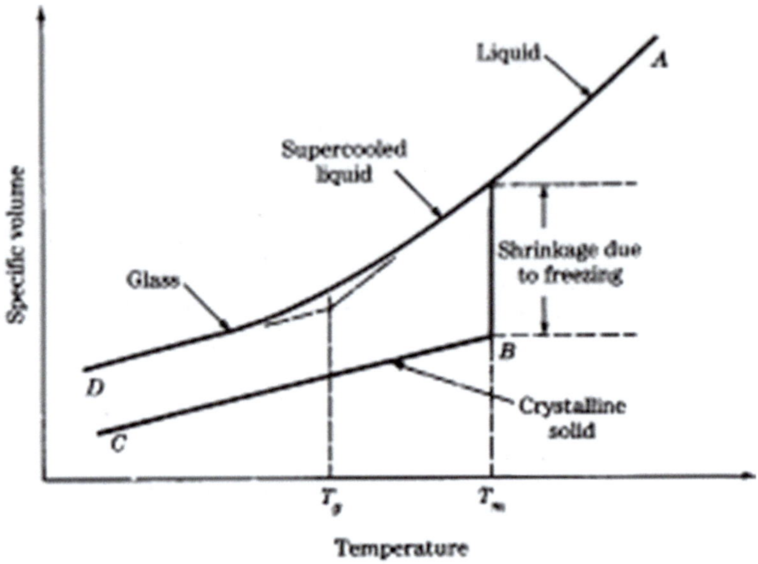

Fig. 1.50 Variation of specific volume with temperature for glass amorphous and crystalline materials

Fig. 1.51 (**a**) Potential well for crystalline lattice. (**b**) The same with a random potential energy at each well, as in Anderson (1958). The density of states (N_E) is also shown

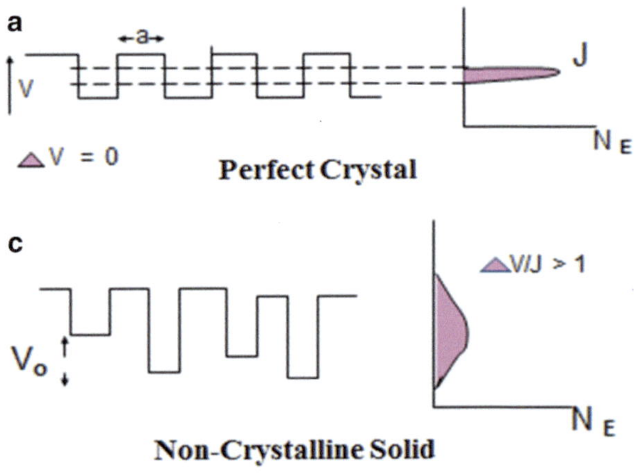

demonstrated the localization of a carrier if the fluctuation in the potential becomes excessive. He examined the effect of variation of fluctuation in the potential well depth V. For $\delta V = 0$ an energy band of width J is formed, where J depends on the overlap of the wavefunction. The wavefunctions in this case are Bloch states and electrons can move freely through the system. If $\delta V/J$ exceeds a critical value of order unity, the electrons no longer move in a band but become localized in the wells and can jump from one well to the next only by phonon cooperation. All the states in the band become localized and this localization of the entire band is termed

"*Anderson localization.*" If the parameter $\delta V/J$ is not large enough to give localization throughout the band, Mott [44] pointed out that states in the band tails would be localized and that energies of localized and extended (non-localized) states would be separated by a sharp energy E_c, known as a mobility edge [19]. For the conduction band of a non-metal, the existence of a mobility edge means that the lowest states have become traps, and conduction will normally be by electrons excited to the mobility edge E_c, so that the conductivity exhibits activated temperature dependence as

$$\sigma = \sigma_o \exp\left[-(E_c - E_f)/kT\right]$$

where E_f is the Fermi level energy, k is the Boltzmann constant.

The absence of long-range order in amorphous semiconductors does not have a major effect on the energy distribution of the electronic levels. Photoemission studies show that the density of states retains a profile similar to that in the crystalline phase and the band gaps survive. Of more importance is the presence of short range order which manifests itself in the form of band tailing and localization of the electronic states therein. The localized states are separated from the extended states in both the conduction and valence bands by mobility edges (Fig. 1.52), so named because conduction through the localized states can occur only by a thermally assisted tunneling process which leads to a low (temperature-activated) mobility. Even beyond the mobility edges the free carrier mobilities are considerably lower than found in good crystalline semiconductors, owing to strong scattering and a short mean free path. A carrier placed in extended states can move through the system without the cooperation of phonons. A localized state on the other hand is characterized by a wave function which falls off exponentially from some origin. A carrier placed in such a state can escape from it only by phonon cooperation. Although some control over the extent of band tailing is possible, the presence of a continuum of localized levels at the band edges is considered as an intrinsic property of amorphous semiconductors. On the other hand, states deeper in

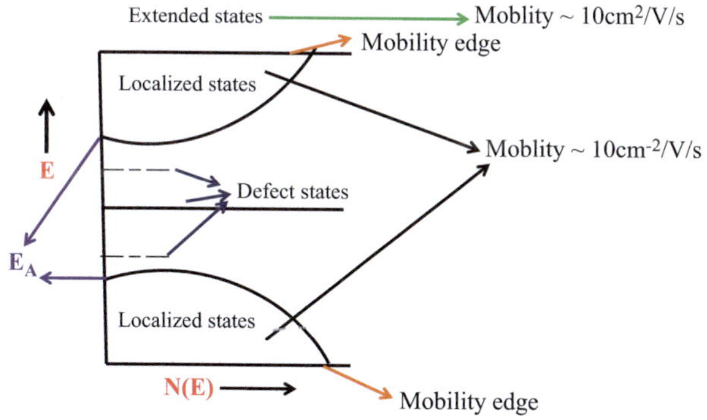

Fig. 1.52 Schematic representation of band structure of amorphous semiconductors

the gap can be classed as extrinsic, in the sense that they arise from point defects whose density is a strong function of preparation conditions or post deposition treatments.

Because in the amorphous semiconductors there is no periodic ordering of atoms, the electron wave function cannot be described meaningfully by Bloch wave functions, and the electron wave vector "k" is no longer a good quantum number. Consequently, no evidence exists as far as the electronic theory of the energy wave vector dispersion relations to describe the band structure of crystalline materials is concerned. Nevertheless, the Schrödinger equation for the solid must have eigenfunctions and eigenvalues so that the concept of density of states is still valid.

In lone-pair, amorphous, chalcogenide semiconductors, of which the chalcogenide glasses are the chief example, a spectrum of localized states tailing into the gap from the valence and conduction bands is introduced by the interaction of lone pairs with each other and with their local environment. Because the lone-pair orbitals are filled at equilibrium, these tail states are paired and thus diamagnetic. Excitation creates empty orbitals (holes) which are compensated by electronic or structural means. In non-compensated, tetrahedral, amorphous materials such as a germanium and silicon, an equilibrium density of dangling bonds should exist, and these should be reflected in paramagnetically detectable, singly occupied states.

Thermodynamic equilibrium does not exist in amorphous materials. They are stable because the thermal energy at room temperature is not sufficient to overcome the potential barrier to change a semiconductor from amorphous to crystalline state. So when we heat an amorphous material, they show a transition to a crystalline state with a release of considerable energy. Amorphous materials can change their structure by external means in such a way that reversible changes may occur in them. The unique properties of amorphous semiconductors and the advantages of producing them in thin film form which do not limit their size and make them adaptable to integration with other solid state technologies, have made these materials the base for a new area of science and technology.

Some of the chalcogen based ternary and quaternary systems such as Ge-Sb-Te, exhibit phase change characteristics exhibiting different electrical and optical properties for different phases of the same composition of the material upon supplying energy using light or heat. On the one hand, they show a pronounced change of optical and electrical properties on crystallization, which is indicative of a significant change in the local atomic arrangement. Nevertheless, crystallization of the amorphous phase proceeds on a very short time scale. This property renders these materials suitable for data storage applications. They are also strong contenders to replace flash memories. Hence, considerable effort has been put forward to identify materials that provide required property combination. Tellurium based compounds such as AgInSbTe and GeSbTe, are fragile glass-formers, and widely used in rewritable optical disks and phase-change memory devices. By applying heat, they can be switched between an amorphous glass and a crystalline state. Thereby they can change their optical and electrical properties and provide the storage of information. Phase-change materials of which $Ge_2Sb_2Te_5$, is a prototypical example, crystallized at a temperature at about 160 °C. The temperature constraint makes the devices not suitable for application when high temperature is required. Hence the search for improved materials and technology is crucial for wider application.

With typical dimensions in the range 1–100 nm, nanocrystals bridge the gap between small molecules and large crystals, displaying discrete electronic transitions reminiscent of isolated atoms and molecules, as well enabling the exploitation of the useful properties of crystalline materials. Bulk semiconductors are characterized by a composition dependent band gap energy (E_g), which is the minimum energy required to excite an electron from the ground state valence band into the vacant conduction energy band. With the absorption of a photon of energy greater than E_g, the excitation of an electron leaves an orbital hole in the valence band. The negatively charged electron and positively charged hole may be mobilized in the presence of an electric field to yield a current, but their lowest energy state is an electrostatically bound electron–hole pair, known as exciton. The exciton has a finite size within the crystal defined by the Bohr exciton diameter which can vary from 1 nm to more than 100 nm depending on the material. If the size of the nanocrystal is smaller than the size of the exciton, the charge carriers become spatially confined, which raises their energy. Therefore, the exciton size delineates the transition between the regime of bulk crystal properties and the quantum confinement regime, in which the optical properties are dependent on the nanocrystal size. In a bulk semiconductor, the conduction electrons are free to move around in the solid, so their energy spectrum is almost continuous, and the density of allowed electron states per energy increases as the square root of energy (Fig. 1.53). If one can reduce the size of the semiconductor so small that the electron feels confined, the continuous spectrum will become discrete and the energy gap will increase as indicated in Fig. 1.53.

Control of size allows engineering of the optical properties of a semiconductor; strong absorption occurs at certain photon energies, at the expense of reduced absorption at other energies. Ideally, the density of states becomes a series of delta functions (narrow spikes) in the quantum dot (QD), so those transitions must be very strong in order to conserve the integrated absorption. Lead sulfide (PbS) quantum dots are attractive candidates for optical refrigeration due to the tunability of their band gap as a function of size and the resulting quantum confinement.

1.7 Defects in Amorphous Chalcogenides

Defects play an important role in the conductivity and photo–induced effects in chalcogenide glasses. Broken bonds resulting in departure from fully coordinated random network are called dangling bonds represented by a "$D^{(0,+,-)}$" in general, by a "T" for tetrahedral, "C" for chalcogenides, "P" for pnictides, etc. An illustration of the three states of defects—D^0, D^+, and D^- is shown in Fig. 1.54. When a

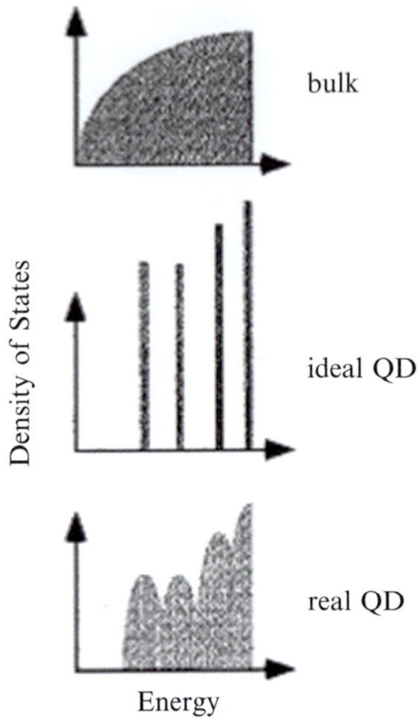

Fig. 1.53 Density of electron states in bulk and size quantized semiconductors (Reprinted with permission [12])

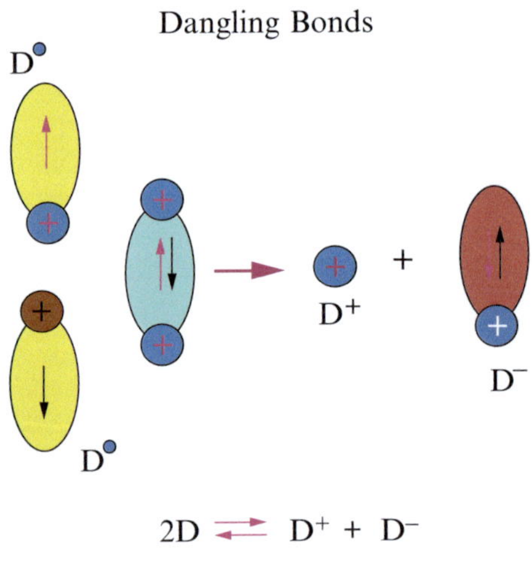

Fig. 1.54 Illustration of three states of dangling bond defects

covalent bond is formed sharing of a pair of electrons from different atoms (D^0 states-single spin, neutral defect), having opposite spins takes place. If due to any reason the bond breaks, it results in the formation of a D^+ center (deficient of an electron) and a D^- center (having an extra spin). In amorphous semiconductors there is a strong tendency for electrons to be paired in bonding configurations, the Coulomb repulsion between electrons at the same site being outweighed by a negative term in the energy due to electron–phonon interaction which leads to configuration changes in the local atomic structure.

Charged Dangling Bond Model:
A common example is that of two dangling bonds at the ends of the selenium chain. When they each contain a single electron defect (D°), transfer of an electron from one chain end to the other (Fig. 1.55a) leads to the formation of two charged defects (D^+ and D^-) according to the reaction [46]:

$$2D^\circ \rightarrow D^+ + D^-$$

Chemically, the reaction can be considered exothermic in nature, since at D^+, an extra bond with a neighboring chain can be formed by utilizing the normally non-bonding lone pair electrons. The coordination of Se atoms at D^+ is therefore

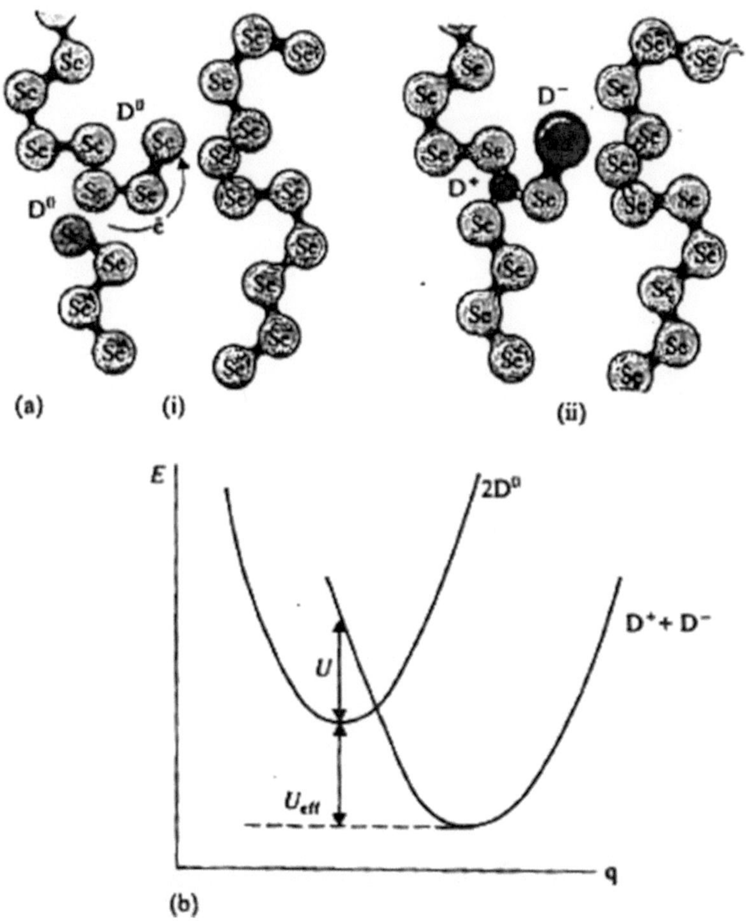

Fig. 1.55 (**a**) Creation of D^+ and D^- defects after charge transfer between two D° defects in a -Se chain [45]. (**b**) Configurational diagram showing negative effective correlation energy during the reaction $2D^\circ \rightarrow D^+ + D^-$ [45]

three, in contrast to that at D^- where it is one and at a normally bonded Se atom where it is two. Mott, Davis, and Street proposed [45] that the lattice distortion at D^- is negligible, at D^+ it is considerable and at D° it is intermediate. The repulsive Hubbard energy U involved in placing an extra electron on the same site to form a D^- center is postulated to be outweighed by the energy gained in forming the extra bond at the D^+ site, rendering the reaction $2D^\circ \rightarrow D^+ + D^-$ exothermic with an overall negative effect correlation energy U_{eff}. On a configurational coordinate diagram as shown in Fig. 1.55b, the positive correlation energy U associated with the two electrons at D^- in the absence of configurational changes become negative (U_{eff}) after lattice relaxation.

The essential feature of the "charged dangling bond" model may be understood by consideration of the monatomic system amorphous Se. the structure of amorphous Se, which is twofold coordinated, is believed to consists mainly of chains (with perhaps a few rings). Any chain end will be the site of a dangling bond orbital, which in the simplest case will contain an unpaired electron and be electrically neutral relative to the bulk; this dangling bond defect can be written as D° or equivalently $C_1{}^\circ$. Mott et al. [47] postulated that, following Anderson [48], electrons residing at D° centers should experience a negative U_{eff} and electron pairing should be energetically favorable as a result of atomic rearrangements.

Electron–lattice coupling is strong in amorphous chalcogenides because of the low atomic coordination resulting in a high degree of network flexibility, but more importantly there exist non-bonding, lone-pair p (pi) orbitals at the chalcogen atoms which may be utilized in forming additional bonds, and which are energetically accessible since they are high lying, forming the top of the valence band [21].

The transfer of an electron from one D° center to another produces one site which has the original dangling bond orbital containing two spin paired electrons and which is consequently negatively charged, and the other which has an empty orbital which is then free to form a dative bond with a lone pair of a fully connected neighboring atom; the defect now becomes threefold coordinated and positively charged. Note that in the latter case the site the defect moves from the originally dangling bond to the atom providing lone-pair electrons for the dative bond which correspondingly becomes over-coordinated; this feature has important ramifications for systems containing more than one type of atom.

The appropriate configurational coordinate, q, in this case may be taken to the sum of the distances between two D° centers and their respective nearest neighbor, but non-directly bonded, atoms, $q_1 + q_2$. Transfer of an electron between D° centers without a change in configurations cost the Hubbard energy U, but the lowest energy state for the two resulting oppositely charged centers lie at a different configurational coordinate because of the atomic relaxation (bonding) that takes place at the D^+ center; the energy for this equilibrium configuration is lower by U_{eff} than that is for the two D° centers.

There are in fact two possible configurations for the neutral defect D° in chalcogenide glasses, one being under-coordinated and the other over-coordinated. Kastner et al. [21] proposed that the symmetric $C_3{}^\circ$ center had the lowest total energy, whereas Mott et al. (1975a) believed that the actual configuration lay between the two, with one elongated bond. However, realistic total-energy calculations by Vanderbilt and Joannopoulos [49] have subsequently shown that C_1° center is the lowest energy configuration, principally because of pi-bonding interactions between dangling-bond orbital on the C_1° and a lone pair on the neighboring C_2° atom.

1.8 Photoinduced Effects in Amorphous Semiconductors

Amorphous semiconductors, being intrinsically metastable in nature, exhibit a wide variety of changes in their physical properties, particularly when irradiated with band gap light. Selenium shows peculiar phenomena when it is illuminated with light comparable to the band gap energy [50, 51]. Some of the important effects observed in chalcogenides are shown in Fig. 1.56. Upon illumination with light, having a photon energy comparable with that of band gap of the material, electron–hole pairs are created, which in turn can contribute to (1) the electrical response of the material (e.g., photoconductivity), (2) recombine, either radiatively (giving rise to photoluminescence (PL)), or (3) non-radiatively. In general, the periodic arrangement of atoms and molecules in crystalline materials imposes constraints on the pathways available to be explored after the creation of electron–hole pair. On the other hand, amorphous semiconductors and insulators exhibit wider variety of photo-induced phenomena because of the freedom and flexibility associated with the atomic structure resulting from the relaxation of crystallographic constraints. At the same time, presence of structural disorder in these materials can lead to localization of electron and hole states at the band edges in the vicinity of the

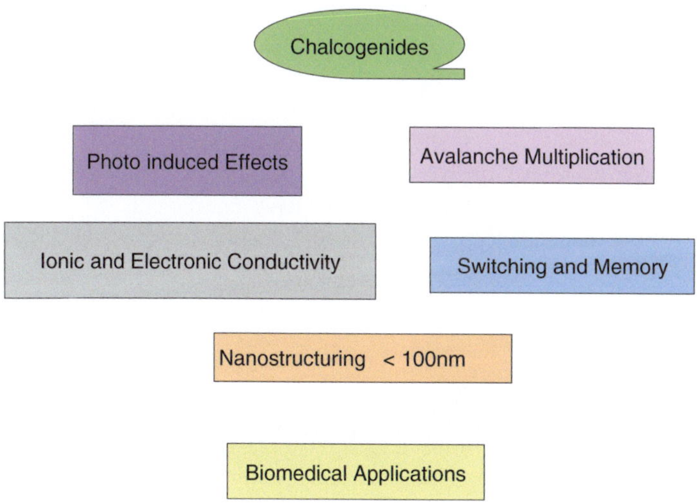

Fig. 1.56 Various effects observed in chalcogenides

gap, which facilitates the likelihood of induced metastable changes. In crystals where such a mechanism is absent, it is only in certain cases (e.g., polaron formation associated with strong electron–phonon interactions) that the otherwise extended wave functions associated with the charge carriers can become sufficiently localized so as to enable local structural changes to occur.

The optical absorption in general may involve either lattice or electronic contributions. The lattice absorption is a result of the interaction of light with optical phonons and is observed in the Infrared region. The electronic absorption involves the electronic states of the energy band. The effect of carriers may be intra- or inter-band in nature. The optical energy gap is calculated by locating the energy at which absorption increases sharply and by considering the nature of the transition (viz., direct or indirect) [52]. Both transitions involve the interaction of an electromagnetic wave with an electron in the valence band, which is raised across the fundamental gap to the conduction band. However, indirect transitions also involve simultaneous interaction with lattice vibrations. Thus, the wave vector of an electron can change in the optical transition, the momentum change being taken or given up by phonons. The radiation imparts negligible momentum to the electron.

If exciton formation (electron–hole interactions) is neglected, the forms of the absorption coefficient "α" as a function of phonon energy $h\omega$ depend on the dependence on energy of $N(E)$ for the bands containing the initial and final states. For simple parabolic bands $N(E) \propto E^{1/2}$ and for direct transitions:

$$\alpha n_o h\nu_o \sim (h\nu - E_o)^n$$

where $n = 1/2$ or $3/2$ depending on whether the transition is allowed or forbidden in the quantum mechanical sense, E_o is the optical gap and no is the refractive index.

For indirect transitions:

$$\alpha n_o \hbar\omega = \frac{(\hbar\omega - E_o + h\nu_{\text{ph}})^n}{\left(\exp\left(\frac{h\nu_{\text{ph}}}{kT}\right) - 1\right)} + \frac{(\hbar\omega - E_o - h\nu_{\text{ph}})^n}{\left(1 - \exp\left(\frac{-h\nu_{\text{ph}}}{kT}\right)\right)}$$

The two terms represent the contributions from transitions involving phonon absorption and emission respectively. And have coefficients of proportionality and temperature dependences. For allowed transitions $n = 2$ and for forbidden transitions $n = 3$.

1.9 Avalanche Multiplication

The phenomenon of avalanche multiplication is commercially utilized in avalanche photodiodes. Stable avalanche multiplication of extrinsic charge carriers has been shown to occur in amorphous selenium (a-Se) at electric fields $>8 \times 10^5$ V/cm [53–

55]. The phenomenon is already commercialized in the Harpicon camera tubes [55, 56].

Avalanche breakdown in semiconductors is caused by the cumulative multiplication of free charge carriers under the action of an applied electric field. Some of the carriers gain enough energy to liberate new electron–hole pairs by impact ionization which in turn can generate further pairs, i.e., an avalanche takes place. The conduction electron in a periodic potential energy function has a traveling Bloch wave function. Phonons in the crystal spatially deform the potential energy and thereby cause the electron wavefunction to change in a manner that is tantamount in scattering. The mean free path of the electron at room temperature is typically of the order of several hundreds of mean interatomic separations and the drift mobility is for silicon (a typical semiconductor) is about 1400 $cm^2\ V\ s^{-1}$. At sufficiently high fields (~$10^5\ V\ m^{-1}$) the electron drift velocity saturates at ~$10^5\ m\ s^{-1}$. As the field increases further, the conduction electron is able to impact ionize the crystal to generate free electron–hole pairs and cause avalanche. The avalanche process occurs when a critical field is reached across the semiconductor. The critical field is the voltage gradient that just causes avalanche in a given specimen of semiconductor and is typical for materials used. The process of avalanche multiplication of charge carriers in amorphous selenium (a-Se), is known since many years [57, 58] and has received a significant attention of the scientific community due to the application of these materials in X-ray imaging devices [56, 57, 59] and in Harpicon tubes—ultrahigh sensitive TV pickup tubes that use avalanche multiplication in a Se to capture images at extremely low light intensities [60–62]. Although many features of the avalanche in a-Se are similar to those in crystalline semiconductors, there are some important distinctions between these phenomena. The most prominent is the temperature dependence of impact ionization rates for electrons "α" and for holes "β" [59]. In crystalline semiconductors, α and β decrease with increasing temperature due to greater phonon scattering at higher temperatures [63]. In contrast, the impact ionization rates in a-Se increase with temperature implying that the mean free path of hot carriers is not controlled by electron–phonon interactions. The lucky-drift (LD) model for impact ionization, earlier proposed by Ridley [64] was extended by Jandieri et al. to account for avalanche phenomena in a-Se [65]. Although the LD model clarifies in general the origin of avalanche multiplication in amorphous media, it is still unclear why a-Se is still the only amorphous semiconductor that clearly evidences reproducible avalanche multiplication [65].

1.10 Ionic and Electronic Conductivity

Solid state Ionic conductors are crucial to a number of major technological developments, particularly in the domains of energy storage and conversion and in environmental monitoring such as battery, fuel cell, and sensor technologies. Solid state ionic membranes based on fast ion conductors potentially provide important advantages over liquid electrolytes, including the elimination of sealing problems and the ability to miniaturize electrochemical devices. Ionic conduction refers to the movement of charges through a semiconductor due to the displacement of ions within the crystal lattice, such movement being maintained by a continuous supply of external energy.

During the last three decades, considerable scientific attention has been paid to the ion-selective electrodes (ISEs) with functional membranes based on chalcogenide glasses. One of the main requirements upon materials used for ISEs is that they have a high ionic conductivity. The latter increases in the order $MeS \rightarrow MeSe \rightarrow MeTe$, which determines the choice of the system. An ion-selective electrode (ISE) is a transducer (or sensor) that converts the activity of a specific ion dissolved in a solution into an electrical potential, which can be measured by a voltmeter or pH meter. The voltage is theoretically dependent on the logarithm of the ionic activity, according to the Nernst equation. The sensing part of the electrode is usually made as an ion-specific membrane, along with a reference electrode.

Chalcogenide glasses have been recommended as exhibiting several advantages compared to the corresponding chalcogenide crystals. For example, they have shown higher stability against dissolution in the electrolyte and chemical attack in corrosive media and better selectivity in the presence of some interfering ions [66–68]. They have therefore been successfully applied as membrane materials for metal ion detection such as Ag^+ [67], Cu^{2+} [69, 70], Pb^{2+} [71], Cd^{2+} [71, 72], Cr^{2+} [73], Br^- [72], Hg^{2+} [74], and Fe^{3+} [75].

ChG exhibit better chemical durability in acidic and redox media [76–79], and in many cases, possess better selectivity and reproducibility of the analytical characteristics, and longer time of life [80] as compared with the polycrystalline chalcogenides, which are broadly used as membrane materials, e.g., insoluble pressed sulfides and selenides [80, 81]. These sensors are simple in construction, and special conditions are not required, neither for their storage, nor for the surface recovering even after continuous measurements. Due to these advantages, ChG became preferred membrane materials in ISEs for Ag^+-, Cd^{2+}-, Pb^{2+}-, and Cu^{2+}-ions determination [77, 82] instead of their crystalline analogous.

In the literature, data on the study of chemical sensors based on chalco-halide glasses are also published [80, 83, 84]. The introduction of halogen atoms in the chalcogenide glass former network enhances the defectiveness and flexibility of the structure [85], which facilitates the ion mobility and the exchange between the membrane surface layer and the investigated solution. The sensitivity mechanism of $AgI–Sb_2S_3$ membranes against heavy metal ions is connected with the formation of insoluble sulfides at the surface active centers [85]. The selectivity of ISEs based on the $AgI–Sb_2S_3$ and other sulfide glasses is closely connected with the solubility of the corresponding crystalline sulfides, i.e., the decrease of the selectivity against metal ions in the order $Cu^{2+} > Pb^{2+} > Cd^{2+} > Zn^{2+} > Ni^{2+}$ corresponds to the increase of the solubility of their sulfides.

1.11 Nanostructuring

Selenium is a highly important inorganic semiconducting material with large Bohr radius. In addition to its interesting physical properties such as thermoelectric and nonlinear optical responses, high conductivity, and piezoelectric effects [86], Se in appropriate amounts is an essential element for living organisms [87]. Se has been shown to prevent cancer in numerous animal model systems when fed at levels exceeding the nutritional requirement [88]. Clarke et al. showed cancer chemopreventive efficacy using a Se supplement in humans [89]. Protein–selenium bioconjugate nanomaterials are reported to be cytotoxic for tumor cells [90]. Se NPs have also been studied for their antioxidant activity [22]. Use of BSA in synthesizing the shape controlled BSA-Se bioconjugate semiconducting nanomaterials are expected to have several advantages over simple Se nanomaterials as far as their biomedical applications are concerned.

We have addressed a very significant and important problem of shape controlled synthesis of ordered bio-nanomaterials in the presence of proteins [33]. Se particles were synthesized by chemical reduction of sodium selenite by hydrazine. In a typical procedure, 10 ml of aqueous BSA ($1–10 \times 10^{-4}$ g/ml) along with 2.5 % hydrazine was taken in a round bottom glass flask. Under constant stirring, sodium selenite (6–25 mM) was added in it. After mixing all the components at room temperature, the reaction mixture was kept in water thermostat bath at 85 °C for 48 h. The color of the solution changed from colorless to deep orange within 2 h and remained same till 48 h. Initial pH of BSA + water solution was 6.7 which increased to 7.8 upon addition of hydrazine. Addition of Na_2SeO_3 further increases the pH = 9 due to the following reaction:

$$N_2H_4(aq) + SeO_3^{2-}(aq) \rightarrow Se(s) + N_2(g) + 2OH^-(aq) + H_2O(l)$$

Within 2 h, pH further rises to 11 due to the denaturation of BSA (at 85 °C) and remained fairly constant for 48 h. The samples were purified by spinning the reaction product at 10,000 rpm for 10 min with repeated washing from distilled water.

Figure 1.57a shows the FESEM image of Se nanobars (NBs) synthesized with $[Na_2SeO_3] = 6$ mM in the presence of BSA $= 1 \times 10^{-4}$ g/ml. BSA macromolecule contains 607 residues with average molecular weight 66432 Da [91]. Assuming all residues of the same nature, each one will contribute about 109.4 mol, and thus $[Na_2SeO_3/BSA]$ mole ratio = 6.7/residue. Bundles of fine multifaceted bars are evident (Fig. 1.57b) with average aspect ratio 4.7 ± 1.8. Figure 1.57c indicates (block arrows) the presence of conjugated BSA on the surface of NBs. A close-up bright field TEM image (inset) further confirms this. It appears that most of the surface of each NB is covered with conjugated BSA. HAADF image (Fig. 1.57d) clearly shows bright surface patches on each NB indicated by block arrows. A dark dotted circle encloses few spots of BSA covered areas of the sample that were damaged by the beam. TEM image of a single bar is shown in Fig. 1.57e which is

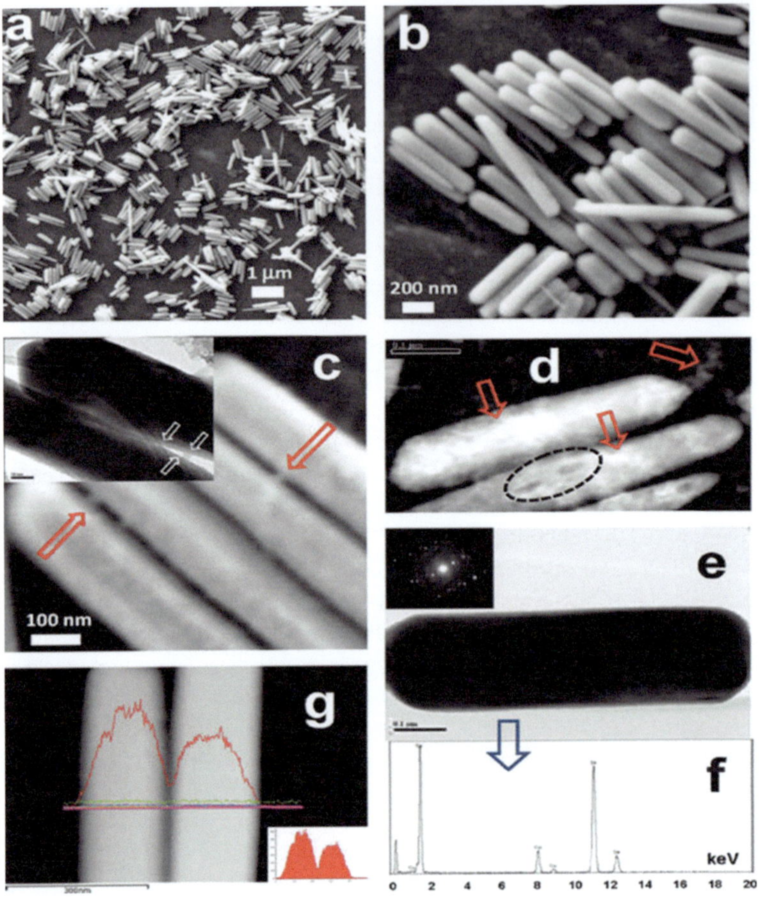

Fig. 1.57 (**a**) Shows low resolution FESEM image of several bundles of Se nanobars. (**b**) Magnified image of several multifaceted nanobars. (**c**) A close-up images shows surface adsorbed BSA and presence of BSA in between the nanobars. *Inset*, dark field image showing the BSA coating. (**d**) HAADF image showing bright patches of adsorbed BSA and few dark patches were created by the exposure to electron beam. (**e**) TEM images of a single nanobar with SAED image (*inset*) and (**f**) its EDX spectrum. (**g**) A line EDX line spectrum across two fused nanobars showing emission due to Se

positioned along the {110} zone as evident from the diffraction image (inset). EDX analysis of this particle further confirms the presence of Se (Fig. 1.57f). A few weak Cu emissions are due to the Cu grid. Se EDX line spectrum (Fig. 1.57g) helps us to understand the extent of fusion among NBs in a bundle which is presumed to be facilitated by the conjugated BSA. Se has a trigonal arrangement due to spiral chains of Se atoms associated with each other through van der Waals interactions in a hexagonal lattice [92]. Such an arrangement provides a unidirectional growth tendency. A thermodynamically stable t-Se is expected to favor the growth along ⟨100⟩ direction and eventually lead to the formation of 1D nanostructures like NBs.

An effective capping ability of BSA helps to attain well-defined rod shape geometries. BSA binds endogenous as well as exogenous substrates in its hydrophobic pockets [93]. The overall shape of BSA macromolecule is oblate ellipsoid in its native state but denaturation is often followed by a massive "unfolding" of the protein. The secondary structure consists of hydrogen bonded α-helices and β-sheets, and is called the large-scale structure. At 85 °C, BSA is considered to be in its unfolded form because the overall denaturation temperature is usually reported to be close to 60 °C depending on different methodologies and detection techniques [94]. Although the exact mechanism is still unclear, the unfolded BSA with predominantly hydrophobic domains might be adsorbed favorably on freshly cleaved Se nucleating centers [22, 90, 95]. Thus, a selective adsorption of unfolded form of BSA on low atomic density {100} crystal planes will direct the crystal growth on {111} planes and will result in the rod shape formation. The adsorption of BSA on Se NBs is further confirmed from XPS studies.

A low magnification FESEM image of such several groups of spheres is shown in Fig. 1.58a. Size distribution histogram computes an average size of a sphere equal to 346 ± 110 nm. A high magnification image (Fig. 1.58b) shows several interconnected spheres in a single group. The spheres are in fact fused together sidewise (Fig. 1.58c). In order to further evaluate the mode of fusion, we carefully selected a pair of spheres and got the HAADF image. This image (Fig. 1.58d) fully confirms the fact that the spheres are indeed fused with each other. EDX line spectrum (Fig. 1.58e) running across the two balls further confirms that the fused part does not contain BSA because full Se emission is visible from the total surface area. Figure 1.58f shows a TEM image of a single ball. Surprisingly, the diffraction image (inset) of this ball gives no diffraction rings suggesting the amorphous nature of spheres. EDX spectrum (Fig. 1.58g) further proves that this ball is entirely made up of Se. Hence, the interparticle fusion of such spheres is predominantly caused by their amorphous nature and not because of BSA capping.

The results conclude that BSA works well in controlling the overall geometry of nanoparticles when [Na_2SeO_3/BSA] mole ratio is ≈6. At this mole ratio, denatured BSA becomes even more hydrophobic due to the neutralization of oppositely charged sites by SeO_3^- ions. A predominantly hydrophobic BSA is a better shape directing agent. But at too high mole ratio = 28, BSA proves to be a poor capping agent due to the presence of too many nucleating centers whose growth cannot be simultaneously controlled by BSA macromolecules. On the other hand, when the mole ratio ≈1, then unfolded BSA macromolecule works as a soft template by accommodating maximum nucleating centers on it and thereby facilitating the Ostwald ripening.

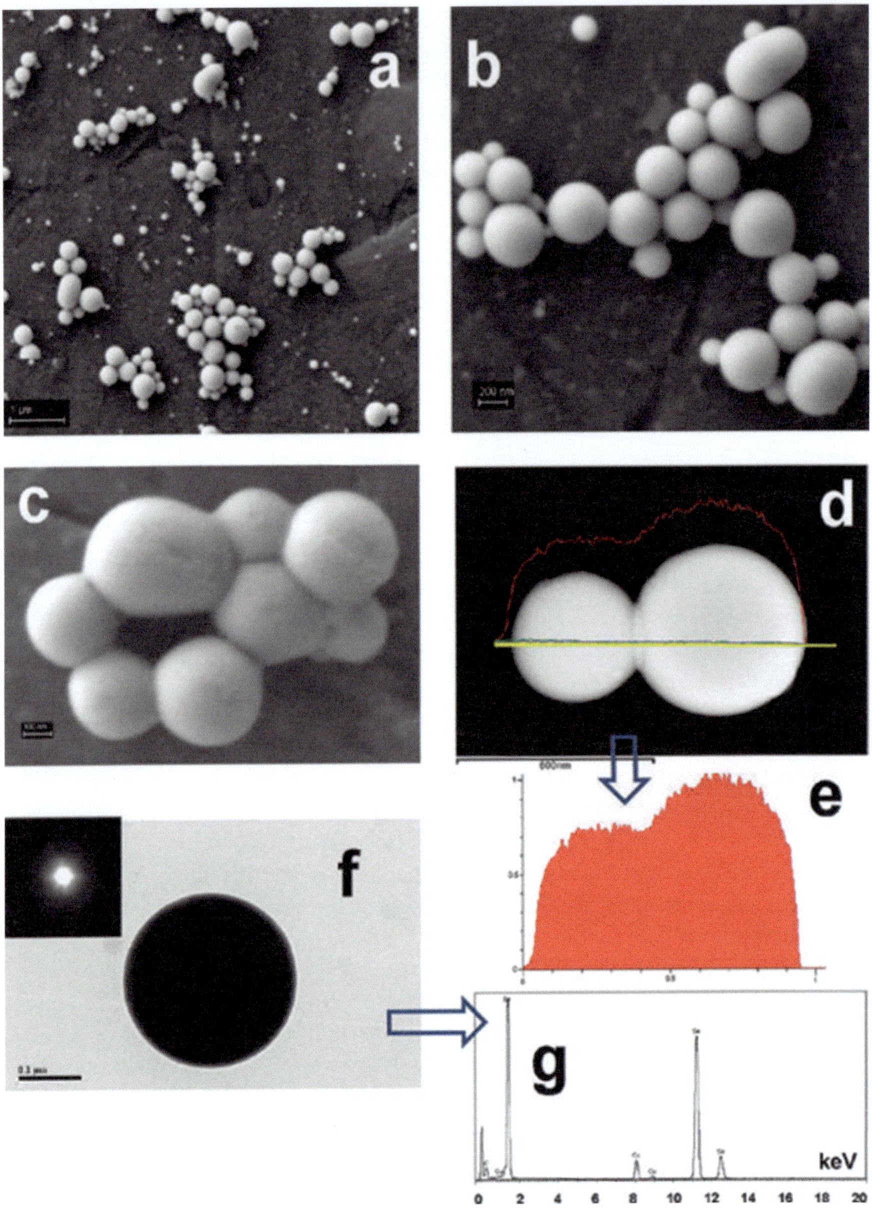

Fig. 1.58 (**a**) Shows low resolution FESEM image of several groups of Se spheres. (**b**) Magnified image of a single group containing several spheres. (**c**) A close-up image showing fused spheres. (**d**) HAADF image showing a pair of fused spheres along with (**e**) line spectrum across them indicating the emission due to Se. (**f**) TEM image of a single sphere with SAED image (*inset*) and (**g**) its EDX spectrum

1.12 Biomedical Applications: Cancer Chemopreventive Effects

Selenium is an essential micronutrient in the diet of many life forms, including humans and other mammals. Significant health benefits have been attributed to this element. It is rapidly becoming recognized as one of the more promising cancer chemopreventive agents [96] and there are strong indications that it has a role in reducing viral expression [97], in preventing heart disease and other cardiovascular and muscle disorders [98], and in delaying the progression of AIDS in human immunodeficiency virus-infected patients [99]. Additional evidence suggests that selenium may have a role in mammalian development [100], in immune function [101], in male reproduction [102], and in slowing the aging process [103]. Despite the many potential health benefits of selenium, the means by which this element promotes better health are only just beginning to be elucidated [104, 105]. There are about 20 known selenium-containing proteins in mammals [106], and it would seem very likely that several of these are mediators of health benefits of dietary selenium.

Selenium compounds exert their biological effects either directly or by being incorporated into enzymes and other bio-active proteins. The main inorganic dietary form of selenium is sodium selenite (Na_2SeO_3). In the organic forms selenomethionine and selenocysteine, a selenium atom is present in the position occupied by a sulfur atom in the amino acids methionine and cysteine. Se-methylselenocysteine is selenocysteine modified by the replacement of the hydrogen atom with a methyl group on the selenium atom.

1.13 Brief Review

Research and Development in the field of Amorphous Chalcogenides [39]:

- 1870—Earliest experimental data on oxygen free glasses reported—Schilz-Sellack
- 1902—Optical properties of vitreous Selenium—Wood, Meyer
- Systematic research on chalcogenide glasses—Middle Twentieth Century
- 1950—As_2S_3—Frerichs—"New optical glasses transparent in infrared up to 12 mm"—Started Se and binary S based glasses
- Winter-Klein—Published papers on numerous Chalcogenides prepared in vitreous state,
- 1955—First research program in the field of ChGs—B.T. Kolomiets, Goryunova, and R.L. Myuller
- 1968—S.R. Ovshinsky—Switching and memory effects in ChGs
- 1969—Amorphon Model for amorphous germanium with little success for non-tetrahedral Chalcogenides—R. Grigorovici

- 1974—Mott, Davis, and Street—Model for states in the gap for non-crystalline chalcogenides, theory of dangling bonds, electronic structure of magnetic and disordered systems
- 1977—Nobel Prize—Sir Nevill Mott, P.W. Anderson, John H. Van Vleck
- 1978—Ordering at not too large scale but more extended than SRO—in order to explain FSDP—Popescu
- 1979—ditto—Phillips

1.14 Effects Discovered in Chalcogenide Glasses [39]

- Elongation effect—1919—Vonwiller
- Switching—Ovshinsky—1959
- Polarization Effect—Kolomiets and Lyubin—1962
- Photo-dissolution and photo-doping—Kostishin et al.—1966
- Photo-crystallization—Dresner and Stringfellow—1968
- Photo-polarization—Andreichin—1970
- Photo-darkening/photo-bleaching—Berkes et al. 1971, Keneman 1971, de Neufville, Moss, Ovshinsky—1973
- Photo-vaporization—Janai and Rudman—1974
- Photo-polymerization—deNeufville—1975
- Photo-expansion—Hamanaka et al.—1976
- Oscillatory transmission—Hajto et al.—1977
- Ultrasonic induced modifications of optical properties—Okano—1978
- Anisotropy of the transmittance—Hajto and Ewen—1979
- Photo-induced softening and hardening—Tanaka—1980
- Photo-contraction—Bhanwar Singh et al.—1982
- Thermal dissolution—Kolobov et al.—1985
- Gamma-radiation induced effect—Shoptyuk et al.—1987
- Photo-amplified oxidation—Tichy et al.—1987
- Photo-induced change of ac transport—Shimakawa et al.—1987
- Photo-anisotropy—Weigert-1920, Lee—1990
- Photo-èlastic birefringence—K. Tanaka—1990
- Mechanically induced anisotropy—Tanaka—1990
- Photo-induced amorphization—Elliott and Kolobov—1991
- Photo-induced girotropy—Lyubin and Tikhomirov—1991
- Photo-induced scattering of Light—Lyubin and Tikhomirov—1991
- Photo-plasticity—Trunov and Anchugin—1992
- Athermal photo-induced transformation—Kolobov et al.—1992
- Acoustic-optical effect—Abdulhalim et al. 1993 Laine and Seddon—1995
- Giant photo-expansion—Hisakuni and Tanaka—1994
- Photo-induced fluidity—Hisakuni and Tanaka—1995
- Optomechanical effect—Krecmer and Elliott—1997
- Anisotropic opto-mechanical effect—Krecmer et al.—1997

- Light stimulated interdiffusion—Kikineshi—2001
- Self-organization effect—Boolchand—2001
- Ion induced interdiffusion—Kikineshi—2001
- Photo-induced ductility—Kastrissios et al.—2002
- Thermo-stimulated inter-diffusion—Malyovanik, Kikineshi et al.—2003
- Deuteron-radiation induced effect—Kokenyesi—2003
- Polarization dependent photo-plasticity—Trunov and Bilanich—2004
- Multistate switching effect—Ovshinsky—2004

References

1. A. Owens, A. Peacock, Compound semiconductor radiation detectors. Nucl. Instrum. Methods. Phys. Res. **531**(p), 18–37 (2004)
2. T.E. Schlesinger, J.E. Toney, H. Yoon, E.Y. Lee, B.A. Brunett, L. Franks, R.B. James, Cadmium zinc telluride and its use as a nuclear radiation detector material. Mater. Sci. Eng. **32**, 103 (2001)
3. P.N. Luke, M. Amman, *15th International Workshop on Room-temperature Semiconductor X- and Gamma-Ray Detectors*, San Diego, CA, 2006
4. J.A. Peters, Z. Liu, B.W. Wessels, I. Androulakis, C.P. Sebastien, H. Li, M.G. Kanatzidis, Alkali metal chalcogenides for radiation detection. Mater. Res. Soc. Symp. Proc. **1341**, 133–138 (2011)
5. A.D. Andreev, A.A. Lipovski, Phys. Rev. B **59**, 15402 (1999)
6. H. Du, C. Chen, R. Krishnan, T.D. Krauss, J.M. Harold, F.W. Wise, M.G. Thomas, J. Silox, Nano Lett. **2**, 1321 (2002)
7. B.W. Wehrenberg, C. Wang, P. Guyot-Sionnest, J. Phys. Chem. B **106**, 10634 (2002)
8. J.M. Petryga, R.D. Schaller, D. Werder, M.H. Stewan, V.I. Klimov, J.A. Hollingsworth, J. Am. Chem. Soc. **126**, 11752 (2004)
9. R.D. Schaller, V.I. Klimov, Phys. Rev. Lett. **92**, 186601–1 (2004)
10. R.D. Schaller, M.A. Petruska, V.I. Klimov, J. Phys. Chem. **107**, 13765 (2003)
11. V.I. Klimov, A.A. Mikhailovsky, S. Xu, A. Malko, J.A. Hollingsworth, C.A. Leatherdale, H.J. Eisler, M.G. Bawendi, Science **290**, 314 (2000)
12. F.W. Wise, Acc. Chem. Res. **33**, 773 (2000)
13. T.D. Krauss, F.W. Wise, D.B. Tanner, Phys. Rev. Lett. **76**, 1376 (1996)
14. T.D. Krauss, F.W. Wise, Phys. Rev. Lett. **79**, 5102 (1997)
15. I. Kang, F.W. Wise, J. Opt. Soc. Am. B **14**, 1632 (1997)
16. A. Olkhovets, R.-C. Hsu, A. Lipovskii, F.W. Wise, Phys. Rev. Lett. **81**, 3539 (1998)
17. N.F. Borrelli, D.W. Smith, J. Non-Cryst. Solids **180**, 25 (1994)
18. A. Lipovskii, E. Kolobkova, V. Petrikov, I. Kang, A. Olkhovets, T.D. Krauss, M. Thomas, J. Silcox, F.W. Wise, Q. Shen, S. Kycia, Appl. Phys. Lett. **71**, 3406 (1997)
19. N.F. Mott, Adv. Physiol. Educ. **16**, 49 (1967)
20. A. Nwachuku, M. Kuhn, Appl. Phys. Letts. **12**, 163 (1968)
21. M. Kastner, Bonding bands, lone pair bands, and impurity states in chalcogenide semiconductors. Phys. Rev. Letts. **28**(6), 355 (1972)
22. R.A. Zingaro, W.C. Cooper (eds.), *Selenium* (Litton Educational Publishing, New York, 1974), p. 25. and 174–217
23. L.I. Berger, *Semiconductor Materials* (CRC Press, Boca Raton, FL, 1997), p. 86
24. Q. Lu, F. Gao, S. Komarneni, Chem. Mater. **18**, 159 (2005)
25. B. Gates, Y. Yin, Y. Xia, J. Am. Chem. Soc. **122**, 12582 (2000)

26. S. Xiong, B. Xi, W. Wang, C. Wang, L. Fei, H. Zhou, Y. Qian, Cryst. Growth Des. **6**, 1711 (2006)
27. Y. Ma, L. Qi, W. Shen, J. Ma, Langmuir **21**, 6161 (2005)
28. Z. Wang, L. Wang, H. Wang, Cryst. Growth Des. **8**, 4415 (2008)
29. Q. Lu, F. Gao, S. Komarneni, Langmuir **21**, 6002 (2005)
30. J. Hu, T.W. Odom, C.M. Lieber, Acc. Chem. Res. **32**, 435 (1999)
31. R.I. Baitser, V.V. Vainberg, S.S. Varshava, J. Phys. IV **6**, C3–429 (1996)
32. B. Mayers, G. Byron, Y. Yin, Y. Xia, Adv. Mater. **13**, 1380 (2001)
33. M.S. Bakshi, V.S. Jaswal, G. Kaur, T.W. Simpson, P.K. Banipal, T.S. Banipal, F. Possmayer, N.O. Petersen, J. Phys. Chem. C **113**, 9121 (2009)
34. T.W. Smith, S.D. Smith, S.S. Badesha, J. Am. Chem. Soc. **106**, 7247 (1984)
35. C. Kittel, *Introduction to Solid State Physics* (John Wiley & Sons Inc., New York, 1996)
36. Y.M. Yiu, M.W. Murphy, L. Liu, Y. Hu, T.K. Sham, AIP Conf. Proc. **1590**, 26 (2014), M.W. Murphy, M.J. Ward, L. Liu, Y.M. Yiu, Y. Hu, J.A. Zapien, T.K. Sham, Examination of the electronic and optical properties of CdS_xSe_{1-x} nanostructures by XANES, XEOL and DFT, J. Appl. Phys. **116**, 193709 (2014)
37. A.D. Becke, E.R. Johnson, J. Chem. Phys. **124**, 221101 (2006). doi:10.1063/1.2213970
38. M.V. Yakusheva, A.V. Mudryib, V.F. Gremenokb, V.B. Zalesskic, P.I. Romanovc, Y.V. Feofanova, R.W. Martina, R.D. Tomlinson, J. Phys. Chem. Solids **64**, 2005–2009 (2003)
39. M. Popescu, Non-crystalline solids, past, present, future. J. Non-Cryst. Solids **352**(9–20), 887–891 (2006)
40. F. Serdouk, M.L. Benkhedir, Physica B **459**(122) (2015)
41. M.L. Benkhedir, M. Brinza, G.J. Adriaenssens, J. Phys. Condens. Matter **16**, S5253 (2004)
42. S. Chand, G.D. Sharma, R.C. Batheja, S. Chandra, Appl. Phys. Lett. **61**, 1915 (1992)
43. R.W.G. Wyckoff, *Crystal Structures*, vol. 1–6 (John Wiley and Sons, Inc., New York, 1963–1971)
44. P.W. Anderson, Absence of diffusion in certain random lattices. Phys. Rev. **109**(5), 1492–1505 (1958)
45. M.H. Brodsky, *Amorphous Semiconductors* (Springer Verlag, New York, 1985), p. 54
46. M.H. Cohen, H. Fritzsche, S.R. Ovshinsky, Phys. Rev. Lett. **22**, 1063 (1969)
47. N.F. Mott, E.A. Davis, R.A. Street, Philos. Mag. **32**, 961 (1975)
48. N.F. Mott, Philos. Mag. **34**, 1101 (1976)
49. P.W. Anderson, Model for the electronic structure of amorphous semiconductors. Phys. Rev. Lett. **34**(15), 14 (1975)
50. D. Vanderbilt, J.D. Joannopoulos, Theory of defect states in glassy selenium. Phys. Rev. B **22**, 2927 (1980)
51. V.G. Zhdanov, B.T. Kolomiets, V.M. Lyubin, V.K. Malinowsky, Phys. Status Sol. A **52**, 621 (1979)
52. K. Shimakawa, A. Kolobov, S.R. Elliott, Adv. Physiol. Educ. **44**, 475 (1995)
53. M. Cardona, in *Semiconductors and Semimetals*, ed. by R.K. Williardson, A.C. Beer (Academic Press, New York, 1967), p. 3
54. G. Juska, K. Arlauskas, Phys. Status Sol. A **59**, 389 (1980)
55. K. Tanioka, J. Yamazaki, K. Shidara, K. Taketoshi, T. Kawamura, S. Ishioka, K. Takasaki, IEEE Electron. Dev. Lett. **EDL-8**, 388 (1987)
56. K. Tsuji, T. Ohshima, T. Hirai, N. Gotoh, K. Tanioka, K. Shidara, Mater. Res. Soc. Symp. Proc. **219**, 507 (1991)
57. F. Okano, J. Kumada, K. Tanioka, SMPTE J. **99**, 612 (1991)
58. N. Hindley, J. Non-Cryst. Solids **5**, 31 (1970)
59. K. Tsuji, Y. Takasaki, T. Hirai, K. Taketoshi, J. Non-Cryst. Solids **114**, 94 (1989)
60. D. Hunt, S. Kirby, L. Rowlands, Med. Phys. **29**, 2464 (2002)
61. S. Kasap, J.A. Rowlands, S.D. Baranovskii, K. Tanioka, J. Appl. Phys. **96**, 2037 (2004)

62. K. Tanioka, J. Yamazaki, K. Shidara, K. Taketoshi, T. Hirai, Y. Takasaki, Adv. Electron. Electron. Phys. **74**, 379 (1988)
63. S. Kasap, J. Rowlands, K. Tanioka, A. Nathan, in *Charge Transport in Disordered Solids*, ed. by S. Baranovskii (Wiley, Chichester, 2006)
64. S.M. Sze, *Physics of Semiconductor Devices*, 2nd edn. (Wiley, New York, 1981)
65. B.K. Ridley, Lucky-drift mechanism for impact ionization in semiconductors. J. Phys. C **16**, 3373–3388 (1983)
66. K. Jandieri, O. Rubel, S.D. Baranovskii, A. Reznik, J.A. Rowlands, S.O. Kasap, Lucky-drift model for impact ionization in amorphous Semiconductors". J. Mater. Sci. Mater. Electron. **20**, S221–S225 (2009)
67. Y.G. Vlassov, E.A. Bychkov, Ion-Sel. Electrode Rev. **9**, 5 (1987)
68. Y.G. Vlassov, E.A. Bychkov, Sens. Actuators **12**, 275 (1987)
69. Y.G. Vlassov, E.A. Bychkov, B.L. Seleznev, Sens. Actuators B **2**, 23 (1990)
70. Y.G. Vlassov, E.A. Bychkov, A.M. Medvedev, Anal. Chim. Acta **185**, 137 (1986)
71. A.E. Owen, J. Non-Cryst. Solids **35–36**, 999 (1980)
72. N. Tohge, M. Tanaka, Chalcogenide glass electrodes sensitive to heavy metal ions. J. Non-Cryst. Solids **80**, 550–556 (1986)
73. Y.G. Vlasov, E.A. Bychkov, Ion-selective chalcogenide glass electrodes. Ion sel. Electrode Rev. **9**(1), 5–93 (1987)
74. J.H. Christensen, P. Clemmensen, G.H. Hansen, K. Lilcorp, J. Mortensen, Response characteristics and applications of chalcogenide glass Cr (VI) selective electrode. Sens. Actuators B Chem. **45**(3), 239–243 (1997)
75. M. Miloshova, E. Bychkov, V. Tsegelnik, V. Strykanov, H. Klewe-Nebenius, M. Bruns, W. Hoffman, P. Papet, J. Saradin, A. Padel, M. Ribes, Tracer and surface spectroscopy studies of sensitive mechanism of mercury ion chalcogenide glass sensors. Sens. Actuators B Chem. **57** (1–3), 171–178 (1999)
76. C.E. Koenig, E.W. Grabner, Electroanalysis **7**(11), 1090 (1995)
77. P.G. Protasova, Phys. Chem. Glasses **16**(5), 831 (1990)
78. Y.G. Vlasov, E.A. Bychkov, Ion-Sel. Electrode Rev. **9**, 3 (1987)
79. Y.G. Vlasov, E.A. Bychkov, A.V. Legin, Electrokhimia **21**, 1444 (1989)
80. Y.G. Vlasov, E.A. Bychkov, A.D. Safarov, P.P. Antonov, M.S. Miloshova, Zh. Anal. Khim. **40**(8), 1438 (1985)
81. J. Koryta, K. Stulik, in *Ion-Selective Electrodes*, ed. by O.M. Petruhin (Mir, Moscow, 1989)
82. R.K. Christova, V.D. Simeonov, in *Potentiometry*, ed. by S. Stanislavova (Nauka i Izkustvo, Sofia, 1984)
83. Y.G. Vlasov, Fresen. Z. Anal. Chem. **335**, 92 (1989)
84. Y.G. Vlasov, E.A. Bychkov, Anal. Lett. **22**(5), 1125 (1989)
85. Y.G. Vlasov, A.V. Legin, A.M. Rudnitskaya, Zh. Anal. Khim **52**(8), 837 (1997)
86. X. Gao, J. Zhang, L. Zhang, Adv. Mater. **14**, 290 (2002)
87. (a) I. Rosenfield, O.A. Beath, *Selenium: Geotoxicity, Biochemistry, Toxicity and Nutrition* (Academic Press, New York, 1964); (b) A. Bock, in *Encyclopedia of Inorganic Chemistry*, vol. 7, ed. by R. Bruce-King (Wiley, New York, 1994), p. 3700; (c) G. Mugesh, W. du Mont, H. Sies, Chem. Rev. **101**, 2125 (2001)
88. C. Ip, Selenium inhibition of chemical carcenogenesis. Fed. Proc. **44**, 2573–2578 (1984)
89. L.C. Clark, G.F. Combs Jr., B.W. Turnbull et al., Effects of selenium supplementation for cancer prevention in patients with carcinoma of skin; a randomized controlled trial. J. Am. Med. Assoc. **276**, 1957–1963 (1996)
90. F. Sieber, J. Daziano, W.H. Gunther, M. Krieg, K. Miyagi, R.W. Sampson, M.D. Ostrowski, G.S. Anderson, I. Tsujino, R.J. Bula, Phosphorus Sulfur Silicon Relat. Elem. **180**, 647 (2005)
91. (a) D.C. Carter, X.J. Ho, Adv. Protein Chem. **45**, 153 (1994); (b) A. Papadopoulos, R.J. Green, R.A.J. Frazier, Agric. Food. Chem. **53**, 158 (2005); (c) C. Honda, H. Kamizono, T. Samejima, K. Endo, Chem. Pharm. Bull. **48**, 464 (2000)

92. T. Peters Jr., *All About Albumin: Biochemistry, Genetics, and Medical Applications* (Academic Press, San Diego, 1996)
93. (a) K. Murayama, M. Tomida, Biochemistry **43**, 11526 (2004); (b) S. Baier, D.J. McClements, J. Agric. Food Chem. **49**, 2600 (2001); (c) Y. Moriyama, K. Takeda, Langmuir **21**, 5524 (2005); (d) V. Militello, V. Vetri, M. Leone, Biophys. Chem. **105**, 133 (2003)
94. A.V. Hippel, J. Chem. Phys. **16**, 372 (1948)
95. B. Mishra, P.A. Hassan, K.I. Priyadarsini, H. Mohan, J. Phys. Chem. B **109**, 12718 (2005)
96. Z.U. Borisova, *Chalcogenide Semiconductive Glasses* (Publ. National University of Leningrad, Leningrad, 1983), p. 242
97. G.F. Combs, L. Lu, Selenium as a cancer preventive agent, in *Selenium: Its Molecular Biology and Role in Human Health*, ed. by D.L. Hatfield (Kluwer Academic Publishers, Norwell, MA, 2001), pp. 205–217
98. M.A. Beck, Selenium as an antiviral agent, in *Selenium: Its Molecular Biology and Role in Human Health*, ed. by D.L. Hatfield (Kluwer Academic Publishers, Norwell, MA, 2001), pp. 235–245
99. R.J. Coppinger, A.M. Diamond, Selenium deficiency and human disease, in *Selenium: Its Molecular Biology and Role in Human Health*, ed. by D.L. Hatfield (Kluwer Academic Publishers, Norwell, MA, 2001), pp. 219–233
100. M.K. Baum, A. Campa, M.J. Miguez-Burbano, X. Burbano, G. Shor-Posner, Role of selenium in HIV/AIDS, in *Selenium: Its Molecular Biology and Role in Human Health*, ed. by D.L. Hatfield (Kluwer Academic Publishers, Norwell, MA, 2001), pp. 247–255
101. J. Kohrle, The deiodinase family: selenoenzymes regulating thyroid hormone availability and action. Cell. Mol. Life Sci. **57**, 1853–1863 (2000)
102. R.C. McKenzie, T.S. Rafferty, G.J. Beckett, J.R. Arthur, Effects of selenium on immunity and aging", in *Selenium: Its Molecular Biology and Role in Human Health*, ed. by D.L. Hatfield (Kluwer Academic Publishers, Norwell, MA, 2001), pp. 257–272
103. L. Flohé, R. Brigelius-Flohé, M. Maiorino, A. Roveri, J. Wissing, F. Ursini, Selenium and male reproduction, in *Selenium: Its Molecular Biology and Role in Human Health*, ed. by D.L. Hatfield (Kluwer Academic Publishers, Norwell, MA, 2001), pp. 273–281
104. V.N. Gladyshev, Selenium in biology and human health: controversies and perspectives, in *Selenium: Its Molecular Biology and Role in Human Health*, ed. by D.L. Hatfield (Kluwer Academic Publishers, Norwell, MA, 2001), pp. 313–317
105. H.J. Thompson, Role of low molecular weight, selenium-containing compounds in human health, in *Selenium: Its Molecular Biology and Role in Human Health*, ed. by D.L. Hatfield (Kluwer Academic Publishers, Norwell, MA, 2001), pp. 283–297
106. V.N. Gladyshev, G.V. Kryukov, Evolution of selenocysteine-containing proteins: significance of identification and functional characterization of selenoproteins. Biofactors **14**, 87–92 (2001)

Chapter 2
Techniques for Structural Investigations (Theory and Experimental)

Yun Mui Yiu

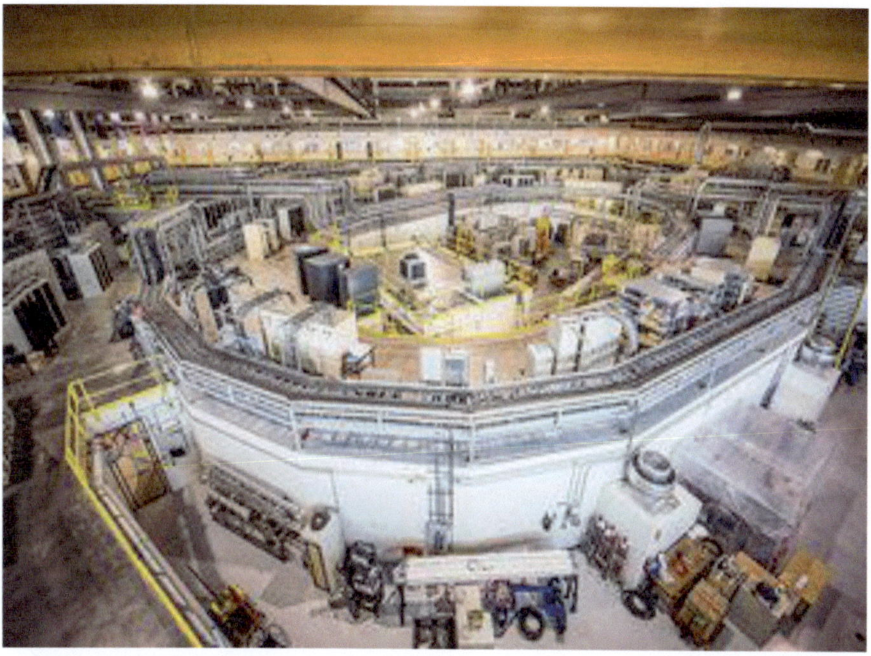

Canadian Light Source (CLS), Synchrotron facility at University of Saskatchewan, Saskatoon, Canada Picture: Courtesy: Canadian Light Source Inc.

This chapter reviews some of the techniques widely used for characterization of structure and properties of materials.

Y.M. Yiu, Ph.D. (✉)
Chemistry Department, University of Western Ontario, 1151 Richmond Street, London, N6A5B7, ON, Canada
e-mail: yyiu@uwo.ca

G.K. Ahluwalia (ed.), *Applications of Chalcogenides: S, Se, and Te*,
DOI 10.1007/978-3-319-41190-3_2

2.1 Introduction

The theory described in this section concerns mainly the X-ray determination of the chalcogenide substances. Surface second harmonic generation using laser is also presented. To characterize a material in microscale to nanoscale, SEM (scanning electron microscopy) and TEM (transmission electron microscopy) are useful laboratory tools. Together with XRD, it can assist with the determination of crystalline states and further determine space group symmetry of the materials under investigation for DFT (density functional theory) calculations. The cluster form of matters with a small scale of order can be investigated by the second program, RSMS (Real Space Multiple Scattering). The RSMS program provides an analysis to the nanoscale properties of materials.

2.1.1 Density Functional Theory (DFT)

Density functional theory (valid for all types of materials) proposes that the electronic charge density, $n(r)$, completely determines all features of the electronic behavior of a system of interacting electrons in the ground state. Energy of the interacting electronic gas is a unique functional (function of a function) of the charge density [1]. Although the functional is not known, the Hohenberg–Kohn theorem allows a variational principle to be established to find the ground state energy: the energy functional takes its minimum value (i.e., the ground state energy) when the charge density $n(r)$, is the true ground state charge density.

Nobel Prize winners, P. Hohenberg and W. Kohn, first introduced the density functional theory [1]. It has been developed using the concept of inversely mapping a "measurable" quantity, the electron density, into Eigen-function in Hilbert space. The recent novel approach highlighted the many-body problem by introducing the exchange-correlation function. DFT program can thus solve the differential equation by the local density approximation (LDA). Generalized gradient approximation (GGA) is later introduced. It assures a first energy derivative correction to the total energy of the system. The calculations then enter into a self-consistent iteration by minimizing the energy to tune into the equilibrium state of the physical system. From a theoretical point of view, the complexity of materials creates a big task for modern computation facilities. The numerical calculation is costly and CPU time demanding for materials possessing complex symmetries.

After the work of Hohenberg et al., the theory was further developed by Kohn and Sham. Its formalism is outlined as follows. The DFT is based on the assumption that there is a relatively homogenous distribution of electrons around the nucleus. Thus, the solution to the differential equation has no singularity in Hilbert space and is then analytical mathematically. For an actual compound, it is a many-body problem. The complicated problem is difficult to solve even for a reasonable system. To circumvent the difficulty, local density approximation (LDA) and generalized gradient approximation (GGA) are used to solve the problem. However, both LDA and GGA underestimate the band gaps of semiconductors. Later on, the introduction of an exchange-correlation potential, mBJ (modified Becke–Johnson) potential, into the program enables the fitting of the calculated band gap energies close to the experimental requirement. These calculations have been revitalized and applied to the complexity of many body systems. The materials are simplified by spatial symmetry of atomic sites using the concept of Brillouin zone and the reciprocal k space. The Brillouin zone depends on the group symmetry of the crystal. Space group theory has been described abundantly in literature, including the book of Crystallographic Space Group [2]. With the advent of supercomputers and computer clusters, introduction of computational software stimulates serious consideration of materials into the nano size region.

2.1.2 Kohn–Sham's Equation

The total energy of the many-body problem is described by the sum of the kinetic energy $T[n]$, the potential energy $v(r)$, the mutual electron–electron interaction energy, and an exchange-correlation energy. To activate the functional of the inverse mapping in the Hilbert space, these energies are approximated as dependent of the electron density $n(r)$. By keeping the number of the total electron density constant, the Coulomb's part of the interaction can be simplified. The self-consistent problem is solved by iteration of the calculations until the energy is minimized to achieve an equilibrium state of matter.

The total energy of the material is given by:

$$E = T[n] + \int v(r)n(r)d^3r + \int\int \frac{n(r)n(r^{'})}{|r - r^{'}|}d^3rd^3r^{'} + E_{\mathrm{xc}}[n] \tag{2.1}$$

The exchange-correlation energy is given by:

$$E_{\mathrm{xc}}[n] = \int e_{\mathrm{xc}}(r)nd^3r + \int e_{\mathrm{xc}}^{\mathrm{GGA}}(n, \nabla n)d^3r + \cdots \tag{2.2}$$

where $e_{\mathrm{xc}}(r)$ is the exchange-correlation energy and $e_{\mathrm{xc}}^{\mathrm{GGA}}(n, \nabla n)$ is the Generalized Gradient Approximated exchange-correlation energy function.

Energy minimization with self-consistent iteration while keeping the total number of electrons constant is done using:

$$\frac{\delta E(n)}{\delta n} = 0, \qquad \forall \mathrm{Z} = \mathrm{Const.} \tag{2.3}$$

where the total number of electrons, Z, is given by,

$$Z = \int n(r) d^3 r \tag{2.4}$$

and the electron density,

$$n = \int \psi^*(r)\psi(r) d^3 r \tag{2.5}$$

2.1.3 Full Potential Augmented Plane Wave Method

The system of the many-body problem following the Kohn–Sham approach can be written as the following form:

$$\left[-\nabla^2 + V_{\mathrm{Ne}} + V_{\mathrm{ee}} + V_{\mathrm{xc}}^{\sigma}\right]\chi_{ik}^{\sigma}(r) = \varepsilon_{ik}^{\sigma}\chi_{ik}^{\sigma}(r) \tag{2.6}$$

where V_{Ne} is the nuclei-electron potential, V_{ee} is the electron–electron potential, and V_{xc}^{σ} is the exchange-correlation function. The wave function is χ_{ik}^{σ} and the energy Eigen-value is $\varepsilon_{ik}^{\sigma}$.

The LAPW (linearized augmented plane wave) method is a procedure for solving the Kohn–Sham equations for the ground state density, total energy, and Kohn–Sham Eigen-values (energy bands) of a many-electron system by introducing a basis set which is especially adapted to the problem. The Augmented Plane Wave method [3] is outlined as the following. The unit cell of the crystal is divided into (1) nonoverlapping atomic spheres, centered at the atomic sites and (2) an interstitial region. In the atomic site S, the wave function φ_k is characterized by the spherical harmonics Y_{lm} and a radial part. The radial wave function is determined by $u_l(r,E_l)$ and its energy derivative $\partial u_l(r,E_l)/\partial E_l$. The interstitial region between S is described by sets of plane waves. The number of plane waves can be set as high as 20,000. The atomic radius is determined by the mean atom–atom distance and denoted as r.

The radial wave is described by u_l and its energy derivative:

$$\varphi_{k_n} = \sum_{lm}\left[A_{lm}u_l(r,\, E_l) + B_{lm}\frac{\partial u_l(r,E_l)}{\partial E_l}\right]Y_{lm}(r), \qquad \forall r \in S \tag{2.7}$$

$$\varphi_{k_n} = \frac{1}{\sqrt{\omega}}e^{-ik_n r}, \qquad \forall r \notin S \tag{2.8}$$

where $k_n = k + K_n$, k is the wave vector in the first Brillouin zone, and K_n is the reciprocal lattice vector. The coefficients A_{lm} and B_{lm} are functions of k_n, u_l and

$\partial u_l(r,E_l)/\partial E_l$ and are obtained by numerical integration of the radial Schrodinger equation on a radial mesh inside the sphere.

The boundary conditions of the self-consistent calculation are briefly shown. The Eigen vectors of the atomic sites should be either matched at the surface sphere of the atomic sites or matched by the first radial derivative of the Eigen function. Boundary Conditions are:

$$\varphi_{k_n}(S)|_{r_s} = \varphi_{k_n}(I)|_{r_s} \tag{2.9}$$

$$\left.\frac{\partial \varphi_{k_n}(S)}{\partial r}\right|_{r_s} = \left.\frac{\partial \varphi_{k_n}(I)}{\partial r}\right|_{r_s} \tag{2.10}$$

The density of states (DOS) can be calculated using the modified tetrahedron method of Bloch et al. [4]. X-ray absorption and emission spectra are determined using Fermi's golden rule and dipole matrix elements, between a core state to valence band or core state to conduction band state [5].

2.1.4 MBJ (Modified Becke–Johnson) Exchange Potential

Both the local density approximation and the generalized gradient approximation tend to underestimate the band gap energy of semiconductors as described in subsequent sections. The inablility of distinguishing semiconductors from metals urges a further investigation. The exchange-correlation energy of materials has been seeking a realistic model. WIEN as described in Sect. 2.2.1 has adopted the evaluation of the modified Becke–Johnson potential [6].

The exchange-correlation function is evaluated as the Becke–Johnson potential $v_{x,\sigma}^{\mathrm{BJ}}(\vec{r})$ and the modified Becke–Johnson exchange-correlation function as $v_{x,\sigma}^{\mathrm{mBJ}}(\vec{r})$. The modified mBJ exchange-correlate energy depends on the BR (Becke–Roussel) potential, $v_{x,\sigma}^{\mathrm{BR}}(\vec{r})$ by adding a term which depends on the kinetic energy of the system:

$$v_{x,\sigma}^{\mathrm{mBJ}}(\vec{r}) = c v_{x,\sigma}^{\mathrm{BR}}(\vec{r}) + (3c-2)\frac{1}{\pi}\sqrt{\frac{5}{12}}\sqrt{\frac{2t_\sigma(\vec{r})}{\rho_\sigma(\vec{r})}} \tag{2.11}$$

where c is described as

$$c = A + B\left[\frac{1}{V_{\mathrm{cell}}}\int_{\mathrm{cell}} \frac{\left|\nabla\rho(\vec{r'})\right|}{(\vec{r'})} d^3r'\right)\right]^e \tag{2.12}$$

The electron density is given by

$$\rho_\sigma = \sum_{i=1}^{N_\sigma} |\psi_{i,\sigma}|^2 \quad (2.13)$$

The kinetic energy

$$t_\sigma = \frac{1}{2}\sum_{i=1}^{N_\sigma} \vec{\nabla}\psi_{i,\sigma}^* \vec{\nabla}\psi_{i,\sigma} \quad (2.14)$$

The Becke–Roussel Potential is

$$v_{x,\sigma}^{\mathrm{BR}}(\vec{r}) = -\frac{1}{b_\sigma(\vec{r})}\left[1 - e^{-x_\sigma(\vec{r})} - \frac{1}{2}x_\sigma(\vec{r})e^{-x_\sigma(\vec{r})}\right] \quad (2.15)$$

where

$$b_\sigma = \left[\frac{x_\sigma^3 e^{-x_\sigma}}{8\pi\rho_\sigma}\right]^{1/3} \quad (2.16)$$

$$\frac{xe^{-2x/3}}{x-2} = \frac{2}{3}\pi^{2/3}\rho_\sigma^{5/3}\left\{\frac{1}{6}\left(\nabla^2\rho_\sigma - 2\gamma\left[\sum_{i=1}^{\sigma}|\nabla\psi_i|^2 - \frac{(\nabla\rho_\sigma)^2}{4\rho_\sigma}\right]\right)\right\}^{-1} \quad (2.17)$$

The parameters are $\gamma = 1$, $A = -0.012$, $B = 1.023$ bohr$^{1/2}$, and $e = 0.5$. These parameters are chosen so that a number of wide band gap materials can be fitted into the category by comparing the predicted band gap energies to the experimental data.

2.1.5 *Real Space Multiple Scattering (RSMS)*

The real space multiple scattering has been introduced mainly by John Rehr's group [7]. The program is user friendly in the latest version. It enables a choice of a platform of Microsoft Windows or Unix/Linux based graphic interface. The calculation is less demanding as compared to the DFT calculation in terms of computer CPU time. With the current Window interface, FEEF 9 is a user friendly endeavor. It can be used in notebook computers. Scientists can bypass the effort of understanding the computer terms using this program. With the computer models of material phases, the electronic properties of matters, including the XANES, EXAFS, and density of states, can be calculated. The validity of the program is justified by comparing the calculated properties of materials to the experimental data and is verified in later section, Sect. 2.6.2.

The Green's function approach has been outlined closely in the following formalism initiated by J. Rehr et al. For one-particle Green's function:

$$G = \frac{1}{E - H - i\Gamma} \tag{2.18}$$

where E is energy, H is Hamiltonian, and Γ is the band width.

In position space, the XAFS has the spectral representations:

$$G\left(\boldsymbol{r}, \boldsymbol{r}'; E\right) = \sum_{f} \frac{\psi_{\mathrm{f}}\left(\boldsymbol{r}'\right)\psi_{\mathrm{f}}^{*}\left(\boldsymbol{r}'\right)}{E - E_{\mathrm{f}} + i\Gamma} \tag{2.19}$$

where ψ_{f} are final sites with energies E_{f} of the effective final-state one-particle Hamiltonian H. H is described by the optical potential $v(\boldsymbol{r})$. Γ is a net lifetime, including effects of extrinsic and intrinsic losses. The absorption from a given core level c is given by:

$$\mu = -\frac{1}{\pi}\mathrm{Im}\left\langle c\middle|\hat{\varepsilon}\cdot\boldsymbol{r}G\left(\boldsymbol{r}, \boldsymbol{r}'; E\right)\hat{\varepsilon}\cdot\boldsymbol{r}'\middle|c\right\rangle \theta_{\Gamma}(E - E_{\mathrm{f}}) \tag{2.20}$$

where θ_{Γ} is a broadened step function at the Fermi energy E_{f}. The local representations of G^{c} and G^{SC} are:

$$G^{\mathrm{c}}\left(r, r'; E\right) = -2k\sum R_L(r_<)H_l(r_>) \tag{2.21}$$

$$G^{\mathrm{SC}}\left(r, r'; E\right) = -2k\sum_{L,L'} R_L(r)R_{L'}\left(r'\right)G^{\mathrm{SC}}_{L'0,L0}(E) \tag{2.22}$$

where $L = (l, m)$. The angular solution to the Dyson equation, the equation of motion of the corresponding Green's function, in quantum theory is

$$R_L(\mathbf{r}) = i^l R_l(\mathbf{r})Y_L(\mathbf{r}) \tag{2.23}$$

The outgoing part is

$$H_L(\boldsymbol{r}) = i^l H_l(r)Y_L(\hat{r}) = N_l(\boldsymbol{r}) - iR_L(\boldsymbol{r}) \tag{2.24}$$

The two-center free propagator is $G^0_{LR,L'R}$.

The dimensionless scattering t matrix is written as following with the phase shift δ:

$$t_{\boldsymbol{LR},L'\boldsymbol{R}'} = e^{i\delta_{i\boldsymbol{R}}}\sin(\delta_{i\boldsymbol{R}})\delta_{\boldsymbol{RR}'}\delta_{LL'} \tag{2.25}$$

The total scattering propagator matrix can be written as the matrix inversion

$$G^{SC} = G^{SC}_{LR,L'R'} = \exp\left(i\delta_l + i\delta_{l'}\right)\frac{G^0}{1 - G^0 t} \tag{2.26}$$

The expression for μ can be written in terms of an atomic background absorption modulated by the XAFS χ as

$$\mu = \mu_0(1 + \chi_l) \tag{2.27}$$

The embedded atomic cross section is given by:

$$\mu_0 = |\langle c|\hat{\varepsilon} \cdot \boldsymbol{r}|R_L\rangle|^2 \tag{2.28}$$

$$\chi_l = \operatorname{Im} e^{2i\delta_l}\left\langle \sum_m \left[\frac{G^0}{1 - G^0 t}\right]_{lm,lm} \right\rangle \tag{2.29}$$

where the brackets denote the thermal and configuration average. G^0 is zero for $\vec{R} = \vec{R}'$. The projected density of states ρ_l' is related to χ and the density of states for an embedded atom by

$$\rho_l' = \rho_{0l}'\left(1 + \chi_l'\right) \tag{2.30}$$

The short range order theory can be obtained by expanding the inverse matrix as a geometric series. For a single path, the contribution to XANES is:

$$\chi_\Gamma = \sum \langle lm|G^0 t_N \ldots G^0 t_2 G^0 t_1 G^0|lm\rangle \tag{2.31}$$

For a multiple path of scattering, the following holds

$$\chi_l = \sum_\Gamma \operatorname{Im}\left[e^{2i\delta_l}\sum_m \langle lm|G^0 t_N \ldots G^0 t_2 G^0 t_1 G^0|lm\rangle\right] \tag{2.32}$$

The scattering t matrix at site i is t_i. The partial wave phase shift is δ_l. The outer brackets indicate thermal and configuration average.

2.1.6 Rehr–Albers Method

The computational difficulties of multiple scattering expansions can be overcome by the Rehr–Albers method. These difficulties include the large angular sites, diversity of multiple scattering paths, and the correlated Debye–Waller factors for finite temperature. The solution involves a rotation matrix rotated along z direction.

An exact and separable representation of the z-axis propagators can thus established as follows.

The exact matrix factorization of the full propagator is given as

$$G_{LR,\,L'\boldsymbol{R}'} = \frac{e^{ik\boldsymbol{R}''}}{k\boldsymbol{R}''} \sum_{\lambda} Y_{L,\lambda} \widetilde{Y}_{\lambda,L'} \tag{2.33}$$

where $\boldsymbol{R}'' = \boldsymbol{R}' - \boldsymbol{R}$ and the generalized spherical harmonics is Y and $\tilde{Y}$.

The XAFS amplitude is given as

$$\chi_{\Gamma}(p) = \operatorname{Im} S_0^2 \frac{e^{i(\rho_1+\rho_2+\cdots\rho_N+2\delta_1)}}{\rho_1\rho_2\cdots\rho_N} e^{-\frac{\sigma_\Gamma^2 p^2}{2}} TrM_l F^N \ldots F^2 F^1 \tag{2.34}$$

The notations are

$$\rho_i = p(\boldsymbol{R}_i - \boldsymbol{R}_{i-1})$$

The photo-electron momentum measured with respect to the muffin-tin zero in Rydberg atomic units

$$p = \sqrt{E - V_{\text{mt}}}$$

The scattering matrix at site i is F^i.

The termination matrix for the final state of angular momentum l is M_l.

The many-body reduction factor is S_0^2.

The mean square variation in total path length $\boldsymbol{R}_{\text{tot}}$ is σ_Γ^2.

The scattering matrix is:

$$F_{\lambda',\lambda} = \sum_{L} \widetilde{Y}_{\lambda',L} t_L Y_{L,\lambda}$$

The standard XAFS formula can be obtained by substituting $k = \sqrt{p^2 - k_F^2}$ and recasting terms as

$$\chi_{\Gamma}(k) = S_0^2 \operatorname{Im} \frac{f_{\text{eff}}}{k\boldsymbol{R}^2} e^{2ki\boldsymbol{R}+2i\delta_l} e^{-2\sigma^2 k^2} \tag{2.35}$$

2.1.7 EXAFS (Extended X-ray Absorption Fine Structure) Classic Theory

Synchrotron radiation facilities around the world provide opportunity for the scientific community to perform experiments using tunable and high-intense X-ray sources. The experimental data of the materials under investigation are abundant. The techniques involve XANES (X-ray absorption near edge spectroscopy), EXAFS (extended X-ray absorption fine structure), XEOL (X-ray excited optical luminescence), resonant XPS (X-ray photoelectron; apectroscopy), etc. It was Stern et al. who first proposed that the oscillatory part of the EXAFS spectra is interpreted as a sinusoidal function of the interatomic distance and phase shift [8]. EXAFS is the extended part of the oscillation part above the adsorption edge of an element, typically only around 50–1000 eV above the adsorption edge of the spectra. It is an element specific characterization of matters. The solid materials under investigation can either be in amorphous form or crystalline state. The oscillation part can be described as a sinusoidal function of kR_j and a phase shift of $\delta_j(k)$. k is the momentum and R_j is the radius of the j atom in real space. By Fourier transformation of the oscillated part, the nearest-neighbor distances can be obtained.

The absorption coefficient of the matter is written as $\mu(E)$. The modulation of the absorption spectrum as a function of energy E is given by $\chi(E)$:

$$\chi(E) = \frac{\mu(E) - \mu_0(E)}{\mu_0(E)} \tag{2.36}$$

where

$$\chi(k) = c\sum_j \frac{N_j f_j(k) e^{-2k^2\sigma_j^2}}{kR_j^2} \sin\left[2kR_j + \delta_j(k)\right] \tag{2.37}$$

The EXAFS spectrum has been added the Debye–Waller factors to address at the finite temperatures of materials under investigation. The modified oscillatory part of the spectrum is written in the following form by the FEFF team:

$$\chi(k) = \sum_j \frac{N_j S_0^2}{kR_j^2} \left| f_j^{\text{eff}}(k, R_j) \right| \sin\left(2kR_j + \varphi_j(k)\right) e^{-2R_j/\lambda} e^{-2\sigma_j^2 k^2} \tag{2.38}$$

The Debye–Waller factors are given by

$$\sigma_j^2 = \frac{1}{4}\left\langle \left[\sum_{i=1}^{n_j} (\vec{u}_i - \vec{u}_{i+}) \hat{R}_{ii+} \right]^2 \right\rangle \tag{2.39}$$

where $\sigma(T)$ is given by

$$\sigma_j^2(T) = \frac{\hbar}{2\mu_j} \int \frac{d\omega}{\omega} \rho_j(\omega) \coth \frac{\hbar\omega}{2k_B T} \tag{2.40}$$

The projected vibration density of states is $\rho_j(\omega)$. The projected reduced mass μ_j is described as the following:

$$\frac{1}{\mu_j} = \sum_{i=1}^{n_j} \frac{1}{M_i} \left(\frac{\hat{R}_{ii-} + \hat{R}_{ii+}}{2} \right)^2 \tag{2.41}$$

2.2 Computer Programs for Electronic Structure Calculations

The useful computer programs for the density functional theory include ABINIT, CPMD, CRYSTAL, Gaussian, VASP, and WIEN2k. WIEN2k and RSFMS (real space full multiple scattering) program of FEFF are described in this chapter.

2.2.1 WIEN Program

The WIEN program [9] is developed and based on DFT theory. The local density approximation or GGA serves as a starting point. It then enters into a self-consistent iteration to obtain a minimization of the energy of a system under consideration. The major algorithm depends on the formalism as outlined in the previous sections.

The program acquires a procedure based on a crystal or crystalline system. The input file can be set as a struct or import from a cif file. The struct file can be used either by a Window (Window X11) based graphic interface or by a unix/linux input file. The cif file is determined by XRD (X-ray diffraction) data. It can be converted to the struct file by the sub-program cif2struct. One can either search through a crystallographic data base or by directly acquiring the laboratory diffractometers to obtain cif files. To use the struct file directly, one has to choose a material with identified atomic sites and group symmetry. The sophistication depends on the input factors. These factors depend on various quantities: number of atomic sites, radius of the atoms, and the group symmetry of the crystalline states. The calculation then starts as it requires the potential to be chosen. It can either use local density approximation or generalized gradient approximation. The system is then entering in the initial stage of providing the calculated electron density.

The calculation procedure includes:

1. Structure generation.
2. Initializing the calculations involves: x nn, x sgroup, x symmetry, x lstart, x xkgen, and x dstart.
3. Run scf: The self-consistent algorithm depends on the seeking of an energy minimization to get the actual equilibrium ground state. The total energy depends on the electron density distribution in the core and valence states. The potential is then generated by lapw0, lapw1, and lapw2. The core level is calculated by lcore. The mixer subortine, mixer, then adds the initial density to the calculated density. Starting the self-consistent loop, it then recalculates the energy. Once the difference of the energy of each iteration is reached to a prescribed value, e.g., 0.0001 Ry or charge to 0.001 e, the calculated system is achieved at the equilibrium stage.

4. Calculation of Properties: To justify the validity of the calculations, the properties of the materials are compared to the experimental data. In particular, it is capable of calculating density of states (DOS), XAFS (X-ray absorption fine spectra), and band structures. The calculated XAFS spectra are described by the electrical dipole transition between the ground state and the excited state and follow Fermi's golden rule. It fulfills the transition requirement of state selection, i.e., $\Delta_l = \pm 1$ for electrical dipole transition, where l is the angular quantum number. The valence band distribution of the electronic state then compares to the XPS spectra. The core excitation can be realized by the calculated density of states, e.g., s to p transition. The calculated XAFS spectra can compare to the experimental XANES spectra. The WIEN2k code is employed by the major WIEN contributors—Peter Blaha and K. Schwarz [9].

2.2.2 FEFF Program

The multiple scattering computer code by the FEFF team has been developed based on the real space calculation through the Dyson equation. The Green's function approach [7] has been added to the evolved programs as described in Sect. 2.1.5.

The input programs and the EXAFS analyses are further evaluated based on the programs Artemis and Athena. Artemis [11] can generate an input file for FEFF from a cif file or a specified input file. It adopts the cif file which is generated by XRD (X-ray Diffraction) as input file. It can also allow a graphic input file specification. The information needed for the input includes the atomic species, atomic sites, space group symmetry, and cluster radius of the investigated model. It then calculates the atomic positions of the specified cluster. With the generated file together with the specific control cards, FEFF program can run self-consistent loop and evaluate XANES or EXAFS for the chosen materials.

Athena [12] can then analyze the output file of FEFF and compare it to the experimental files. It can perform Fourier transformations of EXAFS spectra and evaluate the interatomic distance between the nearest-neighbor atoms.

The primary subroutines of FEFF code are established as the following calculations as proposed by Rehr et al. [7]

1. The atomic potentials and electron densities of the clusters.
2. Generates a muffin-tin scattering potential that includes the excited-state self-energy and calculated energy reference.
3. The complex partial wave phase shifts.
4. The curved-wave scattering amplitudes and phase shifts, and the exact single-scattering curved-wave XAFS spectra.
5. Multiple scattering path expansion.

2.3 XPS (X-ray Photo-electron Spectroscopy)

XPS is also known as ESCA (electron spectroscopy for chemical analysis). X-ray photo-electron scattering (XPS) is a technique which utilizes X-ray photon bombarded samples. The X-ray beam kicks the electrons out from the samples. The outgoing electrons are collected and their energies are monitored. Since the escape depths of electrons are shallow, XPS is a surface-sensitive quantitative spectroscopic technique which measures the elemental composition in the parts per thousand range. Chemical and electronic states of the elements that exist within a material can be detected. XPS spectra are obtained by measuring the kinetic energy and number of electrons that escape from the top 100 Å of the material. Operation of the XPS requires high vacuum ($P \sim 10^{-8}$ mbar) or ultra-high vacuum ($P < 10^{-9}$ mbar) conditions. The energies of the escape electrons can then be fitted to the corresponding core-hole levels and valence band. The oxidation states of the XPS spectra are governed by the binding energy of the samples, following basically Einstein's photo-electric effect. XPS can be either acquired by the lab equipment using monochromatic X-ray source or by synchrotron facilities. With the lab facility, the incident photon energy is confined to a constant depending on the choice of X-ray source, e.g., Al-Kα line. Elaborating the synchrotron facilities around the world, the incoming X-ray can be tuned according to the resonant states of samples. The resonant state can then be manifested by the enhanced resonant electron spectra.

2.4 Analysis of Se-Te System Using XPS, WIEN, and TEM

The appropriate optical and thermal properties and the crystallizing behavior of tellurium based alloys are attractive for phase change memory device applications. Since pure tellurium can crystallize rapidly at about 10 °C, it has to be alloyed with a certain amount of selenium (Se), germanium (Ge), tin (Sn), antimony (Sb), or arsenic (As) in order to stabilize the amorphous phase. In order to investigate these materials for the applications, a detailed knowledge of the local structure is crucial. The structural changes caused by changing the Se–Sb ratio and keeping the tellurium content constant at 20 at.% in the $Se_{80-x}Sb_xTe_{20}$ ($0 \leq x \leq 9$) system has been investigated [13]. XPS of Se-Te, As_2Se_3, and Zn-doped As_2Se_3 systems have been taken at Surface Science Western facility. The XPS core level and valence band spectra were obtained with a Kratos Axis Ultra spectrometer using a monochromatic Al-Kα source (1486.6 eV) in a working pressure lower than 10^{-9} mbar. Spectra were obtained using an analysis area of ~300 × 700 μm and were charge corrected to carbon 1 s at 284.8 eV.

The valence band is hence compared to the computer simulated models using the Ab initio, WIEN program, and the full multiple scattering, FEFF program. The valence band spectra favor the GGA (Generalized Gradient Approximation) even though it tends to underestimate the band gaps of semiconductors. Zero point energy is chosen at the Fermi energy for valence plots of XPS. The calculated valence electron spectra have been compared to the experimental spectra of the samples.

The crystal structure of Se has been modeled as α-Monoclinic Se with space group $P2_1/n$ (#14) [14]. The unit cell dimensions used are $a = 9.054 \pm 0.003$, $b = 9.083 \pm 0.005$, $c = 11.601 \pm 0.006$A, and $\beta = 90.81 \pm 0.05^0$. The four Se sites are at 4e positions: *a*: (0.3209, 0.484, 0.2362), *b*: (0.4254, 0.6625, 0.3569), *c*: (0.3178, 0.6376, 0.5378), and *d*: (0.1343, 0.8186, 0.5529). In order to calculate the Se, Se-Te, and Se-Te-Sb systems, the following simplified models have been made. The Se-Te compound has been modeled as Se occupying the three 4e positions, a, c, and d, and Te being at the b position. The Se-Te-Sb model compound is configured such that selenium atoms are locating at a, and d positions, Te is placing at the b position, and Sb is setting at the c position.

The electron configurations of α-Monoclinic Se and Se_2TeSb crystals along the [100] plane have been plotted in Figs. 2.1 and 2.2 using the WIEN program.

The experimental X-ray photoelectron spectra (XPS) of the valence band of compound $Se_{80-x}Te_{20}Sb_x$ with $x = 2$, 4, and 9 have been graphed as shown in Fig. 2.3 [13]. The theoretically calculated densities of states of the Se_2TeSb crystal using GGA have been plotted together with the XPS valence band of the SeTeSb systems for comparison. It can be seen that as the Sb content increases from 2 to 9 at.%, the valence band of the $Se_{80-x}Te_{20}Sb_x$ system moves towards the Fermi level. The calculated densities of states by GGA are closer to the Fermi level than the experimental valence bands.

The Se and Sb 3d core level spectra of XPS (X-ray photoelectron spectroscopy) for $Se_{80-x}Te_{20}Sb_x$ samples with $x = 4$ are shown in Figs. 2.4 and 2.5. The core level spectra have been de-convoluted into $3d_{5/2}$ and $3d_{3/2}$ Gaussian components. The Se $3d_{5/2}$ and $3d_{3/2}$ spin-orbit split components are located at about 55 eV and 55.8 eV, respectively. The Sb $3d_{5/2}$ and $3d_{3/2}$ components are situated at about 529 eV and 539 eV, respectively. The shoulder around 1.5 eV above the core level is associated with Se-Sb interaction. Since Se is more electronegative, there is charge redistribution and less electron shielding potential around the Sb atoms. It indicates the existence of a bonding between Se and Sb on the surface of the SeTeSb sample. Since no argon sputtering was performed prior to the measurements, the shoulder at about 532 eV is the O 1s peak from the surface oxide.

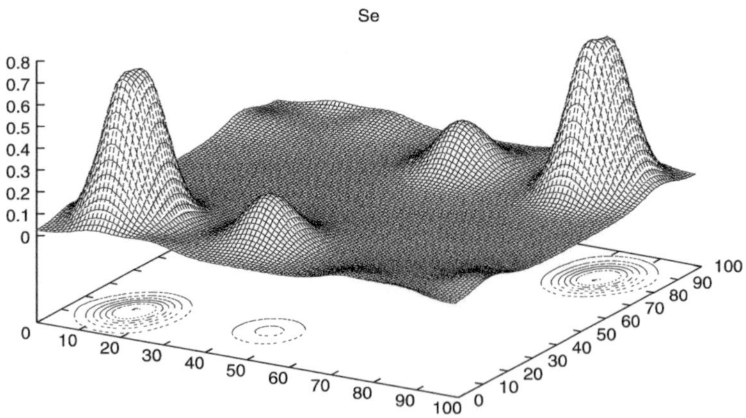

Fig. 2.1 The electron configuration of α-Monoclinic Se crystals along the [100] plane

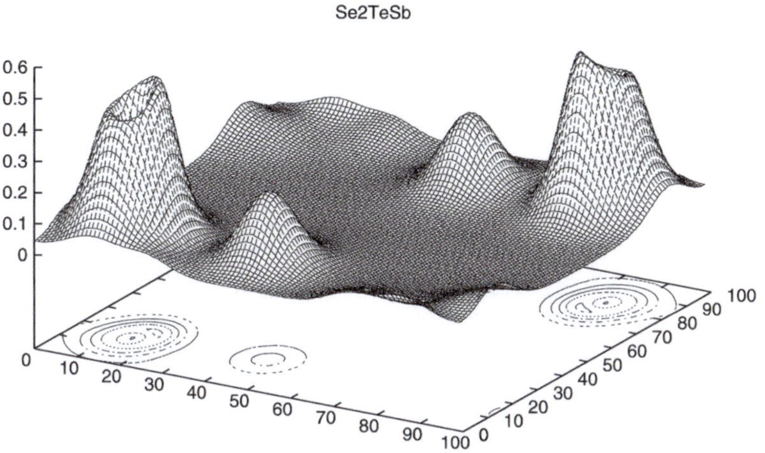

Fig. 2.2 The electron configuration of α-Monoclinic Se_2TeSb crystals along the [100] plane

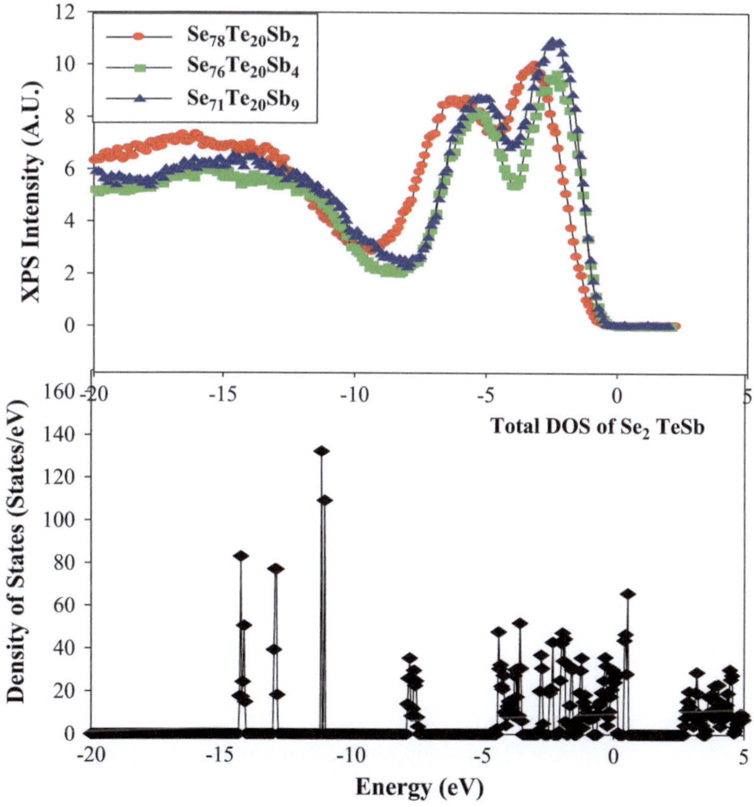

Fig. 2.3 XPS and DOS of the valence band of Se-Te doped with Sb

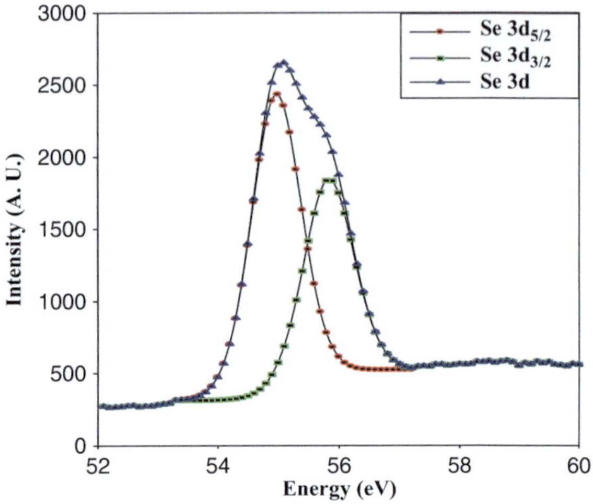

Fig. 2.4 The XPS spectra of Se-Te doped with Sb have been monitored at Se d level

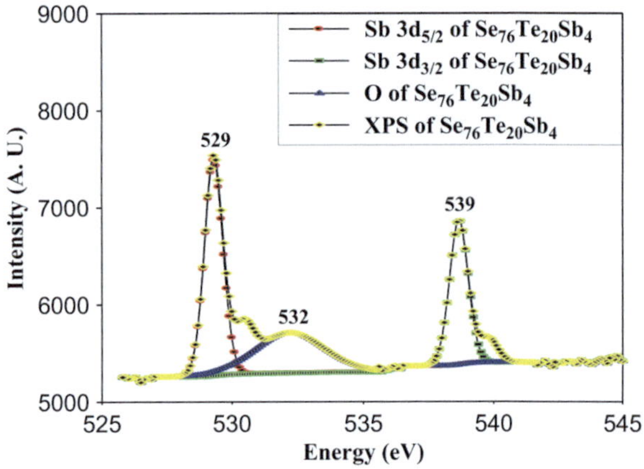

Fig. 2.5 The XPS spectra of Se-Te doped with Sb have been monitored at Sb and oxygen core level

In another study, [15] Se–Te alloy nanocrystals (NCs) were synthesized in aqueous micellar phase at 85 °C by using Na_2SeO_3/Na_2TeO_3 as Se/Te source in the presence of different amounts of hydrazine as reducing agent over the entire mole fraction range of Se (x_{Se}) from $x_{Se} = 0$–1. The shape, structure, and composition of alloy NCs were characterized by scanning electron microscopic (SEM), transmission electron microscopic (TEM), and energy dispersive X-ray spectroscopic (EDS) measurements. Despite many similar characteristic features of elemental Se and Te, a drastic change in the morphology of NCs was observed from $x_{Se} = 0$–1. Amorphous water-soluble Se nanoparticles were obtained at $x_{Se} = 1$ in the excess of hydrazine, whereas long nanoribbons of several micrometers of Te were produced at $x_{Se} = 0$. In the intermediate mole fractions, a smooth transition in the morphology from predominantly rhombohedral in the Se-rich region to long nanoribbons in the Te-rich region of the mixtures was observed. At relatively much lower hydrazine concentration (i.e., 0.1 M), large plate-like morphologies of pure Se were obtained while the morphologies of other mole fractions were slightly distorted especially in the Se-rich region of the mixture. A careful EDS analysis performed on each kind of morphology revealed a homogenous mixing between elemental Se and Te but with a significant nonideal behavior over the whole mole fraction range. Figure 2.6 shows Se-Te alloy long needles prepared with a mole

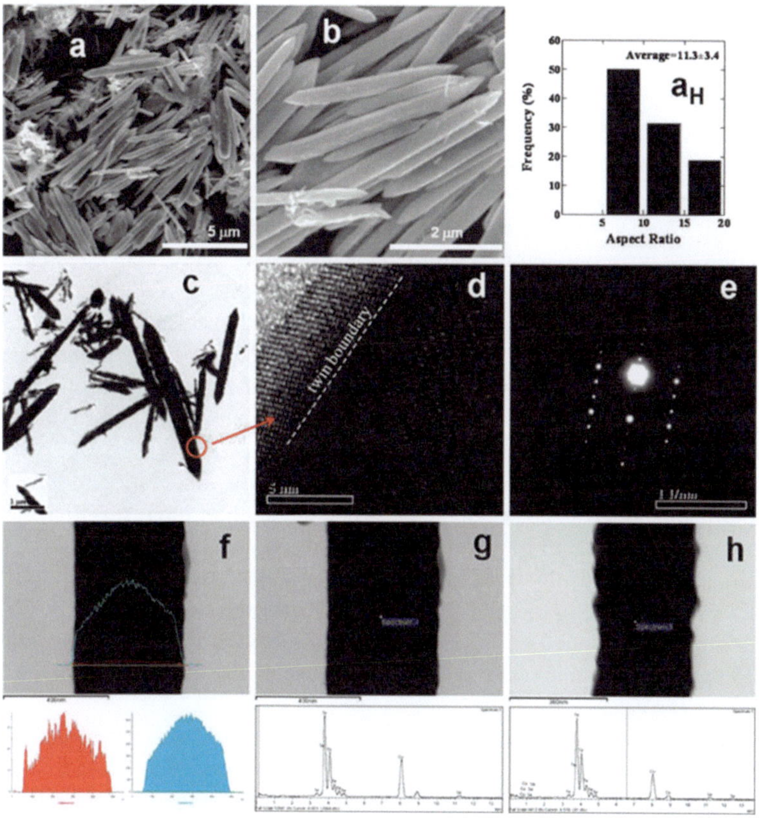

Fig. 2.6 (**a**, **b**) Se-Te alloy long needles prepared with a mole fraction of Se, $x_{Se} = 0.2$, in the presence of 3.6 M hydrazine. The average aspect ratio is calculated from the size distribution histogram shown in (**a$_H$**). (**c**) TEM image of several long needles, and (**d**) a lattice resolved image showing a twin boundary separating {111} and {110} crystal planes from each other. (**e**) Selected area diffraction image taken along the ⟨110⟩ zone axis showing the single crystal nature of a needle. (**f**) EDS line analysis across the short axis of a single needle to compute the atomic % of Se and Te. Note insignificant emission due to elemental Se while each needle is entirely constituted by Te. (**g**, **h**) A single spot EDS analysis on two different needles with corresponding EDS spectra. After [15]

fraction of Se, $x_{Se} = 0.2$, in the presence of 3.6 M hydrazine. NCs were always highly rich in the Te contents even in the Se-rich region of the mixtures and the overall growth was predominantly driven by metalloid (crystalline) nature of Te rather than nonmetallic (amorphous) nature of Se.

2.5 XANES (X-ray Absorption Near-Edge Spectroscopy)

X-ray absorption spectra (XAS) can be measured by synchrotron facilities around the world. These Synchrotron facilities produce incident photons with energies over the range of 200–35,000 eV. X-ray absorption spectroscopy deals with the measurement and interpretation of absorption of light by matter in a given energy region of the electromagnetic spectrum. In the X-ray region, e.g., from ~50 eV (soft X-rays, Li *K*-edge ~56 eV) to 40 keV (I *K*-edge, ~33 keV) is often used. X-ray absorption spectroscopy requires a tunable and intense X-ray source which can be provided by synchrotrons. The dominant physical process of XAS is one where the absorbed photon ejects a core photoelectron from the absorbing atom, leaving behind a core hole. The atom with the core hole is subsequently excited. The ejected photoelectron's energy is equal to that of the absorbed photon minus the binding energy of the initial core state. The ejected photoelectron interacts with electrons in the surrounding non-excited atoms.

X-ray Absorption Fine Structure (XAFS) refers to the modulation of the absorption coefficient above an absorption edge when the atom of interest is in a chemical environment as compared to a free atom. It is often used as a general term by the practitioners of XAS using hard X-rays. XAFS is often divided into two regions, the XANES for the near edge region and the EXAFS for the extended region. XANES spectra represent a range of ~50 eV above the absorption edge of an element. EXAFS spectra extend above the absorption edge to higher energy, ranging from ~50 to ~1000 eV.

XAFS can be used to probe the electronic structure of materials and is associated with X-ray absorption edges of elements. The absorption coefficient increases abruptly producing an edge jump at an energy specific to the excitation threshold of a core level (*K*, 1*s*; *L*, 2*s*, 2*p* and *M*, 3*s*, 3*p*, 3*d*, etc.). Both XANES and EXAFS are element specific. XANES probes the unoccupied density of states in the conduction band and provides a detailed understanding of the oxidation state, coordination, and local chemical environment of the specific element being probed. EXAFS shows the local structure of materials and provides the information of distances between neighboring atoms.

Both XANES and EXAFS spectra can be collected by TEY (total electron yield) and FLY (fluorescence yield) modes. TEY mode collects electrons and is surface sensitive, since electrons have shorter escape depths. The FLY monitors the Fluorescence photons. The sampling depth of FLY is more than about 100 nm below the absorption surfaces depending on the selected samples. It makes the FLY data collected more bulk sensitive.

The local electronic structure at the atomic site has been investigated by the XANES measurements. XANES reflects the local density of unoccupied states and probes the structure of the conduction band. The L_3/L_2-edge XANES gives information about the s or d to p transition. The local density of states has been calculated according to the model compounds.

The theoretical calculations of XANES are based on the two ab initio calculations: DFT and RSMS theories, as described in Sects. 2.1.1 and 2.1.5. DFT calculation is based on a model compound with spatial symmetry. For the DFT calculation, scalar relativistic density function was used with GGA correction. Self-consistent minimization of energy was achieved with 20,000 k points to keep a better accuracy in each calculation. Tetrahedron method [4] for the k space integration was chosen. The theoretical XANES spectra have been calculated using the electrical dipole transition following the selection rule of $\Delta l = \pm 1$ (l is the angular momentum quantum number) and convoluted by instrumental Gaussian and core-hole lifetime Lorentzian broadenings [16]. FMS (Full Multiple Scattering) calculation starts from a cluster of atoms arranged with or without spatial symmetry. For RSMS calculation, the XANES spectra can be obtained after self-consistent loop has been converged to the minimum energy.

2.6 Analysis of Material Systems

2.6.1 CdS, CdSSe, and CdSe

The CdS_xSe_{1-x} nanostructures were prepared in a horizontal tube furnace. The as grown CdS_xSe_{1-x} alloy products formed along a temperature gradient are depicted in Fig. 2.7.

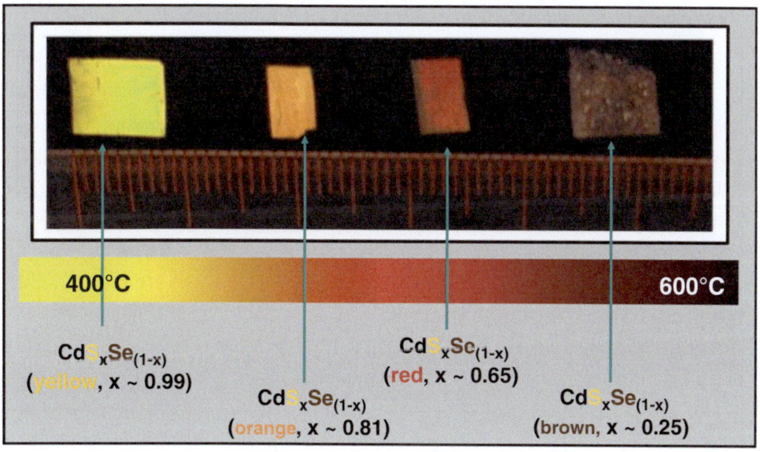

Fig. 2.7 Optical image correlating the composition of the CdS_xSe_{1-x} nanostructures with the growth temperature gradient

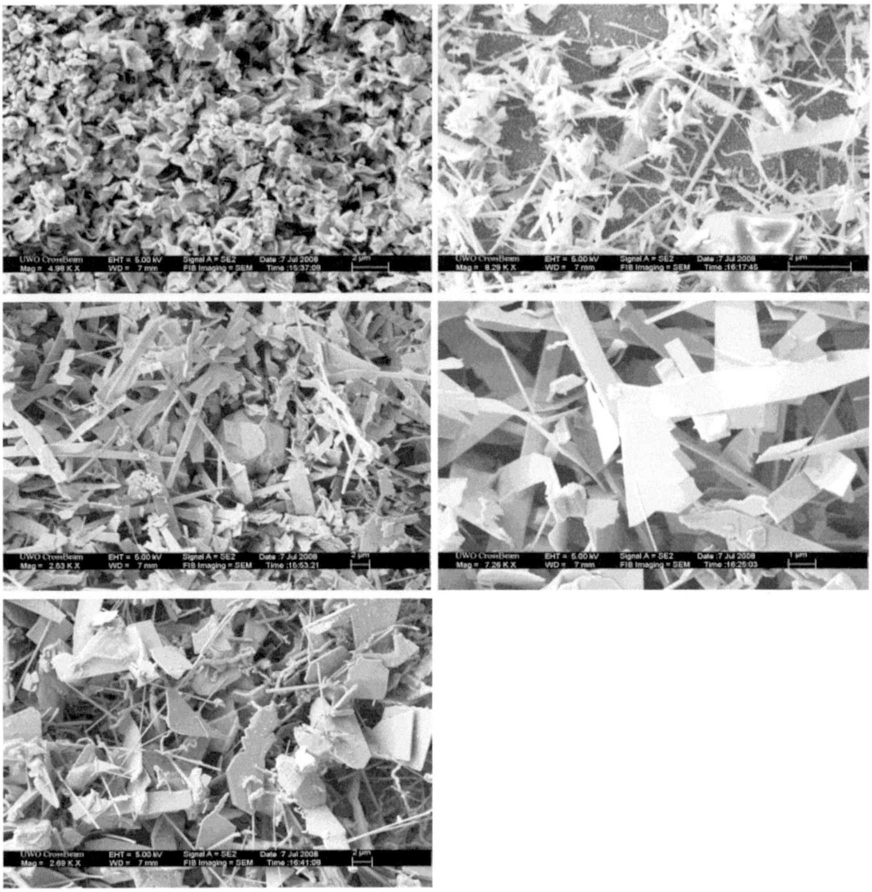

Fig. 2.8 SEM images of CdSSe nanostructures: (**a**) yellow, (**b**) orange, (**c**) red-orange, (**d**) red, (**e**) brown

The compositions for the CdS_xSe_{1-x} obtained from the EDX (energy dispersive X-ray spectroscopy) are $CdS_{0.99}Se_{0.01}$ (Yellow), $CdS_{0.81}Se_{0.19}$ (Orange), $CdS_{0.74}Se_{0.26}$ (Red-Orange), $CdS_{0.65}Se_{0.35}$ (Red), and $CdS_{0.25}Se_{0.75}$ (Brown) denoted by the color of appearance, and referred to hereafter as Yellow, Orange, Red-Orange, Red, and Brown, respectively.

The morphology and composition of the CdS_xSe_{1-x} nanostructures were characterized by Scanning Electron Microscopy (SEM), as shown in Fig. 2.8. The B3 model structure (cubic zinc-blende, Hawleyite) has cubic crystal symmetry of space group, F-43m (#216). The B4 model structure (hexagonal wurtzite, Greenockite) has hexagonal symmetry of $P6_3mc$ (#186). The B3 and B4 structures are shown in Fig. 2.9. For the B3 structures, Cd is at the 4a position of (000), S or Se are at the 4c position of (1/4, 1/4, 1/4). The lattice parameters of CdS and CdSe are 5.832 Å and 6.05 Å, respectively. For the B4 structures, Cd, S, or Se are at the 2b positions of

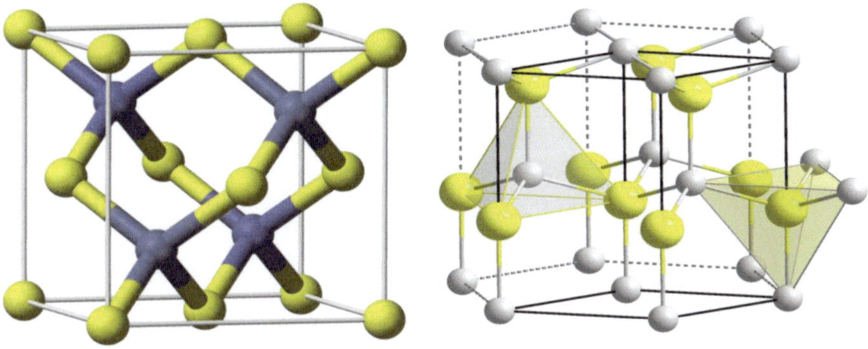

B3 (cubic) B4 (hexagonal)

Fig. 2.9 Crystal structures of B3 (cubic) and B4 (hexagonal) forms of cadmium sulfide

(1/3, 2/3, z) and (2/3, 1/3, $1/2+z$). The lattice constants of CdS and CdSe are $a=4.138$ Å, $c=6.718$ Å and $a=4.302$ Å, $c=7.014$ Å, respectively.

The XAFS (X-ray Absorption Fine Structure) experiments at the Cd and Se K-edge were conducted at Pacific Northwest Consortium and X-ray Science Division (PNC/XSD) of the Advanced Photon Source at Argonne National Laboratories. Cd L_3- and S K edges were performed at the Soft X-ray Micro-characterization beam-line (SXRMB). Se $L_{3,2}$-edge was performed at the spherical grating monochromator (SGM) beam-line and S $L_{3,2}$-edge was done at the variable line spacing plane grating monochromator of the Canadian Light Source at the University of Saskatchewan, Saskatoon, Canada. XANES spectra were recorded in total electron yield (TEY) and total fluorescence yield (FLY), using specimen current and a multichannel plate detector, respectively. The optical-XAFS spectra were recorded in photo-luminescence yield (PLY) using an optical detector [21].

Cd K-edge of CdS and CdSe in FLY mode has been accumulated and is plotted in Fig. 2.10. Cd K-edge involves the dipole transition of a Cd 1s electron to an unoccupied $5p_{3/2,1/2}$ or $np_{3/2,1/2}$ orbital. Cd L-edge is shown in Fig. 2.11. S K and L-edges are monitored and compared to GGA calculated results and are illustrated in Figs. 2.12 and 2.13, respectively. Experimental Se K and L-edges are also compared to GGA calculated results and are illustrated in Figs. 2.14 and 2.15, respectively.

By comparing the experimental FLY spectra with the theoretical calculated spectra of the corresponding B3 (cubic) to B4 (tetragonal) structures of CdS and CdSe, the onset of the absorption edge of the theoretical curves of B3 structure have delays in the rising part of the spectra. This indicates that these Chalcogenide nano materials favor B4 structure.

XAFS spectra at the Cd, S, and Se K and $L_{3,2}$-edges have been used to probe the electronic structures of the CdS_xSe_{1-x} nanostructures which are found to be dependent on the S to Se ratio and display properties that are between that of pure CdS and CdSe. It can clearly be seen from the color, densities of states, and band structures, that there is a tunable red shift in both the band-gap and defect emissions with decreasing S or increasing Se content in the CdS_xSe_{1-x} nanostructures.

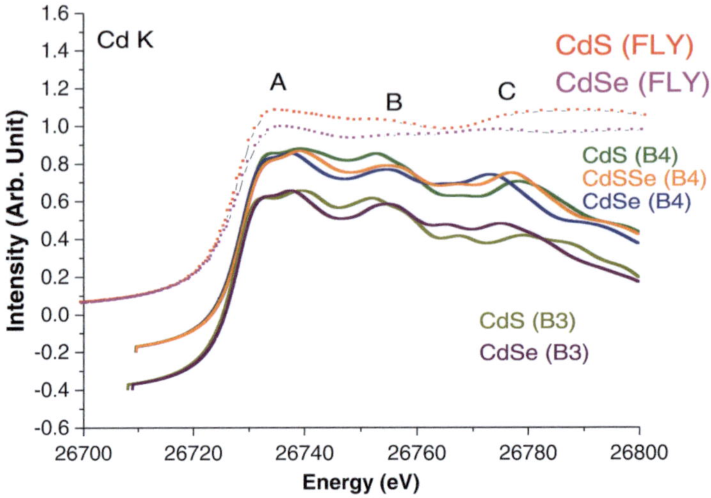

Fig. 2.10 *Cd K*-edges of CdS and CdSe are plotted in B3 and B4 forms together with the experimental FLY

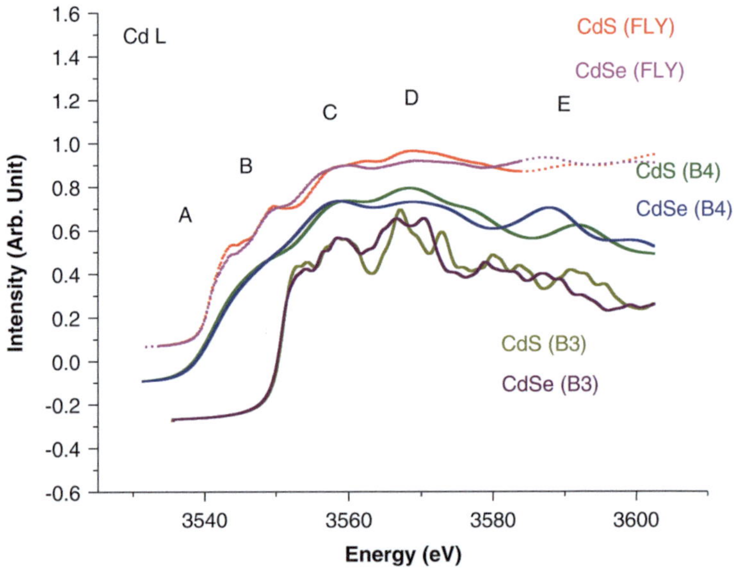

Fig. 2.11 Cd *L*-edges of CdS and CdSe are illustrated in B3 and B4 forms together with experimental FLY

The experimental Cd L_3 edge of both CdS and CdSe favors the B4 (hexagonal wurtzite) structure as the pre-edge regions indicate the present of small peaks and they fit to the theoretical curves of the B4 structure.

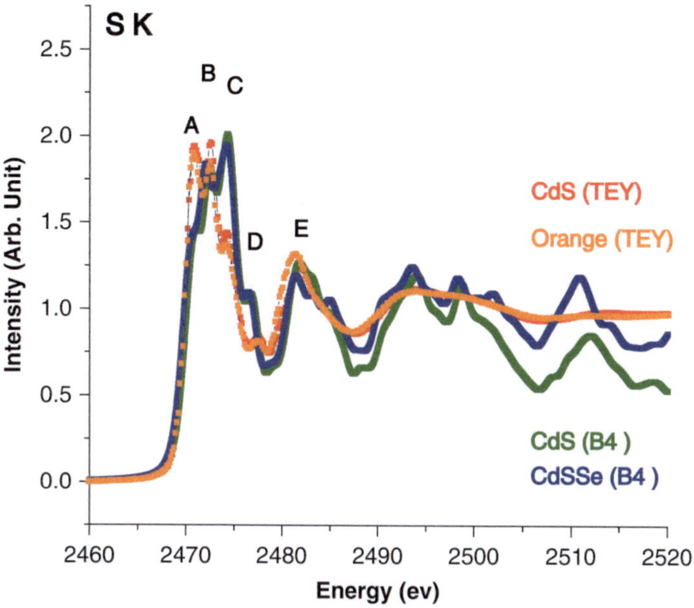

Fig. 2.12 S *K* edges of CdS and CdSe are plotted in B3 and B4 forms comparing the experimental TEY and GGA calculated spectra

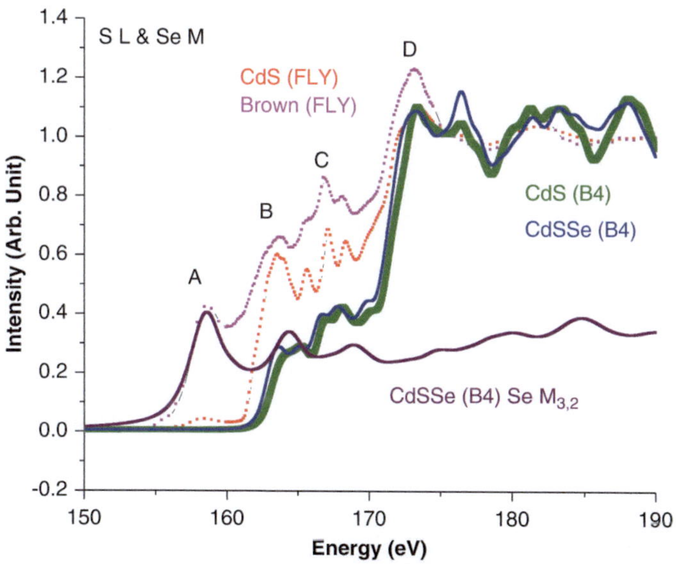

Fig. 2.13 S *L*-edges of CdS and CdSe in B3 and B4 forms are plotted togather with Se *M*-edge. The experimental FLY spectra are compared to the theoretical GGA calculated results

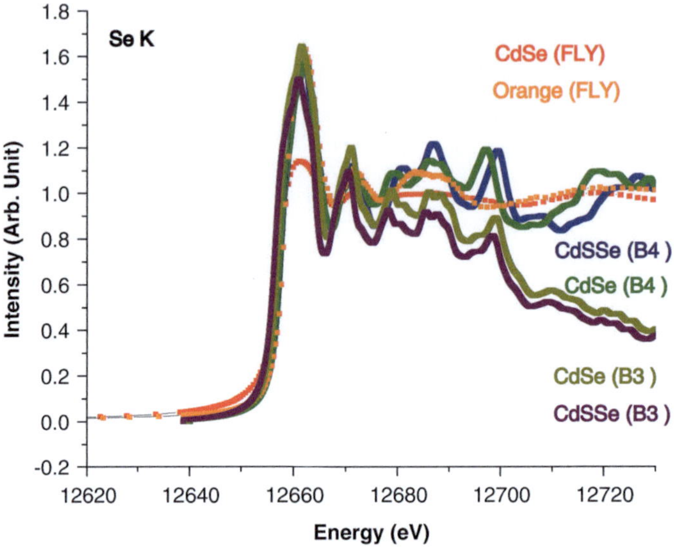

Fig. 2.14 Se *K* edges of CdS and CdSe are plotted in B3 and B4 forms comparing the experimental FLY and GGA calculated spectra

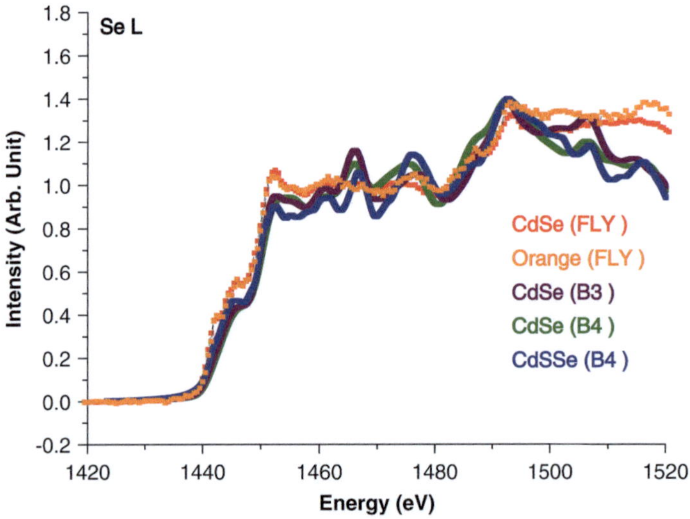

Fig. 2.15 Se *L*-edges of CdS and CdSe are plotted in B3 and B4 forms comparing the experimental FLY and GGA calculated spectra

2.6.2 Se, SeTe, and SeTeSb Systems

Bulk amorphous Se_{80-x}-Te_{20}-Sb_x ($0 < x < 9$) materials were synthesized by the conventional melt quenching technique [13]. TEY (Total Electron Yield) XANES spectra for powdered samples at Selenium L-edge were collected at the Canadian Light Source (CLS) (operation energy: 2.9 GeV). The spherical grating monochromator (SGM) beamline (undulator beamline, energy range: 200–2000 eV, resolution $E/\Delta E$: >5000) was used for Se $L_{3,2}$ edge. Operating current for the measurements was 250 mA at injection; current up to 15 mA was used for the single bunch measurement.

The crystal structure of Se has been modeled as α-Monoclinic Se with space group $P2_1/n$ (#14) and calculated by DFT and RSMS methods. For DFT routine, the core-hole lifetime broadenings of the Se L_3 and L_2-edges have been set at 1.0 eV and 1.13 eV, respectively.

The mono-chalcogenide material under investigation is Se. The Se and Se-Te samples have been characterized. The XANES spectra of these materials have been accumulated. XANES reflects the local density of unoccupied states and probes the structure of the conduction band. The L_3/L_2-edge XANES gives information about the d- and to a lesser extent s-like local density of states. The experimental total electron yield (TEY) spectra near the Se $L_{3,2}$ edge have been collected using the synchrotron radiation beam and plotted in Fig. 2.16 for the Se, Se_3Te and

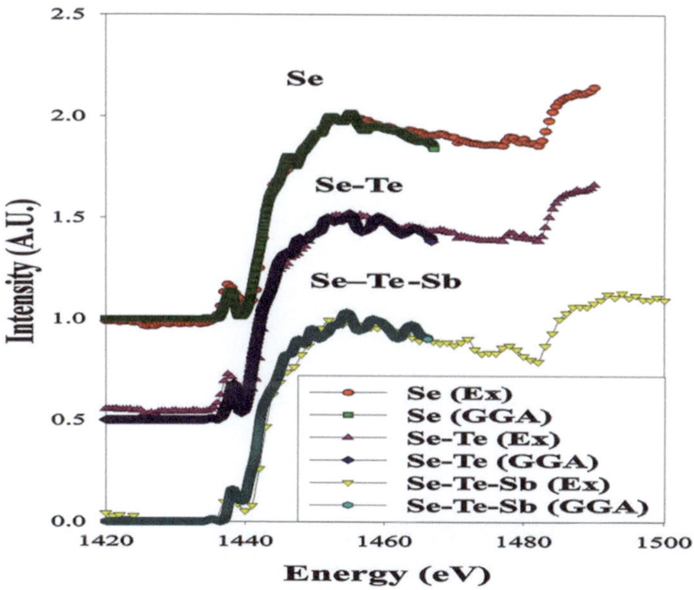

Fig. 2.16 XANES spectra of Se, Se-Te, and Se-Te-SB samples are plotted with GGA theoretical calculated spectra

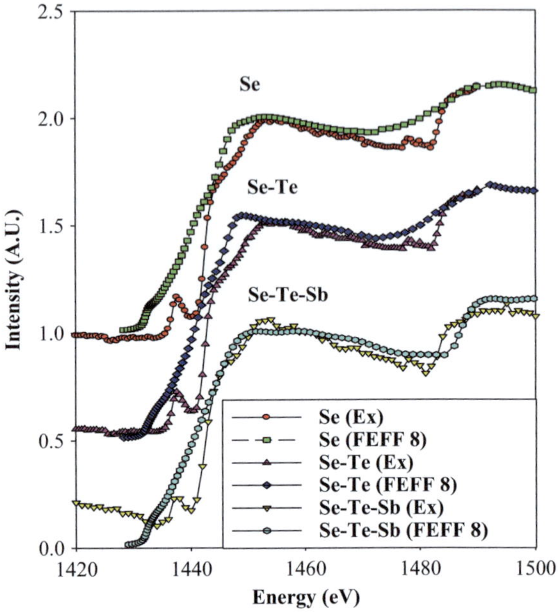

Fig. 2.17 XANES spectra of Se and Se-Te sample are plotted with FEFF theoretical calculated spectra

$Se_{80-x}Te_{20}Sb_x$ systems, together with GGA theoretical spectra. Figure 2.17 reveals the experimental spectra together with FEFF calculation.

The metallic behavior of the Se_2TeSb crystal is due to the presence of Se $4p$ electrons across the Fermi level. The TEY XANES spectra of the Se, SeTe, and SeTeSb systems agree remarkably well with the DFT and GGA calculation, in particular, they possess a small peak at the onset of the L_3 edge as shown in Fig. 2.16. However, the DFT with GGA calculation requires longer computer CPU time in Linux/Unix systems of workstations. DFT method can only calculate around 85 eV above the Se L_3 edge for the Se-based systems.

The RSMS routine is much more efficient and can handle cluster sizes under 1000 atoms in a personal computer platform. The spectra calculated by the RSMS method using the FEFF8 code are in better agreement with the experimental Se, SeTe, and SeTeSb XANES data after the Se L-edge jump. The theoretical L_2 jumps, calculated by RSMS, lag slightly behind the experimental L_2 edges of the Se-Te-Sb systems. Se $L_{3,2}$ edge XANES spectra of $Se_{80-x}Te_{20}Sb_x$ compounds are sensitive to the content x of Sb, where $x = 2, 4, 9$. In particular, there is a glass phase change for 9 at.% of Sb content.

The bulk Se, SeTe, and SeTeSb systems have been modeled as crystals with crystal structures resembled the α-Monoclinic Se structure. The calculated Se L-edge XANES spectra by the density function theory (DFT) with generalized gradient approximation (GGA) fit very well with the experimental TEY of Se, SeTe, and SeTeSb systems, in particular, they possess pre-edge peaks. The Se L-edge XANES spectra calculated by the RSMS method resembles the experimental

Se TEY XANES of the Se, SeTe, and SeTeSb systems after the edge jump of Se. The calculated spectra by RSMS method do not process the pre-edge peaks and make this FEFF calculation deviates slightly from the experimental spectra of the actual samples.

2.6.3 As_2Se_3

Amorphous arsenic tri-selenide is used as a chalcogenide glass for infrared optics by virtue its property of transmitting light from 870 nm to 17.2 μm. Vitreous As-Se films are widely used in xerography due to their high photo-sensitivity and low electrical conductivity [17]. Glassy As_2Se_3 has a property of photo-induced crystallization with potential applications as a computer-memory material. Zinc selenide is a semiconductor material at ambient conditions of technological importance for use as an infrared filter and lenses, and for short wavelength visible light laser devices [18]. Zinc selenide is a semiconductor material. It can be used as an infrared filter, lenses, and short wavelength visible light laser devices. For Zn (II)-ions sensing, one needs detailed structural investigations of these materials.

Crystalline As_2Se_3 has a layered structure based on the $AsSe_3$ spiral chain. The model compound of As_2Se_3 is c-As_2Se_3. The crystal structure of c-As_2Se_3 has a symmetry which follows the space group of $P2_1/c$. The crystal constants are $a = 12.0774$ Å, $b = 9.9037$ Å, $c = 4.2835$ Å, $\beta = 90.458^0$. The As and Se atomic sites are shown in Table 2.1 [19]. Cluster of c-As_2Se_3 is plotted in Fig. 2.18.

The As $L_{3,2}$-edge (L_3: 1323.6 eV and L_2: 1359.1 eV) and Se $L_{3,2}$-edge (L_3: 1433.9 eV and L_2: 1474.3 eV) XANES spectra of the samples were performed at the spherical grating monochromator (SGM, 11ID-1) beam-line of the Canadian Light Source (CLS). A theoretical interpretation of the $L_{3,2}$-edge XANES spectra of un-doped and zinc-doped As_2Se_3 glasses, $L_{3,2}$-edge XANES of glasses with compositions has been obtained [10]. The experimental XANES of powder As_2Se_3 together with the GGA theoretical spectra are illustrated in Fig. 2.19. The pre-edge jumps of the $L_{3,2}$ of As and Se are demonstrated in the GGA theoretical spectra.

Table 2.1 The As and Se atomic sites of crystalline As_2Se_3 in a unit cell are listed

	X	*Y*	*Z*
As (1)	0.2362	0.1967	0.1113
As (2)	0.0108	0.3253	0.3810
Se (1)	0.0977	0.1182	0.4726
Se (2)	0.1534	0.4104	0.0257
Se (3)	0.3785	0.2981	0.4436

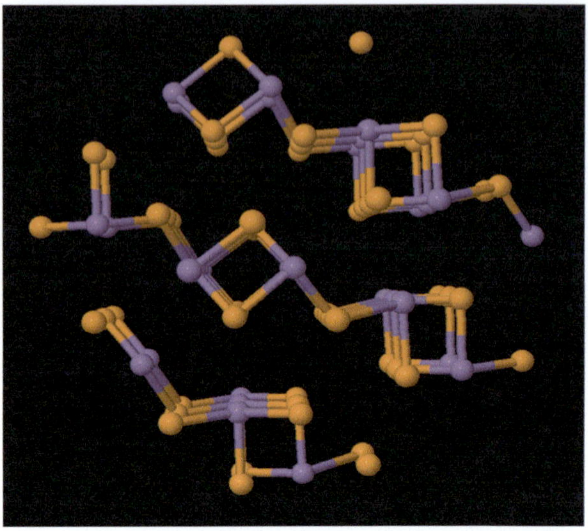

Fig. 2.18 Cluster of c-As_2Se_3 is plotted

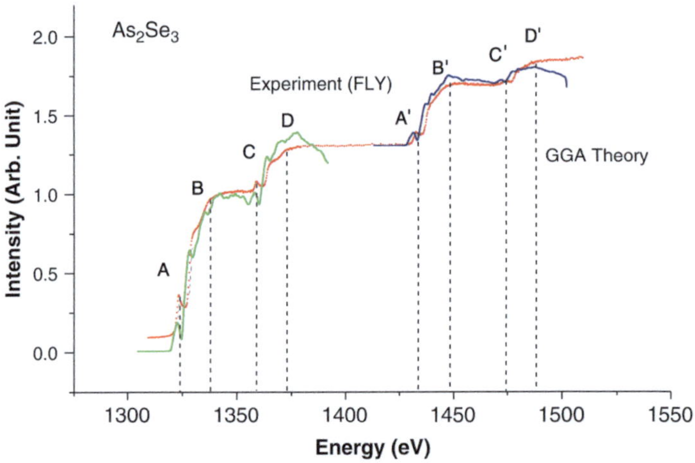

Fig. 2.19 Experimental XANES of powder As_2Se_3 and the theoretical spectra of c-As_2Se_3 calculated by GGA

2.6.4 *ZnS*

The focus mineral sulfide for this section is sphalerite, (Zn, Fe)S. Understanding sphalerite mineral chemistry is important because it is the primary source for virtually all of the world's zinc production. Over 7,000,000 tonnes of zinc each year originate from sulfide ores in which sphalerite (Zn, Fe) S is the dominant zinc mineral [20].

ZnS in general has four crystal structures, B1 (rocksalt), B2 (CsCl structure), B3 (zinc blende structure), and B4 (Wurtzite) structures. The natural crystal, sphalerite, has the zinc blende structure with space group F43m. It has cubic face-centered lattice with Zn at (0,0,0) and S at (1/4,1/4,1/4), and lattice constant of 5.41 Å. The ab initio calculation starts with this model crystal using the WIEN program.

Sulfur $L_{3,2}$ edge XANES spectra were collected using the VLS-PGM beam-line at the Canadian Light Source (CLS), University of Saskatchewan. The spectra are compared with the GGA theoretical spectra in the following Fig. 2.20. High resolution S $L_{3,2}$-edge XANES spectra have been collected from a series of high quality sphalerite samples [21]. Each of the sphalerite samples examined has a characteristic iron concentration and the range is from <0.01 wt.% Fe to 14.79 wt. % Fe. The structures of the S $L_{3,2}$-edges are investigated by comparing ab initio calculated spectra with experimentally measured S $L_{3,2}$-edge spectra collected from synthetic ZnS sphalerite. The theoretical and experimental data provide evidence that S 2*p* electron transitions to empty S 3*s* and S 3*d* states can occur across an

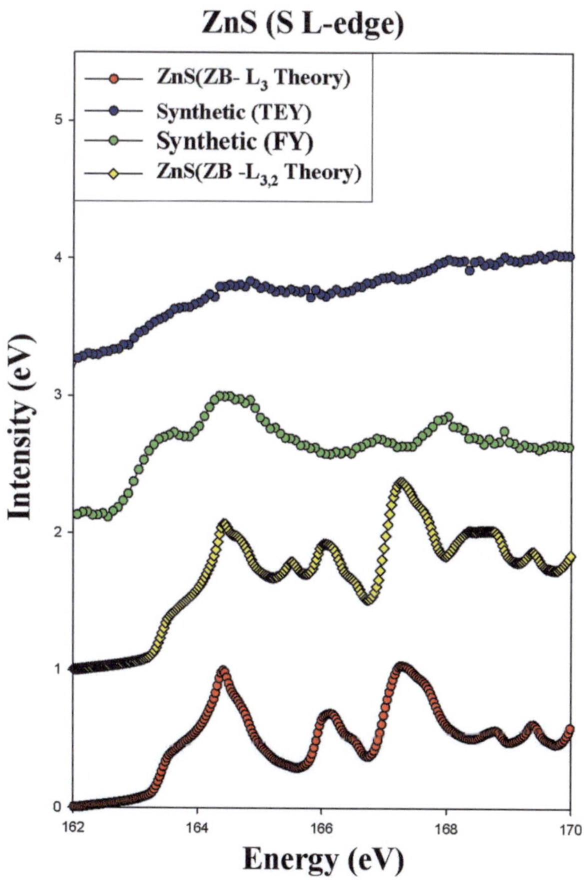

Fig. 2.20 The experimental XANES spectra of ZnS in comparison to the GGA calculated spectra

energy range above the absorption edge. A chemical shift associated with iron to sulfur bonds is identified and for the first time a pre-edge feature associated with S–Fe bonding in sphalerite is observed.

The experimental XANES of ZnS at the S $L_{3,2}$ edges have been plotted in Figs. 2.20 and 2.21. The S L_3 absorption edge is from $2p_{3/2}$ electron transitions to the unoccupied states of *s* or *d* and the L_2 absorption edge results from $2p_{1/2}$ electron transitions to the unoccupied states of *s* or *d*. The first DOS (Density of State) peaks near 3.3 eV above the Fermi level is as shown in Fig. 2.22. Comparison of the DOS peak intensities for S *s* and S *d* shows that the S 2*p* electron transitions to empty S 3*s* states dominate at the initiation of absorption. As the energy difference from the edge is increased, an increase in the S 2*p* electron transitions to empty S 3*d* states is observed. At 4.6 eV above the Fermi level the DOS peaks for S *s* and S *d* have near equal intensities. From this point toward higher energies, the DOS data shows that transitions to both empty S 3*s* and S 3*d* states occur, but it is the transitions to empty S 3*d* states that are dominant. These results show that S 2*p* electron transitions to empty S 3*s* and S 3*d* states occur across an energy range above the absorption edge.

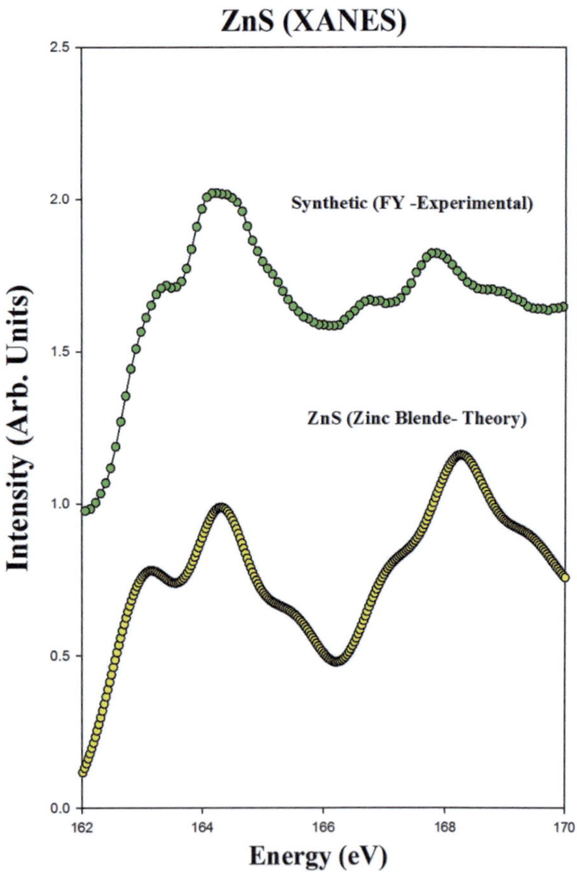

Fig. 2.21 Ab initio calculated L_3 and L_2 edges for ZnS sphalerite, comparing to the experimental FLY spectrum. The line broadening in calculated spectrum is 1.0 eV

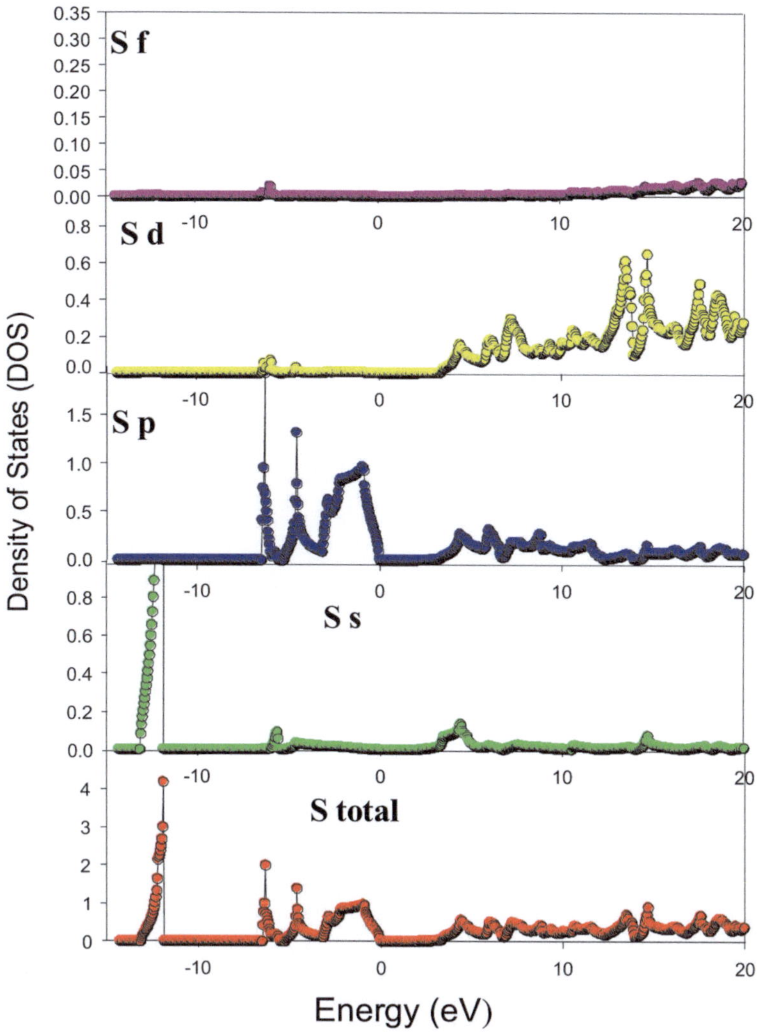

Fig. 2.22 Calculated densities of states for sulfur in ZnS sphalerite using the ideal zinc-blende structure

2.7 EXAFS of CdSSe Nanostructures

EXAFS (extended X-ray absorption fine structure) refers to the oscillation of the absorption coefficients from ~50 to 1000 eV above the absorption edge. The oscillations in the extended region of the XAFS spectrum are due to interference of the outgoing and backscattered photoelectron wave, which modulates the absorption coefficient. EXAFS is the modulation of the absorption coefficient in the extended region to as much as 1000 eV above the threshold. It is represented by

$\chi(k)$ where k is the wave vector. The EXAFS, $\chi(k)$, is the modulation of the absorption coefficent $(\mu - \mu_0)$ in k space relative to the free atom (μ_0),

$$\chi(k) = \frac{\mu - \mu_0}{\mu_0} \tag{2.42}$$

The EXAFS, $\chi(k) = A(k)\Phi(k)$, has two terms, the amplitude $A(k)$ and the phase $\Phi(k)$, describing the oscillation intensity and the periodicity, respectively. Historically, the breakthrough in EXAFS analysis was the use of Fourier transformed (FT) by Sayers, Lytle, and Stern [8] and the detail has been described earlier. The following shows that the FT of the Cd data yield peaks in the real space corresponding to the interatomic distance, peak position $r = R + b$, (R real bond length, b correction term) between the absorbing atom and the neighboring atoms. One can filter the first shell in real space and backtranform it to k space, which gives the amplitude and phase from the back scattering of the nearest atoms.

EXAFS contains information about the interatomic distance between the absorbing atom and the backscattering atom (bond length), R; the coordination number, N_j and the mean square displacement of the interatomic distance due to thermal motion known as the Debye–Waller factor. It has the following expression. EXAFS is additive and is the sum of sine functions for more than one type of bonds.

$$\chi^{(k)} = \sum_j \underbrace{\left[\frac{N_j S_0^2 f(k,\theta) e^{-2R_j/\lambda_e(k)} e^{-2k^2\sigma_j^2}}{kR_j^2}\right]}_{} \mathrm{A}(k) \underbrace{\sin\left[2kR_j + \phi(k)\right]}_{} \tag{2.43}$$

N_j: coordination number of atom type j, e.g., in CdS, Cd has four nearest neighbors, $N_{Cd} = 4$.

S_0^2: amplitude reduction term (shake-up, shake-off loss) due to many-body effects; $S_0^2 \sim 0.8$.

$f(\theta, k)$: atomic scattering amplitude, e.g., $f(\pi, k)$ for single scattering (180° backscattering). It is characteristic of the atomic scatterer, and is k and angle dependent.

$e^{-2R_j/\lambda_e(k)}, e^{-2k^2\sigma_j^2}$: amplitude damping terms due to electron escape depth λ_e and Debye–Waller factor due to thermal motion σ_j, respectively.

R_j: interatomic distance (bond length), most accurately determined parameter from EXAFS.

$\phi(k) = 2\delta_a(k) + \delta_b(k)$: phase function with contributions from the absorber $\delta_a(k)$ and the back-scatterer $\delta_b(k)$.

The samples of CdSSe have been prepared as described in Sect. 2.6.1 [22]. The EXAFS spectra have been acquired by the synchrotron facility as indicated. The spectra are shown in Fig. 2.23. The Fourier transformed curves are evaluated as functions of k-space and real space in Figs. 2.24 and 2.25, respectively. It is revealed in Fig. 2.25 that the nearest neighbors of the Yellow, Brown, and CdSe samples are around 2.5 Å. The next neighbors of these compounds are at around 3.23 Å.

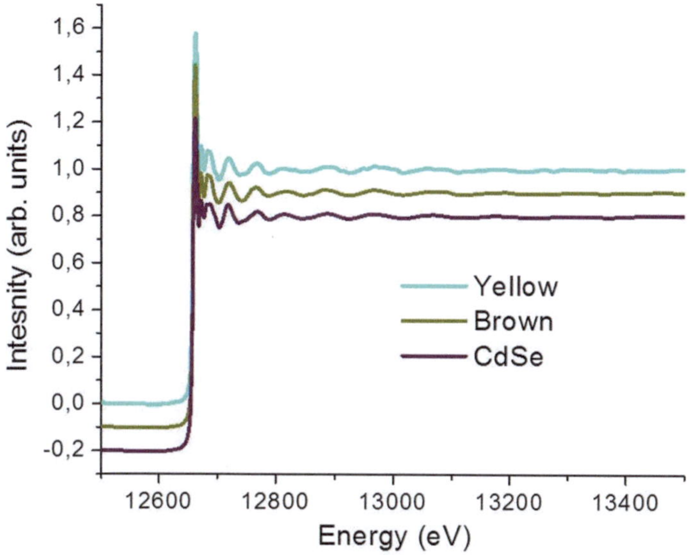

Fig. 2.23 EXAFS spectra of the yellow, brown, and CdSe samples

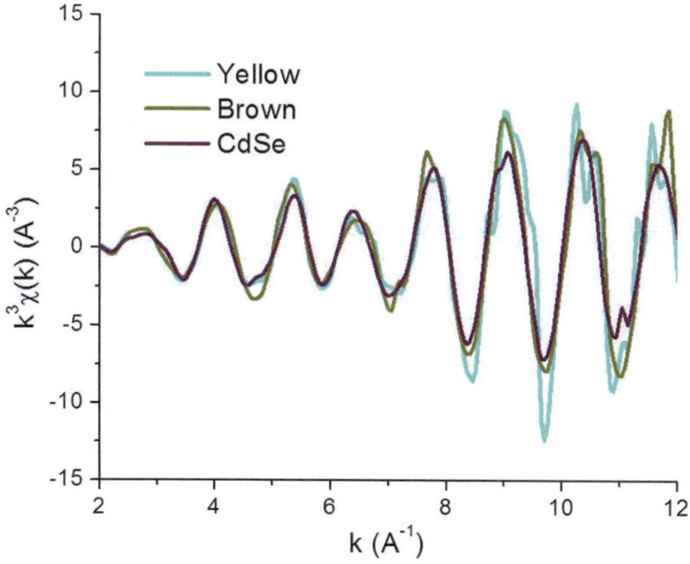

Fig. 2.24 Fourier transform of the EXAFS spectra of the yellow, brown, and CdSe samples have been evaluated as a function of momentum *k*

The EXAFS data was analyzed using the software package IFEFFIT through the graphic interface Arthemis. For the EXAFS fitting, the R-space between 1.4 and 3.4 Å was used for the Se edge with theoretical phase-shifts and amplitudes

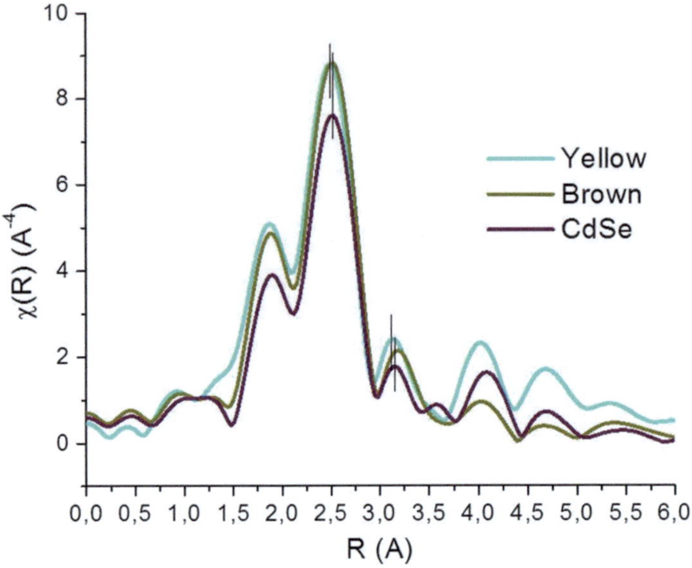

Fig. 2.25 Fourier transform of the EXAFS spectra of the yellow, brown, and CdSe samples in real space *R* representation

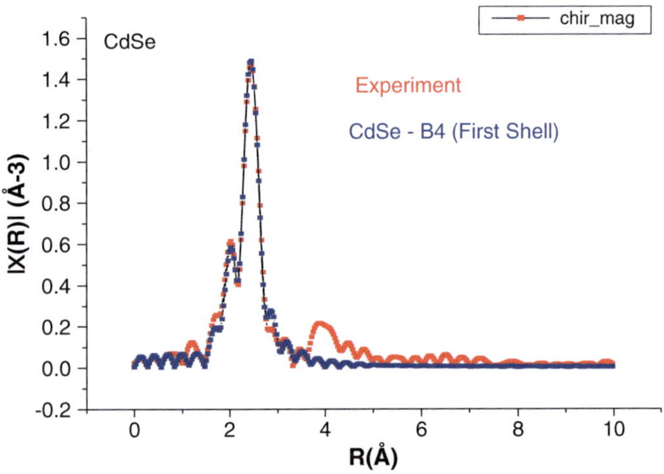

Fig. 2.26 Fourier transform of the EXAFS spectra of CdSe samples and EXAFS first shell fit of CdSe in B4 structure in real space *R* representation

generated by FEFF. Bulk lattice parameters of wurtzite structure of CdSe with first shell coordination numbers of 3 were used to determine the amplitude reduction factor, S_0. The EXAFS single-shell fits in *R*-space of Se *K*-edge for CsSe samples has been plotted in Fig. 2.26.

2.8 XEOL (X-ray Excited Optical Luminescence)

The technique of XEOL has been introduced in the third generation of synchrotron facilities around the world. The incident X-ray beam bombarding the samples under investigation excites the samples. The transitions from the ground state of the samples to the continue states, valence band, and conduction band are monitored. The excited state of the samples then relaxes. The relaxation process involving photons in the visible range is excited optical luminescence and termed as XEOL. XEOL has been shown to be a powerful tool for investigating the local chemical environment of a site that gives rise to a particular luminescence band [23]. In XEOL, one monitors the intensity of the luminescent bands as the X-ray energy is raised above the core level binding energy of a particular element in the system. The photon then excites an electron from a core level to a previously unoccupied state in the conduction band. The decay of the excited state emits to intense luminescence via energy transfer to the optical channel. An increase in signal can be recorded at an energy corresponding to a particular local environment of the probed element.

2.8.1 *XEOL of CdSSe*

XEOL spectra for nano-CdS, CdSe, and CdSSe [22] samples have been accumulated using the synchrotron facility. All the XEOL spectra as shown in Fig. 2.27a–c exhibit two emission bands. The sharper band is the band-gap emission (excitonic), while the broad band is the defect emission of the CdS_xSe_{1-x} nanostructure. From the XEOL spectra in Fig. 2.27, it can be seen that there is a tunable red shift in both the band-gap and defect emissions with decreasing S or increasing Se content in the CdS_xSe_{1-x} nanostructures. The use of a tunable X-ray source (XEOL) to examine the optical emission of CdS_xSe_{1-x} allows us to probe new de-excitation channels which show a defect emission band not previously seen with laser sources [24].

It is known that the band-gap of CdS_xSe_{1-x} alloys can be determined by an interpolation between those of the two binaries (CdS and CdSe) with an additional nonlinear bowing parameter [25] according to Eq. (2.44)

$$E_g(CdS_xSe_{1-x}) = xE_g(CdS) + (1-x)E_g(CdSe) - x(1-x)b \qquad (2.44)$$

where E_g(CdS) is 2.43 eV, E_g(CdSe) is 1.74 eV, and the bowing parameter b is 0.54 for CdS_xSe_{1-x} at room temperature. Using the alloy composition x determined by EDX as shown in Fig. 2.28, we can obtain the theoretical band-gap values of the nanostructures for each composition x. The theoretical band-gap energies (blue triangles) along with the experimental band-gap energies (black squares) and defect emission energies (red circles) are plotted in Fig. 2.27d. The theoretical values as calculated in Eq. (2.44) are in good agreement with the experiment band-gap energies.

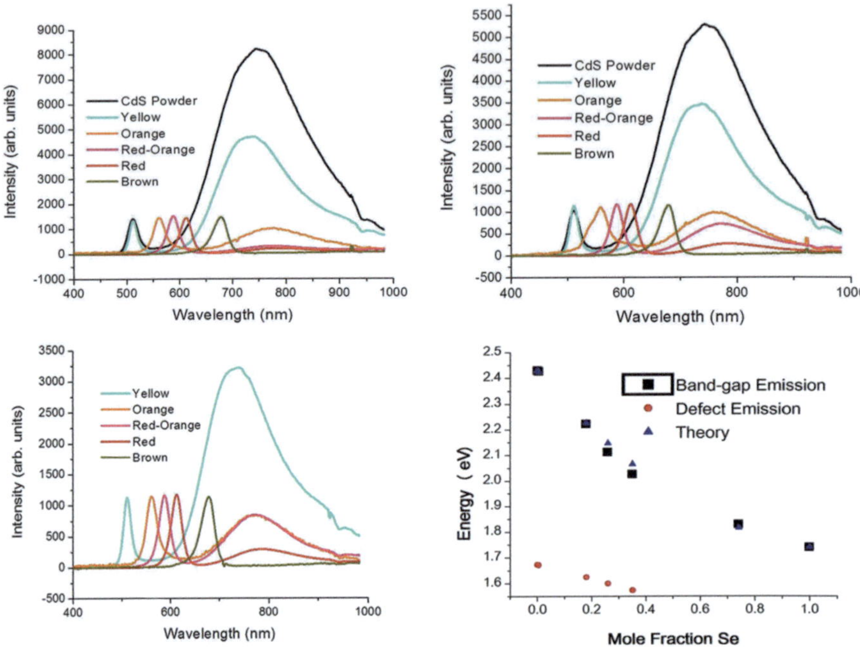

Fig. 2.27 (**a–d**) S L_3-edge XEOL spectra of CdS_xSe_{1-x} normalized to the maximum band-gap intensity when excited at 155 eV (**a**), 163.5 eV (**b**), 170 eV (**c**) and the maximum of band-gap and defect emission energy as a function of mole fraction Se for the 163.5 eV excitation (**d**). Note: The spectral response observed at ~930 nm is an artifact of the detector

The peaks of the band gap emission from the XEOL spectra of CdS, Yellow, Orange, Red-orange, Red, and Brown samples are at 511.32 nm, 551.446 nm, 562.984 nm, 587.541 nm, 612.02 nm, and 678.19 nm, respectively. The peak positions of the band gap emission of these samples do not change when the excitation energy is tuned through the S *L*-edge, changing from 155 eV, 163.5 eV, and 170 eV. The corresponding band gaps of CdS, yellow, orange, red-orange, red, and brown samples are 2.43 eV, 2.25 eV, 2.20 eV, 2.11 eV, 2.03 eV, and 1.83 eV, respectively. Please note that the compositions of these CdSSe nanomaterials are: $CdS_{0.99}Se_{0.01}$ (Yellow), $CdS_{0.81}Se_{0.19}$ (Orange), $CdS_{0.74}Se_{0.26}$ (Red-Orange), $CdS_{0.65}Se_{0.35}$ (Red), and $CdS_{0.25}Se_{0.75}$ (Brown).

The band gaps of CdS, CdSSe, and CdSe have also been calculated using the mBJ (modified Becke–Johnson) exchange-correlation potential [6] within the frame work of WIEN2k. The calculated band gaps are: E_g(CdS-B4) = 2.55 eV, E_g($CdS_{0.5}Se_{0.5}$-B4) = 2.25 eV, E_g(CdSe-B4) = 1.77 eV, E_g(CdS-B3) = 2.42 eV, E_g(CdSe-B3) = 1.74 eV. Both the LDA and GGA underestimate band gaps and the mBJ potential improves the band gaps estimates, and the calculated mBJ band gaps are closer to the experimental data.

Additionally, there is an increase in the branching ratio between the band-gap and defect emissions of the CdS_xSe_{1-x} nanostructures with a decrease in the S to Se ratio. Investigating the change in optical emission across the S L_3-edge (Fig. 2.29),

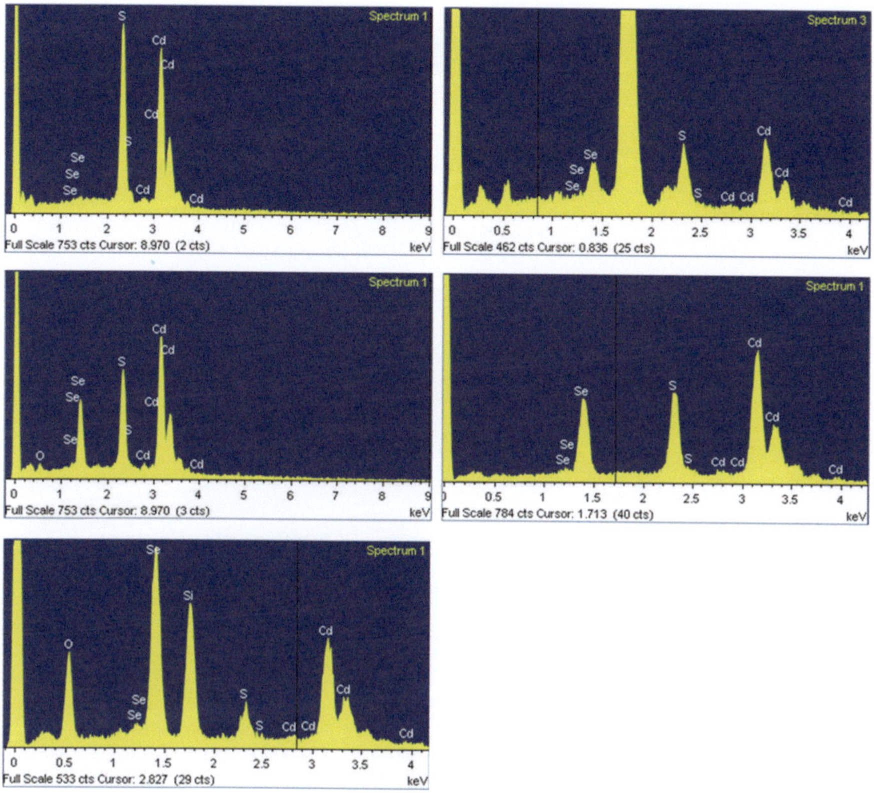

Fig. 2.28 EDX spectra of CdSSe nanostructures: (**a**) yellow, (**b**) orange, (**c**) red-orange, (**d**) red, (**e**) brown

the defect emission decreases across the S L_3-edge for CdS and Yellow, whereas for Orange, Red-Orange and Red the defect emission increases across the S L_3-edge. Moreover, there is a decrease in band-gap intensity moving from below the edge to the edge-jump (163.5 eV) for the Yellow, Orange, Red-Orange, and Red nanostructures and an increase for the Brown. It is evident from these results that with deviation from pure CdS the S takes a more dominant role in defect emission. The decrease in band-gap emission across the S L_3-edge indicates that the band-gap luminescence from the nanostructures is either more likely to originate from the Cd sites than the S sites or the radiative channel is suppressed by the increasing presence of Se [26].

Figure 2.30a–c shows the XEOL spectra taken at excitation energies below the Se L_3-edge (1405 eV) and above the Se L_2-edge (1500 eV) for the Yellow, Orange, and Brown CdS_xSe_{1-x} nanostructures. The peak positions of the band gap emission below the Se L_3-edge (1405 eV) and above the Se L_2-edge (1500 eV) are 510.5 nm, 543 nm, and 657.7 nm for the Yellow, Orange, and Brown samples, respectively. The corresponding band gap energies of the Yellow, Orange, and Brown samples, are 2.43 eV, 2.28 eV, and 1.89 eV, respectively. The peak positions in the XEOL at

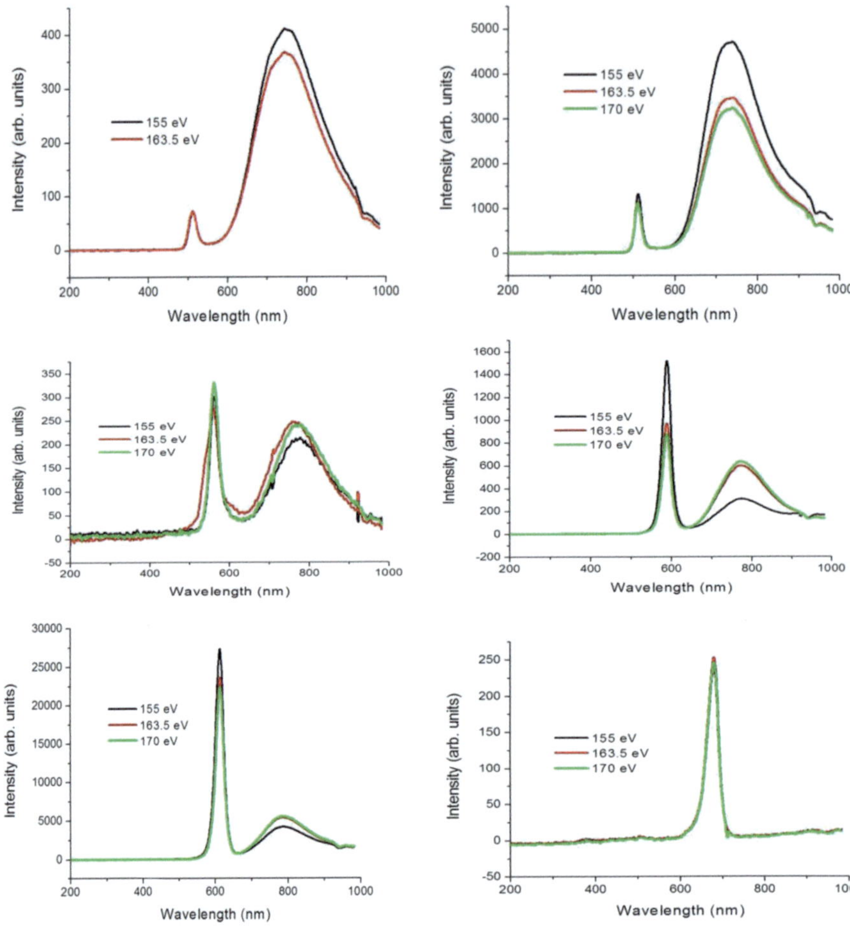

Fig. 2.29 S L_3-edge XEOL spectra for (**a**) CdS, (**b**) yellow, (**c**) orange, (**d**) red-orange, (**e**) red, (**f**) brown. The XEOL spectra are all corrected to the intensity to the intensity of the incoming photons. Note: The spectral response observed at ~930 nm is an artifact of the detector

the Se $L_{3,2}$-edge differ slightly to those at the S $L_{3,2}$-edge because they were recorded at different spots on the respective Yellow, Orange, and Brown samples which have a minor variation in composition along the sample due to the growth gradient. At the Se $L_{3,2}$-edge there is a decrease in defect emission going from below to above the edge. The Yellow and Orange samples, which are closer to CdS, show a decrease in band-gap emission and band-gap to defect branching ratio. The Brown sample, which is closer to CdSe, shows an increase in band-gap emission and band-gap to defect branching ratio. These results demonstrate that when Se is the dominant anion, its excitation will result in a large increase in band-gap emission since energy is transferred directly to the emission sites. However, in CdS_xSe_{1-x} systems in which Se is the nondominant anion, Se excitation will decrease the optical emission by transferring energy away from luminescence sites and de-excitation will occur through non-radiative paths.

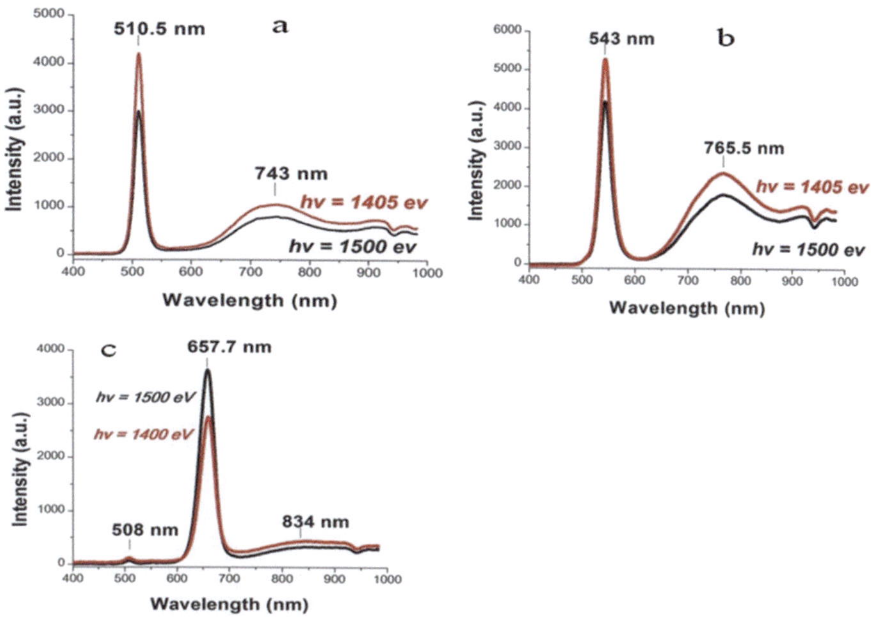

Fig. 2.30 (**a–c**) Se $L_{3,2}$-edge XEOL of CdS_xSe_{1-x} nanostructures (**a**) yellow, (**b**) orange, and (**c**) brown. Note: The spectral response observed around 930 nm is an artifact of the detector

2.9 SSHG (Surface Second Harmonic Generation)

Surface Second Harmonic Generation is a photon in and photon out technique. The incoming photon is provided by lasers which have high enough intensity to be able to generate light with higher order frequency. The incident beam is focused onto the surface layers to achieve the confocal imaging. The outgoing beam is measured by either a photo-detector or a CCD (charge coupled device). Second Harmonic Generation is the generation of an output photon having the double frequency of the input laser beam. For solid or liquid states, the centro-symmetry law dictates that the second harmonic Generation is confined to the surfaces which do not possess centro-symmetry. The third harmonic generation or four-wave mixing is possible in gas phase materials which follow centro-symmetry. Experimental setup for SSHG is shown in Fig. 2.31.

Bulk amorphous $Se_{80-x}Sb_xTe_{20}$ $(0 < x < 9)$ materials were synthesized by the conventional melt quenching technique [27]. WITec confocal microscope system was used for surface second harmonic generation (SSHG) image measurements and Newport high resolution spectrometer (1800 g mm^{-1}) and Andor CCD camera was used for SSHG spectrum analysis. The SSHG spectrum was recorded at each image pixel by an air-cooled (−60 °C) back-illuminated spectroscopic CCD camera (Model DV401-BV, ANDOR) behind a grating (1800 g mm^{-1}) spectrograph

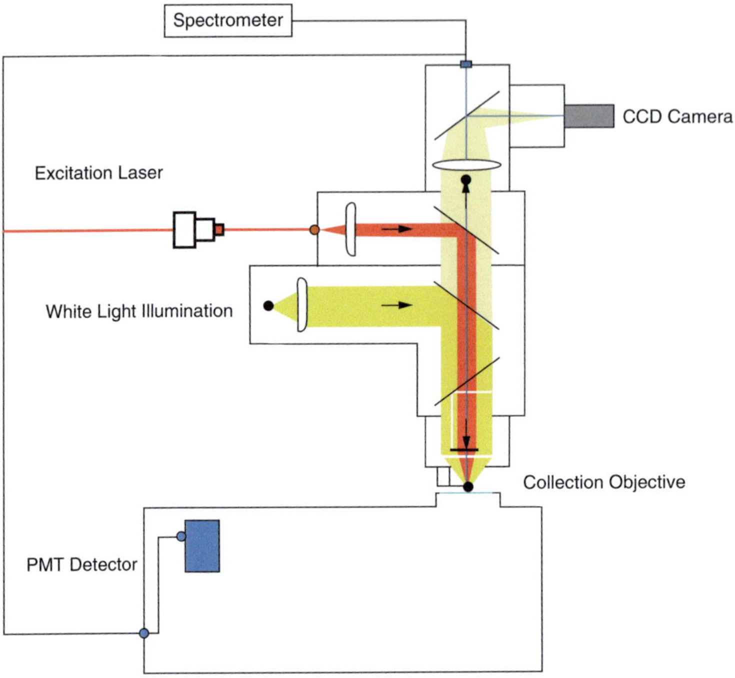

Fig. 2.31 Experimental setup for SSHG

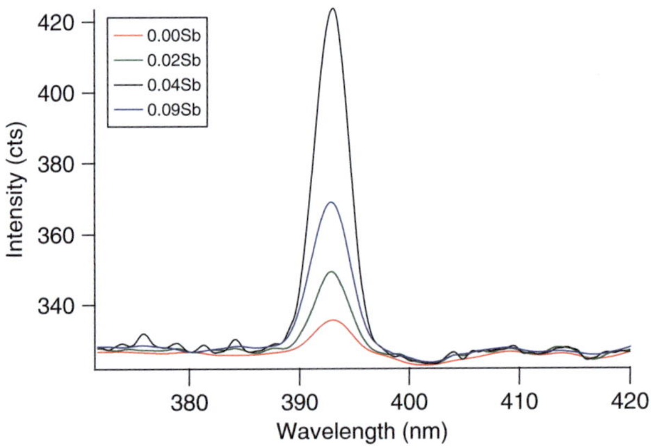

Fig. 2.32 Intensity variation of SSHG signal in $Se_{80-x}Sb_xTe_{20}$ samples

(cornerstone 210). Figure 2.32 shows the variation in intensity of the SHG signal for different compositions.

The following experimental pictures in Fig. 2.33 reveal the 2D configuration of the samples, Se-Te-Sb.

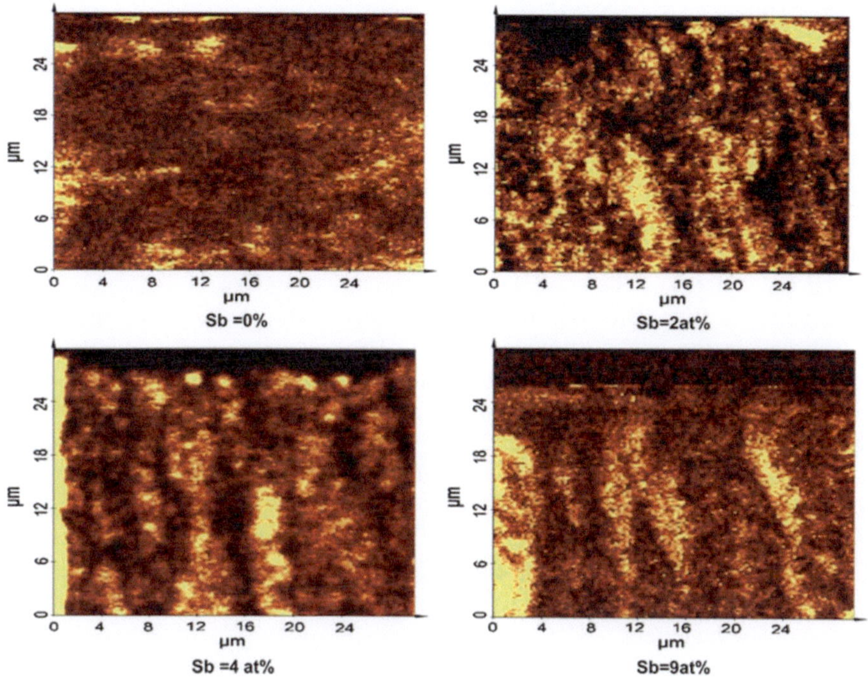

Fig. 2.33 SSHG images of $Se_{80x}Sb_xTe_{20}$ ($0 < x < 9$ at.%) films obtained at 393 nm by the confocal scanning microscope

2.10 Monte Carlo Simulations

Another commonly used computer simulation technique for modeling the structures of materials is the Monte Carlo approach. It uses random numbers in the evaluation of the particle displacements during the course of a simulation run [28]. One popular way of implementing Monte Carlo simulations is to use the Metropolis method. Starting from an initial configuration, say a random distribution of particles (atoms), one atom i is picked randomly and displaced in a random direction by a random amount, from r_i^m to r_i^n, subject to the maximum displacement being the adjustable parameter $\delta r_{\max}$. The change in potential energy of the system, δV_{mn}, resulting from the movement of this atom is then calculated for an assumed form for the interatomic potential. If the atomic displacement is "downhill" in energy terms ($\delta V_{\mathrm{mn}} < 0$), then the new position is accepted unconditionally. If, however, the move is "uphill" in energy ($\delta \mathrm{V}_{\mathrm{mn}} > 0$), then the move is accepted only conditionally, subject to the Boltzmann probability factor, exp ($\delta V_{\mathrm{mn}}/k_{\mathrm{B}}T$).

2.11 Summary

The electronic structures of chalcogenide compounds have been evaluated by comparing the experimental results to the calculated spectra with model compounds. The DFT calculations of WIEN and RSMS of FEFF have been performed. The former is more demanding for computer CPU time.

WIEN program is capable of providing ab initio calculation of band structure for crystalline states having spatial symmetries. It is developed from DFT (Density Functional Theory). The terminology of group symmetry is listed in tables established in literature [2]. Classic materials with crystal symmetry were established in the books of Wyckoff [29]. The calculated density of states of individual atomic sites can be compared to the experimental spectra of XANES and XPS. The band structure describing the propagation along direction of the Brillouin zone can be established. The wide band gap materials underlying the Chalcogenide compounds have been evaluated with the outstanding examples. These examples particularly include CdS, CdSe, CdSSe, Se-Te-Sb, ZnS, ZnSe, and $CuInSe_2$.

FEFF program is based on the RSMS (real space multiple scattering) and can apply to the calculation of materials with clusters conformity. It is especially good for nanoscale materials.

Chalcogenide materials have been reviewed using different computer models. Ab initio calculations have been introduced and compared to the experimental results. The evaluations of the nano properties of mono-chalcogenides have been briefly described. The shortcomings of individual modeling techniques have been discussed. The demands of computer facility are high. It is costly to simulate a simple program for extremely small nanomaterials with complicated structures.

XAFS spectra have been used at the Cd, S, and Se K and $L_{3,2}$-edges to probe the electronic structures of the CdS_xSe_{1-x} nanostructures. These nanostructures are found to be dependent on the S to Se ratio and display properties that are between that of pure CdS and CdSe. The investigation of the color, densities of states, and band structures of CdSSe concludes that there is a tunable red shift in both the band-gap and defect emissions with decreasing S or increasing Se content in the CdS_xSe_{1-x} nanostructures. The experimental Cd L_3 edge of both CdS and CdSe favors the B4 (hexagonal wurtzite) structure as the pre-edge region indicates the presence of small peaks and they compared favorably to the theoretical curves of the B4 structure.

2.12 Discussion

The ab initio calculations of DFT with GGA have shortcoming of underestimating the band gap energy. The band gap is a characteristic of semiconductors. It is confirmed that CdS and CdSe have a direct band gap. The band gap energies are underestimated with LDA (local density approximation) and GGA, in comparison to the experimental results. These experimental results are estimated by the synchrotron results of XAS (X-ray absorption spectra) of XES (X-ray emission spectra) or XEOL (X-ray emission optical luminescence). To correct this deficiency, an exchange-correlation potential, mBJ potential has been introduced. With the program WIEN, the band gap energies can be best fitted to the experimental gap with the introduction of parameters, *A* and *B* of the mBJ potential. The fitted band gap energies of CdS and CdSe systems can be then obtained. Yet the choice of the global parameters A and B is a measure subject to the actual band gaps of materials. It is determined by the application to the majority of wide band gap materials and insulators. Furthermore, exchange-correlation potentials are being developed to investigate novel materials.

Band structure is an essential investigation tool towards the understanding of semiconductors and metals. DFT together with GGA tends to underestimate the band gap energy as shown in the text. With the modified exchange-correlation potential, this difficulty can be by-passed. With the computer models of the outstanding chalcogenide materials, a basic understanding of the nanoscale can be achieved.

References

1. P. Hohenberg, W. Kohn, Phys. Rev. **136**, B864 (1964); W. Kohn, L.J. Sham, Phys. Rev. **140**, A1133 (1965)
2. N.F.M. Henry, K. Lonsdale, *International Tables for X-Ray Crystallography, vol. 1, Symmetry Group* (Kynoch Press, Kingston, ON, 1965)
3. T.L. Loucks, *Augmented Plane Wave Method* (Benjamin, New York, 1967)
4. P.E. Bloch, O. Jepsen, O.K. Andersen, Phys. Rev. B **49**, 16223 (1994)
5. A. Neckel, K. Schwarz, R. Eibler, P. Rastl, Microchim. Acta, Suppl. 6, 257 (1975); K. Schwarz, A. Neckel, J. Nordgren, J. Phys. F: Metal Phys. **9**, 2509 (1979); K. Schwarz, E. Wimmer, J. Phys. F: Metal Phys. **10**, 1001 (1980)
6. A.D. Becke, E.R. Johnson, J. Chem. Phys. **124**, 221101 (2006). doi:10.1063/1.2213970
7. A.L. Ankudinov, B. Ravel, J.J. Rehr, S.D. Conradson, Phys. Rev. **B58**, 7565 (1998); J.J. Rehr, R.C. Albers, Rev. Mod. Phys. **72**, 621, (2000); A.L. Ankudinov, C.E. Bouldin, J.J. Rehr, J. Sims, H. Hung, Phys. Rev. **B65**, 104107 (2002); J.J. Rehr, J.J. Kas, M.P. Prange, A.P. Sorini, Y. Takimoto, F.D. Vila, C. R. Phys. **10**(6), 548–559 (2009); and for EXAFS theory; J.J. Rehr, J.J. Kas, M.P. Prange, A.P. Sorini, Y. Takimoto, F.D. Vila, C. R. Phys. **10**(6), 548–559 (2009); J.J. Rehr, J.J. Kas, F.D. Vila, M.P. Prange, K. Jorissen, Phys. Chem. Chem. Phys. **12**(21), 5503–5513 (2010)
8. E.A. Stern, D.E. Sayers, F.W. Lytle, Phys. Rev. **11**, 4836 (1975); E.A. Stern, J. Synchrotron Rad. **8**, 49–54 (2001)
9. P. Blaha, K. Schwarz, J. Phys. F **17**, 899 (1987); P. Blaha, K. Schwarz, P. Sorantin, S.B. Trickey, Comput. Phys. Commun. **59**, 399 (1990)
10. Y.M. Yiu, G. Kaur, Q. Xiao, T.K. Sham, J Non-Cryst. Solids **364**, 13–19 (2013)
11. M. Newville, J. Synchrotron Rad. **8**, 322–324 (2001)
12. B. Ravel, M. Newville, J. Synchrotron Rad. **12**, 537–541 (2005)
13. Y.M. Yiu, T.K. Sham, G. Kaur, J. Appl. Phys. **104**, 013713 (2008)
14. P. Cherin, P. Unger, Acta Cryst. B **28**, 313 (1972)

15. G. Kaur, M.S. Bakshi, Non-ideal mixing of Se–Te in aqueous micellar phase for nano-alloys over the whole mole mixing range with morphology control from nanoparticles to nanoribbons. J. Phys. Chem. C **114**, 143–154 (2010)
16. J.C. Fuggle, J.E. Inglesfield, *Unoccupied Electronic States: Fundamentals for XANES, EELS, IPS, and BIS* (Springer-Verlag, Berlin Heidelberg, 1992)
17. G. Lucovsky, J.C. Phillips, J. Surf. Sci. Nanotechnol. **7**, 375–380 (2009)
18. V.J. Leppert, S. Mahamuni, N.R. Kumbhojkar, S.H. Risbud, Mater. Sci. Eng. B **52**, 89 (1989); C.E.M. Campos, J.C. de Lima, T.A. Grandi, J.P. Itie, A. Polian, A. Michalowicz, J. Phys. Condens. Matter **17**, 5187 (2005)
19. A.C. Stergious, P.J. Rentzeperis, Aeitschrift Kristallogr. **173**, 185 (1985); A.L. Renninger, B.L. Averbach, Acta Cryst. **B29**, 1583 (1973)
20. J.E. Dutrizac, A.R. Pratt, T.T. Chen, in *Aqueous and Electrochemical Processing*, vol. III. *Metallurgical and Materials Processing: Principles and Technologies*, ed. by F. Kongoli, C. Yamanchi, H.Y. Sohn (TMS (The Minerals, Metals & Materials Society), USA, 2003), p. 139
21. A. Pratt, L. Zuin, Y.M. Yiu, S. Harmer, Can. J. Chem. **85**, 761 (2007)
22. Y.M. Yiu, M.W. Murphy, L. Liu, Y. Hu, T.K. Sham, AIP Conf. Proc. **1590**, 26 (2014); M.W. Murphy, M.J. Ward, L. Liu, Y.M. Yiu Y. Hu, J.A. Zapien, T.K. Sham, Examination of the electronic and optical properties of CdS_xSe_{1-x} nanostructures by XANES, XEOL and DFT, J. Appl. Phys. **116**, 193709 (2014)
23. A. Rogalev, J. Goulon, in *Chemical applications of synchrotron radiation - Part II: X-ray Applications*, edited by T.K. Sham (World Scientific, Singapore, 2002), pp. 707–760
24. Y.K. Liu, J.A. Zapien, Y.Y. Shan, H. Tang, C.S. Lee, S.T. Lee, Nanotechnology **18**, 365606 (2007); J.A. Zapien, Y.K. Liu, Y.Y. Shan, H. Tang, C.S. Lee, S.T. Lee, Appl. Phys. Lett. **90**, 213114 (2007); Y.L. Kim, J.H. Jung, K.H. Kim, H.S. Yoon, M.S. Song, S.H. Bae, Y. Kim, Nanotechnology **20**, 095605 (2009)
25. R. Hill, J. Phys. C:Solid State Phys. **7**, 521 (1974)
26. X.T. Zhou, P.S.G. Kim, T.K. Sham, S.T. Lee, J. Appl. Phys. **98**, 024312 (2005)
27. G. Kaur, F. Wang, Y.M. Yiu, D.W. Shoesmith, M. Zinke-Allmang, T.K. Sham, Z. Ding, J. Mater. Sci. Mater. Electron. **20**, S164–S169 (2009)
28. E. Stephen, *The Physics and Chemistry of Solids* (John Wiley & Sons, New York, 2008), p. 99. reprinted 2008
29. R.W.G. Wyckoff, in *Crystal Structures*, vol. 1–6 (John Wiley and Sons, Inc., 1963–1971)

Chapter 3
Nanostructured Chalcogenides

Mandeep Singh Bakshi and Gurinder Kaur Ahluwalia

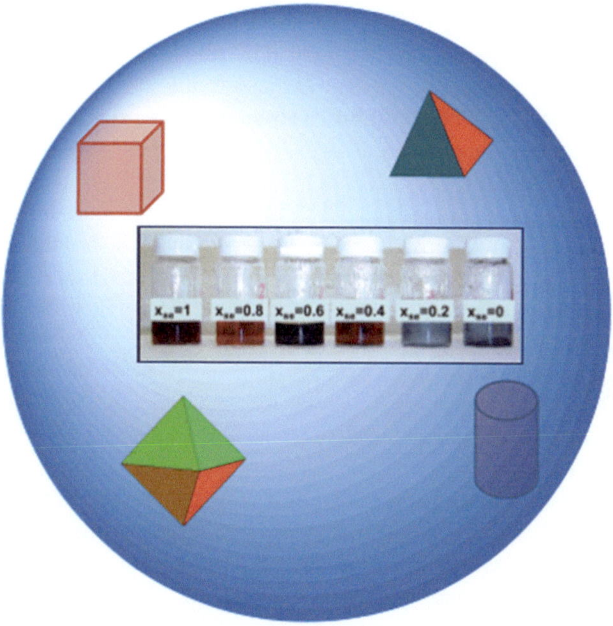

M.S. Bakshi, Ph.D. (✉)
Department of Natural and Applied Sciences, University of Wisconsin - Green Bay, 2420 Nicolet Drive, Green Bay, WI 54311-7001, USA
e-mail: ms_bakshi@yahoo.com

G.K. Ahluwalia, Ph.D
Materials and Nanotechnology Research Laboratory, Department of Physics College of The North Atlantic, Labrador West Campus, 1600 Nichols Adam Highway, Labrador City, A2V 0B8, NL, Canada
e-mail: gurinder.kaur@cna.nl.ca

G.K. Ahluwalia (ed.), *Applications of Chalcogenides: S, Se, and Te*,
DOI 10.1007/978-3-319-41190-3_3

3.1 Introduction

Nanostructuring of chalcogenides is one of the most promising areas of materials science which finds inherent applications from electronic devices to biomarkers for cancer cells detection Colloidal quantum dots (QDs) were first investigated by L. E. Brus et al. [1] in 1983 and have attracted much attention as color saturated, highly luminescent, and robust lumophors, due to their unique optical properties, such as facile bandgap tunability, wide absorption range, spectral purity, and photostability/chemical stability. Among various colloidal QDs based on II–VI (i.e., cadmium sulfide (CdS) [2], cadmium selenide (CdSe) [3–7], cadmium telluride (CdTe) [8, 9], zinc selenide (ZnSe) [10], or zinc oxide (ZnO) [11]), III–V (i.e., indium arsenide (InAs) [12] and indium phosphide (InP) [13]), IV–VI (i.e., lead selenide (PbSe) [14, 15] and lead sulfide (PbS) [16–20]), and IV(i.e., Si [21–24]) compounds, II–VI QDs comprising cadmium and zinc chalcogenides have shown considerable progress in terms of high color purity (full width at half maximum: 20–40 nm) and photoluminescence quantum efficiency (PL QE, above 70 %) in the visible range. Such a success in materials science harnessed the development of colloidal QD-based light-emitting diodes (QD-LEDs) aiming toward economic, stable, and high performance next-generation devices, which will potentially replace conventional inorganic or organic based-LEDs.

From a technological perspective, lead chalcogenide quantum dots (QDs) are of interest because they are among the few materials that can provide tunable electronic transitions at important near-infrared (NIR) wavelengths. PbSe is a convenient choice for inorganic semiconductor QDs for NIR applications ($\lambda > 1$ μm) as the colloidal synthesis is reproducible and yields nanocrystals with narrow size distributions. In addition the large exciton Bohr radius of 46 nm leads to a strong confinement of QD excitons throughout the synthetically accessible range of 2–10 nm (corresponding to absorption peaks $\lambda = 1$ μm (1.2 eV) to $\lambda > 2.4$ μm (<0.5 eV), respectively).

Electroluminescent devices operating at wavelengths important for telecommunications (1.3–1.55 μm) that incorporate PbSe [25, 26] QDs, and water-soluble PbS QDs for in vivo microscopic imaging that absorb and emit in the NIR "biological transparency window" have been reported [27]. In addition, the observation of optical gain and amplified spontaneous emission from PbSe QDs opens the door to possible tunable NIR lasers [28] and multiexciton generation in PbSe QDs may increase the efficiencies of future solar cells [29, 30].

Nanostructuring of germanium chalcogenides such as GeTe and GeSbTe used in data storage devices could potentially pave way for scaling of future devices.

At the nanoscale, a precise synthesis and characterization is the first and most important step to explore such materials. Several methods have been employed to achieve the shape and size control morphologies at nanoscale and among them surfactant-assisted methods are the most promising protocols to be implemented. Surfactants have been very successful in controlling the shape and size of noble metal nanocrystals (NCs) where they selectively adsorb on the lowest energy planes like {100} or {110} to direct the crystal growth in a specified direction. Among conventional semiconductors, lead sulfide (PbS) is particularly important because of its narrow band gap of 0.41 eV in the bulk form, a large excitation Bohr radius of 18 nm, and the diverse morphologies [1]. Most prominent among them include spheres [31a], cubes [31b], rod-like [31a], tubes [31c], wires [31d], truncated octahedrons [31b], dendritic [31e], star-shaped [31f], and flower-shaped structures [31g]. Different synthetic routes such as colloidal solutions [32a], reverse micelles [32b], solvothermal [32c], microwave [32d], and self-assembly [32e] have been employed for the fabrication of PbS nanoparticles. Likewise, lead selenide (PbSe) with narrowest band gap is synthesized by solvothermal [33] and microwave-assisted methods [34]. Copper selenide (CuSe), another semiconducting material is used as a superionic material [35] as well as an optical filter in solar cells [36]. Several methods of synthesis of CuSe NCs include thermolysis of Cu and Se powder mixtures at 400–470 °C in the presence of flowing argon [37], solvothermal process[38], mechanical alloying of Se and Cu in a high energy ball mill [39], and in liquid ammonia [40]. Chalcogenide NCs exhibit unique size and shape dependent physicochemical characteristics [41], which are the outcome of low-dimensional quantum confinement effects and have a wide range of applications as mentioned above. They have also been effectively employed as building blocks for two- and three-dimensional arrays for the development of meta-materials with collective electronic and optical properties [41].

3.2 Size and Shape Controlled Synthesis of Nanoparticles

Conventional surfactants have been widely used for the shape controlled synthesis of noble metal NPs of gold and silver, but few reports are related to their use in the synthesis of PbSe [42] and CuSe [43] NCs. Surfactant-assisted pathways are the simplest in view of the environmental concerns. They are reproducible and best suited for their high yield. Among the ionic surfactants, we have already used cationic gemini surfactants as capping/stabilizing agents for the shape controlled synthesis of PbS NCs [44]. They show remarkable control over crystal growth at nanoscale and produce well defined geometries. Shape controlled morphologies are usual building blocks of nanodevices, and their synthesis in aqueous phase has advantages both from economic and environmental perspectives. While surfactant-assisted synthesis of semiconductor NPs has attracted much attention due to its soft-template effect, reproducibility, and aqueous phase reaction; little is known about the influence of surfactant hydrophobicity on the shape controlled aspects. Along with conventional surfactants, twin-tail cationic surfactants (TTS) shown in Fig. 3.1 can also be used as capping agents in the synthesis of PbS NC. TTS are more hydrophobic in comparison to their monomeric homologous and hence, considered to be even better shape directing agents due to their stronger amphiphilic nature. The capping ability of a shape directing ionic surfactant is directly related to its surface adsorption behavior which in turn is further related to its hydrophobicity as

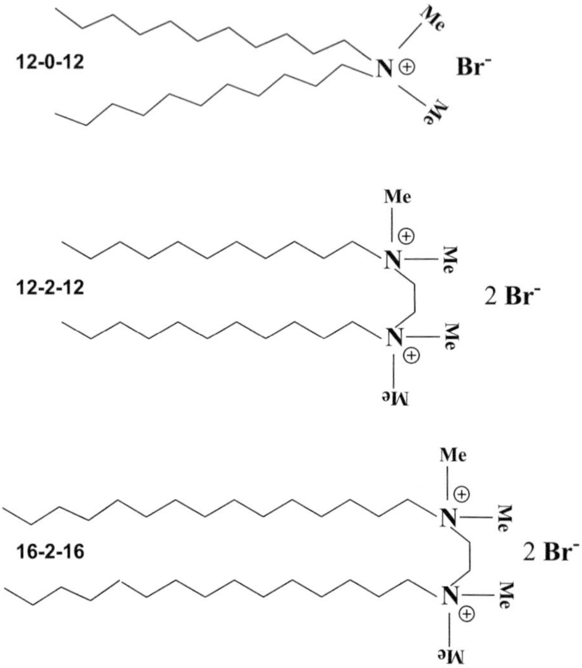

Fig. 3.1 Didodecyl dimethylammonium bromide (12-0-12, critical micelle concentration, cmc = 0.2 mM), dimethylene bis (dodecyldimethyl ammonium bromide) (12-2-12, cmc = 0.84 mM), and dimethylene bis (hexadecyldimethyl ammonium bromide) (16-2-16, cmc = 0.02 mM)

well as the polarity. It has been observed that hydrophobicity is the main contributing factor for a shape directing agent like surfactant to control the morphology of PbS NC.

Apart from surfactants, polystyrene sulfonate (PSS), a polyelectrolyte which when dissolved in water produces polyanions with strong ability to interact with electropositive active sites, can also be used as a shape directing agent. Recently, many studies [45] have demonstrated the formation of thin polyelectrolyte multilayer films capped NCs including PbS NCs. Just like that of surfactants, polyelectrolyte films provide charge and stearic stabilization [46] to the colloidal particles and thereby reduce the possibility of coagulation effects. On the other hand, polyvinylpyrrolidone (PVP) has already been proved to be a strong stabilizing agent as well as a soft template for the synthesis of silver [47a] (Ag) and PbS NCs in different hydrothermal methods. PVP macromolecules show stronger preference for {100} than {111} planes of Ag, thus producing Ag nanowires bound with {100} facets [47b]. A similar mechanism is thought to happen in the case of PbS NCs (Fig. 3.2) synthesized in the presence of PVP. We compare the capping ability of PSS and PVP on the synthesis of shape controlled PbS NCs and try to quantify the differences between the PbS geometries as well as the mechanisms involved in the presence of both polymers.

In addition to PbS and PbSe NCs, both selenium (Se) and tellurium (Te) which belong to sixth group p-block elements and exist in helical chains in crystalline form, can also form semiconducting nanomaterials. Se exhibits photovoltaic action

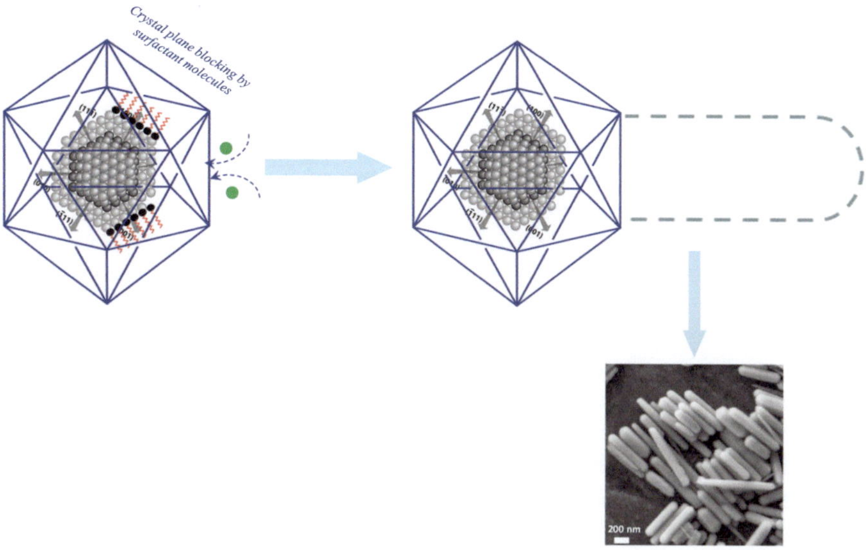

Fig. 3.2 Schematic representation of the fcc nucleation center growth by selective blocking the {100} crystal planes to achieve the formation of nanorods or nanowires

by converting light into electricity through photoconductivity due to its reasonably small band gap of 2.6 eV in the crystalline form. This allows Se to be used in versatile photocells and solar cells [48, 49]. Te also exhibits photoconductivity to a lesser extent. Both Se and Te are p-type semiconductors and hence used in electronics and solid-state applications. Independently, both predominantly form 1D nanostructures [49–54] due to a preferential growth along the c-axis of trigonal hexagonal geometry and exist in the spiral chains. 1D geometries have potential ramifications in electronic device formation [55]. Facile production of such morphologies under economical favorable conditions such as in aqueous phase and at relatively low temperature (<100 °C) is always required. Recently, Se–Te alloys have been found to demonstrate interesting temperature dependence of electrical resistivity [56]. In addition, fine homogenous nearly monodisperse Se–Te nanorods have also been synthesized simply by following hydrazine reduction [57]. Hydrazine is a relatively weaker reducing agent than ammonia but it is best suited for a slow reduction of Se or Te precursors into their atomic forms. The following reaction is expected to take place:

$$N_2H_4(aq) + SeO_3{}^{2-}/TeO_3{}^{2-}(aq) \rightarrow Se/Te(s) + N_2(g) + 2OH^-(aq) + H_2O(l)$$

In our recent work [58] related to the synthesis and characterization of PbSe and CdSe nanocrystals (NCs), we have also observed the formation of Se nanorods and microrods in both cases. It shows that Se is more readily reduced in comparison to Pb and Cd which is understood from its nonmetallic nature as well as its location in the periodic table. The same can be said about Te. Since Se–Te alloy formation [59, 60] is emerging as a potential field with applications in nanoelectronics, and therefore, a comprehensive investigation of their simultaneous reduction requires an attention in view of both reduction chemistry and crystallization kinetic perspectives.

Semiconductor NCs when synthesized in less than 10 nm size, exhibit very interesting luminescence properties and hence, they are called fluorescent quantum dots (QDs). Zero-dimensional nanomaterials have attracted wide attention during the past two decades due to their unique physical and chemical properties. QDs demonstrate high emission quantum yields, high photobleaching thresholds, and excellent photostability [61–63], and hence, they are used as excellent biomarkers for biological assays [64–67]. QDs of cadmium and zinc chalcogenides are high in color purity and photoluminescence quantum efficiency usually above 70 % in the visible range, and hence, they are best suited nanomaterials for light-emitting diodes (QD-LEDs) for high performance next-generation devices. Unique electronic properties of QDs mainly attributed to the structural and compositional nature based on valence and conduction band which are located relatively lower than the lowest unoccupied molecular orbital (LUMO) and highest occupied molecular orbital (HOMO) of organic semiconductors. PbS, PbSe, and PbTe QDs constitute an interesting class of materials because of strong quantum confinement effects due to large Bohr radius of 46 nm in the case of PbSe, small Stokes shift, and bright luminescence. PbSe QDs have been synthesized in different shapes including spheres, cubes, multipods, rods, octahedrons, wires, and rings.

3.3 Methods of Synthesis

3.3.1 Preparation of PbS Nanocrystals (NCs)

The synthesis of PbS NCs can be done by applying the following procedure [44]. In a typical procedure; 30 ml of distilled water was taken in a round bottom glass flask. Under constant stirring, 5 ml of aqueous TTS solution (post-micellar concentration) was added in the flask. Or if the polymers are used for the shape controlled synthesis, then take 30 ml of aqueous PSS/PVP (0.5 mM monomolar basis) in a round bottom glass flask. This was followed by the addition of 4 ml of 1 M aqueous acetic acid. After this, 2 ml of 0.5 M aqueous lead acetate, and 2 ml of aqueous 0.5 M thioacetamide were introduced into the above solution under constant stirring at room temperature. After mixing all the components, the reaction mixture was kept in an oil bath at a temperature of 80 °C for 48 h under static conditions. That led to the formation of a black colloidal solution indicating the formation of PbS NC. The colloidal suspension of PbS was collected and washed twice with water followed by several washing with methanol.

3.3.2 Synthesis of Surfactant Core-Shell PbSe and CuSe NCs

PbSe NCs can be obtained through a reaction between lead acetate and sodium selenite in the presence of aqueous hydrazine [68]. In a typical reaction, 10 ml of aqueous surfactant (12-0-12/12-2-12/16-2-16) solution of concentration 4, 8, or 16 mM containing 2.5 % hydrazine was taken in a round bottom glass flask. A fresh hydrazine should be used. Then, 1.25 mM of lead acetate was added in it. This was followed by the addition of 0.125 ml of 1.25 mM aqueous sodium selenite solution under constant stirring.

$$Pb(CH_3COO)_2(s) \rightarrow Pb^{2+}(aq) + 2CH_3COO^-(aq) \tag{3.1}$$

$$Na_2SeO_3(s) \rightarrow 2Na^+(aq) + SeO^{3-}(aq) \tag{3.2}$$

$$3N_2H_4(aq) + SeO_3{}^{2-}(aq) \rightarrow 2Se^{2-}(aq) + 3N_2(g) + 6H_2O(l) \tag{3.3}$$

$$Pb^{2+}(aq) + Se^{2-}(aq) \rightarrow PbSe(s) \tag{3.4}$$

After mixing all the components at room temperature, the reaction mixture was kept in a water thermostat bath at a fixed temperature of 85 ± 1 °C for 48 h. The color of the solution changed from slightly turbid white due to the low solubility of surfactant initially to clear dark brown and finally it turned black within 4 h and remained same till 48 h. A colloidal black solution thus obtained indicates the presence of PbSe NCs. The purified sample was collected by centrifugation at 10,000 rpm for 5 min after washing it 2–3 times with distilled water. The same procedure can be adopted for the preparation of CuSe nanoparticles by taking copper acetate in place of lead acetate. The color of the solution containing CuSe colloidal suspension changed from light black to dark black within 2 h and remained same for 48 h.

3.3.3 Synthesis of Se–Te Alloy NCs

For the synthesis of Se, Te, or their alloy hydrothermal method is the easiest one to follow in the presence of surfactants [69]. Se–Te NCs were synthesized by a simultaneous hydrazine reduction of both aqueous Na_2SeO_3 and Na_2TeO_3 in a micellar solution. First of all, two aqueous stock solutions of $Na_2SeO_3 = 25$ mM and $Na_2TeO_3 = 25$ mM were prepared separately in the presence of cetyltrimethylammonium bromide (CTAB) = 50 mM Required quantities of both stock solutions were mixed to produce different Se mole fractions, $x_{Se} = 0$, 0.2, 0.4, 0.6, 0.8, and 1, with total $[Na_2SeO_3 + Na_2TeO_3] = 25$ mM, along with a constant addition of 3.6 M hydrazine in each case. It is to be mentioned that all reactions were carried out by using fresh hydrazine. The reaction bottles with final volume of 10 ml of each mole fraction were then kept in an oil bath maintained at 85 °C for 24 h under static conditions. The choice of 85 °C provides better prospects of getting crystalline nanoparticles. The color of the solution changed from colorless to reddish-

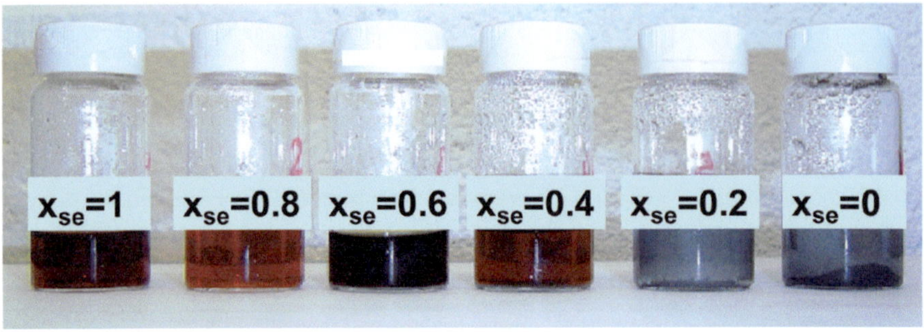

Fig. 3.3 Photos of as prepared reactions over the whole mole fraction range of Se + Te mixtures in the presence of 3.6 M hydrazine (Reproduced with permission from [69])

brown for the mole fractions lying in the Se rich-region, while colorless to grey-black for those lying in the Te rich-region of the mixtures within 1–2 h and remained same till 48 h. Figure 3.3 shows a series of bottles with different mole fractions from $x_{Se} = 0–1$ prepared with hydrazine = 3.6 M. The colloidal suspensions thus obtained at other mole fractions were cooled to room temperature. NCs were collected after repeated washing with pure water (at least three times) to remove surfactant followed by spinning at 10,000 rpm for 5 min in each case.

3.3.4 Synthesis of QDs

The most versatile method for the synthesis is the "Hot-injection method" (Fig. 3.4). For a typical synthesis of PbSe QDs, 0.1 g (0.26 mmol) of lead acetate, 0.5 mL of L-121 (block polymer as stabilizer), and 5 mL of diphenyl ether were introduced to a 50 mL three-necked flask connected to a condenser circulating with cold water. The mixture was then vigorously stirred at 100 °C and purged with argon for 1 h. On the other hand, 0.2 g (2.5 mmol) of Se was dissolved in 1.6 mL of trioctyl phosphine (TOP). The Pb: Se molar ratio was 1:10. To initiate the synthesis, the temperature of the lead salt mixture was adjusted to the desired temperature (100 °C), and the TOP-Se mixture was quickly injected. Upon injection of TOP-Se, the colorless mixture turned to light brown. As the reaction proceeded, the color intensified to dark brown because of the nucleation and growth of PbSe QDs stabilized in the presence of L-121 (block co-polymer) and TOP. With increasing growth temperature, the size of the resulting PbSe QDs increases. The slow growth at relatively high temperatures allows the QDs to anneal and to form nearly defect-free wurtzite lattices, which are identical to the bulk lattice. Similar method can be used for synthesis of ternary QDs.

The following procedure can be implemented for the synthesis of a typical core–shell QDs, for example, PbSe and PbSe/PbS core/shell. The first step was the preparation of small PbSe core QDs 2.0–2.5 nm in diameter. The main reaction mixture was composed of lead oleate and hexadecane. Another solution of selenium

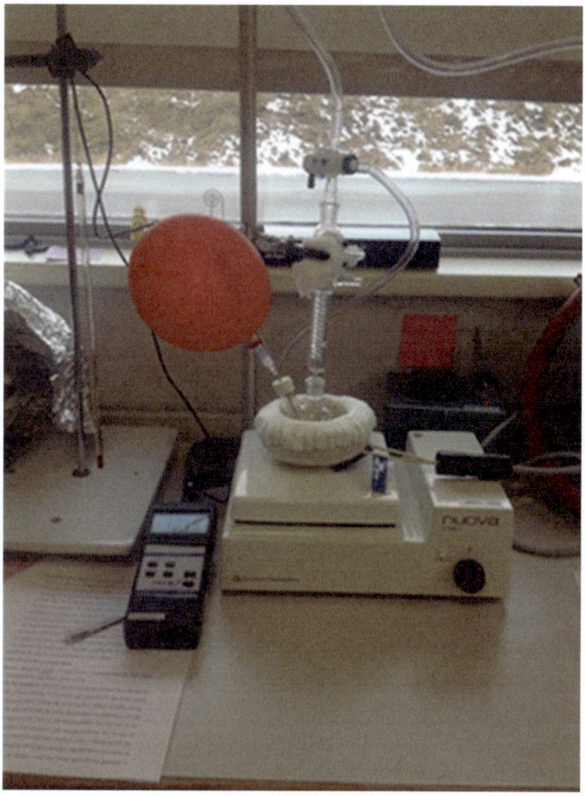

Fig. 3.4 A typical experimental setup for the synthesis of QDs by hot injection method

precursor containing TOP, diphenylphosphine, and hexadecane, was heated under vacuum for 1 h. This solution was injected into the main reaction mixture. QDs thus obtained were separated by centrifuging. At the second step of the procedure, the shell growth was carried out. Diluted bis(trimethylsilyl) sulfide was added dropwise, at 70 °C, into the reaction mixture, containing PbSe QDs, lead oleate, and diphenylether which resulted in the formation of core/shell QDs.

3.4 Shape and Size Characterization of Nanocrystals

3.4.1 Lead Sulfide (PbS)

3.4.1.1 Star Shaped Morphology

In this section we show different morphologies of PbS NCs synthesized under different reaction conditions. A typical fern like *star shaped* PbS micro crystals are produced in the presence of 1 mM (Fig. 3.5) dodecyltrimethylammonium bromide (DTAB) with average size distribution close to 5 μm.

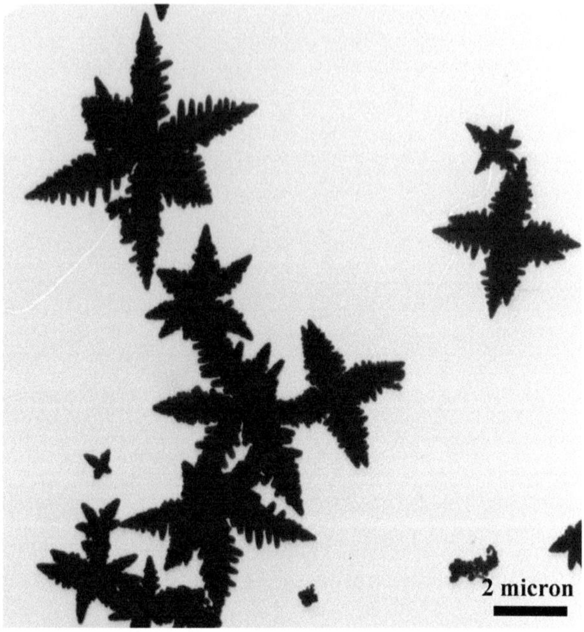

Fig. 3.5 TEM micrographs of PbS micro crystal controlled by DTAB (Reproduced with permission from [44])

No appreciable change in the morphology of PbS micro crystals is observed with the increase in DTAB concentration from 0.1 to 2 mM. DTAB is a strong ionic surfactant and is expected to adsorb electrostatically at {111} facets of fcc geometry of PbS (Pb^{+}/S^{-}) due to the presence of greater charge driven by high atomic density [70]. This would leave low atomic density {100} facets poorly capped or uncapped with surfactant molecules and remain available for further nucleation. This would consequently result in a preferential growth at {100} facets in comparison to {111}. On the contrary, in the case of noble metal nanoparticles (NPs) such as Au and Ag, surfactant molecules (CTAB and sodium dodecylsulfate, SDS) prefer to adsorb at {100} facets of fcc geometry in comparison to {111}. This preference practically comes from a relatively low atomic density at {100} facets with appropriate atomic spacing suitable for the size of surfactant head groups to interact. Thus, PbS being semiconductor possesses insufficient electron density at {100} facets to attract surfactant molecules in comparison to that of Au or Ag. This fundamental difference of surfactant preferential adsorption at {111} facets in the case of PbS semiconductor from that at {100} facets of Au or Ag, tells us why mostly cubic or star shaped geometries are obtained in the former case in comparison to nanorods/nanowires in the latter case [71]. However, if another tail is introduced into a DTAB molecule to make it 12-0-12 (Fig. 3.1), the shape and size of PbS MC dramatically changed. 0.5 mM 12-0-12 (Fig. 3.6) produces fine monodisperse crystalline cubic NCs with average size distribution 64.7 ± 5.4 nm.

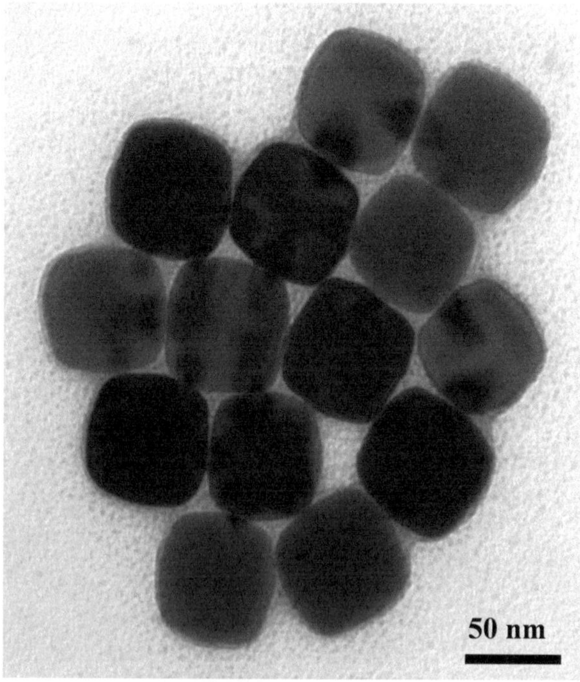

Fig. 3.6 TEM micrographs of PbS NCs controlled by 12-0-12 (Reproduced with permission from [44])

3.4.1.2 Hexagonal

When a spacer of two methylene groups is introduced in the head group of 12-0-12, it is converted into a gemini surfactant i.e., 12-2-12. A use of 12-2-12 as capping agent produces star shaped structures with average size distribution close to 5 μm at 12-2-12 = 0.1 mM. At 12-2-12 = 0.5 mM, the star shaped morphologies are converted into typical fern like structures. A systematic increase in the amount of 12-2-12 from 0.1 to 2 mM transfers the star shaped micro crystals into *monodisperse hexagonal* NCs (Fig. 3.7).

3.4.1.3 Cubic

Consequently, if we increase the hydrophobicity further as in the case of 14-2-14, again different morphologies, i.e., star shaped and well-defined cubic NCs, are observed but at much lower concentrations than those observed in the case of 12-2-12. At 14-2-14 = 0.01 mM star shaped NCs are obtained, while at 14-2-14 = 0.02 mM, the number of star NCs decreases and plate like geometries appear which eventually seem to transform into predominantly large cubic NCs with average size distribution 144 ± 17.8 nm at 14-2-14 = 0.05 mM. Further increase in the concentration, i.e., 0.1 and 0.2 mM, produces well-defined *monodisperse cubic* NCs (Fig. 3.8) with average size distribution 99 ± 14.1 nm. When similar PbS

Fig. 3.7 TEM micrographs of PbS NCs controlled by 12-2-12 (Reproduced with permission from [44])

Fig. 3.8 TEM micrographs of PbS NCs controlled by 14-2-14 (Reproduced with permission from [44])

synthesis carried out in the presence 10-2-10 which is less hydrophobic in comparison to 12-2-12 and 14-2-14 due to smaller twin tails, only star shaped NC were obtained at all 10-2-10 concentrations with appreciable anisotropic growth.

In order to check the surfactant specific shape controlled behavior, the addition of SDS in the case of 12-0-12 leads to the formation of star shaped micro crystals rather than well-defined spherical NC. Similar addition in the case of 14-2-14 eliminates the presence of well-defined PbS nanocubes and produces several micron sized rod shaped structures along with the polyhedral geometries. Thus, the induction of negatively charged anionic surfactant like SDS alters the capping ability of 12-0-12 and 14-2-14 by undergoing strong electrostatic interactions[41] with them and that consequently changes the morphology of micro crystals. Apart from this, if P103, a non-ionic block polymer/surfactant, is used as a capping agent instead of twin tail surfactants (TTS), then no ordered morphology is obtained except large PbS flake type structures. Therefore, the absence of any ordered morphology of PbS NCs in the presence of mixed surfactant systems as well as P103 ultimately suggests a high degree of TTS selectivity in controlling the overall shape of PbS NCs.

All results from different studies indicate that TTS are efficient capping agents and hence effectively control the shape and size of PbS NCs. Hydrophobicity is considered to be the main shape controlling factor as evident from the TEM micrographs. A stronger hydrophobic surfactant has lesser solubility in aqueous phase in comparison to a weaker hydrophobic surfactant. Thus, the former will have higher interfacial adsorption in comparison to the latter. Hence, a stronger hydrophobicity acts as a driving force for a better surface adsorption of surfactant molecules [71]. This will push greater number of strongly hydrophobic surfactant molecules from aqueous bulk at water-PbS NC interface where they would undergo preferential adsorption at {111} planes with greater charge density. Even, a greater amount of a strongly hydrophobic surfactant will have greater interfacial adsorption and would effectively control the shape evolution during the crystal growth.

3.4.1.4 Octahedron

Other interesting geometries have also been reported in the literature [72]. A large-scale synthesis of well-defined truncated *octahedron* with highly geometrical symmetry has been reported. Each truncated octahedron has a composition of six {100} squares and eight {111} hexagons, and the overall structure shares 24 identical edges (Fig. 3.9). This kind of nanopattern of truncated octahedron crystals with its 14 well-defined symmetries could find potential applications because of their unique structure such as a nanometer-sized construction element or a new kind of potential optical device that is independent or coupled with other materials. These high-quality single crystalline PbS polyhedrons represent a new platform for further studies of nanoscale phenomena as well as for applications in various fields of nanotechnology.

Fig. 3.9 TEM micrographs of PbS Octahedrons controlled by CTAB (Reproduced with permission from [72])

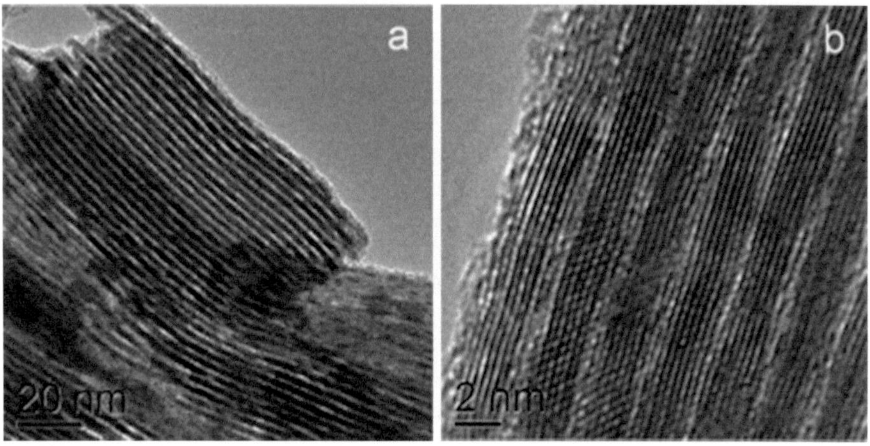

Fig. 3.10 (**a**) TEM and (**b**) HRTEM images of ultranarrow PbS wires self-assembled into 2D super crystalline arrays (Reproduced with permission from [73])

3.4.1.5 Nanowires/Nanothreads/Nanosheets

One-dimensional (1D) semiconductors are essential components for various applications as high-performance optoelectronic devices, field-effect transistors, logic circuits, non-volatile memories, and biosensors. However, the success of such potential applications eventually depends on the quality and control over the nanowire composition and morphology. Ultra narrow (1.8 nm) PbS nanowires (Fig. 3.10) are synthesized in a single step, under benchtop conditions at relatively low temperature (90 °C) [73]. The nanowires exhibit a nearly perfect crystal lattice, high width uniformity, and tight side-by-side registry. Two-dimensional (2D) assembly over large areas (>15 μm^2) is achieved using the Langmuir–Blodgett method. The wire width can be readily controlled in the range

1.8–10 nm by a surface pressure-induced coalescence reaction, as monitored by transmission electron microscopy and Raman spectroscopy. The fluorescence of the 2D assembly shows strong polarization dependence along the long axis of the wires, making the system potentially suitable for orientation sensitive devices.

Nanowires can also be obtained by simple hydrothermal method by using the water soluble polymer as shape directing agents. The formation of nanothreads and nanosheets at higher polystyrene sulfonate concentration might be due to the soft template effect of some self-assembled state of polystyrene sulfonate (PSS) [74] generated in the presence of lead acetate, i.e., due to ionic effect. Although it is not clear especially what kind of self-assembled state facilitates the formation of nanothreads and nanosheets, however, recent studies have demonstrated that poly-electrolyte and mesoporous silica based films are best templates for organized assemblies of PbS [75, 76]. Usually, nanorods bound with {100} crystal planes obtained in the polymer films due to their lamellar structures, while nanoparticles formation takes place only in the solution. [PSS] = 4 mM may also lead to the formation of such self-assembled layered structures which may act as soft templates to produce *nanothreads* and *nanosheets* (Fig. 3.11). The soft template effect of PSS is considered to be the most likely reason for the 1D nanostructure. This is not the case with polyvinylpyrrolidone (PVP) which produces mainly cubic NCs. PVP is expected to predominantly adsorb at {111} planes due to the electrostatic interactions, which originate from lone pair of oxygen atom on the carbonyl group of pyrrolidone moiety. It directs the crystal growth at {100} crystal planes to produce perfect cubic nanocrystals as the amount of PVP increases. A perfect cubic geometry bound with {100} planes suggests a complete coverage of {111} crystal planes by PVP molecules. Although PSS should have even much stronger adsorption at {111} facets due to polyanionic nature, the presence of roughly cubic geometries NCs along with the nanothreads and nanosheets (Fig. 3.10a) indicates a somewhat different mechanism where competing growth at {111} and {100} facets is observed. It means that PSS has not only preferential surface adsorption for {111} crystal planes, but it has some tendency to adsorb at even {100} planes due to its polyanionic nature. Otherwise a clear cubic geometry might have emerged from each polyhedral amorphous nanoparticle just like that in the presence of PVP.

3.4.1.6 Quantum Dots (QDs)

QDs are inorganic nanocrystals made from group III to V semiconductors with a size in the range of 1–10 nm. Small size results in quantum confinement effects of the nanoparticles, which results in size-tunable absorption and emission wavelengths. The large band gap energy provides smaller emission and absorption wavelengths compared to the bulk material. Due to the many advantages over common fluorophores, use of QDs in biological applications has achieved a considerable importance [77, 78]. As *biosensors*, [79], QDs must be extremely sensitive, very selective, and offer high spatial resolution, and the measurement should

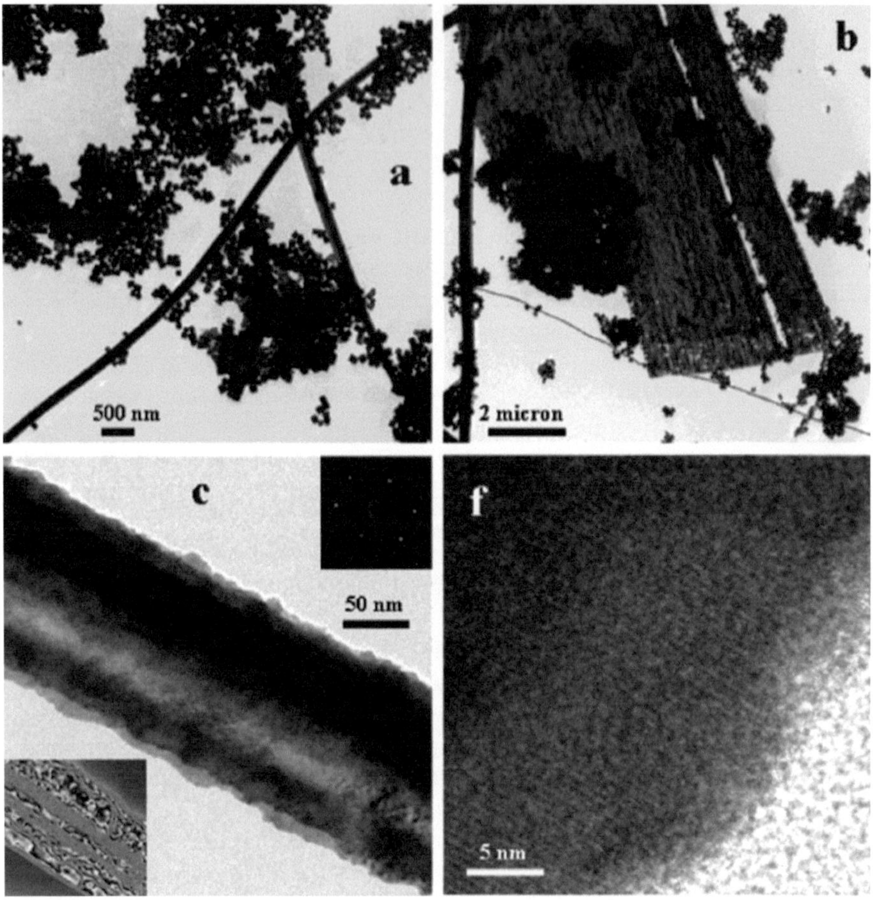

Fig. 3.11 (**a**) TEM image of roughly cubic PbS NCs along with PbS nanothreads of several micrometers in length synthesized in the presence of [PSS] = 4 mM. (**b**) The presence of threads in the form of nanosheets along with the roughly cubic PbS NCs of the same sample. (**c**) A section of a single thread coated with a thin film of PSS, and (**d**) HRTEM showing crystal planes of single threads (Reproduced with permission from [74])

be reproducible, fast, inexpensive, and miniaturized. For all of these requirements, QDs possess specific advantages compared to other fluorescent markers. QDs are usually synthesized by following the hot injection method where lead precursor (a typical lead salt) in oleylamine (OLA) is taken under N_2 flow. The mixture is heated to 100 °C in vacuum followed by a temperature increased to 120 °C for 30 min under vigorous stirring. In this mixture, elemental S in OLA is injected at room temperature at 100 °C and kept for ~7–9 h during the growth and subsequently rapidly quenched into cold hexane followed by an addition of small amount of ethanol to induce precipitation of the nanocrystals. The solution is centrifuged, the supernatant is discarded, and the precipitate is dispersed in hexane.

3.4.1.7 Superlattices

The quantum dot *superlattices* (QDSLs) [80] (Fig. 3.12) also demonstrate interesting optical properties. The QDSLs fabricated by self-organization of PbS quantum dots depict the period of the QDSLs is proportional to the QD diameter. The ligand layer thickness increases from 1.1 to 3.4 nm with an increase in the QD diameter from 2.8 to 8.9 nm (Fig. 3.13).

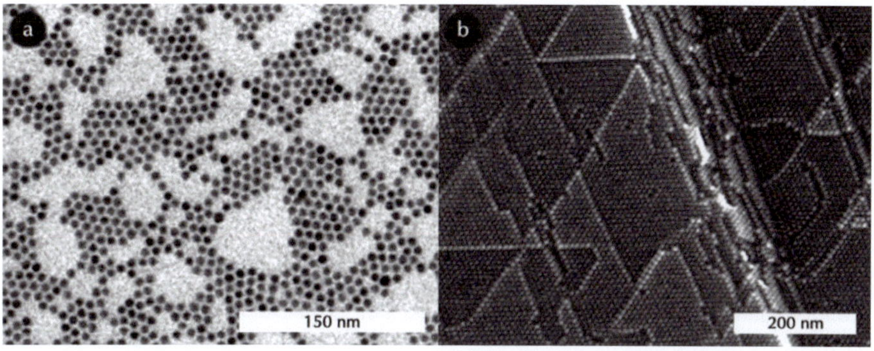

Fig. 3.12 Characterization of PbS nanocrystals: (**a**) TEM image of a PbS nanocrystal; (**b**) HRSEM image of a PbS nanocrystal superlattice (Reproduced with permission from [80])

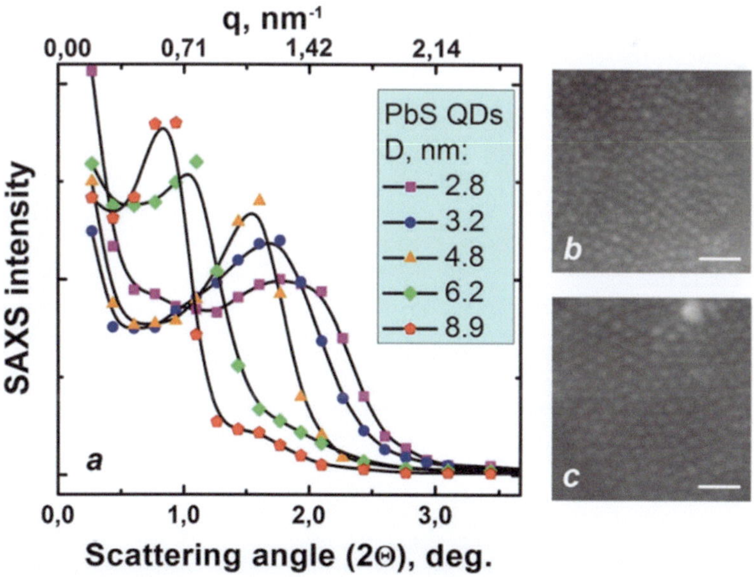

Fig. 3.13 (**a**) Typical SAXS patterns for PbS QDs of different diameters (*D*), deposited on the cover slides. (**b**, **c**) SEM images of the QDSLs formed from QDs with sizes of 6.2 and 8.9 nm, respectively; scale bars of 20 nm (Reproduced with permission from [32e])

The average QD photoluminescence (PL) lifetime of QDSL decreases sharply whereas the absorption or PL spectra of the QDs does not significantly change, this is due to the appearance of additional nonradiative channels of dissipation of photoexcited electrons and holes. In the QDSLs, the life time depends on the QD size. It decreases with the increase in size. The QDLS impairs when the amount of ligands on the small QDs surface reduces while inter-dot spacing remains almost unchanged, but for the large QDs, ligand removal leads to a strong reduction in the inter-dot distances resulting in the improvement of the ordering of the superlattice. The small inter-dot distances allow fast charge carrier transport through the QDSLs enabling potential applications such as photodetectors and efficient photovoltaic devices.

3.4.2 PbSe

3.4.2.1 Nanocrystals (NCs)

PbSe NCs are usually having the same shape and size when cationic surfactants are used for their synthesis as that of PbS NCs. Shape controlled reaction of 12-0-12 cationic surfactant produces mostly cubic crystalline geometries of 39.1 ± 9.1 nm with fine shell of 12-0-12 around each NC. One can find a surfactant monolayer of ~1.5 nm thick corresponding to tail length of C12 (Fig. 3.14). This value is in good agreement with the radius of DTAB micelle with C12 carbon tail length, i.e., 1.8 nm [81]. A slightly larger latter value can be attributed to the hydrated state of the

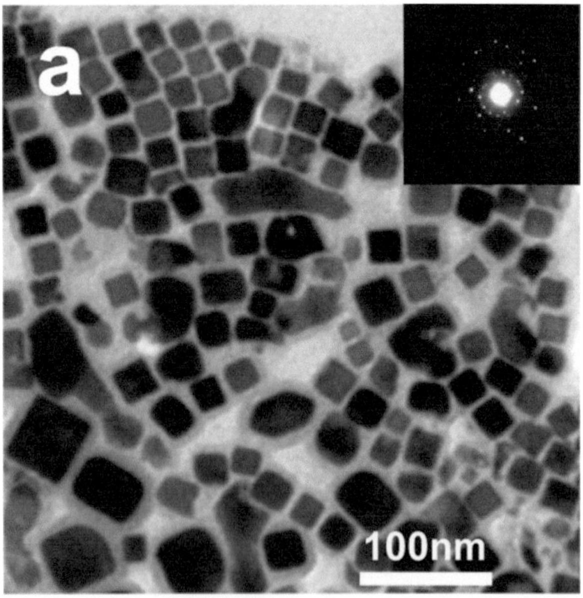

Fig. 3.14 (**a**) Plate like core-shell morphologies of predominantly cubic geometries of PbSe particles with a clear surfactant coating of a sample synthesized in the presence of 12-0-12 = 16 mM (Reproduced with permission from [68])

Fig. 3.15 TEM images of a size-selected PbSe nanocube assembly and HRTEM image of a single PbSe nanocube (Reproduced with permission from [82])

micelle in comparison to the dried monolayer on the surface of NC. The presence of surfactant monolayer around each NC was further confirmed from the gel electrophoresis. All samples show displacement towards the negatively charged electrode indicating the fact that PbSe NCs are positively charged due to the presence of monolayer of cationic surfactants.

3.4.2.2 Nano Cubes

If the reactions are conducted at high-temperature by using a growth technique of dynamic injection, fine nanocubes can be produced (Fig. 3.15) [82]. It can be done by rapidly injecting a phenyl ether solution of lead acetate and trioctylphosphine selenide into a vigorously stirred hot phenyl ether at 180 or 230 °C, in the presence of oleic acid. Small PbSe clusters immediately nucleated and started to grow with stabilization provided by the capping ligands. The shape evolution of PbSe NCs is dependent on the growth time, whereas the crystalline size can be tuned by varying the growth temperature. The higher growth rate in the {111} direction compared to that in the {100} direction results in the formation of nanocubes.

3.4.2.3 Nano Wires (NWs)

Nanowires of PbSe of much better quality than reported above for PbS can be synthesized through oriented attachment of NPs (Fig. 3.16) [83]. Semiconductor nanowires may one day be employed in high-performance field-effect transistors (FETs) [84, 85], logic circuits [86], non-volatile memories [87], and biosensors [88]. Nanowire-based applications demand nanowires with controlled composition, diameter, length, and morphology. Crystalline PbSe nanowires can be obtained by oriented attachment of collections of nanocrystals that attach and fuse along identical crystal faces forming oriented chains. Aqueous phase reaction temperature of 190–250 °C is required for efficient annealing of crystalline defects in the presence of surfactant mixtures to mediate growth and achieve nanowires uniform

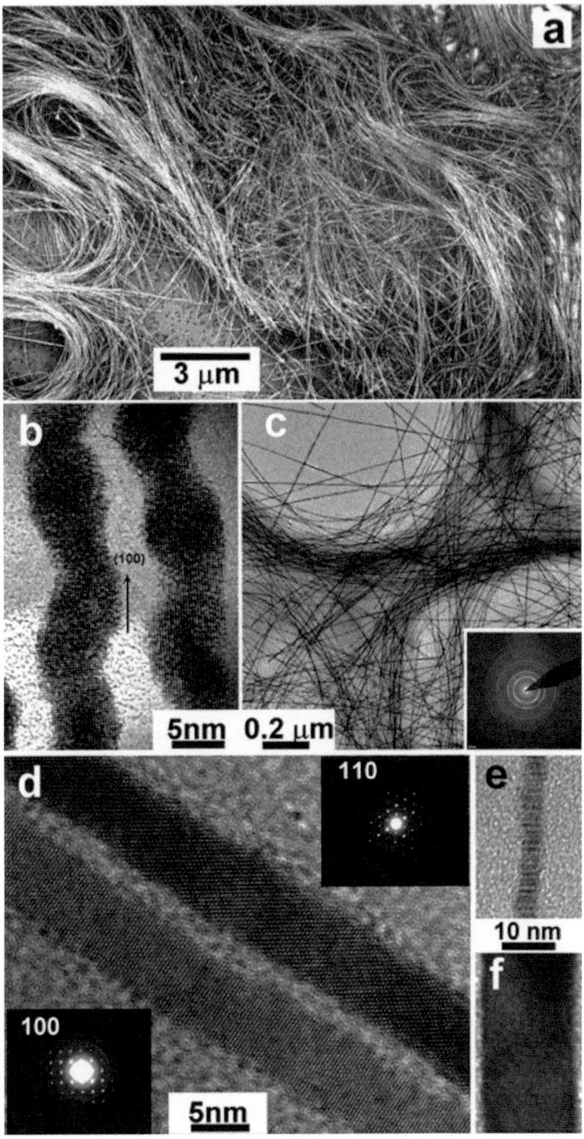

Fig. 3.16 High-resolution (**a**) SEM and (**b**) TEM images of PbSe nanowires grown in solution in the presence of oleic acid. (**c**) Overview and (**d–f**) high-resolution TEM images of PbSe nanowires formed in the presence of oleic acid and *n*-tetradecylphosphonic acid (Reproduced with permission from [83])

in diameter. The entire process of nanowire formation and growth is fast. The nanowires form stable colloidal dispersions. In addition to straight nanowires, zigzag, helical, branched, and tapered nanowires as well as single-crystal nanorings could all be prepared by adjustment of the reaction conditions. Straight nanowires are useful for high carrier mobilities necessary for high-performance FETs, nanowire-based sensors [89], and photovoltaic devices [90].

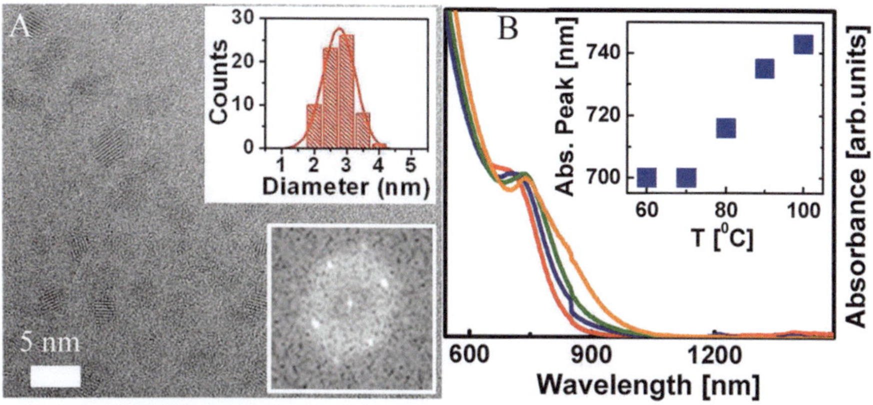

Fig. 3.17 (**a**) HRTEM image of 2.5 ± 0.4 nm PbSe CQDs. *Bottom inset*: the corresponding FFT pattern of the HRTEM image. *Top inset*: Histogram of the CQD size distribution. (**b**) Absorption spectra of 2 ± 0.5 nm PbSe CQDs recorded after thermal annealing at $60-120$ °C. *Inset*: dependence of the lowest absorption band on the annealing temperature (Reproduced with permission from [91])

3.4.2.4 PbSe/PbS QDs

The small-sized PbSe/PbS core/shell QDs with the band-edge energy in the range of 1.1–1.4 eV are promising materials usually synthesized by the hot injection method [91]. First small-size PbSe QDs are synthesized, and then a shell of PbS is created by using a desired amount of precursors. PbSe QDs show first exciton absorption band at 70 °C. Above 70 °C, the red shift of the first excitonic transition increases with temperature due to the Ostwald ripening process (Fig. 3.17). The PbS shell coating onto ultrasmall PbSe is performed by using bis (trimethylsilyl) sulfide (TMS_2S) as a sulfur precursor in the form of a monolayer. In the core/shell QDs, the PL intensity significantly increases with increasing temperature. The small-sized PbSe/PbS QDs offer significant benefits for their implementation in optoelectronic devices and in QD-based photovoltaic cells.

The effect of core/shell nonconcentricity on electronic structure, [92] optical properties, and on the stability of PbSe/CdSe QD heterostructures has been studied particularly with the shells that are limited to the outermost layers of the QD [93, 94]. Such a distribution predicts potential application of the investigated materials in various optoelectronic devices. This causes excitonic initial blue shift of $E_{1Se-1Sh}$ with respect to that of the parent PbSe cores and after ~25 min, a retracting red shift of ~30 meV takes place (Fig. 3.18). The XRD patterns suggest a (202) peak is associated with the existence of a Pb_3O_4 phase in the XRD pattern of the parent PbSe QDs. The intensity of this peak is reduced and totally absent in the XRD pattern of the core/shell QDs while that of the (111), (220), and (222) peaks are enhanced and are associated with the existence of the CdSe (zb) phase. These signatures designate the formation of the CdSe domain, presumably grown in a specific facet of the parent cores. The results showed a blue shift of a band gap upon

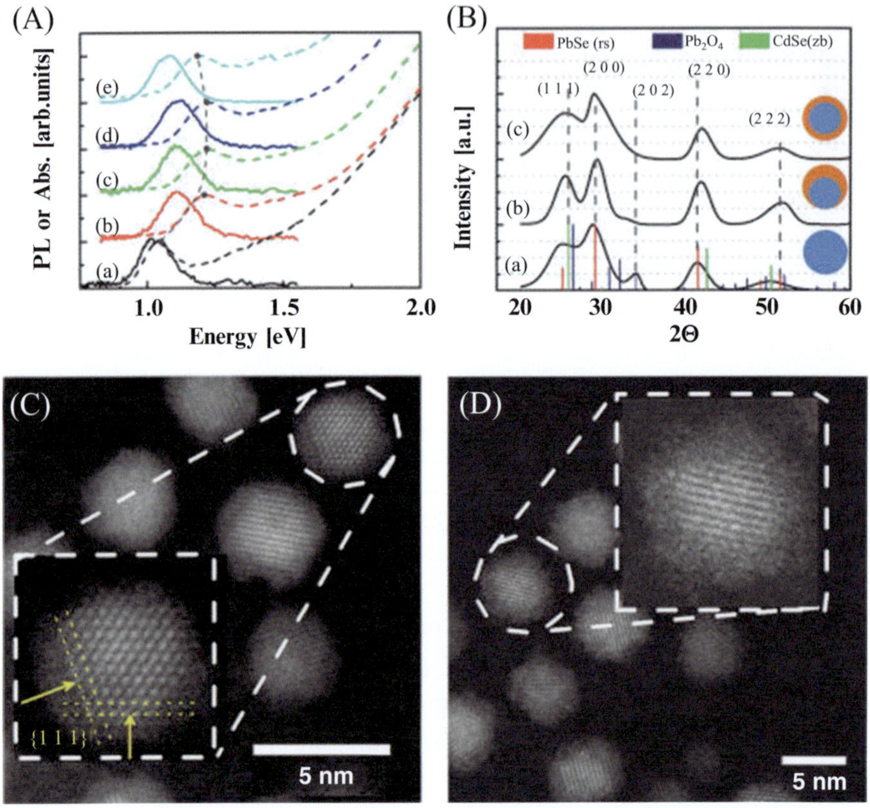

Fig. 3.18 (**A**) The absorbance (*dashed*) and PL (*solid*) spectra of (*a*) PbSe and PbSe/CdSe after (*b*) 10, (*c*) 15, (*d*) 20, and (*e*) 25 min of exchange reaction. (**B**) The XRD patterns of (*a*) PbSe and PbSe/CdSe after (*b*) 30 and (*c*) 60 min of exchange reaction. (**C**, **D**) Representative HRSTEM images of nonconcentric (**C**) and concentric (**D**) PbSe/CdSe QDs. The *dashed box* outlines a magnified image of a selected QD. The *dashed yellow lines* indicate the crystallographic facets, and the reaction direction is shown by the *arrows* (Reproduced with permission from [92])

exchange, which is counterbalanced by a red shift, due to nonconcentricity at the preliminary stages of the exchange process. Prolonging the reaction time induced the formation of a thin concentric shell that provided the QDs with a higher emission quantum yield (QY), longer radiative lifetime, and, important of all, tolerance to oxidation.

3.5 Cadmium Sulfide (CdS)

CdS and CdSe are the extensively investigated semiconducting materials due to their unique size- and structure-dependent optical and electronic properties which have found a variety of applications. In terms of shape-controlled synthesis,

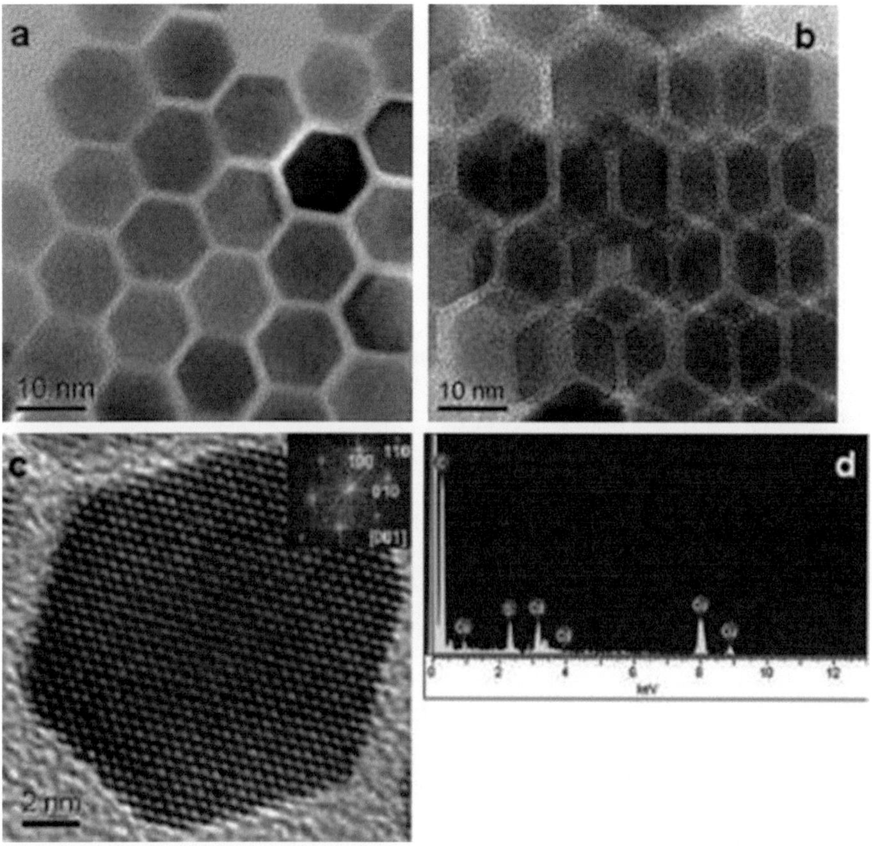

Fig. 3.19 (**a**, **b**) TEM micrographs of 2-D and 3-D assembly of hexagonal CdS nanocrystals; (**c**) HRTEM image of a single hexagonal CdS nanocrystal, the *inset* is the corresponding FFT pattern; and (**d**) EDS spectrum from the hexagonal CdS nanocrystals (Reproduced with permission from [97])

cadmium chalcogenides are considered to be the most important materials which find their wide potential applications in photoluminescence device and solar cells. These materials can be molded into a variety of different faceted geometries, including rods, tetrapods, hexagons, cubes, and pyramids [95–98] (Fig. 3.19).

Traditionally, CdS NPs are synthesized by using Cd $(CH_3COO)_{2.}2H_2O$. It is dissolved in dodecylamine to form a homogeneous solution followed by the addition of S powder. The mixture is then transferred into a Teflon-lined stainless steel autoclave and quickly heated and maintained at 220 °C for 10 h in an oven. It leads to the shaped controlled morphologies. In the reaction, cadmium acetate forms a cadmium–dodecylamine complex, and the presence of dodecylamine offers an alkaline environment needed for the disproportionating reaction of element S to release S^{2-}. The long-chain alkylamine acts as a ligand and slows the initial growth of the nuclei, which favor the size distribution of nanoparticles.

Fig. 3.20 Typical evolution of the CdSe/CdS multipod shape when large molar fractions of $CdCl_2$ are introduced in the synthesis of core-shell CdSe/CdS NCs (Reproduced with permission from [103])

The resulting CdS NPs are regular hexagon plates of around 10 nm that can form a 2-D close-packed super-lattice assembly (Fig. 3.20). A shape control between hexagon and pyramid is achieved by manipulating the precursor Cd/S molar ratio in the synthesis. Structurally, the formation of the flat, hexagonal CdS nanoparticles means a higher growth rate in the {110} directions than that in the perpendicular {001} directions.

3.5.1 *CdS/CdSe/CdTe Heterostructures*

CdS nanomaterials also known for their anisotropic different shape controlled morphologies with unique photophysical properties. The presence of unknown impurities is that of anisotropic NCs, sometime leads to the formation of nanorods, nanowires, and branched NCs, whose synthesis often depends on a delicate balance of parameters. A possible source of impurity is represented by halide ions present in trace amounts in the surfactants used in the synthesis, in the solvents, as residual species in NCs for which metal halides were used as precursors, and as species adsorbed on glassware or even on the magnetic stir bars. Halide ions interact strongly with cadmium chalcogenides [99–103]. A strong binding between Cd^{2+} and halide ions reduces the reactivity of the precursors and is responsible for the decrease in the number of nuclei formed in the nucleation stage. It results in the increase of average size and in phase transformation of the CdS NCs. A unique branched CdSe/CdS core/shell NCs can be obtained in the presence of chloride ions. By adding chloride ions in the reaction, via the introduction of the Cd precursor as a mixture of CdO and $CdCl_2$, the synthesis of octapods becomes fully reproducible when a fairly low amount of $CdCl_2$ is used. A relatively high amount of $CdCl_2$ causes a shape evolution from octapods to tetrapods through asymmetric octapods. Monodispersed octapods with tunable pod aspect ratio is indeed of major importance when their assemblies in tailored superstructures (1D, 2D, or 3D) are desired for specific applications [104, 105]. Asymmetric octapods (Fig. 3.21), with four pods having dimensions and shapes much different from the other four pods, might uncover interesting opportunities for their shape-dependent optical and electronic properties and for their self-assembly behavior.

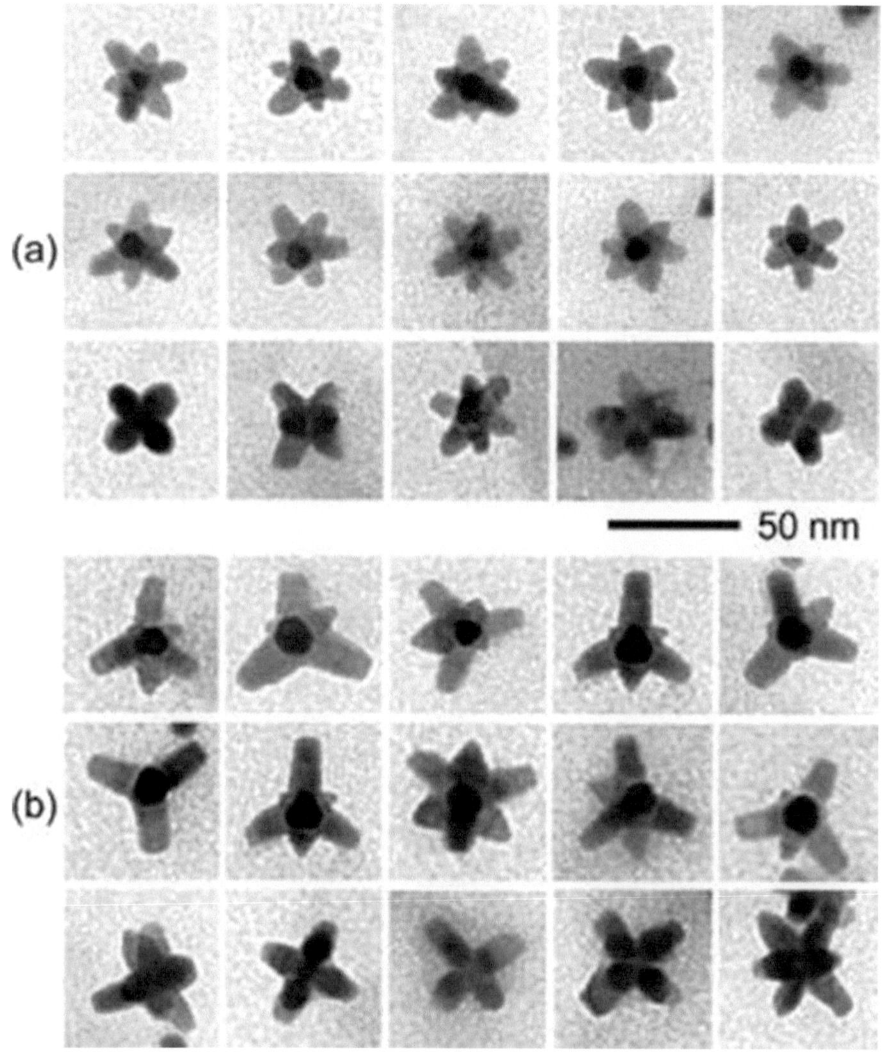

Fig. 3.21 Aliquots of the synthesis run at $fCdCl_2 = 0.32$, taken at 30 s (**a**) and 1 min (**b**) after the injection. After 1 min, several "asymmetric" octapods were present. In these TEM images, most of the octapods are oriented such that two pods, one pointing upward and one pointing downward, are parallel to the electron beam: this orientation is possible for these particles due to the uneven length of opposite pods (i.e., they lay on the support film with three of the longer pods touching it) (Reproduced with permission from [103])

3.5.2 CdS-NWs/Nanorods

One-dimensional (1D) CdS nanowires (NWs), nanotubes (NTs), and nanoribbons (NRs) are critical building blocks toward understanding fundamental properties of low-dimensional materials [106–110]. They are produced under the dislocation-driven

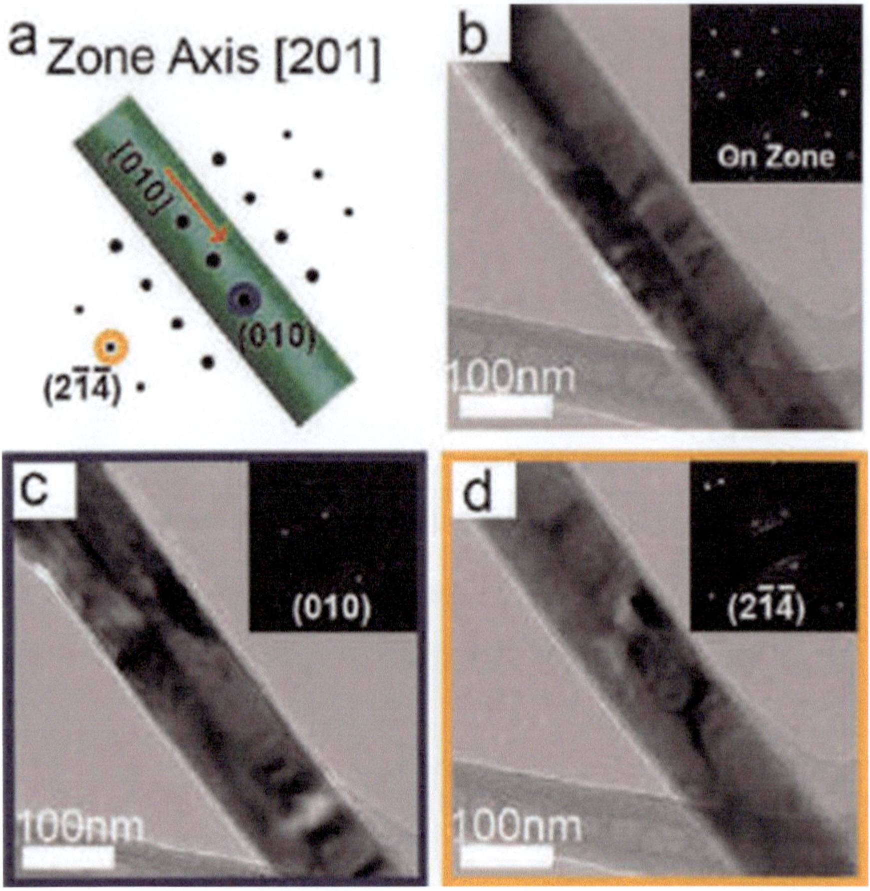

Fig. 3.22 Diffraction contrast TEM imaging of the dislocation in a CdS NW under two-beam conditions. (**a**) Schematic illustration of the superposition of real space and reciprocal lattice in CdS NW along the [201] zone axis. (**b**) Zero-beam TEM image of a representative CdS NW on the [201] zone axis with a dislocation contrast line seen along the NW's axial direction. The growth direction of this NW was [010], and the corresponding SAED pattern is shown in the *inset*. (**c**, **d**) TEM images with corresponding SAED patterns under strong two-beam conditions of (010) and (214), respectively (Reproduced with permission from [113])

growth mechanism [111–113], where axial screw dislocations provide self-perpetuating growth spirals to facilitate anisotropic crystal growth. The dislocation driven growth mechanism may generally exist in a wide variety of 1D nanomaterials, such as CdS and CdSe NWs, both of which possess the dislocation-prone wurtzite crystal structure [113]. Dislocation and stacking fault defects (Fig. 3.22) were observed in millimeter-scale whiskers, ribbons, and platelets of CdS in the 1960s, and it was suggested that dislocation drove the initial growth of these crystals [114]. CdS/CdSe NWs and nanorods (NRs) are successfully grown with uniform size distribution in the presence of a gold nanoparticle catalyst.

Without the catalyst, a variety of morphologies such as nanopods and nanocones are obtained at high system pressure, while NWs are still observed to grow at relatively low pressure. Dislocation contrast is observed for a NW along the [201] zone axis (SAED shown in Fig. 3.22b inset). The zero-beam TEM image reveals a clear dark contrast line along the NW's axial direction (Fig. 3.22b). When the (010) spots are excited, this dark contrast is visible (Fig. 3.22c). When the (214) spots perpendicular to the NW growth direction are excited, the dark contrast along the NW axis disappears (Fig. 3.22d). This experiment suggests that the observed dark contrast is due to a dislocation with screw character and possibly a screw dislocation along the NW growth axis. Gold nanoparticles provide nucleation sites to promote and initiate the growth of CdS/CdSe NWs, which further leads to dislocation sites that are responsible for the dislocation-driven growth. At a low supersaturation level, these dislocation sites continue to drive and sustain the 1D growth.

Heterostructure growth in colloidal nanocrystals (Fig. 3.23) has so far been limited and permits more complexity in their properties which are fully tunable. Anisotropic colloidal heterostructures are fabricated by sequential growth of semiconductor dots and rods of different materials. Branching is introduced through

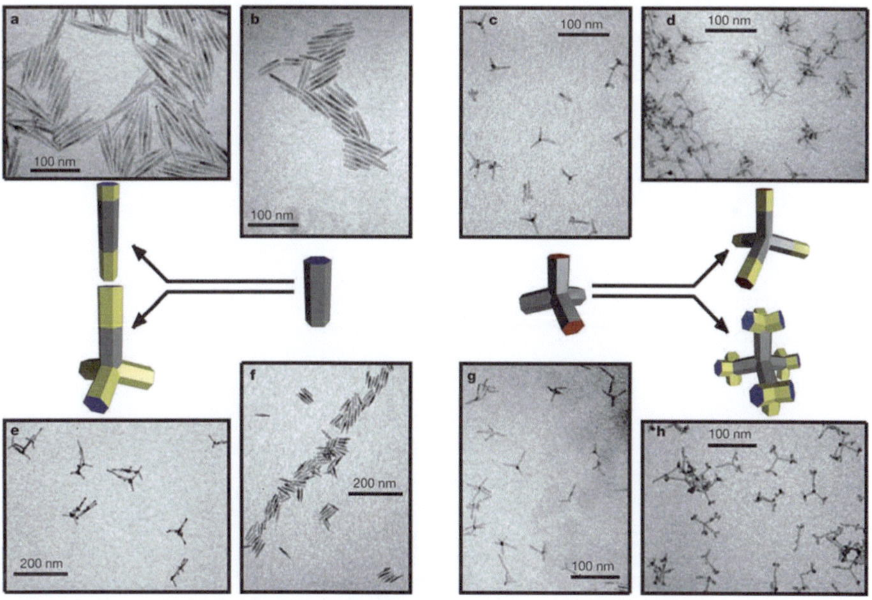

Fig. 3.23 Survey of nanocrystal heterostructures: Extended rods (**a**) were formed first by growing CdS nanorods (**b**), then adding CdSe extensions to each end. Branched rods (**e**) result from nucleation of CdTe on either end of CdSe nanorods (**f**). A CdTe zincblende region at one end creates the branch point. CdSe tetrapods (**c, g**) comprise a zincblende core and four wurtzite arms. Extending each arm linearly with CdTe yields extended tetrapods (**d**). Branched tetrapods (**h**) result from nucleation of CdTe zincblende branch points on the end of each arm. The model sequences illustrate the growth processes. In each case, the first generation is shown in *grey* and the second in *yellow*, while *red* and *blue* indicate the unique crystal faces of the wurtzite structure (Reproduced with permission from [115])

crystal phase control [115–117] leading to a zincblende–wurtzite polytypism [118]. In all cases, cadmium oxide complexed by alkylphosphonic acids is used as the cation precursor [116]. The anion precursor [elemental Se, S or Te dissolved in tri (*n*-alkylphosphine)] is injected at a temperature between 280 and 320 °C to initiate growth. Linear heterojunctions are formed when precursors for a second material are added to a growth solution containing nanorods or tetrapods. Branched junctions are introduced preferentially at high supersaturation of these precursors. The second generation is typically grown without isolation of the nanocrystals by using an excess of cadmium in the first step and injecting the anion precursor for the second material. Thus, all the reported heterostructures have cadmium as a common cation. Linearly extended rods are synthesized with CdS rods and CdSe extensions, while the other structures were synthesized with CdSe in the first and CdTe in the second generation. Similar results are achieved by isolation of the first-generation nanocrystals and reintroduction of cadmium precursor before injecting the second anion precursor to produce heterostructures that do not share common ions.

3.5.3 CdSSe

Using a single injection of premixed trioctylphosphine sulfur (TOPS) and trioctylphosphine selenium (TOPSe) to a bis(phosphonate)cadmium [$Cd(ODPA)_2$] complex at 320 °C, axially anisotropic $CdS_{1-x}Se_x$ nanorods with a thick "head" and a thin "tail" (Fig. 3.24) can be synthesized [119]. Nanorod heads are CdSe rich,

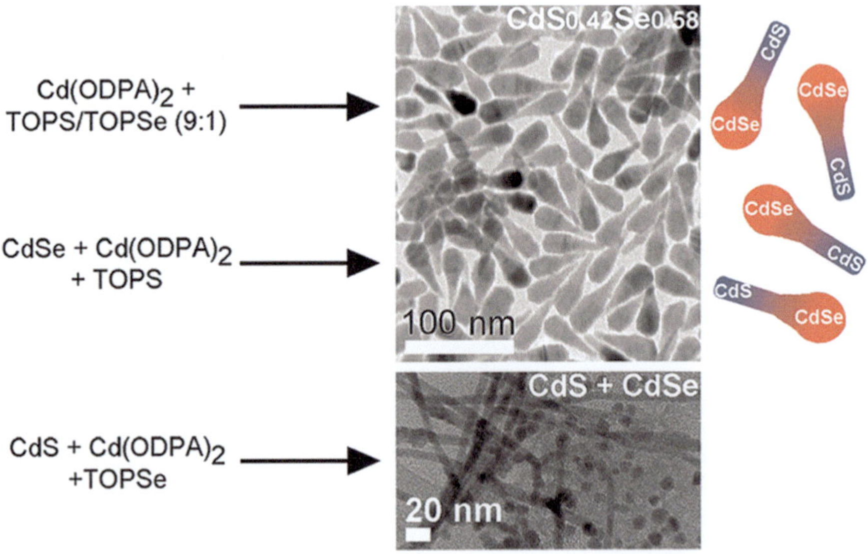

Fig. 3.24 $CdS_{1-x}Se_x$ nanorods with a thick "head" and a thin "tail" (Reproduced with permission from [120])

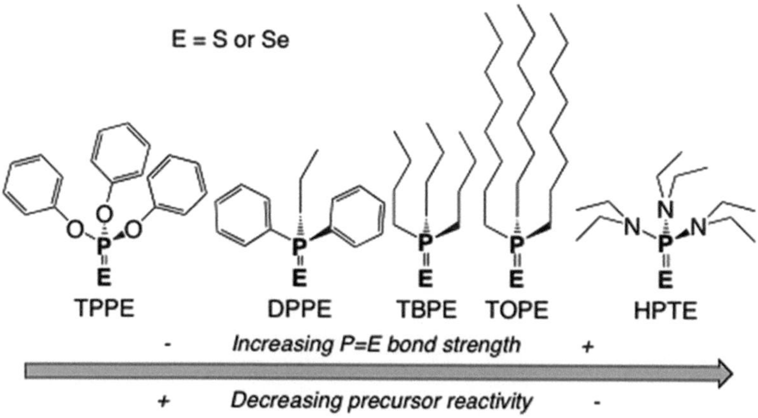

Fig. 3.25 Phosphine-chalcogenide precursors (Reproduced with permission from [120])

whereas nanorod tails are CdS-rich [120]. Time evolution and mechanistic studies showed that $CdS_{1-x}Se_x$ nanorods form sequentially, starting with quick CdSe head nucleation (<20 min), followed by slow CdS tail growth (~85 min) (Fig. 3.25). The relative ease of formation between these two nanorod segments is due to the relative TOPS versus TOPSe precursor reactivity. By tuning sterics and electronics and therefore altering molecular precursor reactivity, one could control relative formation rates of different nanocrystalline phases. A comparative reactivity of different chalcogenide (sulfide and selenide) derivatives, $R_3P{=}E$ where 'E' is a chalcogen (S or Se), of five commercially available phosphines, R_3P: Triphenylphosphite (TPP), diphenylpropylphosphine (DPP), tributylphosphine (TBP), trioctylphosphine (TOP), and hexaethylphosphorustriamide (HPT) (Fig. 3.25) can be evaluated. All of the phosphines in this series are liquid at room temperature (RT), which facilitates precursor preparation by chemical "dissolution" of chalcogen (sulfur or selenium). More electron-donating groups would lead to better stabilization of a partial positive charge on phosphorus, thus to stronger phosphorus-chalcogen (P=E) bonds and less reactive precursors (Fig. 3.25). A decrease in phosphorus-chalcogen (P=E, E=S or Se) bond strength, that is, lower P=E bond energy, increases phosphine-chalcogenide precursor reactivity. More reactive R_3PE precursors react faster and more uncontrollably with Cd $(ODPA)_2$, leading to lower selectivity for anisotropic (unidirectional) 1D growth along wurtzite's z-axis and resulting in smaller nanorod aspect ratios.

3.5.4 CdSe/CdS

A comparison can be made for w (wurtzite)-CdSe and zb (zinc-blende)-CdSe nanocrystals in the seeded growth of CdSe/CdS nanohererostructures with different morphologies. Combining CdSe and CdS in a single nanostructure creates a material with heterogeneous carrier confinement or "mixed dimensionality" where holes are confined to CdSe while electrons can move freely between CdSe and CdS phases, spreading over the entire nanostructure [121, 122]. Both w-CdSe and zb-CdSe nanocrystals form stable colloidal solutions in nonpolar solvents like hexane, toluene, and TOP; and show narrow PL bands associated with the recombination of photoexcited carriers from 1Sh and 1Se quantum-confined states. Both w-CdSe and zb-CdSe nanocrystals are injected into the reaction mixture containing TOPO, TOP, *n*-octadecylphosphonic acid (ODPA), *n*-propylphosphonic acid (PPA), a Cd-ODPA complex and trioctylphosphine sulfide (TOPS) in concentrations optimized for synthesis of high-quality wurtzite-phase CdS (w-CdS) nanorods with long (001) axes. CdS easily nucleates at the surface of CdSe seeds [91, 92]. The structure of the CdSe seeds determined the morphology of CdSe/CdS nano-heterostructures. w-CdSe nanocrystals seeded the formation of CdSe/CdS nanorods (Fig. 3.25a–c), whereas zb-CdSe seeds initiated growth of CdSe/CdS nanotetrapods (Fig. 3.26d–f). The phosphonic acids present in the reaction mixture selectively bind to the {100} type facets of w-CdS and w-CdSe, slowing the nanocrystal growth along these directions and forcing w-CdS to grow along the *c*-axis. The w-CdS phase will primarily nucleate on "polar" {001} and {100} facets of w-CdSe seeds, which are structurally identical to fast growing {001} and {100} facets of w-CdS. The small mismatch in lattice constants of CdSe and CdS (~3.8 % for the {001} planes) is favorable for the epitaxial relation between the CdSe and CdS parts of the nanostructure.

CdSe/CdS nanorods present the appealing characteristics of strong and tunable light emission from green to red, and they can be spontaneously self-assembled on substrates, and easily organized in close-packed ordered arrays over large areas that are reminiscent of liquid-crystalline phases (Fig. 3.27). The formation of these types of assemblies is entropically driven and has been observed already in several colloidal nanorod solutions [123, 124]. Also, ordered arrays of close-packed, vertically aligned nanorods can be obtained by evaporation-mediated self-assembly of nanorods on substrates from more concentrated nanorod solutions. Larger areas, often of the order of several square micrometers, can be achieved when the process is carried out on the surface of water [125]. Let the solvent evaporate from a concentrated solution of nanocrystals in toluene that was deposited on the surface of water in a glass beaker, and then was collected part of the floating nanocrystal film with a carbon-coated TEM grid. Regular patterns of vertically stacked nanorods can be seen, which are organized either in hexagonally close-packed monolayers or in ordered multilayer structures.

The lateral alignment of nanorods can be extended over larger areas and controlled in direction if the evaporation- mediated assembly is influenced by an external perturbation like an electric field [126–128]. The presence of a significant dipole in this type of nanorods can laterally align them on planar surfaces using external electric fields (Fig. 3.28a). Interdigitated electrode devices (electrode separation 4 μm, electrode length 600 μm), which were fabricated by optical

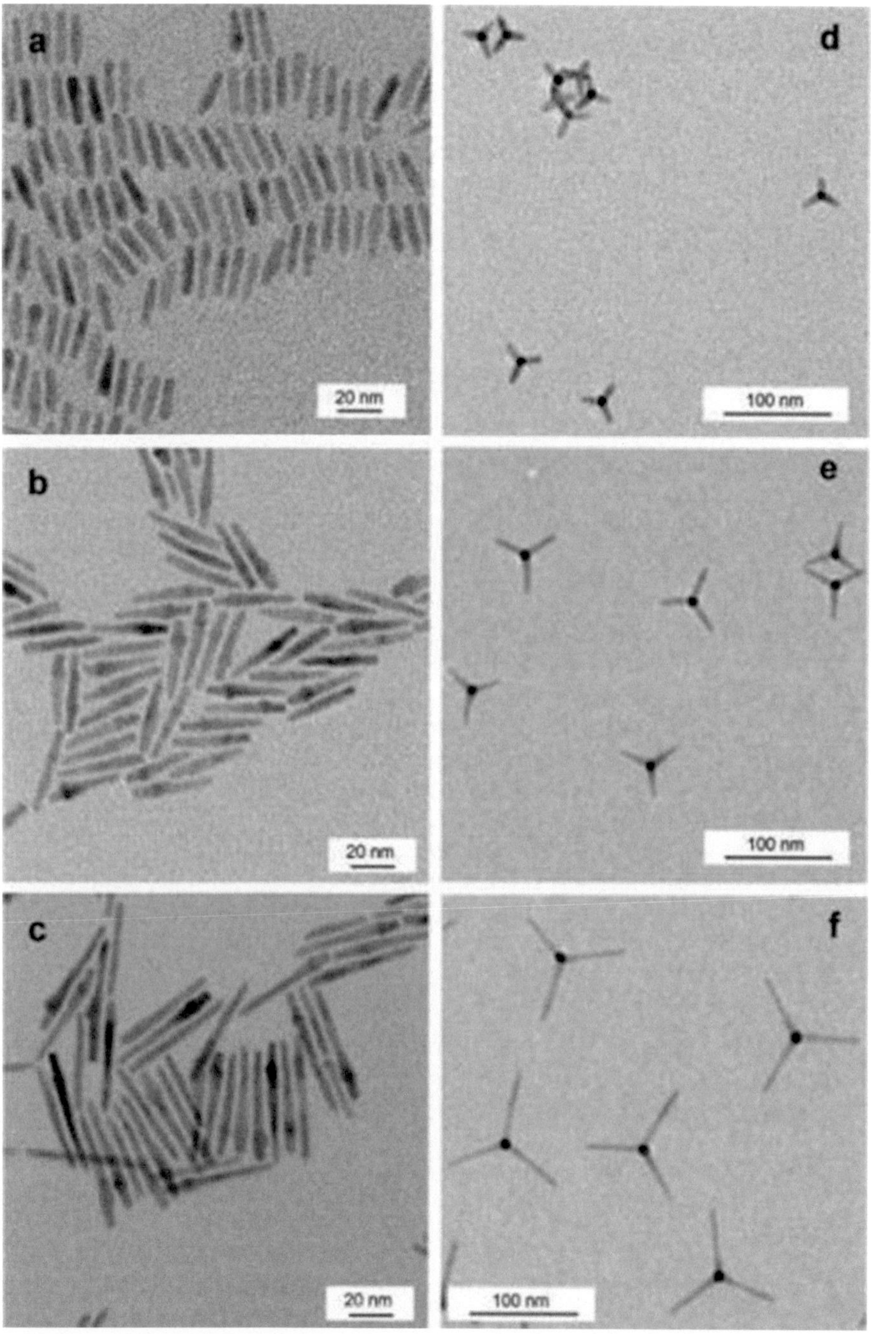

Fig. 3.26 (**a–c**) TEM images of CdSe/CdS nanorods with different aspect ratios, all grown from 4.4 nm w-CdSe seeds. (**d–f**) TEM images of CdSe/CdS nanotetrapods with different lengths of the arms grown from 4.0 nm zb-CdSe seeds (Reproduced with permission from [122])

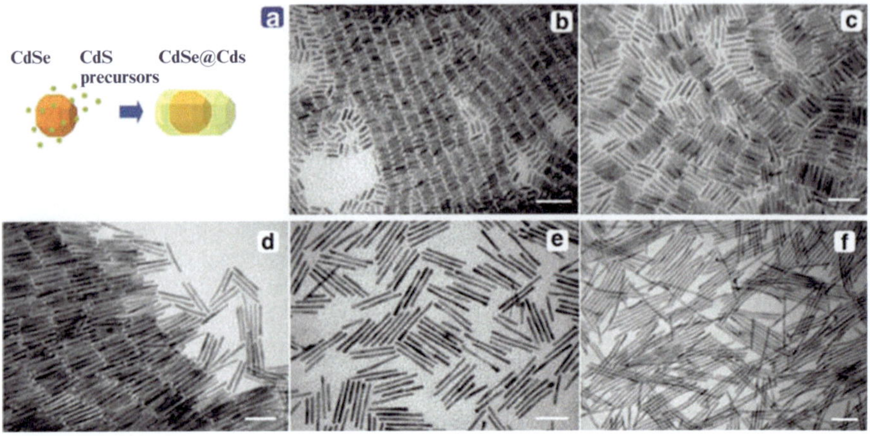

Fig. 3.27 Evaporation-mediated self-assembly of CdSe/CdS nanorods prepared by seeded growth. (**a**) Sketch of the seeded growth approach. (**b–f**) TEM of self-assembled CdSe/CdS nanorods. As soon as the solvent, in which the CdSe/CdS are dissolved (toluene), is allowed to evaporate from the grid on which the drop of solution is deposited, the nanorods start self-assembling. Short aspect ratio (rod length/rod diameter, AR) rods (**b–d**) tend to form locally ordered 2D smectic phases, while longer rods tend to form either disordered assemblies, ribbons, or locally ordered 2D nematic phases (Reproduced with permission from [123])

lithography and subsequent metal evaporation (500 nm Al) on Si/SiO_2 (100 nm oxide thickness) substrates can be employed. A quantity of typically 50 μL of nanorod solution is drop-casted onto the electrode device under ambient conditions, and a dc electric field of 2.5 105 V/cm is applied during solvent evaporation. After the alignment process, the electrode devices are inspected by scanning electron microscopy (SEM) imaging. The rods aligned uniformly over areas on the scale of several tens of square micrometers by following the field streamlines throughout the electrode gap. Magnifications of randomly picked areas in between the electrodes showed indeed an almost perfect lateral alignment of rods in monolayers or multilayered structures, with a type of organization that resembles either the nematic (Fig. 3.28b) or the smectic liquid-crystal phases (Fig. 3.28c). The polarized emission from the regions of aligned nanorods under ambient conditions is recorded with a conventional micro-photoluminescence setup. The sample is excited by a circularly polarized He-Cd laser at 325 nm. The PL of such large arrays of oriented nanorods is strongly polarized along the alignment direction.

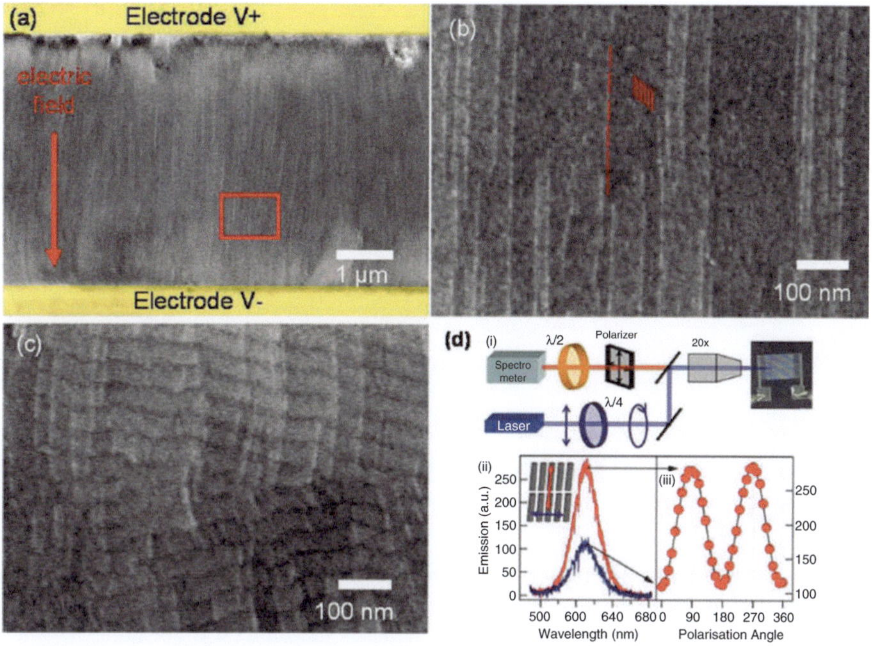

Fig. 3.28 Lateral alignment of nanorods with electric fields and polarized emission from aligned nanorods. (**a–c**) SEM images showing aligned arrays of nanorods with an aspect ratio of 10. The *red arrow* indicates the direction of the electric field that was applied during the evaporation of the nanorod solution, and the red square marks the region that is displayed in (**b**). (**b**) Micrometer size area where the individual rods can be resolved. Some rods are highlighted in *red* as a guide to the eye. (**c**) Zoom that displays an area of nanorods assembled in ribbon like structures. (**d**) (*i*) Schematic illustration of the experimental setup. (*ii*) Nanorod emission spectra, recorded from the micrometer size regions of aligned nanorods under ambient conditions, polarized parallel (*red*) and perpendicular (*blue*) to the long axis of the nanorods, as illustrated by the schematic inset. (*iii*) Red dots show the PL intensity dependence on the polarization angle and can be fitted to a cos 2 function (*black line*) (Reproduced with permission from [123])

3.6 Zinc Oxide/Zinc Sulfide/Zinc Selenide (ZnO/ZnS/ZnSe)

3.6.1 ZnO

Another important category of versatile semiconductor materials comprises of ZnO, ZnS, and ZnSe. ZnO is a biofriendly, cheap, wide band gap II–VI semiconductor, with a range of applications in everyday products as well as high tech applications such as piezoelectric sensors and mechanical actuators [129, 130], solar cells [131, 132], varistors [133], and thermoelectrics. ZnO is most commonly found in the wurtzite crystal structure, which is non-centrosymmetric leading to pyroelectric and piezoelectric properties. Hydrothermal synthesis of ZnO is simple, fast, flexible, scalable, and even green as it is free of organic solvents [134, 135].

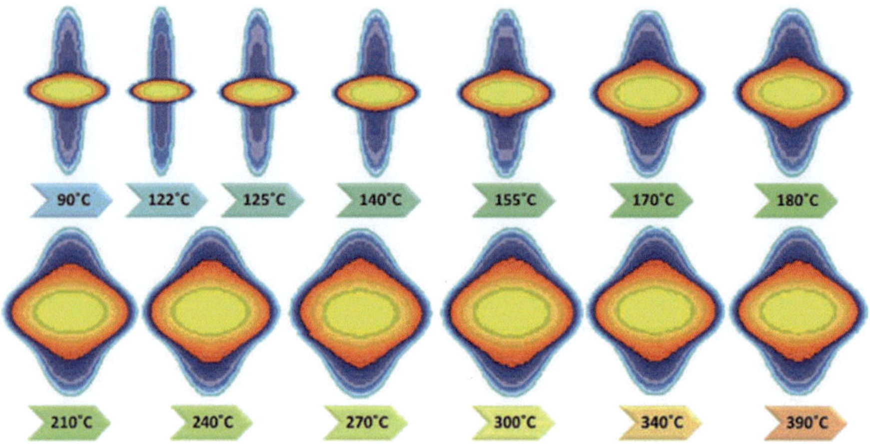

Fig. 3.29 Average morphology of particles synthesized at different reactor temperatures (Reproduced with permission from [136])

At low temperatures the particles appear to have an elongated diamond shape, whereas higher temperatures lead to a more isotropic crystallite domain shape (Fig. 3.29). As the reaction temperature increases the dielectric constant of water decreases and a higher degree of supersaturation is achieved. The fewer nucleation sites at lower temperatures and lower supersaturation favor growth by diffusion on steps and kinks and possibly by Ostwald ripening, where smaller particles dissolve and grow onto larger particles [136, 137]. Below 210 °C, the dominant particle appearance is large double-cone structures with rough surfaces indicative of diffusion controlled growth.

3.6.1.1 Spheres

Colloidal crystalline spherical ZnO NPs with smooth surfaces have been synthesized by nanosecond pulsed laser irradiation [138, 139]. Before laser irradiation, ZnO NPs are smaller than 100 nm with a hexagonal phase. Irradiating of the raw nanoparticles dispersed in water for 10 min using a 355 nm laser transforms the irregular NPs into spherical which still possess a hexagonal phase. It happens when ZnO spherical particles grow from raw nanoparticles (Fig. 3.30a). For 30 s irradiation, a small number of spherical particles with diameters below 100 nm are obtained which are mainly dominated by raw nanoparticles. If the laser irradiation time is increased to 60 s, the number of raw nanoparticles is significantly decreased (Fig. 3.30b). Careful examination of the surfaces of the formed spherical particles at this stage reveals that the spherical particles are combined from several particles with irregular shape, which can be evidenced by the apparent crystal boundaries on the surfaces. When further prolonging the irradiation time to 120 s, the product is dominated by solidified spherical particles (Fig. 3.30c). After laser irradiation for

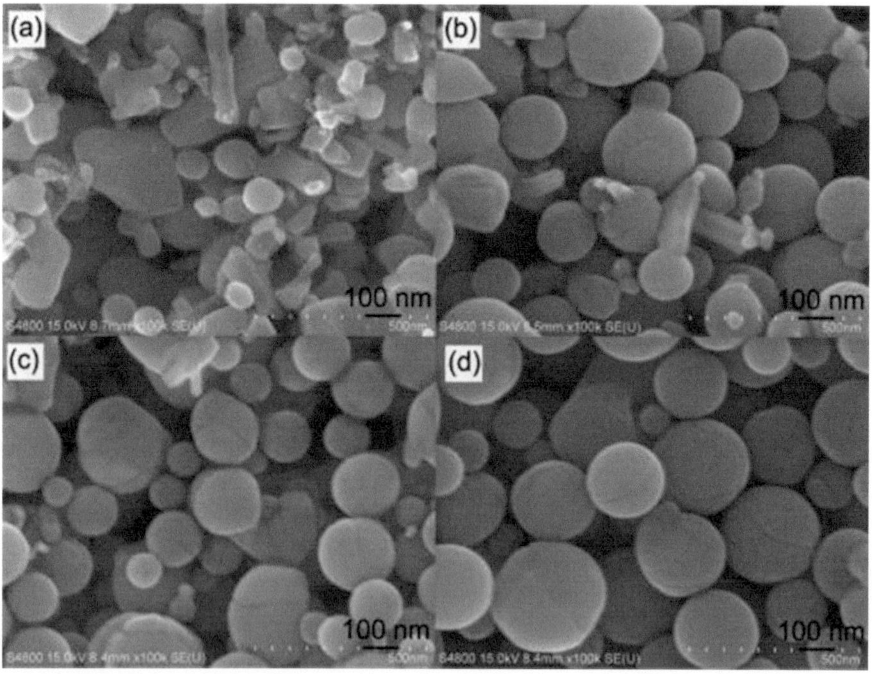

Fig. 3.30 FESEM images of ZnO obtained by laser irradiation of commercial ZnO nanoparticles with different laser irradiation time (third harmonic, in water, 133 mJ/(pulse cm^2)): (**a**) 30 s, (**b**) 60 s, (**c**) 120 s, (**d**) 5 min (Reproduced with permission from [139])

5 min, the product exhibits more homogeneous size distribution (Fig. 3.30d). Finally, after laser irradiation for 10 min, single crystalline spherical particles with smooth surfaces are obtained.

3.6.1.2 Nanowires

ZnO nanowires and nanobelts have received attention due to the potential applications as UV lasers, [140], light-emitting diodes [141], gas sensors [142, 143], and UV photodetectors [144]. Because of their wide band gap, low cost, strong radiation hardness, and ease of manufacturing, ZnO nanowires and nanobelts are regarded as one of the most promising candidates for UV photodetectors [145, 146]. The effect of Au nanoparticles on the performance of ZnO single nanowire (Fig. 3.31) photodetectors demonstrates that the dark current decreases by 2 orders of magnitude after covering ZnO nanowires with Au nanoparticles. Large enhancements of photoresponse and response speed have been observed due to the presence of Au nanoparticles at the same time.

Schematics of ZnO nanowire-based photodetectors with and without Au nanoparticles and a representative SEM image of this device are shown in Fig. 3.32a–c. The Au electrodes are 100 nm thick, and the distance between two

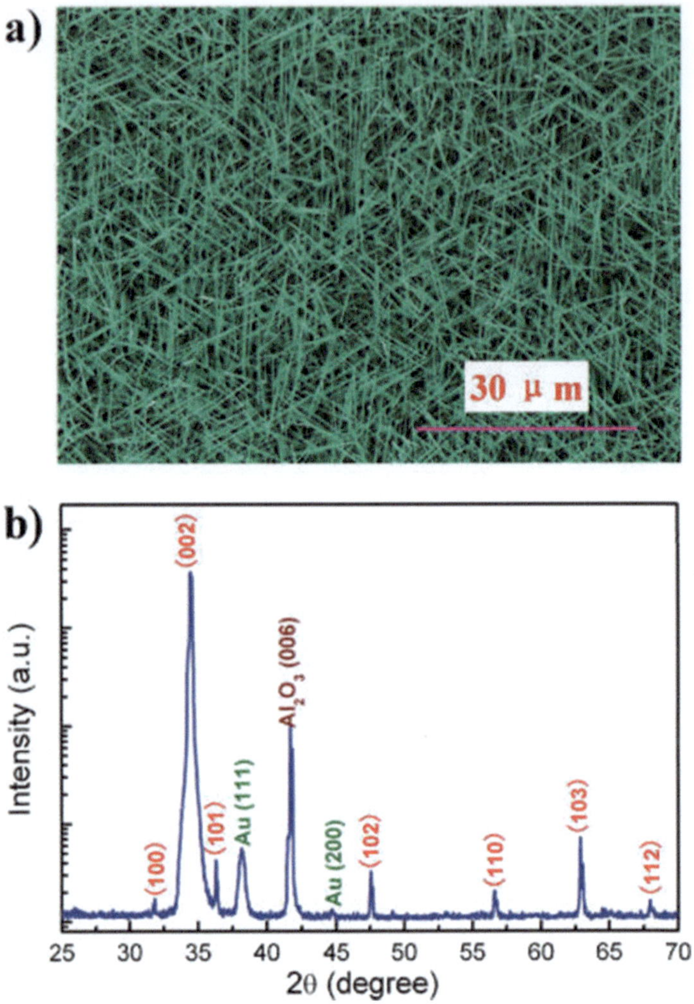

Fig. 3.31 SEM image (**a**) and XRD pattern (**b**) of as-grown ZnO nanowires on sapphire substrate (Reproduced with permission from [145])

electrodes is 2 μm. The I–V curves in dark for two devices are nonlinear, which indicates that Au electrodes and ZnO nanowires are typical Schottky contacts. When the devices are illuminated by 350 nm UV light, the currents increase largely due to the photon energy (350 nm, 3.54 eV) is larger than E_g of ZnO (3.4 eV) and they can produce electron-hole pair, thereby changing the electrical conductivity of ZnO. The ratio of photo current to dark current of bare ZnO photodetector is around 10^3. As for the device with Au nanoparticles, this value can be increased as large as 5×10^6. These results indicated that covering with Au nanoparticles on the surface of devices can decrease the dark current and increase the photocurrent.

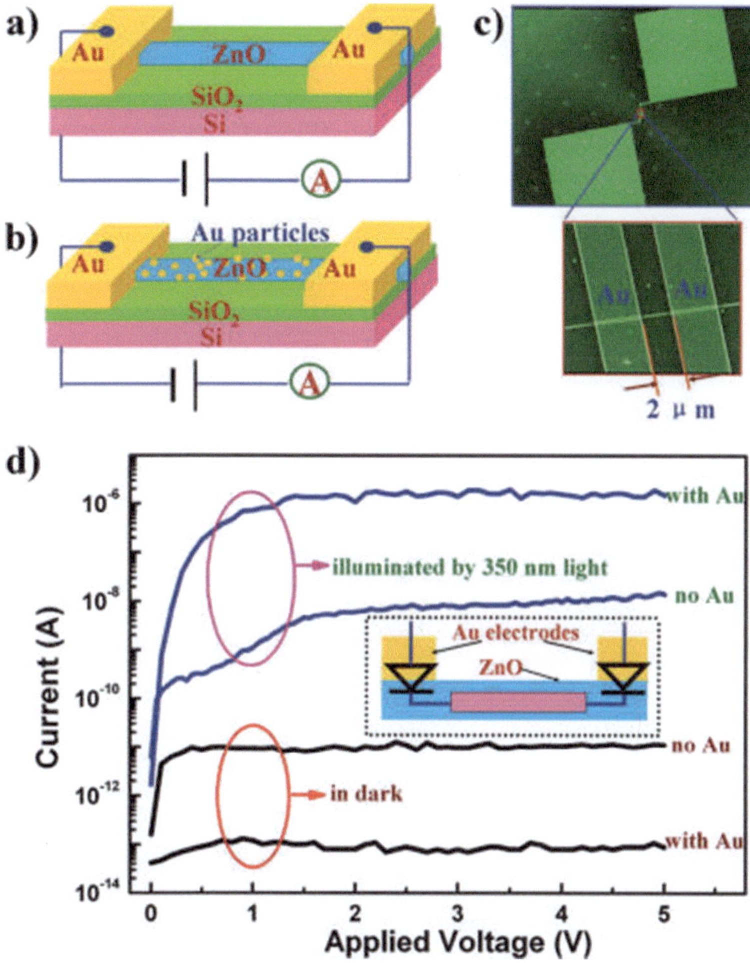

Fig. 3.32 Schematic of ZnO nanowire photodetectors without (**a**) and with (**b**) Au nanoparticles. (**c**) SEM image of the device. (**d**) *I–V* characteristics of ZnO nanowire photodetectors both in dark and under 350 nm UV light illumination. The *inset* of "**d**" is the equivalent circuit for the devices (Reproduced with permission from [145])

3.6.2 ZnS

Like ZnO, ZnS has also attracted considerable research interest due to its extensive usage in thin-film electroluminescent devices, infrared windows, flat panel displays, sensors, and lasers [147]. Single crystalline tower-like complex ZnS structures (Fig. 3.33) can be formed via a ZnS nanoparticles-assisted vapor deposition route [148]. The growth of nanotowers is initiated by the preferred nucleation on the ZnS nanoparticles precoated on Si substrate, followed by alternate two-dimensional (2D) growth of hexagonal plates/layers inside nanotowers and one-dimensional

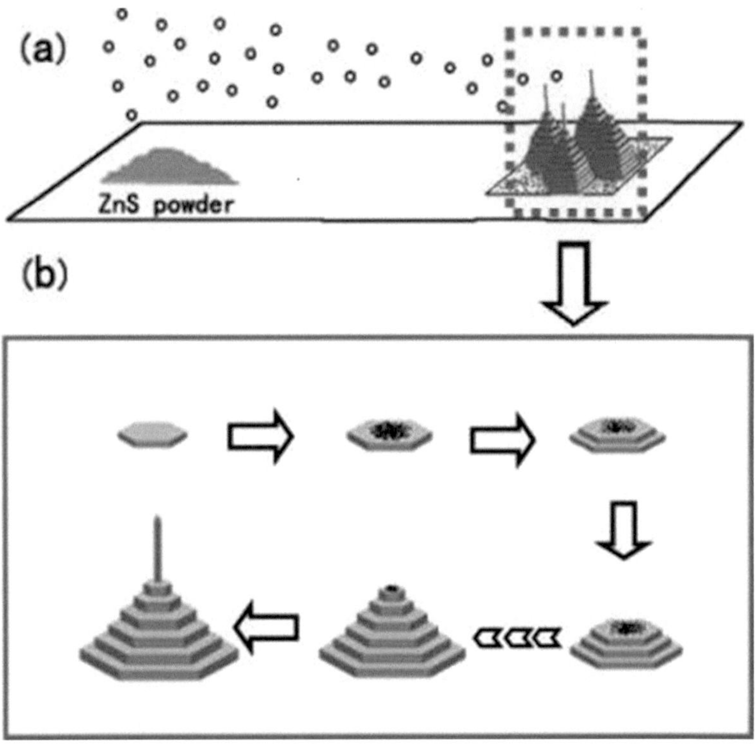

Fig. 3.33 (**a**, **b**) Schematic illustration for the growth mechanism of the ZnS nanotowers (Reproduced with permission from [148])

(1D) stacking in the direction perpendicular to the plates/layers via vapor–solid (VS) growth processes. Such ZnS nanotowers display new photoluminescence emission bands and anomalous Raman scattering modes compared with nanoplates prepared without the assistance of ZnS nanoparticles. The tower-like structures are likely to be made up in a layer-by-layer fashion and consist of quasi-hexagonal plates. The photoluminescence spectra of the nanotowers show two strong emission bands located at 394 and 592 nm. In Raman spectra, two anomalous modes at 233 and 488 cm^{-1} are observed.

In addition to ZnS, monodisperse ZnSe NPs can be self-assembled in supraparticles via "in situ aggregation" through a simple hot-injection method [149]. The "in situ aggregation" strategy can be extended to tune the shapes of ZnSe supraparticles (Fig. 3.34), and introduce novel functional magnetic and luminescence properties. Shape control is manipulated with oleic acid as ligands, which balances the attractive interparticles van der Waals forces and steric repulsive forces from the ligands. With the increase of oleic acid concentration, a morphology change from microspheres to asymmetrical multimer and three-dimensional nanoflowers is observed. "Doping" preformed Fe_3O_4 nanoparticles into ZnSe supraparticles endow them with magnetic properties. The magnetism of these Fe_3O_4@ZnSe supraparticles depends on the dosage of dopant. Doping of preformed CdS nanocrystals was also

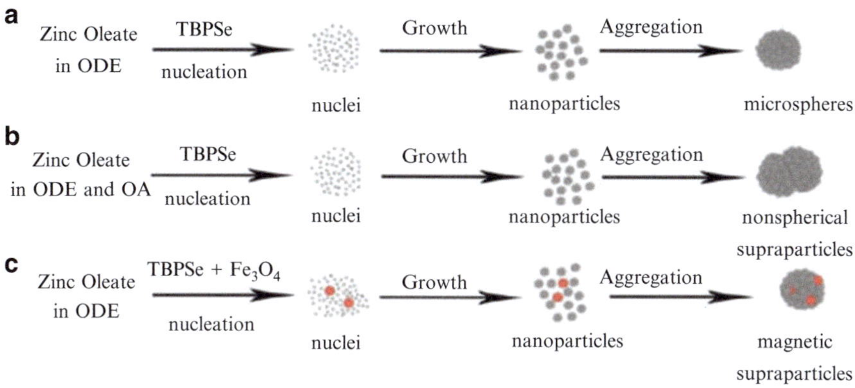

Fig. 3.34 Diagram of the formation processes of (**a**) Spherical ZnSe supraparticles, (**b**) Nonspherical ZnSe supraparticles, and (**c**) Preformed NP Doped ZnSe supraparticles (Reproduced with permission from [149])

studied, resulting in emissive hybrid CdS@ZnSe supraparticles with diameters of 50–100 nm. It is noted that the doping of Fe_3O_4 and CdS nanoparticles show differing morphologies. The differences can be explained by variance in the lattice mismatches which leads to differing potentials for crystal growth.

3.7 Se/Te Nanocrystals

In this section, we present fine crystalline Se–Te alloy morphologies which can be obtained over the whole composition range of Se and Te mixtures [69]. The shape, structure, and elemental composition of different morphologies depend very much on the amount of reducing agent (i.e., hydrazine). Their different mixing ratios produce different morphologies with regular transitions from rhombohedral NCs in the Se rich-region to spindle shaped NCs close to equimolar proportions, and then to fine crystalline needles in the Te rich-region of the mixtures. All morphologies show a complete homogenous mixing between elemental Se and Te in Se–Te alloy but with a marked nonideal behavior with respect to their stoichiometric mole fractions over the total mixing range. All morphologies show a significantly low Se and high Te contents than their ideal amounts.

The results are compared in a phase diagram (Fig. 3.35) which shows a drastic variation in the morphology of Se–Te alloy NCs over the whole composition range. Pure Se ($x_{Se} = 1$) always produced small predominantly amorphous nanoparticles in the presence of 3.6 and 2.2 M of hydrazine, while micrometer sized flakes (*image 1*) are formed with 0.1 M hydrazine. Thus, amount of hydrazine has a significant influence on pure Se nano-morphologies. This is all related to the nucleation kinetics because excess of hydrazine will reduce maximum number of SeO_3^{2-} ions into atoms with little possibility of further nucleation, but low hydrazine will leave many SeO_3^{2-} ions for further nucleation. Hence, the former nucleating

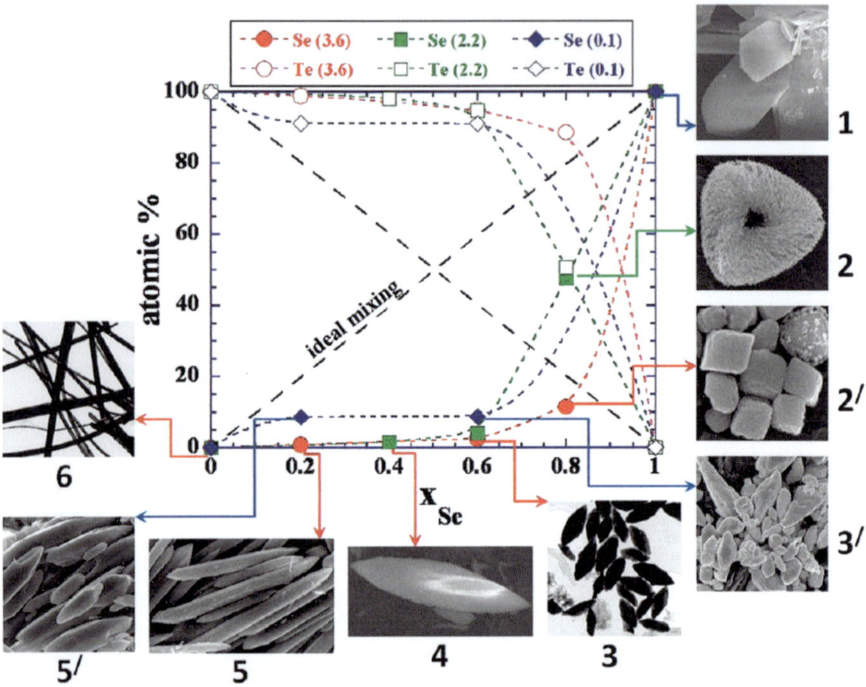

Fig. 3.35 A plot of atomic % of Se and Te in Se-Te alloy NCs versus stoichiometric mole fraction of Se (x_{Se}) over the whole composition range of Se + Te mixtures in the presence of 3.6 M (*red symbols*), 2.2 M (*green symbols*), and 0.1 M (*blue symbols*) hydrazine. *Empty* and *filled symbols* belong to Te and Se atomic %, respectively. Straight dotted lines represent the ideal mixing between elemental Se and Te over the whole mixing range of Se + Te mixtures. SEM or TEM images (1 through 6) of different Se-Te alloy morphologies belong to different mole fractions. A *color* of the *arrow* indicates to which hydrazine concentration that mole fraction belongs. For instance, images 1 and 6 are indicated by the "*blue*" and "*red*" arrows, respectively. The "*blue*" colored symbols belong to 0.1 M hydrazine and "*red*" colored symbols to 3.6 M hydrazine (see legends of the graph). (Reproduced with permission from [69])

centers will remain in aqueous phase stabilized with CTAB but latter ones will continue the autocatalytic process to produce petal-like morphologies (*image 1*). As the amount of Se reduces at $x_{Se} = 0.8$, there are also 20 % Te nucleating centers which possess predominantly metallic properties with little solubility in the aqueous phase. A contrasting solubility preference between Te and Se nucleating centers does not allow their stoichiometric mixing to pass on to heterogeneous crystalline phase. A greater aqueous solubility preference of Se nucleating centers will leave much greater amount of Te rather than Se in heterogeneous crystalline phase even in the Se-rich region of the mixture. But as the amount of hydrazine reduces to 2.2 M, greater number of unreduced SeO_3^{2-} ions will undergo autocatalytic reduction on already available Te nucleating centers and hence will increase the atomic % of Se in crystalline phase along with Te to produce Se–Te nano-alloy. Thus, different atomic % of Se induces a change in the overall shape and structure of resulting NCs. Note *image 2* with greater amount of Se shows polyhedral morphologies but *image 2′* with much lesser amount of Se shows predominantly rhombohedral morphologies. Same is true with $x_{Se} = 0.6$. Low atomic % Se produces fine spindles (*image 3*) but relatively greater atomic % distorts the same morphology (*image 3′*). Likewise, $x_{Se} = 0.2$ produces fine hexagonal needles (*image 5*) with low atomic % of Se but distorted ones (*image 5′*) with relatively high atomic %. In addition, there is a smooth transformation in the rhombohedral morphology to spindle shape from $x_{Se} = 0.8$–0.6, short spindles to long spindles from $x_{Se} = 0.6$–0.4, long spindles to long needles from $x_{Se} = 0.4$–0.2, and finally long needles to long ribbons from $x_{Se} = 0.2$–0.

This sequence can very well be related to both predominantly amorphous nature of Se and its amount in Se–Te alloy NCs. Thus, right hand side (Se-rich region) of the phase diagram prefers polyhedral or rhombohedral shape while left hand side (Te-rich region) favors 1D nano-morphologies. Equimolar region of the mixtures clearly identifies the transitions between these two extremes.

3.8 Selenium-Nanocrystals and Nanowires

Amorphous selenium with a mean particle size of 20 Å has been synthesized in aqueous solution by Dimitrijevic et al. [150]. The solutions, however, are subject to extreme photo-corrosion and are stable for only a few hours. Selenium has been incorporated into various forms of zeolite and identified by optical and Raman spectroscopy and anomalous x-ray scattering [151, 152].The diameter of the zeolite supercages is about 13 Å, which restricts the size of the selenium entities. Chains and/or rings of selenium are produced, depending on the nature of the counteraction and the zeolite used.

Johnson et al. [153] have synthesized nanoparticles of selenium using a reverse-micelle medium on an intermediate length scale of 20–200 Å. The micelle structure was examined with small-angle neutron scattering (SANS). Sodium bis (2-ethylhexyl) sulfosuccinate usually called AOT (obtained from Aldrich and used without further purification) was dissolved in heptane to make a 0.1 M solution. Three values of w (=$[H_2O]/[AOT]$) were used, namely 5, 10, and 20. The appropriate amount of Nanopure water required to give the desired w was added to the heptane–AOT solution, and the mixture was stirred for several hours to ensure equilibrium. The solution of heptane, AOT, and water was fully deoxygenated by bubbling ultrahigh-purity argon through the mixture. Selenious acid (H_2SeO_3) was added to half of the solution to give a concentration of 2.0×10^{-3} M, and the same molar amount of hydrazine was added to the other half; both chemicals were pre-dissolved in deoxygenated nanopure water. After several hours, these two solutions were combined, initiating the reaction inside the water pool.

These operations were carried out inside a nitrogen-filled glovebox. In all the experiments, an orange-red precipitate was obtained that was identified as selenium by energy-dispersive X-ray spectrometry (EDS). The color implied that its form was either amorphous or monoclinic since trigonal selenium is black. It should be noted that nanoparticles could not be isolated by traditional capping techniques[1] but no changes in color or X-ray patterns were noticed in samples aged up to 2 weeks. The solution went from colorless to pale yellow to orange during the first 15 min of reaction time, and samples were removed from solution at several times (1, 2, 5, 10, 15, and 30 min) after the reaction initiation and placed on carbon grids for examination by electron microscopy. The rate of color change varied with the concentration of selenious acid and hydrazine and the value of *w*. After about 2 h, a red-orange precipitate was deposited at the bottom of the flask. The solution was then centrifuged to extract the particles, which were washed in heptane and methanol to remove any residual AOT and water. Drying with petroleum ether and under vacuum changed the color of the particles from red to black so instead, the methanol was allowed to evaporate slowly. However, in samples prepared at high pH (by adding 5 mL of 10 M NaOH immediately after the solutions containing selenious acid and hydrazine was mixed), the color change was much slower to appear. TEM pictures of the Se NPs thus synthesized is shown in Fig. 3.36. The crystalline particles were found to be spherical with a monoclinic structure and a size distribution on a nanometer scale.

Synthesis of one-dimensional nanostructures of Se and Te using chemical methods has been reported by Mayers et al. [154]. Diameter-selected synthesis of single crystalline trigonal selenium nanowires has been reported by Changhua et al. [155]. They designed a low temperature hydrothermal synthesis route to the direct production of monocrystalline trigonal selenium (t-Se) nanowires, and their diameters are dependent on the reductants used. Sodium selenite (Na_2SeO_3) was used as oxidant to provide selenium source, and $Na_2S_2O_3$ or $(NH_4)_2S_2O_3$ was chosen as reductant to prepare t-Se nanowires with different diameters, respectively. In a typical procedure, Na_2SeO_3 (0.01 mol), $Na_2S_2O_3$ (0.01 mol), or $(NH_4)_2S_2O_3$ (0.01 mol) were added into a Teflon-lined autoclave with the capacity of 50 ml, which was filled with 40 ml distilled water to form a homogeneous

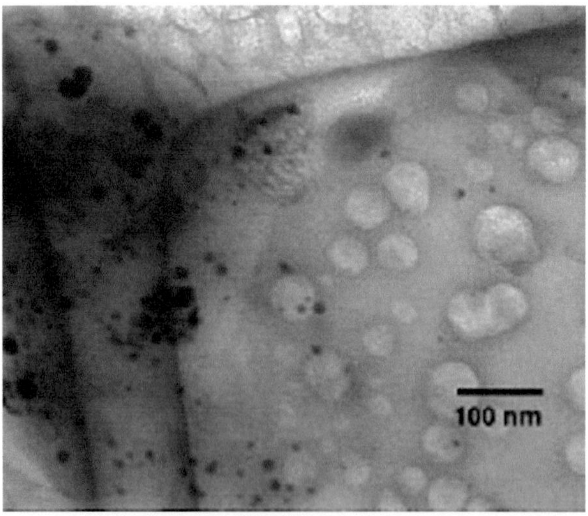

Fig. 3.36 TEM of selenium particles. Reproduced with permission from [153]

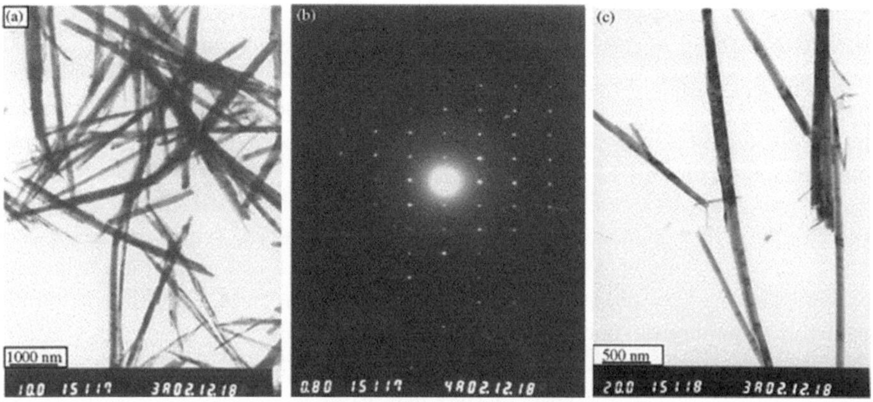

Fig. 3.37 Low (**a**) and high magnification (**c**) TEM images and SAED pattern (**b**) of the sample prepared with $NH_4S_2O_3$ as reducing agent. Reproduced with permission from [155]

mixture. Then the autoclave was sealed and maintained at 180 °C for 12 h. After the heating treatment, the autoclave was allowed to cool for about 6 h to room temperature naturally. The precipitates were filtered and washed with distilled water, absolute alcohol several times. After drying in vacuum at 50 °C for 4 h, black powders were collected for characterization. TEM and selected area electronic diffraction (SAED) of the NWs produced are shown in Fig. 3.37a, c, the t-Se nanowires (in 100 % morphological yield) existed in the sample.

The calculated mean diameter and length of these wires are 60 nm and 3 μm, respectively. The average aspect ratio of the nanowires is about 50, which is much smaller that of the sample prepared with $Na_2S_2O_3$ as a reducing agent. Figure 3.37b gives the SAED pattern of a single nanowire, exhibiting the single-crystalline nature of nanowires. The twofold, rotational symmetry associated with this pattern indicates that these t-Se nanowires had predominately grown along the [0 0 1] direction (*c*-axis), with the helical chains of Se atoms parallel to the longitudinal axis [156]. Because the diffraction patterns taken from various regions of an individual nanowire were essentially the same, it could be concluded that the nanowires synthesized using the present procedure were single crystalline in structure.

3.9 Germanium Antimony Telluride (GeSbTe): GST

The family of Ge-Sb-Te (GST) materials is one of the most important systems for phase change memory devices due to their rapid reversible change between crystalline and amorphous states at temperatures compatible with those of device applications (for $Ge_2Sb_2Te_5$, melting point 610 °C, glass transition temperature 350 °C) [157]. Germanium telluride and Sb_2Te_3 are the two important members of this family of materials and function as the basis to fabricate GST ternary alloys. They have melting points of 724 and 630 °C [158] and amorphous phase crystallization temperatures of 145 °C [159] and 77 °C [160] respectively.

Among the phase-change chalcogenide alloys, $Ge_2Sb_2Te_5$ (GST) is the most widely used material and exhibits the best data storage performance when used in nonvolatile resistive switching random access memory. GST exhibits excellent thermal stability, high crystallization rate at elevated temperatures (can be recrystallized in less than 50 ns by laser heating pulse), and extremely good durability (more than 10^{10} write/erase cycles). However, the large thermal threshold energy needed to generate structural phase transition in thin GST films is a bottleneck in further increasing the storage density. Recent studies on Ge-doped SbTe "line cell" fabricated by thin film deposition and photolithography showed nanosecond-level phase switch at very low power dissipation, suggesting that the nanowire configuration could be ideal for data storage devices.

Colloidal synthesis of GeTe nanowires has been reported by several research groups [161, 162]. The synthesis of GST nanowires as reported by Sun et al. was carried out [163] in a high-vacuum two-zone furnace system. In this setup, the source evaporation temperature high and the nanowire growth temperature low are separately controlled. The reactor tube was evacuated to a base pressure of 10^{-2} Torr prior to the growth experiment. A mixture of 5 N purity GeTe powder, and Sb_2Te_3 powder 99 with a mole ratio of 2:1 was placed in the high temperature zone. The SiO_2-coated Si 100 substrate, dispersed with gold nanoparticles 20 nm as catalyst, was loaded downstream in the low-temperature zone as the target substrate for GST NW growth. The argon carrier gas was introduced at a flow rate of 25 SCCM (SCCM denotes cubic centimeter per minute at STP and pressure of 200 Torr). The temperatures of the source material and the substrate were ramped up to 690 and 450 °C, respectively, and maintained during the 1 h growth. GST nanowires were dispersed onto the SiO_x-film-coated copper TEM grids for TEM imaging. The SEM image in Fig. 3.38 shows high-yield nanowire growth with NWs of 20–40 nm in diameter when 20 nm Au nanoparticles were applied as catalyst.

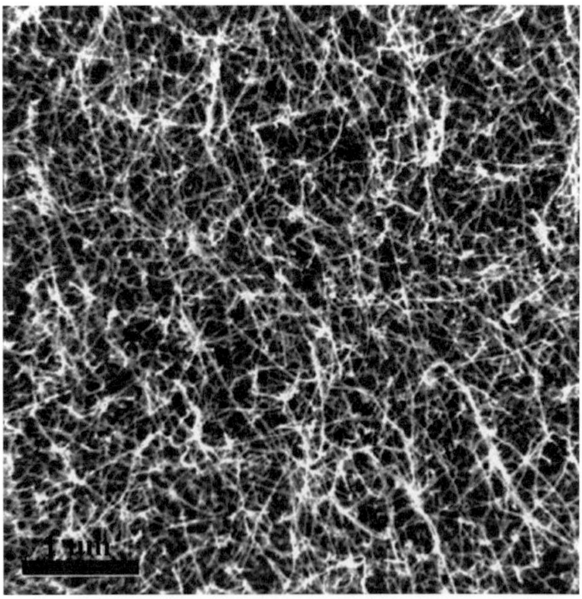

Fig. 3.38 SEM image of the as-synthesized GST nanowires grown on the surface of SiO_2 substrate. Scale bar is 1 mm. Reproduced with permission from [163]

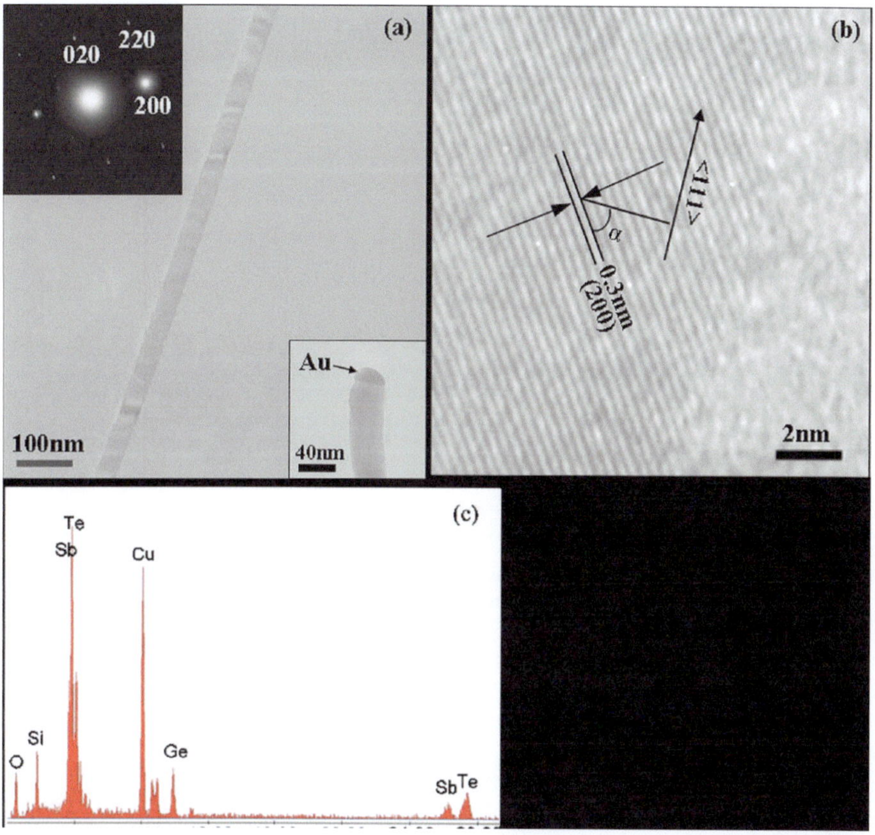

Fig. 3.39 1D Phase-change $Ge_2Sb_2Te_5$ nanowire. (**a**) TEM image of an individual NW grown on SiO_2 using 20 nm Au particle as catalyst. The NW is 36 nm in diameter and over 5 mm in length. *Top left inset*: SAED pattern of an fcc cubic structure. *Bottom right inset*: Au particle at the NW tip. (**b**) HRTEM image of the (200) planes of fcc GST with an interplanar distance of ~0.3 nm. The angle ($a = 54.7°$) in the image matches the one between the (200) and the $\langle 111 \rangle$ planes, indicating a preferential GST NW growth in the $\langle 111 \rangle$ orientation. (**c**) EDX spectrum of the $Ge_2Sb_2Te_5$ nanowire shown in (**a**). Reproduced with permission from [163]

A representative TEM image Fig. 3.39 shows that the GST nanowire has a smooth surface with a clearly noticeable metal particle on the tip bottom inset, Fig. 3.39a indicating that the growth process followed the VLS mechanism. Local EDX analysis of the individual nanowires discloses that Ge, Sb, and Te are present approximately in an atomic ratio of 2:2:5 (Fig. 3.39c). The Si, O, and Cu signals in the EDX spectrum are from the SiO_x-film-coated copper TEM grid. Both SAED image (top inset, Fig. 3.39a) and HRTEM image (Fig. 3.39b) confirm the single crystalline nature of the GST NWs. The SAED pattern shows regular spot patterns, which further confirms the perfect single crystalline nature of the NWs. The diffraction spots in the SAED pattern can be indexed to cubic fcc structure with a

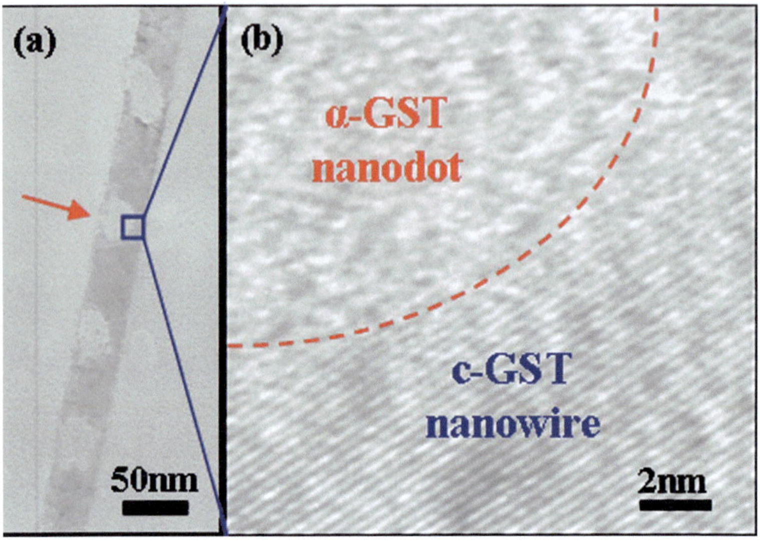

Fig. 3.40 Thermally induced data encoding observed on an individual 1D GST nanowire with scanning focused electron beam. (**a**) A series of α-GST nanodots was created by highly localized thermal heating with an e-beam spot of ~25 nm in diameter. (**b**) HRTEM image of the boundary between the α-GST nanodot and the as-synthesized c-GST nanowire. The amorphous-to-crystalline interface is marked by the *dash line*. Reproduced with permission from [163]

lattice constant *a* of ~6.0 Å, and the growth direction of the NW is along the $\langle 111 \rangle$ direction. The HRTEM image (Fig. 3.39b) shows the planes with a lattice spacing of ~3.0 Å that corresponds to *d* spacing of the (200) plane of GST fcc structure. The angle (=54.7°) in the image fits the one between the $\langle 200 \rangle$ plane and $\langle 111 \rangle$ plane, which indicates that the nanowire preferentially grows in the $\langle 111 \rangle$ direction.

Local *c*-to-a phase transition was observed in the GST nanowires at room temperature under local thermal stimulus. High-resolution TEM image (Fig. 3.40b) shows the clear interface between the amorphous dot and the original crystalline nanowire. The dot size and the interdot spacing are fully controllable by the beam geometry of the scanning focused electron beam of TEM used for the purpose and may be further reduced by using scanning probes [164, 165] such as scanning tunneling microscopy (STM) or atomic force microscopy (AFM).

3.10 Future Outlook and Summary

The shape-controlled morphology of semiconductor nanomaterials is the key to the advancements in electronic industry. New diverse morphologies as required by the devices can be obtained by employing new synthetic methods and strategies. Since crystalline phase is required to have ordered morphologies, and therefore, controlled nucleation under carefully designed reaction protocols is an important aspect to be implemented. This chapter sheds light on some of the important methods of synthesis for designing different shapes and sizes of semiconductor nanocrystals of different types required for different applications.

References

1. R. Rossetti, S. Nakahara, L.E. Brus, Quantum size effects in the redox potentials, resonance Raman spectra, and electronic spectra of CdS crystallites in aqueous solution. J. Chem. Phys. **79**(2), 1086–1088 (1983)
2. W.W. Yu, X. Peng, Formation of high-quality CdS and other II-VI semiconductor nanocrystals in noncoordinating solvents: tunable reactivity of monomers. Angew. Chem. Int. Ed. Engl. **41**(13), 2368–2371 (2002)
3. X. Chen, A.Y. Nazzal, M. Xiao, Z.A. Peng, X. Peng, Photoluminescence from single CdSe quantum rods. J. Lumin. **97**(3–4), 205–211 (2002)
4. L. Qu, Z.A. Peng, X. Peng, Alternative routes toward high quality CdSe nanocrystals. Nano Lett. **1**(6), 333–337 (2001)
5. L. Qu, W.W. Yu, X. Peng, In situ observation of the nucleation and growth of CdSe nanocrystals. Nano Lett. **4**(3), 465–469 (2004)
6. X. Wang, L. Qu, J. Zhang, X. Peng, M. Xiao, Surface-related emission in highly luminescent CdSe quantum dots. Nano Lett. **3**(8), 1103–1106 (2003)
7. B.K.H. Yen, N.E. Stott, K.F. Jensen, M.G. Bawendi, A continuous-flow microcapillary reactor for the preparation of a size series of CdSe nanocrystals. Adv. Mater. (Deerfield Beach, FL) **15**(21), 1858–1862 (2003)
8. G. Morello, M. De Giorgi, S. Kudera, L. Manna, R. Cingolani, M. Anni, Temperature and size dependence of nonradiative relaxation and exciton–phonon coupling in colloidal CdTe quantum dots. J. Phys. Chem. C **111**(16), 5846–5849 (2007)
9. Z.A. Peng, X. Peng, Formation of high-quality CdTe, CdSe, and CdS nanocrystals using CdO as precursor. J. Am. Chem. Soc. **123**(1), 183–184 (2001)
10. M.A. Hines, P. Guyot-Sionnest, Bright UV-blue luminescent colloidal ZnSe nanocrystals. J. Phys. Chem. B **102**(19), 3655–3657 (1998)
11. M. Shim, P. Guyot-Sionnest, Organic-capped ZnO nanocrystals: synthesis and n-type character. J. Am. Chem. Soc. **123**(47), 11651–11654 (2001)
12. D. Battaglia, X. Peng, Formation of high quality InP and InAs nanocrystals in a noncoordinating solvent. Nano Lett. **2**(9), 1027–1030 (2002)
13. O.I. Micic, C.J. Curtis, K.M. Jones, J.R. Sprague, A.J. Nozik, Synthesis and characterization of InP quantum dots. J. Phys. Chem. **98**(19), 4966–4969 (1994)
14. J. Xu, J.-P. Ge, Y.-D. Li, Solvothermal synthesis of monodisperse PbSe nanocrystals. J. Phys. Chem. B **110**(6), 2497–2501 (2006)
15. W.W. Yu, J.C. Falkner, B.S. Shih, V.L. Colvin, Preparation and characterization of monodisperse PbSe semiconductor nanocrystals in a noncoordinating solvent. Chem. Mater. **16** (17), 3318–3322 (2004)
16. K.A. Abel, J. Shan, J.-C. Boyer, F. Harris, F.C.J.M. van Veggel, Highly photoluminescent PbS nanocrystals: the beneficial effect of trioctylphosphine. Chem. Mater. **20**(12), 3794–3796 (2008)
17. L. Cademartiri, J. Bertolotti, R. Sapienza, D.S. Wiersma, G. von Freymann, G.A. Ozin, Multigram scale, solventless, and diffusion-controlled route to highly monodisperse PbS nanocrystals. J. Phys. Chem. B **110**(2), 671–673 (2006)

18. M.A. Hines, G.D. Scholes, Colloidal PbS nanocrystals with size-tunable near-infrared emission: observation of post-synthesis self-narrowing of the particle size distribution. Adv. Mater. (Deerfield Beach, FL) **15**(21), 1844–1849 (2003)
19. S.-M. Lee, Y.W. Jun, S.-N. Cho, J. Cheon, Single-crystalline star-shaped nanocrystals and their evolution: programming the geometry of nano-building blocks. J. Am. Chem. Soc. **124** (38), 11244–11245 (2002)
20. W. Lin, K. Fritz, G. Guerin, G.R. Bardajee, S. Hinds, V. Sukhovatkin, E.H. Sargent, G.D. Scholes, M.A. Winnik, Highly luminescent lead sulfide nanocrystals in organic solvents and water through ligand exchange with poly(acrylic acid). Langmuir **24**(15), 8215–8219 (2008)
21. D.S. English, L.E. Pell, Z. Yu, P.F. Barbara, B.A. Korgel, Size tunable visible luminescence from individual organic monolayer stabilized silicon nanocrystal quantum dots. Nano Lett. **2** (7), 681–685 (2002)
22. J.D. Holmes, K.J. Ziegler, R.C. Doty, L.E. Pell, K.P. Johnston, B.A. Korgel, Highly luminescent silicon nanocrystals with discrete optical transitions. J. Am. Chem. Soc. **123**(16), 3743–3748 (2001)
23. Z. Kang, Y. Liu, C.H.A. Tsang, D.D.D. Ma, X. Fan, N.-B. Wong, S.-T. Lee, Water-soluble silicon quantum dots with wavelength-tunable photoluminescence. Adv. Mater. (Deerfield Beach, FL) **21**(6), 661–664 (2009)
24. J. Zou, R.K. Baldwin, K.A. Pettigrew, S.M. Kauzlarich, Solution synthesis of ultrastable luminescent siloxane-coated silicon nanoparticles. Nano Lett. **4**(7), 1181–1186 (2004)
25. J.S. Steckel, S. Coe-Sullivan, V. Bulvic, M.B. Bawendi, 1.3 mm to 1.55 mm tunable electroluminescence from PbSe quantum dots embedded within an organic device. Adv. Mater. **15**, 1862 (2003)
26. L. Bakueva, S. Musikhin, M.A. Hines, T.W.F. Chang, M. Tzolov, G.D. Scholes, E.H. Sargent, Size-tunable infrared (1000–1600 nm) electroluminescence from PbS quantum dot nanocrystals in a semiconducting polymer. Appl. Phys. Lett. **82**, 2895 (2003)
27. L. Bakueva, I. Gorelikov, S. Musikhin, X.S. Zhao, E.H. Sargent, E. Kumacheva, One-stage aqueous synthesis of tunable 800-1300 nm quantum dots with stable efficient luminescence. Adv. Mater. **16**, 926 (2004)
28. R.D. Schaller, M.A. Petruska, V.I. Klimov, Tunable near-infrared optical gain and amplified spontaneous emission using PbSe nanocrystals. J. Phys. Chem. **107**, 13765 (2003)
29. R.D. Schaller, V.I. Klimov, High efficiency carrier multiplication in PbSe nanocrystals: implications for solar energy conversion. Phys. Rev. Lett. **92**, 186601 (2004)
30. R.J. Ellingson, M.C. Beard, J.C. Johnson, P. Yu, O.I. Micic, A.J. Nozic, A. Shabaev, A.L. Efros, Nano Lett. **5**, 865 (2005)
31. (a) S. Wang, S. Yang, Langmuir **16**, 389 (2000); (b) S.M. Lee, Y.W. Jun, S.N. Cho, J.W. Cheon, J. Am. Chem. Soc. **124**, 11244 (2002); (c) E. Leontidis, M. Orphanou, T. -Kyprianidou-Leondidou, F. Krumeich, W. Caseri, Nano Lett. **3**, 569 (2003); (d) H. Dai, E.W. Wong, Y.Z. Lu, S. Fan, C.M. Lieber, Nature **375**, 769 (1995); (e) D. Kuang, A. Xu, Y. Fang, H. Liu, C. Frommen, D. Fenske, Adv. Mater. **15**, 1747 (2003); (f) Y. Ma, L. Qi, J. Ma, H. Cheng, J. Cryst. Growth Des. **4**, 351 (2004); (g) Y. Ni, H. Liu, F. Wang, Y. Liang, J. Hong, X. Ma, Z. Xu, J. Cryst. Growth Des. **4**, 759 (2004)
32. (a) R. Rossetti, R. Hull, J.M. Gibson, L.E. Brus, J. Chem. Phys. **83**, 1406 (1985); (b) P.S. Khiew, S. Radiman, N.M. Huang, Md. S. Ahmad, J. Cryst. Growth. **254**, 235 (2003); (c) D. Wang, D. Yu, M. Shao, X. Liu, W. Yu, Y. Qian, Cryst. Growth Des. **257**, 384 (2003); (d) Y. Ni, H. Liu, F. Wang, Y. Liang, J. Hong, X. Ma, Z. Xu, Cryst. Res. Technol. **39**, 200 (2004); (e) A.V. Baranov, E.V. Ushakova, V.V. Golubkov, A.P. Litvin, P.S. Parfenov, A.V. Fedorov, K. Berwick, Self-organization of colloidal PbS quantum dots into highly ordered superlattices, Langmuir **31**(1), 506–513 (2015)
33. S.-H. Yu, J. Yang, Y.-S. Wu, Z.-H. Han, J. Lu, Y. Xie, Y.-T. Qiana, J. Mater. Chem. **8**, 1949 (1998)
34. Y. Zhang, Z.-P. Qiao, X.-M. Chen, J. Mater. Chem. **12**, 2747 (2002)

35. (a) A.A. Korzhuev, Fiz. Chim. Obrab. Mater. **3**, 131 (1991); (b) H. Toyoji, Y. Hirohsi, Jpn. Kokai Tokkyo Koho: JP 02 **173**, 622 (1990)
36. (a) S.T. Lakshmikvar, Sol. Energ. Mater. Sol. Cells **32**, 7 (1994); (b) W. Wang, P. Yan, F. Liu, Y. Xie, Y.Geng, Y. Qian, J. Mater. Chem. **8**, 2321 (1998)
37. T. Ohtani, M. Motoki, Mater. Res. Bull. **30**, 195 (1995)
38. G. Henshaw, I.P. Parkin, G. Shaw, Chem. Commun. 1095 (1996)
39. G. Henshaw, I.P. Parkin, G. Shaw, J. Chem. Soc.: Dalton Trans. 231 (1997)
40. (a) X. Wang, Y. Li, Chem. Commun. 2901 (2007); (b) H. Fan, Chem. Commun. 1383 (2008)
41. (a) A.H. Latham, M.E. Williams, Acc. Chem. Res. 41, 411–420 (2008); (b) J.O. Joswig, M. Springborg, G. Seifert, J. Phys. Chem. B **104**, 2617–2622 (2000)
42. D.V. Talapin, H. Yu, E.V. Shevchenko, A. Lobo, C.B. Murray, J. Phys. Chem. C **111**, 14049–14054 (2007)
43. Y. Xie, X. Zheng, X. Jiang, J. Lu, L. Zhu, Inorg. Chem. **41**(2), 387–392 (2002)
44. M.S. Bakshi, P. Thakur, S. Sachar, G. Kaur, T.S. Banipal, F. Possmayer, N.O. Petersen, J. Phys. Chem. C **111**, 18087 (2007)
45. (a) P. Schuetz, F. Caruso, Chem. Mater. **16**, 3066 (2004); (b) H. Yang, P.H. Holloway, J. Phys. Chem. B **107**, 9705 (2003)
46. A. Roucoux, J. Schulz, H. Patin, Chem. Rev. **102**, 3757 (2002)
47. (a) J. Bai, Y. Qin, C. Jiang, L. Qi, Chem. Mater. **19**, 3367 (2007); (b) B.J. Wiley, Y. Chen, J.M. McLellan, Y. Xiong, Z.-Y. Li, D. Ginger, Y. Xia, Nano Lett. **7**, 1032 (2007)
48. R.A. Zingaro, W.C. Cooper (eds.), *Selenium* (Litton Educational Publishing, New York, 1974), p. 25. and 174–217
49. L.I. Berger, *Semiconductor Materials* (CRC Press, Boca Raton, FL, 1997), p. 86
50. Q. Lu, F. Gao, S. Komarneni, Chem. Mater. **18**, 159 (2005)
51. B. Gates, Y. Yin, Y. Xia, J. Am. Chem. Soc. **122**, 12582 (2000)
52. S. Xiong, B. Xi, W. Wang, C. Wang, L. Fei, H. Zhou, Y. Qian, Cryst. Growth Des. **6**, 1711 (2006)
53. Y. Ma, L. Qi, W. Shen, J. Ma, Langmuir **21**, 6161 (2005)
54. Z. Wang, L. Wang, H. Wang, Cryst. Growth Des. **8**, 4415 (2008)
55. Q. Lu, F. Gao, S. Komarneni, Langmuir **21**, 6002 (2005)
56. J. Hu, T.W. Odom, C.M. Lieber, Acc. Chem. Res. **32**, 435 (1999)
57. R.I. Baitser, V.V. Vainberg, S.S. Varshava, J. Phys. IV **6**, C3–429 (1996)
58. M.S. Bakshi, V.S. Jaswal, G. Kaur, T.W. Simpson, P.K. Banipal, T.S. Banipal, F. Possmayer, N.O. Petersen, J. Phys. Chem. C **113**, 9121 (2009)
59. B. Mayers, G. Byron, Y. Yin, Y. Xia, Adv. Mater. **13**, 1380 (2001)
60. T.W. Smith, S.D. Smith, S.S. Badesha, J. Am. Chem. Soc. **106**, 7247 (1984)
61. X. Wu, H. Liu, J. Liu, K.N. Haley, J.A. Treadway, J.P. Larson, N. Ge, F. Peale, M.P. Bruchez, Nat. Biotechnol. **21**, 41–46 (2003)
62. A.P. Alivisatos, Science **271**, 933–937 (1996)
63. P. Alivisatos, Nat. Biotechnol. **22**, 47–52 (2004)
64. W.C.W. Chan, S.M. Nie, Science **281**, 2016–2018 (1998)
65. R. Gill, L. Bahshi, R. Freeman, I. Willne, Angew. Chem. **57**, 1676–1679 (2008)
66. R. C. Mulrooney, N. Singh, N. Kaur, J. F. Callan, Chem. Commun. 686–688 (2009)
67. C.Y. Chen, C.T. Cheng, C.W. Lai, P.W. Wu, K.C. Wu, P.T. Chou, Y.H. Chou, H.T. Chiu, Chem. Commun. 263–265 (2006)
68. M.S. Bakshi, P. Thakur, P. Khullar, G. Kaur, T.S. Banipal, Cryst. Growth Des. **10**, 1813–1822 (2010)
69. G. Kaur, M.S. Bakshi, Non-ideal mixing of Se–Te in aqueous micellar phase for nano-alloys over the whole mole mixing range with morphology control from nanoparticles to nanoribbons. J Phys. Chem. C **114**, 143–154 (2010)
70. (a) N. Zhao, L. Qi, Adv. Mater. **18**, 359 (2006); (b) G. Zhou, M. Lu, Z. Xiu, S. Wang, H. Zhang, Y. Zhou, S. Wang, J. Phy. Chem. **110**, 6543 (2006)

71. (a) M.S. Bakshi, G. Kaur, P. Thakur, T.S. Banipal, F. Possmayer, N.O. Petersen, J. Phys. Chem. C **111**, 5932 (2007); (b) M.S. Bakshi, F. Possmayer, N.O. Petersen, Chem. Mater. **19**, 1257 (2007); (c) M.S. Bakshi, A. Kaura, P. Bhandari, G. Kaur, K. Torigoe, K. Esumi, J. Nanosci. Nanotechnol. **6**, 1405 (2006); (d) M.S. Bakshi, P. Sharma, T.S. Banipal, G. Kaur, K. Torigoe, N.O. Petersen, F. Possmayer, J. Nanosci. Nanotechnol. **7**, 916 (2007)
72. N. Wang, X. Cao, G. Lin, S. Yang, Z. Wu, Facile synthesis of PbS truncated octahedron crystals with high symmetry and their large-scale assembly into regular patterns by a simple solution route. ACS Nano **2**, 184–190 (2008)
73. I. Patla, S. Acharya, L. Zeiri, J. Israelachvili, S. Efrima, Y. Golan, Synthesis, two-dimensional assembly, and surface pressure-induced coalescence of ultranarrow PbS nanowires. Nano Lett. **7**(6), 1459–1462 (2007)
74. M.S. Bakshi, G. Kaur, F. Possmayer, N.O. Petersen, Shape controlled synthesis of PSS and PVP capped lead sulfide nano-cubes, bars, and threads. J. Phys. Chem. C **112**, 4948–4953 (2008)
75. (a) S. Wang., S. Yang, Langmuir **16**, 389 (2000); (b) A.A. Patel, F. Wu, J.Z. Zhang, C.L. Torres-Martinez, R.K. Mehra, Y. Yang, S.H. Risbud, J. Phys. Chem. B **104**, 11598 (2000)
76. S. Wang, S. Yang, Langmuir **16**, 389 (2000)
77. R. Gill, M. Zayats, I. Willner, Semiconductor quantum dots for bioanalysis. Angew. Chem. Int. Ed. **47**, 7602–7625 (2008)
78. P. Zrazhevskiy, M. Sena, X.H. Gao, Designing multifunctional quantum dots for bioimaging, detection, and drug delivery. Chem. Soc. Rev. **39**, 4326–4354 (2010)
79. N. Hildebrandt, Biofunctional quantum dots: controlled conjugation for multiplexed biosensors. ACS Nano **5**, 5286–5290 (2011)
80. L. Cademartiri, E. Montanari, G. Calestani, A. Migliori, A. Guagliardi, G.A. Ozin, Size-dependent extinction coefficients of PbS quantum dots. J. Am. Chem. Soc. **128**, 10337–10346 (2006)
81. T. Imae, S. Ikeda, J. Phys. Chem. **90**, 5216 (1986)
82. W. Lu, J. Fang, Y. Ding, Z.L. Wang, Formation of PbSe nanocrystals: a growth toward nanocubes. J. Phys. Chem. B **109**, 19219–19222 (2005)
83. K.-S. Cho, D.V. Talapin, W. Gaschler, C.B. Murray, Designing PbSe nanowires and nanorings through oriented attachment of nanoparticles. J. Am. Chem. Soc. **127**, 7140–7147 (2005)
84. Y. Cui, Z. Zhong, D. Wang, W.U. Wang, C.M. Lieber, Nano Lett. **3**, 149–152 (2003)
85. X. Duan, C. Niu, V. Sahi, J. Chen, J.W. Parce, S. Empedocles, J.L. Goldman, Nature **425**, 274–278 (2003)
86. Y. Huang, X. Duan, Y. Cui, L.J. Lauhon, K.-H. Kim, C.M. Lieber, Science **294**, 1313–1317 (2001)
87. X. Duan, Y. Huang, C.M. Leiber, Nano Lett. **2**, 487–490 (2002)
88. F. Patolsky, G. Zheng, O. Hayden, M. Lakadamyali, X. Zhuang, C.M. Lieber, Proc. Natl. Acad. Sci. U. S. A. **101**, 14017–14022 (2004)
89. J.-I. Hahm, C.M. Lieber, Nano Lett. **4**, 51–54 (2004)
90. W.U. Huynh, J.J. Dittmer, A.P. Alivisatos, Science **295**, 2425–2427 (2002)
91. D. Yanover, R.K. Capek, A. Rubin-Brusilovski, R. Vaxenburg, N. Grumbach, G.I. Maikov, O. Solomeshch, A. Sashchiuk, E. Lifshitz, Small-sized PbSe/PbS core/shell colloidal quantum dots. Chem. Mater. **24**, 4417–4423 (2012)
92. G. Zaiats, A. Shapiro, D. Yanover, Y. Kauffmann, A. Sashchiuk, E. Lifshitz, Optical and electronic properties of nonconcentric PbSe/cdse colloidal quantum dots. J. Phys. Chem. Lett. **6**, 2444–2448 (2015)
93. B. De Geyter, Y. Justo, I. Moreels, K. Lambert, P.F. Smet, D. Van Thourhout, A.J. Houtepen, D. Grodzinska, C. de Mello Donega, A. Meijerink, D. Vanmaekelbergh, Z. Hens, The different nature of band edge absorption and emission in colloidal PbSe/CdSe core/shell quantum dots. ACS Nano **5**, 58–66 (2010)

94. K.A. Abel, H. Qiao, J.F. Young, F.C.J.M. van Veggel, Four-fold enhancement of the activation energy for nonradiative decay of excitons in PbSe/CdSe core/shell versus PbSe colloidal quantum dots. J. Phys. Chem. Lett. **1**, 2334–2338 (2010)
95. Z.A. Peng, X. Peng, J. Am. Chem. Soc. **123**, 1389 (2001)
96. Y. Jun, S. Lee, N. Kang, J. Cheon, J. Am. Chem. Soc. **123**, 5150 (2001)
97. Y. Cheng, Y. Wang, F. Bao, D. Chen, Shape control of monodisperse CdS nanocrystals: hexagon and pyramid. J. Phys. Chem. B **110**(19), 9448–9451 (2006)
98. J.H. Warner, R.D. Tilley, Adv. Mater. **17**, 2997 (2005)
99. Y. Zou, D.S. Li, D.R. Yang, Shape and phase control of CdS nanocrystals using cationic surfactant in noninjection synthesis. Nanoscale Res. Lett. **6**, 374 (2011)
100. S.J. Lim, W. Kim, S. Jung, J. Seo, S.K. Shin, Anisotropic etching of semiconductor nanocrystals. Chem. Mater. **23**, 5029–5036 (2011)
101. S.J. Lim, W. Kim, S.K. Shin, Surface-dependent, ligand-mediated photochemical etching of CdSe nanoplatelets. J. Am. Chem. Soc. **134**, 7576–7579 (2012)
102. M. Saruyama, M. Kanehara, T. Teranishi, Drastic structural transformation of cadmium chalcogenide nanoparticles using chloride ions and surfactants. J. Am. Chem. Soc. **132**, 3280–3282 (2010)
103. M.R. Kim, K. Miszta, M. Povia, R. Brescia, S. Christodoulou, M. Prato, S. Marras, L. Manna, Influence of chloride ions on the synthesis of colloidal branched CdSe/CdS nanocrystals by seeded growth. ACS Nano **6**(12), 11088–11096 (2012)
104. K. Miszta, J. de Graaf, G. Bertoni, D. Dorfs, R. Brescia, S. Marras, L. Ceseracciu, R. Cingolani, R. van Roij, M. Dijkstra et al., Hierarchical self-assembly of suspended branched colloidal nanocrystals into superlattice structures. Nat. Mater. **10**, 872–876 (2011)
105. W. Qi, J. de Graaf, F. Qiao, S. Marras, L. Manna, M. Dijkstra, Ordered two-dimensional superstructures of colloidal octapod-shaped nanocrystals on flat substrates. Nano Lett. **12**, 5299–5303 (2012)
106. A.L. Schmitt, J.M. Higgins, J.R. Szczech, S. Jin, Synthesis and applications of metal silicide nanowires. J. Mater. Chem. **20**, 223–235 (2010)
107. A.I. Hochbaum, P. Yang, Semiconductor nanowires for energy conversion. Chem. Rev. **110**, 527–546 (2009)
108. R. Yan, D. Gargas, P. Yang, Nanowire photonics. Nat. Photon. **3**, 569–576 (2009)
109. B. Tian, T.J. Kempa, C.M. Lieber, Single nanowire photovoltaics. Chem. Soc. Rev. **38**, 16–24 (2009)
110. W. Lu, C.M. Lieber, Nanoelectronics from the bottom up. Nat. Mater. **6**, 841–850 (2007)
111. M.J. Bierman, Y.K.A. Lau, A.V. Kvit, A.L. Schmitt, S. Jin, Dislocation-driven nanowire growth and Eshelby twist. Science **320**, 1060–1063 (2008)
112. S.A. Morin, M.J. Bierman, J. Tong, S. Jin, Mechanism and kinetics of spontaneous nanotube growth driven by screw dislocations. Science **328**, 476–480 (2010)
113. H. Wu, F. Meng, L. Li, S. Jin, G. Zheng, Dislocation-driven CdS and CdSe nanowire growth. ACS Nano **6**(5), 4461–4468 (2012)
114. J. Chikawa, T. Nakayama, Dislocation structure and growth mechanism of cadmium sulfide crystals. J. Appl. Phys. **35**, 2493–2501 (1964)
115. D.J. Milliron, S.M. Hughes, Y. Cui, L. Manna, J. Li, L.-W. Wang, A. Paul Alivisatos, Colloidal nanocrystal heterostructures with linear and branched topology. Nature **430**, 190–195 (2004)
116. Z.A. Peng, X. Peng, Nearly monodisperse and shape-controlled CdSe nanocrystals via alternative routes: nucleation and growth. J. Am. Chem. Soc. **124**, 3343–3353 (2002)
117. L. Manna, D.J. Milliron, A. Meisel, E.C. Scher, A.P. Alivisatos, Controlled growth of tetrapod-branched inorganic nanocrystals. Nat. Mater. **2**, 382–385 (2003)
118. C.Y. Yeh, Z.W. Lu, S. Froyen, A. Zunger, Zincblende–wurtzite polytypism in semiconductors. Phys. Rev. B **46**, 10086–10097 (1992)
119. G.S. Hammond, A correlation of reaction rates. J. Am. Chem. Soc. **77**, 334–338 (1955)

120. T. Purnima, A. Ruberu, H.R. Albright, B. Callis, B. Ward, J. Cisneros, H.-J. Fan, J. Vela, Molecular control of the nanoscale: effect of phosphine–chalcogenide reactivity on CdS–CdSe nanocrystal composition and morphology. ACS Nano **6**(6), 5348–5359 (2012)
121. X. Peng, M.C. Schlamp, A. Kadavanich, A.P. Alivisatos, J. Am. Chem. Soc. **119**, 7019 (1997)
122. D.V. Talapin, J.H. Nelson, E.V. Shevchenko, S. Aloni, B. Sadtler, P. Alivisatos, Seeded growth of highly luminescent CdSe/CdS nanoheterostructures with rod and tetrapod morphologies. Nano Lett. **7**(10), 2951–2959 (2007)
123. L. Carbone, C. Nobile, M. De Giorgi, F.D. Sala, G. Morello, P. Pompa, M. Hytch, E. Snoeck, A. Fiore, I.R. Franchini, M. Nadasan, A.F. Silvestre, L. Chiodo, S. Kudera, R. Cingolani, R. Krahne, L. Manna, Synthesis and micrometer-scale assembly of colloidal CdSe/CdS nanorods prepared by a seeded growth approach. Nano Lett. **7**, 2942–2950 (2007)
124. K. An, N. Lee, J. Park, S.C. Kim, Y. Hwang, J.G. Park, J.-Y. Kim, J.-H. Park, M.J. Han, J. Yu, T. Hyeon, J. Am. Chem. Soc. **128**, 9753–9760 (2006)
125. K.M. Ryan, A. Mastroianni, K.A. Stancil, H.T. Liu, A.P. Alivisatos, Nano Lett. **6**, 1479–1482 (2006)
126. S. Gupta, Q. Zhang, T. Emrick, T.P. Russell, Nano Lett. **6**, 2066–2069 (2006)
127. Z. Hu, M.D. Fischbein, C. Querner, M. Drndić, Nano Lett. **6**, 2585–2591 (2006)
128. C. Nobile, V.A. Fonoberov, S. Kudera, A. Della Torre, A. Ruffino, G. Chilla, T. Kipp, D. Heitmann, L. Manna, R. Cingolani, A.A. Balandin, R. Krahne, Nano Lett. **7**, 476–479 (2007)
129. D.C. Look, Mater. Sci. Eng. B **80**, 383–387 (2001)
130. Z.L. Wang, J. Phys. Condens. Mater. **16**, 829–858 (2004)
131. U. Ozgur, Y.I. Alivov, C. Liu, A. Teke, M.A. Reshchikov, S. Dogan, V. Avrutin, S.J. Cho, H. Morkoc, J. Appl. Phys. **98**, 103 (2005)
132. M. Gratzel, Nature **414**, 338–344 (2001)
133. S.C. Pillai, J.M. Kelly, D.E. McCormack, P. O'Brien, R. Ramesh, J. Mater. Chem. **13**, 2586–2590 (2003)
134. T. Adschiri, K. Kanazawa, K. Arai, J. Am. Ceram. Soc. **75**, 1019–1022 (1992)
135. P. Hald, J. Becker, M. Bremholm, J.S. Pedersen, J. Chevallier, S.B. Iversen, B.B. Iversen, J. Solid State Chem. **179**, 2674–2680 (2006)
136. M. Sondergaard, E.D. Bojesen, M. Christensen, B.B. Iversen, Size and morphology dependence of ZnO nanoparticles synthesized by a fast continuous flow hydrothermal method. Cryst. Growth Des. **11**, 4027–4033 (2011)
137. M. Kahlweit, Adv. Colloid Interface Sci. **5**, 1–35 (1975)
138. H.Q. Wang, A. Pyatenko, K. Kawaguchi, X. Li, Z. Swiatkowska-Warkocka, N. Koshizaki, Selective pulsed heating for the synthesis of semiconductor and metal submicrometer spheres. Angew. Chem. Int. Ed. **49**, 6361–6364 (2010)
139. H. Wang, A. Pyatenko, N. Koshizaki, H. Moehwald, D. Shchukin, Single-crystalline ZnO spherical particles by pulsed laser irradiation of colloidal nanoparticles for ultraviolet photodetection. ACS Appl. Mater. Interfaces **6**, 2241–2247 (2014)
140. M.H. Huang, S. Mao, H. Feick, H.Q. Yan, Y.Y. Wu, H. Kind, E. Weber, R. Russo, P.D. Yang, Science **292**, 1897 (2001)
141. C.H. Liu, J.A. Zapien, Y. Yao, Z.M. Meng, C.S. Lee, S.S. Fan, Y. Lifshitz, S.T. Lee, Adv. Mater. (Weinheim, Ger.) **15**, 838 (2003)
142. S. Das, J. Kar, J. Choi, T. Lee, K. Moon, J. Myoung, J. Phys. Chem. C **114**, 1689 (2010)
143. Q. Wan, Q.H. Li, Y.J. Chen, T.H. Wang, X.L. He, J.P. Li, C.L. Lin, Appl. Phys. Lett. **84**, 3654 (2004)
144. Y. Zeng, T. Zhang, L. Wang, R. Wang, W. Fu, H. Yang, J. Phys. Chem. C **113**, 3442 (2010)
145. K. Liu, M. Sakurai, M. Liao, M. Aono, Giant improvement of the performance of ZnO nanowire photodetectors by Au nanoparticles. J. Phys. Chem. C **114**, 19835–19839 (2010)
146. W. Yang, R.D. Vispute, S. Choopun, R.P. Sharma, T. Venkatesan, H. Shen, Appl. Phys. Lett. **78**, 2787 (2001)
147. C. Falcony, M. Garcia, A. Ortiz, J.C. Alonso, J. Appl. Phys. **72**, 1525 (1992)

148. Y. Zhang, F. Lu, Z. Wang, H. Wang, M. Kong, X. Zhu, L. Zhang, ZnS nanoparticle-assisted synthesis and optical properties of ZnS nanotowers. Cryst. Growth Des. **7**(8), 1459–1462 (2007)
149. G. Yang, H. Zhong, R. Liu, Y. Li, B. Zou, In situ aggregation of ZnSe nanoparticles into supraparticles: shape control and doping effects. Langmuir **29**, 1970–1976 (2013)
150. N.M. Dimitrijevic, P.V. Kamat, Langmuir **4**, 782 (1988)
151. P. Armand, M.L. Saboungi, D.L. Price, L. Iton, C. Cramer, M. Grimsditch, Phys. Rev. Lett. **79**(11), 2061 (1997)
152. A. Goldbach, L. Iton, M. Grimsditch, M.L. Saboungi, J. Am. Chem. Soc. **118**(8), 2004 (1996)
153. J.A. Johnson, M.-L. Saboungi, P. Thiyagarajan, R. Csencsits, D. Meisel, Selenium nanoparticles: a small-angle neutron scattering study. J. Phys. Chem. B **103**, 59–63 (1999)
154. B. Mayers, B. Gates, Y. Xia, One dimensional nanostructures of chalcogens and chalcogenides. Int. J. Nanotechnol. **1**(1/2), 86–104 (2004)
155. C. Ana, S. Wang, Diameter-selected synthesis of single crystalline trigonal selenium nanowires. Mater. Chem. Phys. **101**, 357–361 (2007)
156. J. Stuke, in *Selenium*, ed. by R.A. Zingaro, W.C. Cooper (Van Nostrand Reinhold, New York, 1994)
157. S. Hudgens, B. Johnson, MRS Bull. **11**, 829 (2004)
158. A. Prince, G.V. Raynor, D.S. Evans, *Phase Diagrams of Ternary Gold Alloys* (Institute of Metals, Brookfield, VT, 1990)
159. N. Yamada, E. Ohno, K. Nishiuchi, N. Akahira, J. Appl. Phys. **69**, 2849 (1991)
160. V. Damodara Das, N. Soundararajan, M. Pattabi, J. Mater. Sci. **22**, 3522 (1987)
161. M.A. Caldwell, S. Raoux, R.Y. Wang, H.-S. Philip Wong, D.J. Milliron, Synthesis and size-dependent crystallization of colloidal germanium telluride nanoparticles. J. Mater. Chem. **20**, 1285–1291 (2010)
162. S. Meister, H. Peng, K. McIlwrath, K. Jarausch, X.F. Zhang, Y. Cui, Synthesis and characterization of phase-change nanowires. Nano Lett. **6**(7), 1514–1517 (2006)
163. X. Sun, B. Yu, M. Meyyappan, Synthesis and nanoscale thermal encoding of phase-change nanowires. Appl. Phys. Lett. **90**, 183116 (2007)
164. H. Satoh, K. Sugawara, K. Tanaka, J. Appl. Phys. **99**, 024306 (2006)
165. H.F. Hamann, M. O'Boyle, Y.C. Martin, M. Rooks, K. Wickramasinghe, Nat. Mater. **5**, 383 (2006)

Part II
Sulfur

Chapter 4
Optical Fibers

Gurinder Kaur Ahluwalia

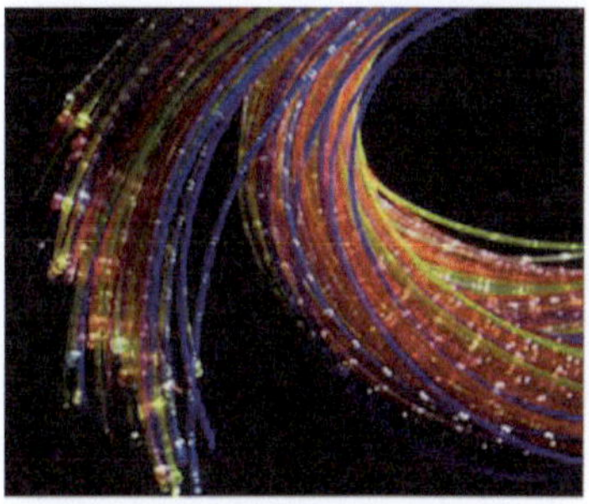

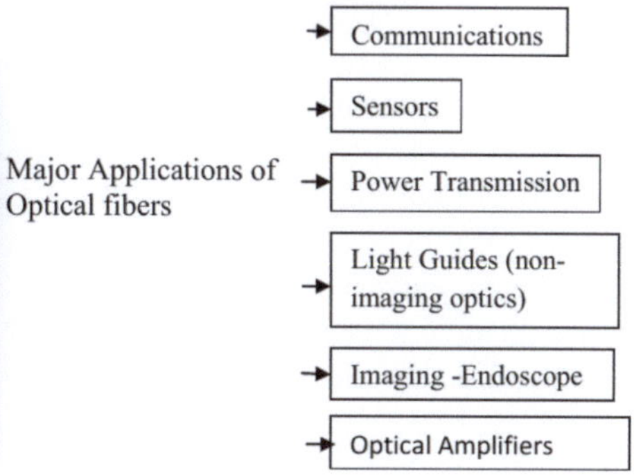

Compact, diode-pumped Pr^{3+}-doped chalcogenide fiber source with broadband emission in the 3–5 μm wavelength region

G.K. Ahluwalia, Ph.D (✉)
Materials and Nanotechnology Research Laboratory, Department of Physics College of The North Atlantic, Labrador West Campus, 1600 Nichols Adam Highway, Labrador City, A2V 0B8, NL, Canada
e-mail: gurinder.kaur@cna.nl.ca

G.K. Ahluwalia (ed.), *Applications of Chalcogenides: S, Se, and Te*,
DOI 10.1007/978-3-319-41190-3_4

4.1 Introduction

Applications of low loss optical fibers as waveguides for light propagation include a wide spectrum such as telecommunications, fiber imaging bundles, fiber gyro, fiber amplifiers, fiber resonators, delay lines, and fiber couplers. Optical methods based on the use of nonlinearity in materials are expected to create large storage devices with convenient signal processing capabilities. These include: information communication, information recording, display, I/O equipment, power transmission, manufacturing technology, sensing, and telecommunication. Much work is in progress in search of new materials and exploiting the nonlinear photonic effects in polymers, glasses, and semiconductors in order to achieve desired characteristics in devices and systems.

Glasses containing chalcogens are among the most nonlinear glasses and much effort has been devoted to their study. Chalcogenide glasses are characterized by high transmission in the infrared (IR) region of the spectrum (greater than ~2 μm), making them particularly suitable for active and passive IR devices in comparison to fluoride glasses. Although various III–V semiconducting materials are being used for applications in IR, there are some important drawbacks which limit the use of those materials, namely, the complex etching process required for micro fabrication and cost. Amorphous chalcogenides, are less toxic, easy to handle, and can be fabricated over large areas. In addition, extra functionality for optical fabrication and performance are provided by their unique set of photo induced effects. Taking advantage of the possibility of modifying the optical band gap and the index of refraction by illumination, optical elements for applications in communications, interconnections, and data storage have been patterned in chalcogenide glasses [1].

The first optical fibers for IR transmission were drawn in the mid-1960s from arsenic-based chalcogenide glasses such as As_2Se_3 and As_2S_3 with losses in excess of 10 dB/m [2]. In general, chalcogenide fibers can be classified into three categories: sulfide, selenide, and telluride. One or more chalcogen elements are mixed with one or more elements such as As, Ge, P, Sb, Ga, Si, and Al to form a glass. Examples of stable glasses include As_2S_3 [3], $Ge_{20}S_{40}Br_{40}$ [4], As_2Se_3 [3], and $Ge_{30}As_{10}Se_{30}Te_{30}$ [5]. Glass systems with rare doping include Ge-Ga-S [6], Ge-As-Ga-S [7], Ga-La-S [8], Ga-Na-S [9], Ge-S-I [10], and Ge-As-Se [11].

Although chalcogen-based optic fibers are more delicate in nature than the oxide glass fibers, they are finding many applications in chemical and temperature sensor systems and as IR imaging bundles. Unlike the fluoride glasses which contain five or more components, most chalcogenide glasses have two or three elemental components, which make them easy to synthesize. Typical characteristics of chalcogenide glasses which make them suitable for applications in fiber optics are:

- High refractive index (~3).
- Do not transmit well in the visible region.
- Low phonon energy, thus allowing many rare-earth transitions to be active that are quenched in silica glass (400–450 cm^{-1} for sulfide and ~350 cm^{-1} for selenide glasses). Large Kerr nonlinearities.

- Ability to be directly patterned by exposure to near band-gap light.
- Many of the compositions are stable enough for fiberization.
- Unlike silica glass fibers, chalcogenide fibers have good transmission in the IR beyond 1.5 μm.
- Many chalcogenide glass compositions are non-hygroscopic and have good chemical resistance.
- As hosts for rare earth ions, the low phonon energy of chalcogenide glasses results in a smaller multiphonon transition probability allowing many mid-IR and long-wave IR rare earth transitions that are normally quenched in silica and fluoride glasses to become active. This makes them potential candidates for mid-IR and long-wave IR rare earth doped fiber lasers.
- The high nonlinearity of several chalcogenide glass compositions allows the potential for mid-IR and long-wave IR Raman fiber lasers. It has been shown that the Raman gain coefficient of As-Se fiber is more than 300 times greater than the Raman gain coefficient of silica at a wavelength of 1.5 μm.

More recent interest in chalcogen-based optical fibers has been generated due to their high IR transmission between 2 and 12 μm, low phonon energy and the fact that these glasses can be fiberized. The fibers can be used to transmit laser power such as CO (5.25 μm) and CO_2 (10.6 μm) light for remote cutting and welding. Laser sources in the mid-infrared are of great interest due to many potential applications in such areas as chemical sensing and laser surgery. For example, studies have shown that the best wavelength to ablate soft tissue with minimal collateral damage is 6.45-μm [12]. The large Raman gain coefficient of As-Se in addition to its low loss in the mid-IR makes this glass composition a promising material for a mid-IR fiber Raman laser. In addition, most molecular species possess characteristic IR vibrational signatures between 2 and 12 μm. Hence, chalcogenide glass fibers can be used for remote spectroscopy and chemical analysis. Other applications include temperature monitoring, medical applications, and IR imaging using fiber bundles.

The basic principle of refraction which makes transmission of light possible through glass fiber was first demonstrated by Daniel Colladon and Jacques Babinet in the early 1840s. Colladon first described the "light fountain" or "light pipe" in an article titled "On the reflections of a ray of light inside a parabolic liquid stream" in 1842. One of the first practical applications was use in close examinations for dental and medical diagnostics in the early twentieth century. The first glass fiber optic was proposed in 1926 by C.W. Hansell [13]. Till the 1950s, the driving force behind fiber-optic transmission was the need for a flexible image bundle for industrial and medical applications [14]. Following developments included fiber bundles for image transmission, fiber-optic semi flexible gastroscope, television cameras for space applications, and telecommunications. Use of optical fiber is especially advantageous for long distance communication, because light propagates through the fiber with little attenuation compared to electrical cables. This allows long distances to be spanned with few repeaters. The per-channel light signals propagating in the fiber have been modulated at rates as high as 111 gigabits per second

(Gbit/s) by NTT, although 10 or 40 Gbit/s is typical in deployed systems. Fibers are often also used for short-distance connections between devices. Most high definition televisions (HDTV) offer a digital audio optical connection. This allows the streaming of audio over light, using the TOSLINK protocol. In addition to the low manufacturing and materials cost, advantages of using optical fibers over conventional copper wiring include (1) a broad bandwidth—a single optical fiber can carry 3,000,000 full-duplex voice calls or 90,000 TV channels, (2) immunity to electromagnetic interference—light transmission through optical fibers is not affected by electromagnetic radiation from any other nearby sources, (3) it does not act as an antenna to pick up electromagnetic signals due to its electrically insulating nature; at the same time, this also prevents possible problems due to conduction of lightening, (4) low attenuation loss over long distances—attenuation loss can be as low as 0.2 dB/km in optical fiber cables, which allows transmission over long distances with no need for repeaters, (5) security of information passed down the fiber-optic cable as compared to copper and other metals which can be tapped with very little chance of detection. Optical fibers can be classified into three broad categories on the basis of their physical structure as shown in Fig. 4.1. Table 4.1 lists the suitability of different kinds for specific applications.

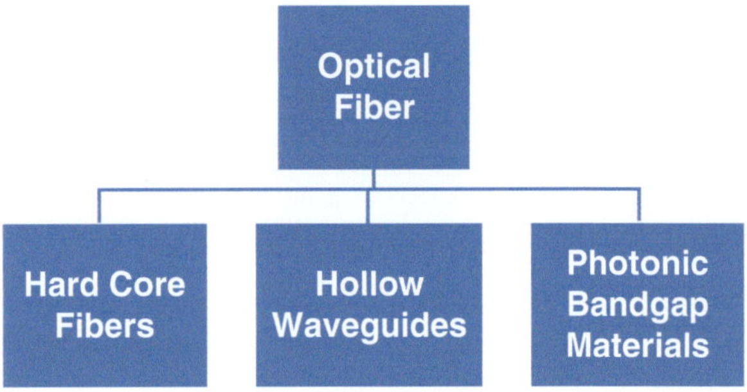

Fig. 4.1 Classification of optical fibers based on structure

Table 4.1 IR fiber candidates for various sensor and power delivery applications [14]

Application	Comments	Suitable IR fibers
1. Fiber-optic chemical sensors	Evanescent wave principle—liquids	AgBrCl, sapphire, chalcogenide, HMFG
2. Fiber-optic chemical sensors	Hollow core waveguides—gases	Hollow glass waveguides
3. Radiometry	Blackbody radiation, temperature measurements	Hollow glass waveguides, AgBrCl, chalcogenide, sapphire
4. Er:YAG laser power delivery	3 μm transmitting fibers with high damage threshold	Hollow glass waveguides, sapphire, germinate glass
5. CO_2 laser power delivery	10 μm transmitting fibers with high damage threshold	Hollow glass waveguides
6. Thermal imaging	Coherent bundles	HMFG, chalcogenide
7. Fiber amplifiers and lasers	Doped IR glass fibers	HMFG, chalcogenide

4.2 Basic Principle: Phenomena of Total Internal Reflection

An optical fiber typically consists of a transparent core surrounded by a transparent cladding material with a lower index of refraction. The boundary between the core and cladding may be abrupt as in step-index fibers, or gradual as in graded index fiber. Light is restricted to the core by the phenomena of total internal reflection, which causes the fiber to act as a waveguide. When a ray of light passes from an optically denser material (i.e., having a higher refractive index, such as glass) to an optically rarer medium (such as air); $n_1 > n_2$, for the incident angle α_i the light emanating from the interface is bent partly reflected and partly transmitted (Fig. 4.2a) at angles α_r and α_t respectively. When the incident angle is increased sufficiently, the transmitted angle (in air) reaches 90°. It is at this point no light is transmitted into air (Fig. 4.2b). This phenomenon is called *total internal reflection* and occurs when a propagating wave strikes a medium boundary at an angle larger than a particular critical angle with respect to the normal to the surface. If the refractive index is lower on the other side of the boundary and the incident angle is greater than the critical angle, the wave cannot pass through and is entirely reflected.

The *critical angle* is the angle above which the total internal reflection occurs and is determined by Snell's law. The angle of incidence is measured with respect to the normal at the refractive boundary. The critical angle is given by Snell's law

$$n_1 \sin \alpha_i = n_2 \sin \alpha_t \tag{4.1}$$

Rearranging Snell's law, we get

$$\sin \alpha_i = \frac{n_2}{n_1} \sin \alpha_t \tag{4.2}$$

To find the critical angle, we find the value for α_i; when $\alpha_t = 90°$ and thus $\sin \alpha_t = 1$. The resulting value of α_i is equal to the critical angle α_c. The sine of α_c is the numerical aperture (NA) of the fiber. The critical angle determines the acceptance

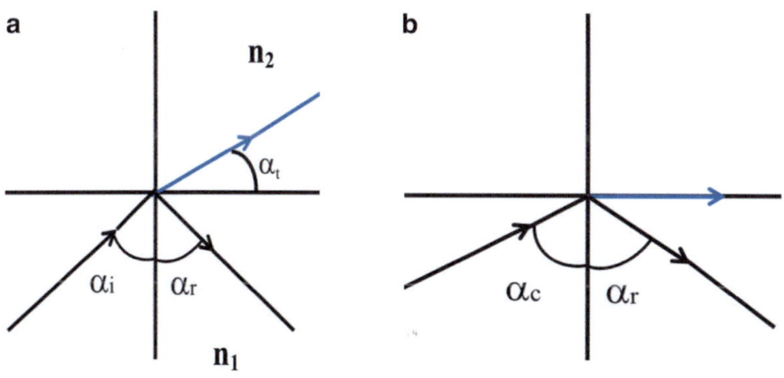

Fig. 4.2 Basic principle for transmission of light through optical fibers—phenomena of total internal reflection

angle of the fiber, and consequently the NA. *Single mode* fibers have a small NA. A high NA for the case of *multimode* fibers allows light to propagate down the fiber in rays close to the fiber axis and at various angles, allowing efficient coupling of light into the fiber. However, at the same time this high value increases the amount of dispersion, which is not a desirable feature.

4.3 Parameters for Optical Fiber Applications

4.3.1 *Refractive Index*

The *refractive index* is one of the most important optical parameters characterizing a dielectric material. In the simplest terms refractive index is the dimensionless ratio of the velocity of light in vacuum to that in the dielectric medium, or $n = c/v$. It is essential to know the refractive indices of core and cladding materials to many significant figures in order to accurately determine the dispersive properties of the fiber. Since the IR fiber applications are short haul, the relative refractive index difference between core and cladding is reasonably large.

4.3.2 *Core Size and Numerical Aperture*

In the design of the optical fiber, two important properties are the *core size* and the *numerical aperture* that together determine the mode field size. Since the Raman gain depends on the intensities of the pump and Stokes waves, a small mode field diameter is important. Additionally, it is important for the fiber to be single mode at the relevant wavelengths to maximize the mode overlap between the pump and Stokes waves. The single mode cutoff wavelength is given by:

$$\lambda_c = \frac{2\pi a}{2.405} \cdot \mathrm{NA} \tag{4.3}$$

where the value, α, is the core radius and NA is the numerical aperture. The cutoff wavelength is chosen to be below both the pump and Stokes wavelengths. With the cutoff wavelength set, either the core diameter or the numerical aperture can be chosen and the above equation will determine the other value [15]. For instance, the NA can be set to 0.60, which requires the core radius to be 3.2-μm to maintain a single mode cut-off wavelength of 5 μm. The mode field radius can then be determined from the following equation:

$$w_0 = a\left(0.65 + \frac{1.619}{V^{1.5}} + \frac{2.879}{V^6}\right)$$

where the value V is the normalized frequency or V-number, $V = (2\pi a/\lambda)$ NA. In this case, for a NA of 0.60 and a core radius of 3.2-μm, the mode field radius is 4.0-μm.

4.4 Loss Mechanisms in Optical Fibers

Optical fibers exhibit three main categories of losses as shown in Fig. 4.3. Infrared fibers in general possess considerably higher losses than the minimum loss of ~0.14 dB/km for silica fiber. This restricts their usage up to ~10 m.

The loss in optic fibers is specified in terms of the absorption coefficient "α" given by

$$A = \frac{10}{l}\log(1/T) = \frac{10}{l}\log(P_{\text{in}}/P_{\text{out}}) \tag{4.5}$$

where T is the transmission, P is the power, and l is the length of the fiber. The total absorption coefficient is the sum of the *intrinsic* and *extrinsic* absorption mechanisms.

Intrinsic losses are the ones arising due to the inherent absorption mechanisms of pure transparent solids and cannot be lowered below a certain limit. For silica this loss is 0.14 dB/km at 1.55 μm [14].

The intrinsic loss mechanisms for any dielectric material include: (1) UV absorption, (2) scattering, and (3) IR absorption. These may be written as:

$$\alpha = A\mathrm{e}^{a/\lambda} + B/\lambda^4 + C\mathrm{e}^{-c/\lambda} \tag{4.6}$$

The first term describes the exponentially increasing absorption in the UV or short wavelength regime as the wavelength decreases and is a result of excitonic or electronic absorption. The second term represents scattering losses resulting from Rayleigh (elastic) and Brillouin (inelastic) scattering. Rayleigh scattering dominates in glassy materials and arises due to density or compositional fluctuations in the refractive index. For crystalline fibers having perfect order, there is no Rayleigh scattering and only Brillouin scattering remains. The final term in Eq. (4.6) describes the exponential increase in attenuation with wavelength that characterizes infrared or multiphonon absorption. Strong IR or multiphonon absorption results from the coupling of light (photons) to lattice vibrations (phonons).

Extrinsic losses in fibers arise as a result of imperfections which cause impurity absorption and extrinsic scattering in the IR material. In general, these losses can be eliminated by taking appropriate steps in the fiber fabrication process. Some common extrinsic losses are listed in Table 4.2 along with an example of an IR fiber where these losses are most likely to occur. In the chalcogenide glasses, hydrogen present in the batching process bonds very strongly with sulfur and

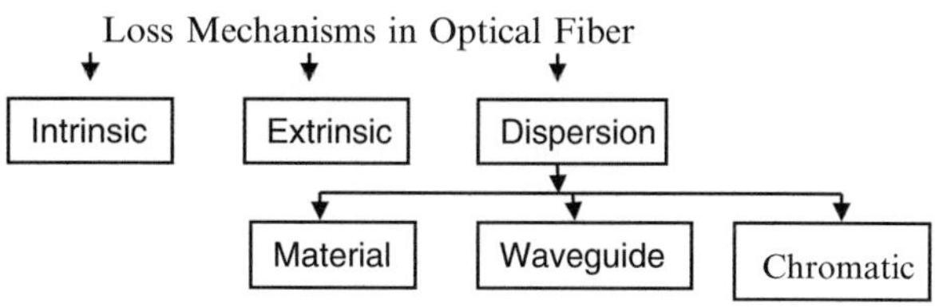

Fig. 4.3 Loss mechanisms in solid core optical fibers

Table 4.2 Sources of extrinsic losses in IR fibers

Extrinsic loss	Source	IR fiber
Impurity absorption	OH–, H_2O	PC, chalcogenide, HMFG
	SeH, SH	Chalcogenide
	Transition metals, e.g., Cu, Fe, Ni	HMFG
	Rare earth ions, e.g., Nd^{3+}, Cr^{3+}	HMFG
	CO^{3-}	PC
Scattering	Bubbles	HMFG
	Crystallites	HMFG
	Grain boundaries	PC
	Core/clad irregularities	HMFG, PC

Reproduced with permission from [14]
PC polycrystalline, *HMFG* heavy metal fluoride glass

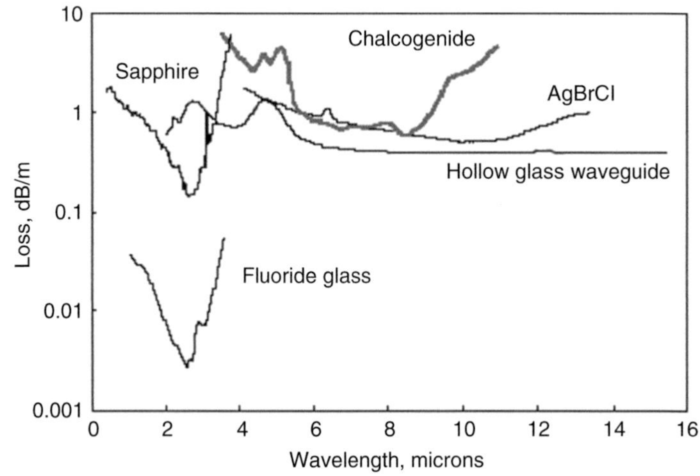

Fig. 4.4 Composite loss spectra for some common IR fiber optics: ZBLAN fluoride glass [16], SC sapphire [17], chalcogenide glass [18], PC AgBrCl [19] and hollow glass waveguide [20]. Reproduced with permission from [21]

selenium giving rise to quite intense absorption bands in the 4.0–4.3 μm region. At the same time, IR fibers often suffer from irregularities in the core–clad interface and other drawing defects which arise during preform preparation and fiber fabrication processes.

In the absence of extrinsic absorption and scattering, the minimum attenuation for $As_{40}S_{60}$ is predicted to be 0.05 dB/km at 5 μm; however, the presence of extrinsic absorption and scattering gives rise to the practically observed attenuation. It is well known that extrinsic scattering can be a limiting factor to achieving low loss. Composite loss spectra for some common IR fiber optics such as ZBLAN fluoride glass [16] SC sapphire [17], chalcogenide glass [18], PC AgBrCl [19] and hollow glass waveguide [20] have been reported in [21] (Fig. 4.4).

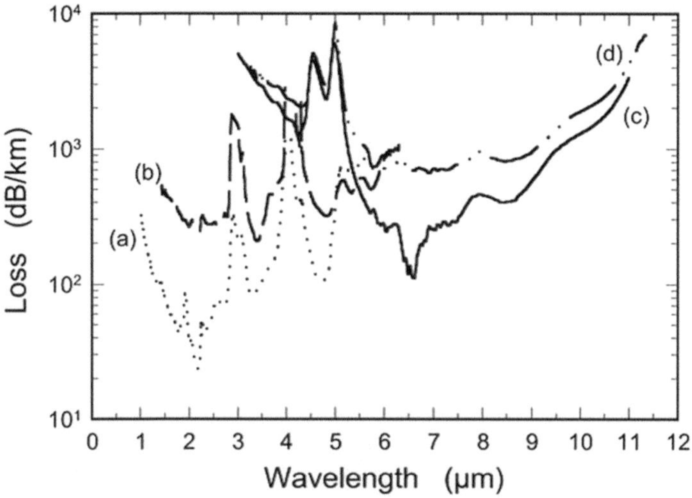

Fig. 4.5 Transmission loss spectra of (**a**) lowest loss sulfide fiber, (**b**) typical sulfide fiber, (**c**) lowest loss telluride fiber, and (**d**) typical telluride fiber. Reproduced with permission from [22]

Figure 4.5 shows comparison of transmission loss spectra for sulfide and telluride glasses.

Dispersion and attenuation are the key optical properties which determine the ultimate performance of a single mode fiber. The important dispersion mechanisms are: (1) material dispersion, (2) waveguide dispersion, (3) nonlinear effects, and (4) polarization mode dispersion. The material and waveguide dispersion taken together are referred to as chromatic dispersion and has a very low effect for silica fibers. Material dispersion has the greatest impact on pulse dispersion which may occur in IR fiber amplifiers.

4.5 Materials for Optical Fibers

Depending on the requirement of specific applications optical fibers can be made from glasses such as silica; pure or doped with germanate, fluorozirconate, fluoroaluminate, and chalcogenide glasses as well as crystalline materials like sapphire. Chalcogen-based optical fibers are used for longer-wavelength infrared or other specialized applications. Silica and fluoride glasses usually have refractive indices of about 1.5, whereas chalcogenides can have indices as high as 3. Typically the index difference between core and cladding is less than one percent. Plastic optical fibers (POF) are commonly step-index multi-mode fibers with a core diameter of 0.5 mm or larger. POF typically have higher attenuation coefficients than glass fibers, 1 dB/m or higher, and this high attenuation limits the range of

Table 4.3 Selected physical properties of key IR fibers compared to silica fiber (permission obtained) [14]

Property	Glass			Crystal		Hollow
	Silica	HMFG ZBLAN	Chalcogenide AsGeSeTe	PC AgBrCl	SC Sapphire	Hollow Silica Waveguide
Glass transition or melting point (°C)	1175	265	245	412	2030	150 (useable T)
Thermal conductivity (W/m °C)	1.38	0.628	0.2	1.1	36	1.38
Thermal expansion coefficient (10^{-6} °C^{-1})	0.55	17.2	15	30	5	0.55
Young's modulus (GPa)	70.0	58.3	21.5	0.14	430	70.0
Density (g/cm^3)	2.20	4.33	4.88	6.39	3.97	2.20
Refractive index (λ, μm)	1.455 (0.70)	1.499 (0.589)	2.9 (10.6)	2.2 (10.6)	1.71 (3.0)	NA
dn/dT, 10^{-5} °C^{-1} (λ, μm)	+1.2 (1.06)	−1.5 (1.06)	+10 (10.6)	−1.5 (10.6)	+1.4 (1.06)	NA
Fiber transmission, range (μm)	0.24–2.0	0.25–4.0	4–11	3–16	0.5–3.1	0.9–25
Loss[a] at 2.94 μm, (dB/m)	~800	0.08	5	3	0.4	0.5
Loss[a] at 10.6 μm (dB/m)	NA	NA	2	0.5	NA	0.4

[a]Typical measured loss
NA stands for Not applicable

POF-based systems. Table 4.3 gives a comparison of selected physical properties of key IR fibers in comparison to the conventional silica fiber.

Table 4.4 lists some physical, mechanical, and optical properties of two chalcogenide glasses used for making optical fibers [22, 24]. Compared to the more traditional oxide glasses, they can be described as having lower glass transition temperature (T_g), higher coefficient of thermal expansion (CTE), lower hardness and higher indices of refraction.

4.6 Chalcogenide Glass Optical Fibers

The transmission range for chalcogenide fibers depends heavily on the mass of the constituent elements. The lighter element glasses such as arsenic trisulfide (As_2S_3) have transmission range from 0.7 to ~6 μm [25]. As_2S_3 and some phosphorus containing and Ge-S-based glasses are the only ones transmitting visible radiation [18]. Longer wavelength transmission is possible upon addition of heavier elements such as Se and Te as can be seen in Fig. 4.6. Depending upon composition, the

Table 4.4 Some physical, mechanical and optical properties of chalcogenide glasses for making optical fibers

	$As_{40}S_{60}$	$Ge_{30}As_{10}S_{30}Te_{30}$
Physical properties		
Tg (°C)[a]	197	265
CTE (10^{-6}/°C)[b]	21.4	14.4
Thermal conductivity (W/m °C)	0.17	~0.2
Mechanical properties		
Denisty (g/cm^3)	3.20	4.88
Knoop hardness (kg/mm^2)	109	205
Fracture toughness (MPa m$^{1/2}$)	~0.2	~ 0.2
Poisson's ratio	0.24	~ 0.26
Youngs modulus (GPa)	16.0	21.9
Optical properties		
Refractive index[c]	2.415 (3.0)	2.80 (10.6)
d*n*/d*T* (10^{-5} °C^{-1})[c,d]	+0.9 (5.4)	+10.0 (10.6)
Bulk transmission (μm)	0.6–10.0	1.0–17.0
Fiber transmission (μm)	0.8–6.5	3.0–11.0
Lowest loss (dB/km)[c]	23 (2.3)	110 (6.6)
Typical loss (dB/km)[c]	100–200 (2.2–5.0)	500–1000 (6.0–9.0)
Estimated minimum loss (dB/km)	1.0	nd

Reproduced with permission from [23]
nd not determined
[a]Tg is the glass transition temperature
[b]CTE is the coefficient of thermal expansion
[c]Wavelength in μm given in parenthesis
[d]d*n*/d*T* is the change in refractive index with temperature

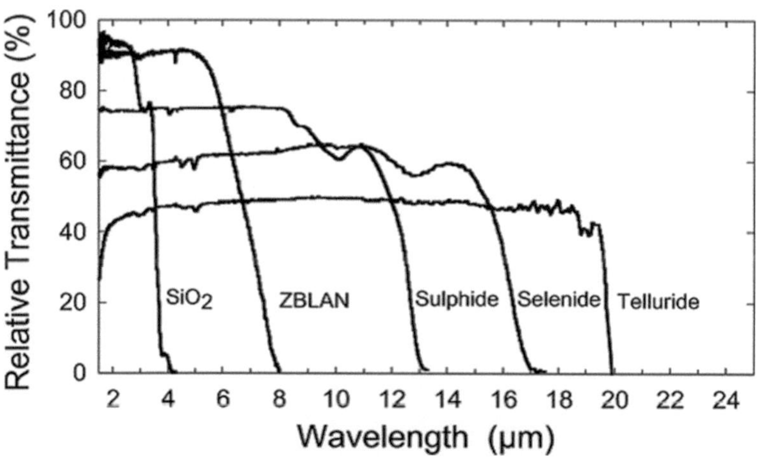

Fig. 4.6 Transmission spectra for several glasses (thickness of about 2–3 mm). Reproduced with permission from [22]

sulfide, selenide, and telluride-based fibers transmit between about 0.8–7 μm, 1–10 μm, and 2–12 μm respectively. Since the chalcogenide glasses transmit to longer wavelengths in the IR than silica and fluoride glasses (Fig. 4.6), there are numerous potential applications in the civil, medical, and military areas. These can be essentially divided into two groups, namely "passive" and "active" applications [22].

4.6.1 Passive Applications

The fibers are used as a light conduit from one location to another without interacting with the light, other than that due to scattering, absorption and end face reflection losses associated with the fiber. Some of the important passive applications are listed in Table 4.5.

4.6.1.1 Laser Power Delivery

Transmitting laser power through fibers enables remote operation. High power Carbon monoxide (CO) and carbon dioxide (CO_2) lasers operating at 5.4 μm and 10.6 μm respectively are readily available for industrial welding and cutting

Table 4.5 Passive applications of IR transmitting chalcogenide glass fibers

Applications	References
Laser power delivery	
5.4 μm	[18, 26]
10.6 μm	[18, 26]
Atmospheric 2–5 μm region	[27, 28]
Medical free electron laser	[29, 30]
Antireflection (AR) coatings	[18, 27]
Chemical sensing	
Aqueous, nonaqueous, toxic chemicals	[30–34]
Polymers, paints, pharmaceuticals	[34–36]
Condition-based maintenance (CBM)	[34]
Cone penetrometer system	[37]
Active coatings	[38]
Bio-medical	[29, 39]
Temperature monitoring	
Grinding ceramics	[40]
Thermal imaging and hyperspectral imaging	
Coherent fiber bundles	[41–44]
Near field microscopy	
Imaging and spectroscopy	[45–47]
Fiber multiplexing	
Fiber couplers	[48]

Reproduced with permission from [23]

applications. Fibers having core diameters <200 μm have demonstrated tolerance to power densities of ~125 kW/cm^2 at 5.4 μm and 54 kW/cm^2 at 10.6 μm without damage [26]. Tellurium glass fiber losses at 10.6 μm have been reported in the range of 1.5–3 dB/m depending on composition and purity [5, 18]. Figures 4.7a, b show the output power versus input power for CO and CO_2 laser transmission and indicate the absence of nonlinear effects in the 1–2 m fiber sections. Telluride glass fiber losses at 10.6 μm lie in the range of 1.5– 3 dB/m, depending upon composition and purity The arsenic sulfide fibers transmit in the atmospheric 2–5 μm region and can be used for transmission of laser power in this region for military applications such as in infrared countermeasures (IRCM), where the fiber guides the IR laser power from a remotely located laser to a Jam Head mounted on the surface of the plane/helicopter [27]. Figure 4.7c shows a plot of output power versus input power for a sulfide fiber using a pulsed laser operating in the 2- 5 μm region [23].

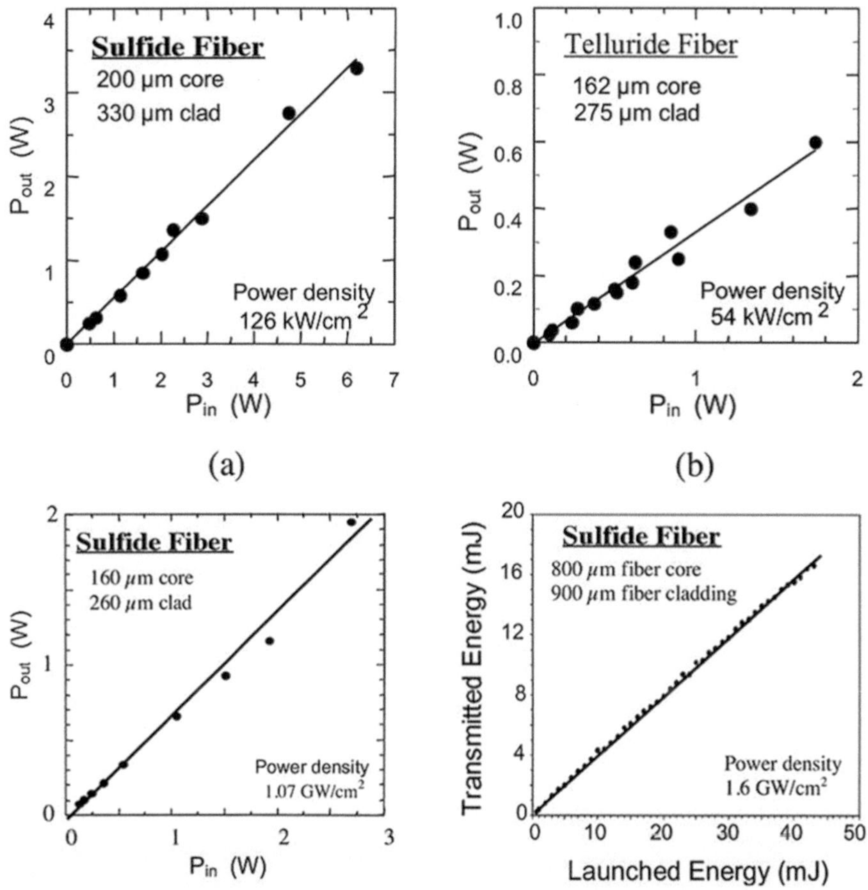

Fig. 4.7 (**a**) CO laser transmission; (**b**) CO_2 laser transmission; (**c**) pulsed high energy laser transmission in the 2–5 μm region. Reproduced with permission from [23]

The average power is about 2.69 W but the peak power is 26.9 kW, which corresponds to a peak power density of 1.07 GW/cm^2 without fiber damage for up to 1.5×10^7 pulses. This remarkable threshold to damage is close to the predicted value of about 3.0 GW/cm^2 due to dielectric breakdown at the surface. This threshold to damage is obtained by control of fiber polishing; otherwise, the fiber end face undergoes damage at relatively lower powers.

4.6.1.2 Chemical Sensing

Chalcogenide fibers are well suited for chemical sensing applications since practically most molecular species vibrate in the infrared region. The chalcogenide fiber-optic chemical sensor systems based on techniques such as evanescent/ATR (attenuated total reflectance), diffuse reflectance, and absorption spectroscopy can be used for quantitative remote detection and identification of chemicals in mixtures. Applications of chalcogen-based fibers for chemical sensing are discussed in Chap. 8.

4.6.1.3 Temperature Monitoring, Thermal Imaging, and Hyperspectral Imaging

As-S fibers with a Teflon cladding have been used by Ueda et al. to measure temperature increases of up to 200 °C on the surface layer of ceramic plates during grinding [40]. Use of chalcogenide fibers for thermal imaging was first suggested by Kapany and Simms [41]. Saito et al. [42] recorded the image of an electric iron at 773 K, with some degree of coherency through a 1000 fiber bundle. Nishii et al. [43] fabricated a flexible fiber bundle containing 8400 Teflon coated fibers and recorded the thermal image of an operating integrated circuit in 3–5.4 μm region using an InSb detector. The area of hyperspectral imaging can be exploited by coupling coherent fiber bundles to focal plane array (FPA) detectors based on InSb (2–5.4 μm) or MCT (3–11 μm). The focal plane array detectors are sensitive enough to be used for performing both spatial and spectral analysis in the infrared.

4.6.1.4 Near Field Microscopy

Sub-diffraction limit resolution has been demonstrated using single mode silica fibers by pulling and/or etching the fiber ends from a diameter of 125 μm to below 100 nm at the fiber tip. Preliminary results were obtained using multimode chalcogenide fiber micro tips with about 1 μm resolution [45]. High quality single mode and multimode chalcogenide fibers have been used to demonstrate 100 nm resolution for both topographic and spectroscopic analysis of polycrystalline diamond [46] and 20 nm topographic resolution of pancreatic cells [47].

4.6.1.5 Fiber Multiplexing

Fused fiber couplers are important since they enable fiber multiplexing. For example, they might be used in IR fiber-optic chemical sensing systems and data transmission systems. A preliminary fused taper fiber coupler has been fabricated using a multimode arsenic sulfide fiber with a coupling ratio of 3:1 at 2.65 μm, but it should be possible to use single mode fibers.

4.6.2 Active Applications

The light propagating through the fiber is modified by a process other than that due to scattering; absorption and end face reflection losses associated with the fiber. Examples of these include fiber lasers, amplifiers, bright sources, gratings, and nonlinear effects (Table 4.6).

4.6.2.1 Rare Earth (RE) Doped Fibers

Rare-earth ions possess characteristic electronic energy levels that are only slightly influenced by the host matrix due to the screening effect of the d electrons. When pumped with the appropriate energy, the electrons are excited into upper levels from which they can subsequently decay to lower levels. Certain transitions become increasingly more efficient in longer wavelength transmitting hosts such as the chalcogenide glasses due to less multiphonon quenching and IR fluorescence

Table 4.6 Active applications of IR transmitting chalcogenide glass fibers

Applications	References
Rare earth doped fibers	
Fiber lasers—1.08 μm (Nd)	[49]
Amplifers	
1.08 mm (Nd)	[50]
1.34 mm (Pr)	[9]
1.34 mm (Dy)	[51]
Infrared scene simulation (IRSS)	[52]
Chemical sensing	[29]
Gratings—1.5 mm	[53]
Nonlinear	
Optical switching	[54]
Second harmonic generation	[55]
Frequency mixing	–
Electrical poling	–

Reproduced with permission from [23]

Table 4.7 IR emission wavelengths in chalcogenide glasses and wavelengths that have exhibited laser oscillations and amplification in chalcogenide fibers

Rare earth ion	IR emission wavelength (μm)
Nd	0.786, 0.919, 1.08, 1.37
Er	0.822, 0.860, 0.987, 1.54, 2.7, 3.5, 4.5
Tm	1.21, 1.45, 1.81, 2.35
Ho	0.76, 0.91, 1.2, 0.9, 3.9
Pr	1.3, 1.6, 2.9, 3.4, 4.5, 4.8, 4.9, 7.2
Dy	1.3, 1.8, 2.3, 4.3
Tb	3.0, 4.8, 8.0

As reported in [23]. Reproduced with permission [23]

emissions beyond 2 μm are only seen in chalcogenide glasses and not in silica. Rare-earth doped fibers possess several emission bands that can be harnessed for 2–4 μm laser emission. IR emission wavelengths found in chalcogenide glasses are listed in Table 4.7 [55–64].

4.6.2.2 Fiber Lasers

The development of fiber-based laser sources operating in the mid-infrared (IR) has been fueled by a wide variety of potential applications [65]. Laser surgery would greatly benefit from high-power, high-beam quality sources near the 3 μm water absorption peak [66]. Laser oscillation in chalcogenide fibers has been reported from Nd at 1.08 μm [49] while amplification has been demonstrated using Nd at 1.08 μm [50] and Pr at 1.34 μm [9]. These results were obtained in sulfide glass hosts. The 1.3-μm wavelength is of interest for telecommunications [67]. The gain coefficient obtained from the single-mode sulfide (Ga–Na–S) fiber at 1.34 μm was a remarkable 0.8 dB m W^{-1} and the efficiency was about 30 %. The fiber loss at 1.34 μm was about 1 dB m^{-1} [67].

Stimulated Raman scattering as an alternative to produce a high-power laser emission from a fiber is being actively investigated by researchers [68]. It allows efficient laser operation on almost any desired wavelength, and therefore it can bridge the gap between the few available RE emission bands in the mid-IR. Cascaded Raman laser cavities based on silicate fibers have been perfected in the near-IR to cover the broad spectral range from 1.1 to 1.5 μm. Recently, the power of these devices has been scaled significantly and reached a maximum of 300 W (CW) at 1.48 μm [69]. Using silicate fibers with high GeO_2 concentrations, this cascade approach was used to extend the wavelength coverage of Ramuan fiber lasers (RFLs) slightly beyond 2 μm [70]. An RFL emission was also recently reported in fluoride glass fibers with a maximum continuous output power of 3.7 W at 2.23 μm [71]. Unfortunately, fluoride glass fibers exhibit a relatively low Raman gain coefficient [72], which hinders their use at longer wavelengths, especially since the Raman gain coefficient scales inversely with the wavelength

[73]. An efficient RFL operating at 3.34 μm [74] based on a chalcogenide (As_2S_3) glass fiber, exhibiting a high nonlinearity and a Raman gain coefficient about 50 times larger than that of a fluoride glass [72, 75] was also reported. A theoretical optimization of such laser was presented in [76] and predicted that As_2S_3 RFLs could, in theory, be developed at any wavelength in the range 3–4 μm.

Solid state lasers such as Ho: YAG and Er: YAG/YSGG are now commercially available and offer high power levels near 2.1 μm and 2.8 to 2.9 μm, respectively [77–79]. In addition, Cr^{2+}:ZnS/Se and Fe^{2+}:ZnSe-doped crystals can be pumped by well-developed sources, i.e., fiber and solid state lasers, to deliver a broad range of long emission wavelengths between 2 and 5 μm [80] Quantum cascade lasers are also good candidates for mid-IR laser emission in the 3.5–10 μm spectral region. However, the heat management of these devices is currently the main challenge that limits their power scaling [81]. Fiber lasers (FL) present several advantageous features over other technologies: a great compactness, good beam quality, efficiency and ruggedness. Nevertheless, aside from Tm- and Ho-doped FLs emitting near 2 μm [80, 82] the number of commercial FLs having a wavelength over 2.1 μm is still limited. This limit actually corresponds to the long wavelength edge of silicate fibers transparency window. There is thus a growing need for developing FLs for the mid-IR based on other glass materials.

An important step toward the development of compact laser sources emitting in the 3–5 μm atmospheric transparency windows has been investigated by Bernier et al. [83]. The longest emission wavelength produced by a fiber laser operating at room temperature is reported.

The 3.77 μm output wavelength was generated by cascaded Fabry–Perot cavities formed by nested pairs of fiber Bragg Gratings (FBGs) written in a 2.8 m segment of low-loss single-mode As_2S_3 fiber. The schematic diagram of the experimental setup used is shown in Fig. 4.8. A maximum output peak power of 112 mW was measured with a laser efficiency of 8.3 % with respect to the launched pump power. The first demonstration of a chalcogenide-based RFL [84], which relies on the use of a specially designed As_2S_3 fiber presenting optical losses below 100 dB/km at 3.4 μm, paves the way for all-FLs operating between 3 and 4 μm.

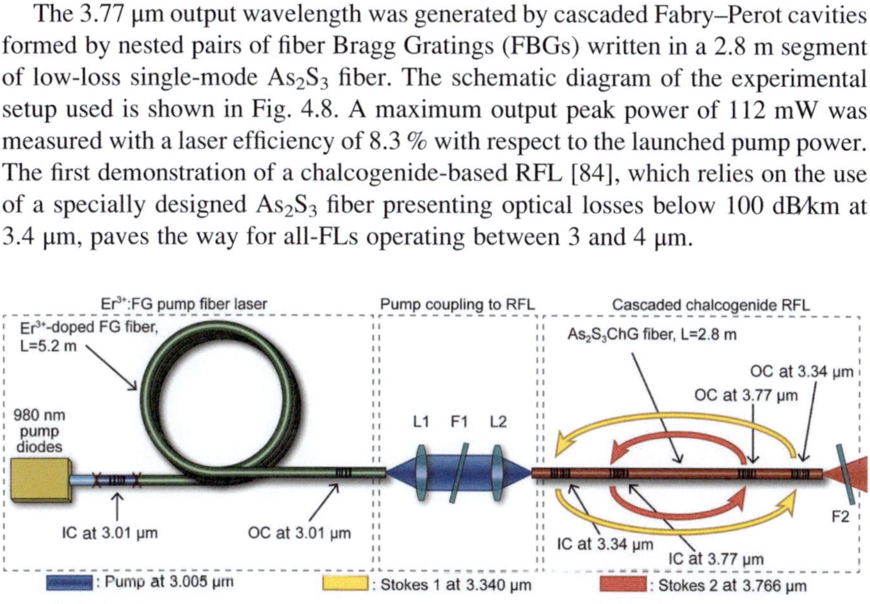

Fig. 4.8 Experimental setup of the As_2S_3-based 3.77 μm cascaded RFL. *F1* long-pass filter at 2.95 μm, *F2* long-pass filter at 3.50 μm, *L1* and *L2* aspheric lenses, *ICs* cavity input couplers, *OCs* cavity output couplers. Permission obtained. After [83]

Since the rare earth doped chalcogenide fibers emit in the 2–5-μm region [29], they are an attractive alternative to blackbody sources for many applications. For example, arrays of the fibers can be used for infrared scene simulation (IRSS) for characterization of focal plane array detectors (e.g., InSb) [52]. Prototype bundles have been fabricated and mid-IR emission measured from the pixels [52]. Black body temperatures of up to 2400 K have been simulated in single pixel pumping indicating that these fibers are capable of providing bright sources in the IR.

4.6.2.3 Fiber Amplifiers (FAs)

Erbium doped fiber amplifier (EDFA) has enabled the development of worldwide fiber-optic communication networks allowing to abandon the use of optoelectronic and electro optical conversions of signals. Prior to the advent of optical amplifiers, the attenuation of optical signals was compensated by electronic regenerators along the optical fiber cable. This slow and expensive electronic regenerating process not only limited the full usage of large transmission capacity and bandwidth which are inherently associated with optical fiber communication networks, but also restricts its wide deployment due to high cost. The usage of optical fiber amplifiers removes the bottleneck of optical fiber communication networks, resulting in rapid increase in the capacity of the system because of their high gain, wide optical bandwidth, high output saturation, near quantum-limited noise, low insertion losses, high reliability and compactness, polarization independence, immunity to saturation-induced and to crosstalk, and possibility of choosing the pumping laser diode at 980 or 1480 nm wavelengths. In addition, the FAs can also be used to increase the received signal power to level that is suitable for detection (pre-amplifier), and boost the signal immediately from transmitter and modulator to a level suitable for transmission (power amplifier). Optical amplifier can also be used for compensating the intrinsic losses of passive devices, such as splitter and switch array. Although this kind of technology is mature and widely employed, further research is needed to obtain amplifiers with higher gain efficiency. The optimization of fiber transversal section is crucial to improve the amplifier performance in terms of gain, noise, and output power characteristics as well as device compactness and pump power consumption. In fact, in the rare-earth-doped devices, the fiber geometry strongly affects the pump intensity, the overlap of the pump, and the signal propagation modes with the doped core.

The large Raman gain coefficient of chalcogenide glass coupled with its large IR transparency show promise for lasers and amplifiers in the near and mid-IR. Stimulated Raman scattering attracts considerable interest, since it provides broadband light amplification. Stimulated Raman scattering can be regarded as a kind of nondegenerate two-photon process. However, in contrast to the two-photon absorption, in the Raman scattering, one photon is absorbed by virtual states and simultaneously one photon is emitted. When the incident light becomes intense, in a similar way to the conventional stimulated light emission, the Stokes-shifted light could be amplified. Related nonlinear polarization is written as $P^{(3)}(\omega_s) \propto i\chi_R^{(3)} I(\omega_L) E(\omega_s)$, or emitted light intensity being $\Delta I(\omega_s) \propto \chi_R^{(3)} I(\omega_L) I(\omega_s)$, where ω_L and ω_s ($<\omega_L$) are an exciting laser frequency and a Stokes-shifted frequency.

$As_{40}Se_{60}$ glass has a much narrower Raman line (~60 cm^{-1}) than silica glass (~250 cm^{-1}) [85]. In addition, the Raman shift for $As_{40}Se_{60}$ glass is much smaller (~240 cm^{-1}) than the Raman shift of silica glass (~440 cm^{-1}) due to the heavier atoms present in the chalcogenide glass. Previous studies have looked at stimulated Raman scattering in $As_{40}S_{60}$ glass, a very similar glass system to $As_{40}Se_{60}$ [86]. These studies found the Raman gain coefficient of $As_{40}S_{60}$ to be almost two orders of magnitude higher than that of silica. It was also found that this enhancement in the Raman gain roughly corresponded to the enhancement in the nonlinear index, n_2. Consequently, one might expect to see an even larger Raman gain coefficient in $As_{40}Se_{60}$ since the selenide glass has shown a larger nonlinearity and also a narrower Raman spectrum. For stimulated Brillouin scattering, the backward wave propagating through the fiber has been measured for different pump powers [87] as shown in Fig. 4.9.

Raman amplification at 1.55 μm has been demonstrated in small core As-Se fiber [15, 87]. The results of the Raman amplification experiment are shown in shown in Fig. 4.10. Over ~23 dB of gain was achieved in a 1.1-m length of fiber pumped by a nanosecond pulse of ~10.8 W peak power at 1.50 μm. The peak of the Raman gain was shifted by ~230 cm^{-1} to 1.56 μm. The Raman gain coefficient was estimated to be ~300 times silica in this experiment. More recent measurements of the Raman

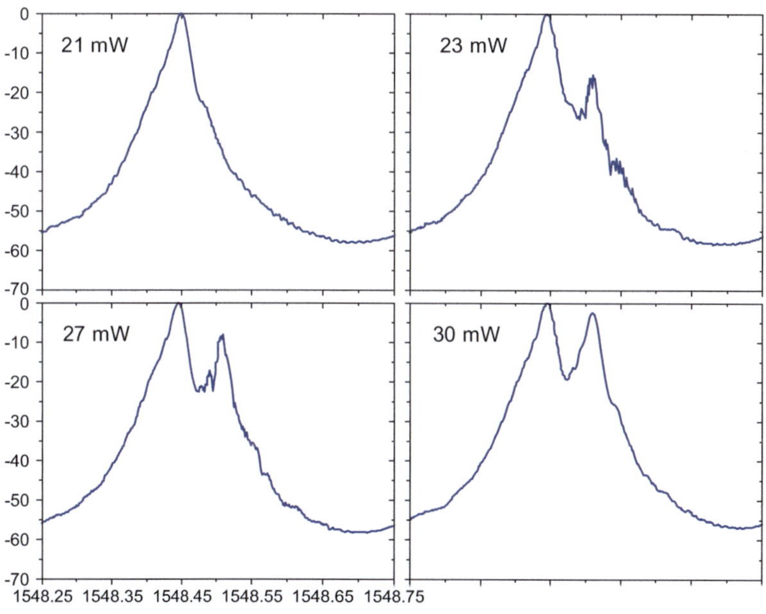

Fig. 4.9 Typical spectra of the reflected light for different launched pump powers into the As_2S_3 fiber core. Fiber length was 10.0 m. Estimated SBS threshold: (27 ± 3) mW. Tick labels shown only on one plot for clarity (the span on the *x*-axis is 0.5 nm). With permission [87]

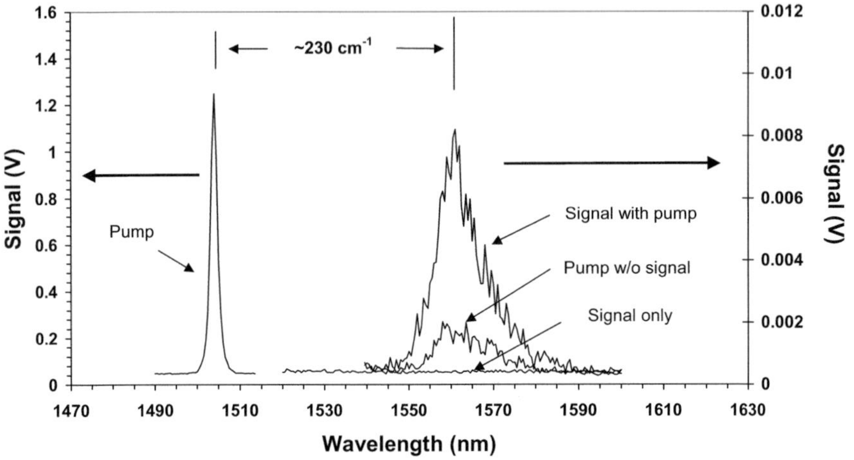

Fig. 4.10 Raman amplification in As-Se fiber. Shown is amplifier output with signal and no pump, pump and no signal (showing background stimulated Raman scattering (SRS) resulting from pump), and amplified signal with pump. With permission After [87]

Table 4.8 Figure of merit for Raman amplification in As–Se fiber at 4 μm-compared Raman amplification in silica fiber at 1.5 μm

	λ (μm)	g_R (cm/W)	Loss (dB/km)	α (cm^{-1})	FOM (10^{-6} W^{-1})
Silica fiber	1.5	0.65×10^{-12}	0.2–0.3	$\sim 6 \times 10^{-7}$	1.1
As–Se fiber	4	1.7×10^{-10}	200	5×10^{-4}	0.34
			3	7.5×10^{-6}	23

After [88] Open access
The loss value of 200 dB/km (a) for As-Se is typical of a "champion" loss value. The loss value of 3 dB/km (b) is theoretical loss

gain coefficient show a value of about 780× greater than that of silica [88] (Table 4.8).

As a consequence, it can lead to the suppression of the amplified spontaneous emission (ASE), the power scaling and the maximum inversion of rare-earth ions as well as to the reduction of the fiber length. Sophisticated design methods and fabrication techniques have been developed to construct single-mode optical fiber amplifiers. To this aim, a fine control of refractive index profile of both core and cladding as well as more design flexibility of fiber cross section is needed. The conventional optical fibers are not able to completely respond to these requirements, while the photonic crystal fiber (PCF) technology seems to be an attractive solution [89]. A PCF is typically characterized by a transverse crystal lattice (usually periodically arranged) containing either air holes or glass strand running along the fiber axis. Figure 4.11 shows cross section of a PCF preform made from sulfide glass and the photonic bandgap (PBG) of a chalcogenide preform [87].

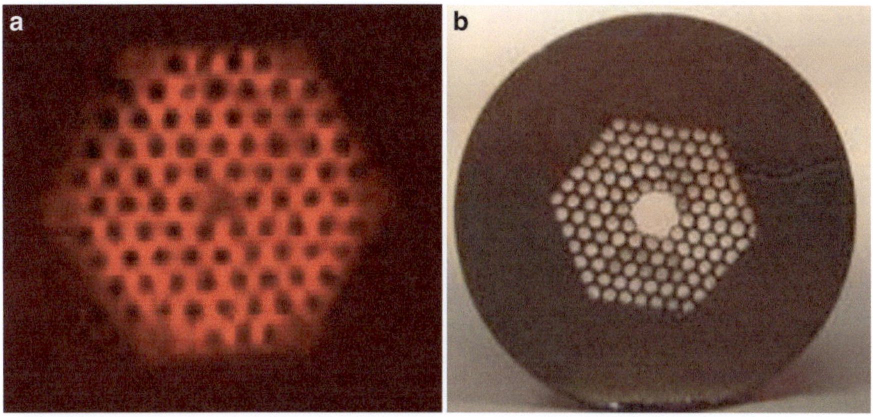

Fig. 4.11 Cross section of (**a**) PCF from sulfide glass and (**b**) PBG chalcogenide glass preform Reproduced with permission [87]

As a consequence, these fibers are different with respect to the conventional ones, and the main difference is in the refractive index profile of both core and cladding regions.

The feasibility of Er^{3+}-doped, $Ga_5Ge_{20}Sb_{10}S_{65}$ (2S2G) chalcogenide glass has been demonstrated in order to obtain a high performance optical amplifier in the third band of fiber-optic communication [90]. The calculated optical gain of the optimized PCF amplifier, 2.79 m long, is close to 23 dB at the signal wavelength of 1.538 μm, by using a pump power of 200 mW and a signal power of 0.1 μW. This result indicates that the proposed fiber amplifier could be a good candidate in optical communication networks and systems. Therefore, by considering that (1) the high refractive index and the low phonon energy of the chalcogenide glass increase the efficiency of transitions among rare-earth energy levels; (2) the excellent rare-earth solubility in chalcogenide glass allows high dopant concentration, without ion clustering and concentration quenching effects, fiber amplifiers based on chalcogenide glasses, operating at 1.5 μm, exhibit higher pump conversion efficiency and shorter fiber lengths than those based on silicate glasses.

A detailed spectroscopic study of Er^{3+}-doped 2S2G glass has been illustrated in [91]. This doped glass (1) facilitates the optical fiber drawing, since it presents suitable thermomechanical properties, (2) provides a better solubilization, since it contains gallium, and (3) enables efficient $^4I_{11/2} \rightarrow {}^4I_{13/2}$ (2.7 μm) and $^4I_{9/2} \rightarrow {}^4I_{11/2}$ (4.5 μm) mid-IR transitions. The radiative lifetime of $^4F_{9/2}$, $^4I_{9/2}$, $^4I_{11/2}$, and $^4I_{13/2}$ energy levels and the related branching ratios are determined. The estimation of Er^{3+} emission cross section in the mid-IR spectral range was 2.85×10^{-25} m^2 at 4.6 μm. The $^4I_{9/2}$ radiative quantum efficiency was estimated to be 64 %. The propagation of fluorescence signals around 2.7 μm and, for the first time, around 4.6 μm was clearly observed in two 2G2S fibers doped with erbium ions under laser pumping at 804 nm.

4.6.2.4 Fiber Bragg Gratings (FBGs)

A key enabler for the integration of efficient fiber lasers is the ability to write highly reflective and low-loss fiber Bragg gratings (FBGs) [92]. A fiber Bragg grating (FBG) is a type of distributed Bragg reflector constructed in a short segment of the optical fiber that reflects particular wavelengths of light and transmits all others. This is achieved by creating a periodic variation in the refractive index of the fiber core, which generates a wavelength-specific dielectric mirror (Fig. 4.12). A fiber Bragg grating can therefore be used as an inline optical filter to block certain wavelengths, or as a wavelength-specific reflector. In the past few years, only a few attempts to inscribe FBGs in As_2S_3 ChG fibers were reported. The first report in 1996 was based on the use of sub-bandgap exposition at 633 nm from a He–Ne laser using the transverse holographic method [86]. An FBG with a peak reflectivity of 80 % was inscribed at 1550 nm following an exposure of about 15 min.

In subsequent reports, the transverse holographic method was replaced with a phase-mask inscription setup which is more mechanically stable [94, 95]. The spectral responses of the resulting FBGs, still inscribed with 633 nm radiation, were relatively poor and the required exposure time to obtain a significant reflectivity were up to an hour long. During the past decade, a new approach was demonstrated to write FBGs in various dielectric and crystalline materials based on the use of high intensity femtosecond (fs) pulses at 800 nm and a phase-mask [96]. The possibility to write FBGs through the polymer jacket of silica fibers was also exploited based on the fact that polymers used to coat optical fibers are usually transparent at 800 nm [97]. The usefulness of the FBG written using the

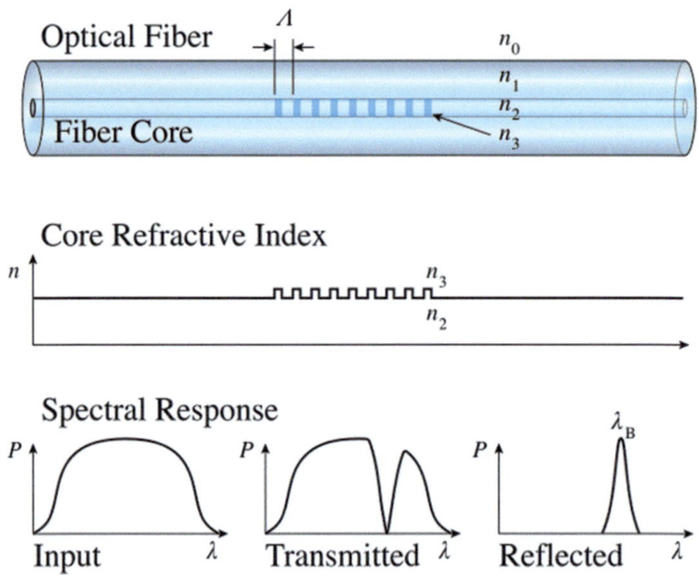

Fig. 4.12 Fiber grating engraved in an optical fiber and its spectral response [93]

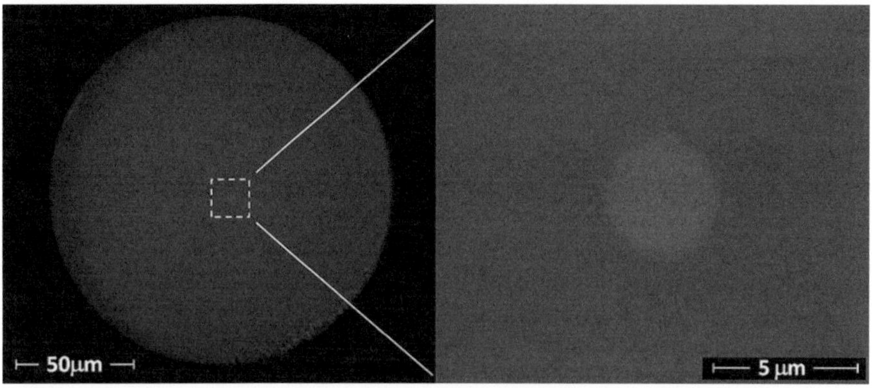

Fig. 4.13 Microscope images of the single-mode As_2S_3 fiber [92]

femtosecond (fs) approach was recently demonstrated for the inscription of strong and permanent first-order FBGs in fluoride glass fibers [98]. This approach has also been used with As_2S_3 fibers but the reported results were limited to microscope images of periodic structures inscribed transversally to the fibers using either the point-by-point [99] or the phase-mask [100] techniques. A first demonstration of FBG writing through the polymer jacket of low loss single-mode As_2S_3 (image in Fig. 4.13) fibers using 800 nm fs laser pulses and a phase mask has been reported [92].

A peak reflectivity in excess of 99 % is demonstrated at 3440 nm after 300 s exposure. The resulting FBG maintained a peak reflectivity of 90 % after 64 min of thermal annealing at 100 °C. Characterization of the loss spectrum of a single-mode ChG fiber in the mid-IR up to 4 μm has been done using a fluoride fiber-based supercontinuum source.

4.6.3 Nonlinear Effects

Materials exhibiting nonlinear optical properties are attractive candidates for future high bit rate optical communication. Glasses in general exhibit several advantages over their crystalline counterparts for applications involving nonlinear behavior. First, the atomic composition of glass can be tailored continuously. For instance, it is possible to obtain nonlinear optical glass with any refractive index in the range of 1.4–3.2 at a wavelength of ~1 μm [101]. Second, the atomic structure can be modified reasonably easily using for example light beams. Such modifications may be regarded as photoinduced phenomena which can be employed in order to add second-order nonlinearity to selected regions. In addition, macroscopic shapes can be changed into arbitrary bulk forms, fibers, thin layers, and microparticles.

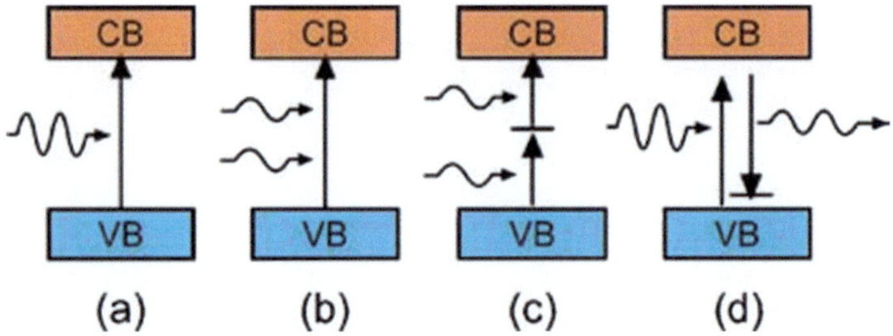

Fig. 4.14 Schematic illustrations of (**a**) one-photon absorption, (**b**) two photon absorption, (**c**) two-step absorption via a mid gap state, and (**d**) Raman-scattering process in a semiconductor with a valence (VB) and a conduction band (CB) [105]

The relation between polarization, P, and electric field, E, is given by [102–104]

$$P = \chi^{(1)}E + \chi^{(2)}E^2 + \chi^{(3)}E^3 + \cdots$$

where $\chi^{(1)}$ depicts the conventional linear susceptibility, $\chi^{(2)}$ and $\chi^{(3)}$ represent send and third order nonlinear susceptibilities. $\chi^{(1)}$ is related to the linear refractive index n_o via $n_o = \{1 + 4\pi\chi^{(1)}\}^{1/2}$. The second term $\chi^{(2)}$ provides time dependence such as $\sin\{(\omega t \pm 2\omega)\}t$ so that it could produce dc and second overtone (2ω) signals. In a similar way the third term could modify the fundamental (ω) signal and generate a third overtone (3ω). Microscopically, the nonlinear terms arise through several mechanisms such as electronic, atomic (including molecular motions), electrostatic, and thermal processes. Among these, the electronic process can provide the fastest response with fs–ps time scales, which will be needed for optical information technologies. A simplified picture of the various mechanisms is given in Fig. 4.14.

For chalcogenide glasses, due to the presence of anions, such as S^{2-}, Se^{2-}, and Te^{2-}, with lone electron pairs, the third order optical susceptibility $\chi^{(3)}$ is several orders higher in comparison to SiO_2 (silica) glasses. Silica-based optical fibers are one of the most promising media for all-optical switches [106, 107]. However, these switches demand relatively high switching power or fiber lengths because of the small nonlinear refractive index ($n_2 \sim 2.8 \times 10^{-16}$ cm^2/W) of silica glasses. In an effort to solve this problem, other glasses with large nonlinearities are being studied [108, 109]. Chalcogenide glasses are promising candidates for these applications due to their high optical nonlinearity. Several chalcogenide compositions have been shown to exhibit large values of nonlinear refractive index, n_2, together with low absorption in the 1–1.5 μm telecommunications window [110]. For As–S–Se glasses investigated by Asobe et al. [86, 111] values of n_2 up to 400 times that of silica have been reported. Such large values have been correlated with the presence of covalent, homopolar Se–Se bonds in the glass. A number of compounds such as $Ge_{0.25}Se_{0.75}$, $Ge_{0.25}Se_{0.65}Te_{0.10}$, $Ge_{0.25}Se_{0.60}Sb_{0.12}$ and As_2Se_3 with Kerr nonlinearities hundreds of times larger than silica have been reported by Lenz et al. [112]. It is possible that compositions with even stronger nonlinearity may exist and a method for predicting such compositions would be useful [113].

4.6.4 Super Continuum Generation

Over the past decade, there has been much interest in the generation of supercontinuum (SC) with broad spectral bandwidth [114–116]. A supercontinuum is formed when a collection of nonlinear processes act together upon a pump beam in order to cause severe spectral broadening of the original pump beam, for example using a microstructured optical fiber. The result is a smooth spectral continuum. Broadband SC sources have been found to have many applications in diverse fields [117], including frequency metrology [118], optical coherence tomography [119], and molecular spectroscopy [120]. The amount of spectral broadening depends on the fiber type, pulse duration, and pulse energies used. The use of tapered fibers has reduced the need for high power and allowed for SC generation over short lengths. Chalcogenide (ChG) glasses, such as As_2S_3 and As_2Se_3 are excellent candidates for broadband SC generation due to their high nonlinearity. Most experiments have been done using optical fibers [121, 122] for obtaining long interaction lengths. For instance, as illustrated in Fig. 4.15, a 350-m single-mode fiber, which is excited by a 2.22-W CW laser with a wavelength of 1.483 μm, can emit a 2.1-W output with a broad spectrum extending 1.43–1.53 μm [123].

The amount of spectral broadening depends on the fiber type, pulse duration, and pulse energies used. The use of tapered fibers has reduced the need for high power and allowed for SC generation over short lengths. Chalcogenide (ChG) glasses, such as As_2S_3 and As_2Se_3 are excellent candidates for broadband SC generation due to their high nonlinearity. Tapering ChG fibers has led to efficient SC generation by shifting the zero dispersion wavelength (ZDW) from the mid-IR to the operating wavelengths of available commercial laser sources. A SC covering an octave wavelength bandwidth was generated using As_2S_3 wires from pulses with a peak

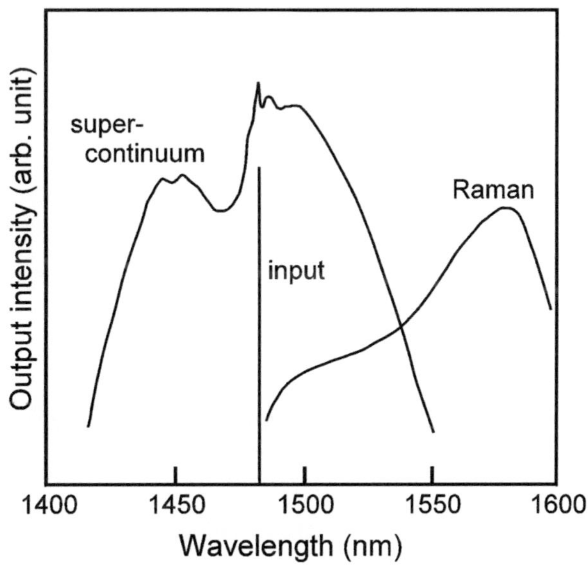

Fig. 4.15 Output spectra of a Raman fiber amplifier [53] and a supercontinuum fiber generator [54], which are excited by 1483 nm light (schematic illustration) [123]

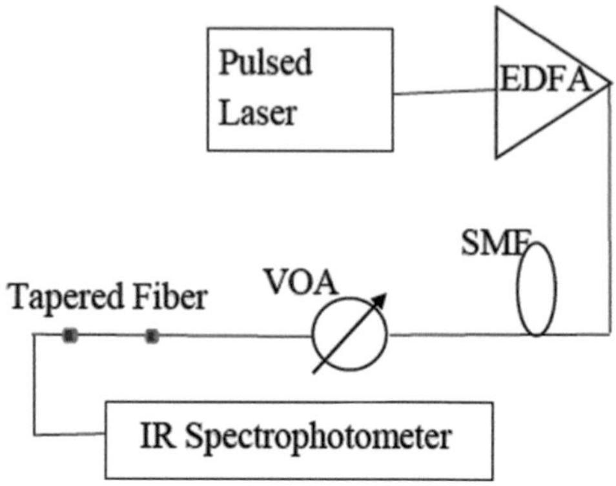

Fig. 4.16 Experimental setup for SC generation. *EDFA* Er-doped fiber amplifier, *SMF* single-mode fiber, *VOA* variable optical attenuator

power of 824 W [124]. However, As_2Se_3 is a more optimal material for SC generation at low peak powers, as it possesses a larger nonlinear refractive index, i.e., up to 1000× that of silica and ~4× that of As_2S_3 [125]. The tapering of As_2Se_3 fibers to subwavelength dimensions has yielded wires with waveguide nonlinearities of $\gamma = 60\ W^{-1}\ m^{-1}$ [126] reaching as high as $\gamma = 176\ W^{-1}\ m^{-1}$ [127]. Using such ChG wires, a SC spanning from 1450 to 1650 nm (−20 dB bandwidth) was achieved at peak powers of only 7.8 W [128]. Figure 4.16 shows the experimental setup for supercontinuum generation.

Two-photon absorption (TPA) is an unavoidable absorption process in As_2Se_3 that is small but nonnegligible at a wavelength of 1550 nm and decreases with increasing wavelength [129–131]. TPA has been shown to limit the efficiency of nonlinear processes including SC generation [128], guided-wave switching [132], Raman gain amplification [133], intrapulse Raman scattering [134], and parametric processes [135]. Quantitatively, the nonlinear absorption (β_{TPA}) for As_2Se_3 drops by a factor of 5 and becomes negligible when the pump wavelength (λ_p) is shifted from 1550 ($\beta_{TPA} = 0.14$ cm/GW) to 1775 nm ($\beta_{TPA} = 0.028$ cm/GW). Therefore, the efficiency of the nonlinear processes could be enhanced by pumping As_2Se_3 at ~1775 nm wavelength where the Kerr nonlinear response is free from a significant TPA [136].

Supercontinuum generation has been demonstrated between 2 and 3 μm in small core sulfide and selenide fibers [87] as well as preliminary photonic crystal selenide fibers [88]. The 1 meter length of fibers were pumped with a Ti:sapphire laser at a wavelength of 2.5 μm using 100 fs pulses and 100 pJ/pulse. The outputs from the fibers are shown in Fig. 4.17.

4.7 Experimental Techniques for Making Optical Fiber [8]

Chalcogenide glasses are made by combining highly purified (>99.9999 purity) raw elements in sealed (glass/quartz) ampoules or in vitreous carbon crucibles located within quartz ampoules and heated and mixed under constant rocking

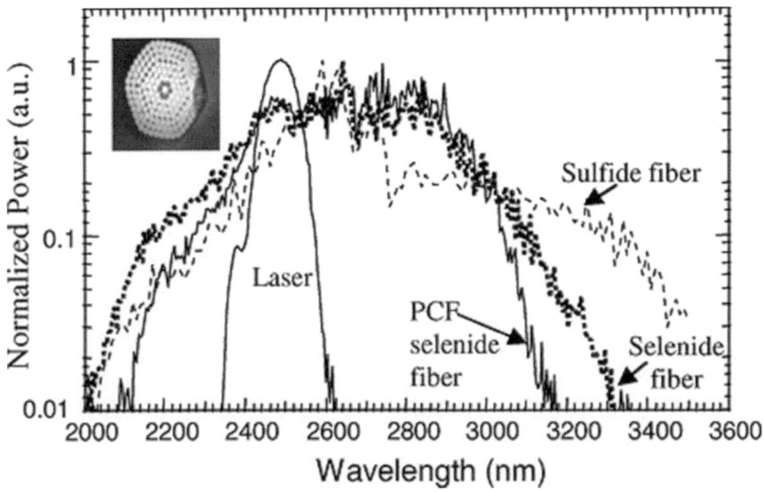

Fig. 4.17 Supercontinuum generation in small core chalcogenide fibers. The *insert* shows the cross-sectional view of the selenide PCF fiber [87]

Fig. 4.18 Schematic representation of fiberization using rod-in-tube technique. Reproduced with permission from [23]

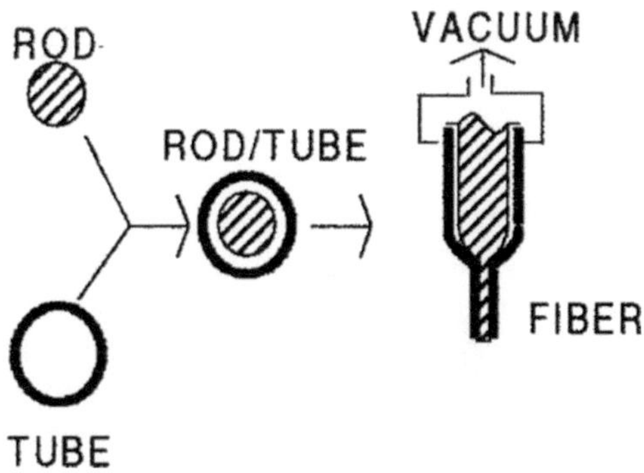

conditions. Typical melt temperatures range from 600 to 1100 °C, depending upon composition. The liquids are quenched and the glass rods annealed at temperatures around the appropriate softening temperatures. The optical fibers are obtained by heating preforms fabricated via rod-in-tube type processes [35, 137] or by double crucible (DC) processes (Figs. 4.18 and 4.19) [35, 138, 139]. The cladding tubes can be obtained via an in situ casting process, which is preferred due to less contamination and better quality surfaces, or by core drilling from larger samples. The preforms can also be obtained by extrusion of core and cladding glass billets [9]. The DC process enables adjustments to be made in the core/clad diameter ratio during fiber drawing by independent pressure control above each melt. Therefore both multimode and single mode fibers can be drawn with relatively fewer processing steps using the DC process. A buffer polymer coating is applied over the cladding using a UV acrylate or by applying a Teflon heat-shrink tube over the preform and then drawing.

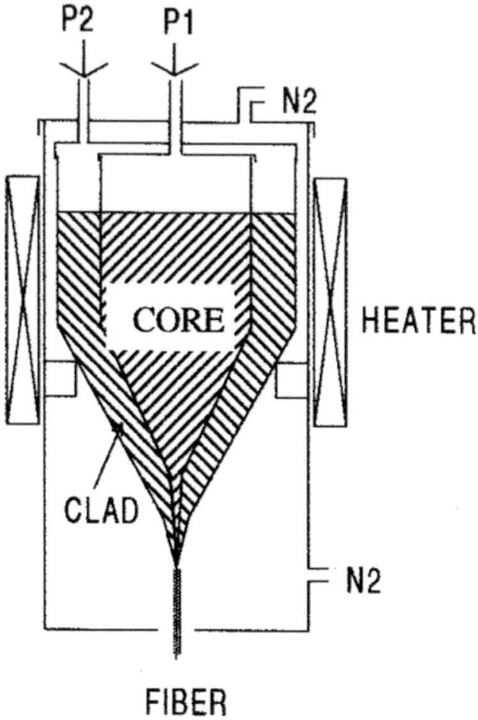

Fig. 4.19 Schematic representation of fiberization using double crucible process. Reproduced with permission from [23]

The first chalcogenide glass microstructured optical fibers (MOF) was fabricated in 2000 [140]. Efforts are still ongoing to meet the applications requirements. Two main techniques have been used to fabricate chalcogenide glass MOFs: (1) capillary-stacking technique, which is the most commonly used to manufacture silica MOFs, and is also widely used for chalcogenide glass MOFs fabrication [141, 142]. The main advantage of this technique is the ability to manufacture microstructured preform with highly complex periodic geometry (>60 holes), but this necessitates a huge amount of tricky handling, and (2) extrusion technique, which seems simpler than the stack and draw one, but which is still quite challenging when aiming for fabrication of preforms with very complex microstructure pattern, like the triangular holey structure with more than three rings of holes [143]. More recently, in view of large scale production, fabrication of di-arsenic tri-sulfide (As_2S_3) glass (MOFs) using a mechanical drilling has been reported [144]. According to this, preforms can be manufactured with various microstructures from simple ones to more complex ones. These fibers could be the base for the fabrication of optical amplifiers, chemical sensors, laser radiation transmission devices, IR fiber radiometers and for the development of new nonlinear fiber optics such as for example new supercontinuum sources exhibiting broadband generation in the infrared spectral range above 2 μm.

Impurities (metal and nonmetals such as C, O, and H) can have significant effect on the optical absorption spectra of the glasses and the precursors for glass formation need to be purified before use. For instance, in the case of arsenic, oxygen present in the raw material is chemically combined and forms arsenic oxides. Carbon is usually present in As in the form of micro inclusions whose size lies from 0.05 μm up to 0.2 μm, with a concentration in commercial high purity arsenic (5 N) reaching 10^7–10^8 cm^{-3} [145–147]. Sublimation is conventionally used for arsenic purification although efficiency of this technique is low. At standard evaporation rates an accumulation of carbon particles takes place at the surface of the sublimed arsenic [146]. To improve this efficiency, M. El-Amraoui et al. [144] tried a very low sublimation rate, up to 10^{-6} cm^3/cm^2 s. However, this low evaporation rate lead to an increasing arsenic/silica ampoule contact time, resulting in further contamination of purified arsenic.

In another approach tried for the purpose, arsenic mono-sulfide (As_4S_4) was chosen as a starting material, instead of arsenic. Arsenic mono-sulfide seemed to be a suitable compound in terms of distillation under vacuum and purification efficiency [144]. The melt of As_4S_4 is a low melting, ($T_{melting} = 320\ °C$) polycrystalline compound having a low viscosity comparable to that of water. This leads to an improvement of purification efficiency. High purity arsenic sulfide glasses can thus be prepared using the following reaction [146]:

$$As_4S_4 + S_x \rightarrow As_4S_{4+x}$$

By adding suitable quantities of pure sulfur, various vitreous compositions of As_xS_y glasses can be prepared. This technique of pure As-S glass preparation was worked out for the first time in 1972 [147]. It led to the design of optical fibers whose losses were 23 dB/km at 2.4 μm. Optical losses of As-S fibers have been reduced to 12 dB/km at 3 μm [148]. After melting of high purity As_2S_3 glass, cylindrical preforms are prepared by ordinary cooling of the liquid batch. The resulting cylinders are then annealed in a conventional way in order to remove mechanical strains which could lead to glass breakdown during the MOF preform mechanical shaping step.

4.7.1 Preforms and Microstructured Fibers Fabrication

Mechanical drilling was performed for the preparation of preforms with a variety of geometrical patterns in view of MOFs drawing [144]. By optimization of the drilling parameters, especially rotation speed and applied force, drilling of a regular pattern of holes in a glass rod was done with quite thin matter bridges between the holes. Versatile patterned preforms were prepared for applications such as—simple rods intended to mono-index fiber drawing, preforms for standard core clad fibers, preforms devoted to "traditional" microstructured fibers with several rings of holes in a triangular pattern (18 holes, 36 holes and 64 holes), preforms for other microstructured fibers profiles with various numbers of holes, and preforms for suspended core fibers with three, four or six holes around a solid core. For all these preforms the outer diameter was 16 mm, the diameter *d* of holes was equal to

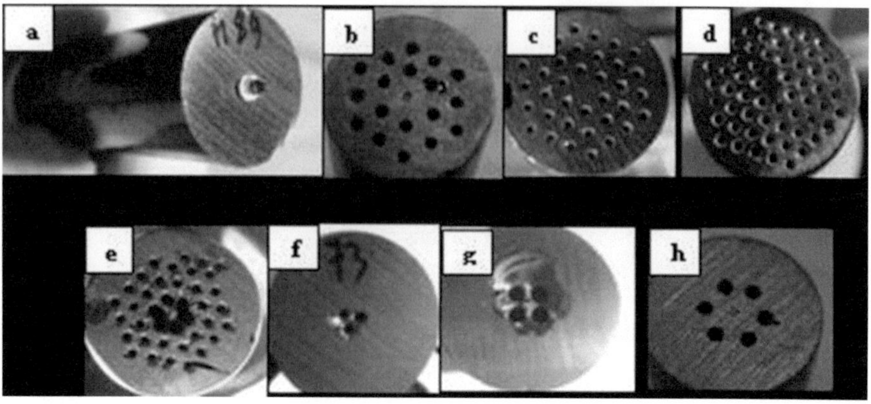

Fig. 4.20 Pictures of 16 mm outer diameter preforms elaborated by mechanical drilling. Preform for core clad fiber (**a**: 1 hole); Preforms for microstructured fibers (**b**: 2 rings/18 holes; **c**: 3 rings/36 holes; **d**: 4 rings/64 holes; **e**: 3 rings/34 holes + 2 big holes); Preforms for suspended core fibers (**f**: 3 holes, **g**: 4 holes, **h**: 6 holes). Permission obtained [144]

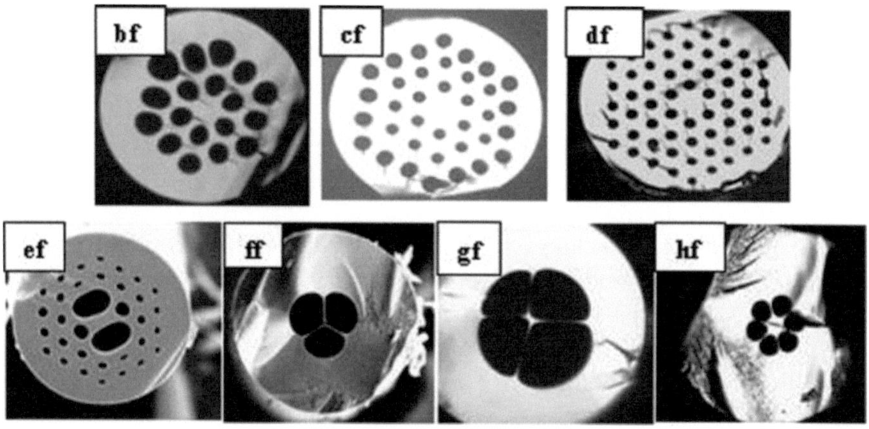

Fig. 4.21 Geometrical profiles of microstructured optical fibers drawn from performs depicted in Fig. 4.20. (**bf**, **cf**, **df**, **ef**, **ff**, **gf**, and **hf** correspond to the preforms **b**, **c**, **d**, **e**, **f**, **g**, and **h**, respectively). Depending on fibers, the external diameter varies from 120 μm up to 160 μm. Permission obtained [144]

0.8 mm and their length about 30 mm. Figure 4.20 shows some pictures of the obtained preforms.

Parameters that need to be controlled during MOFs drawing include (1) preform temperature, (2) translation speed, (3) fiber drawing speed, (4) pressure in the holes, and (5) flow rate of inert gas circulating along the preform. By selecting suitable sets of parameters various kinds of holey microstructured and suspended core fibers can be drawn. Figure 4.21 shows the pictures of the sections of several MOFs that have been drawn from the preforms obtained using the mechanical drilling technique illustrated in Fig. 4.20.

4.8 Summary

Among the family of chalcogenide glasses, As_2S_3 and As_2Se_3 are important (IR) transparent materials for a variety of applications such as IR sensors, waveguides, and photonic crystals. Broadly tunable near and mid-infrared lasers are of interest for applications such as high-resolution spectroscopy, meterology, pumping of nonlinear optical frequency converters, and chemical sensing. Unlike silica glass fibers, chalcogenide fibers have good transmission in the IR beyond 1.5 μm. Sulfur-based chalcogenide glass compositions are non-hygroscopic and have good chemical resistance. As hosts for rare earth ions, the low phonon energy of chalcogenide glasses results in a smaller multiphonon transition probability, allowing many mid-IR and long-wave IR rare earth transitions to become active, making them potential candidates for mid-IR and long-wave IR fiber lasers. The high nonlinearity of chalcogenide glasses allows the potential for mid-IR and long-wave IR Raman fiber lasers.

References

1. H. Nishihara, M. Haruna, T. Suhara, in *Optical and Electro-optical Engineering Series*, ed. by R.E. Fisher, W.E. Smith, vol. 215 (McGraw-Hill, New York, 1989), p. 151
2. N.S. Kapany, R.J. Simms, Recent developments of infrared fiber optics. Infrared Phys. **5**, 69–75 (1965)
3. Z.U. Borisova, *Glassy Semiconductors* (Plenum Press, New York, 1981)
4. J.S. Sanghera, J. Heo, J.D. Mackenzie, J. Non-Cryst. Solids **103**, 155 (1988)
5. J.S. Sanghera, V.Q. Nguyen, P.C. Pureza, F.H. Kung, R. Miklos, I.D. Aggarwal, J. Lightwave Technol. **12**, 737 (1994)
6. E. Snitzer, K. Wei, US Patent 5,379, 1995, 149
7. B. Aitken, M.A. Newhouse, US Patent 5,389, 1995, 584
8. D.W. Hewak, R.S. Deol, J. Wang, G. Wylangowski, J.A. Mederios Neto, B.N. Samson, R.I. Laming, W.S. Brocklesby, D.N. Payne, A. Jha, M. Poulain, S. Otero, S. Surinach, M.D. Baro, Electron. Lett. **29**, 237 (1993)
9. H. Tawarayama, E. Ishikawa, K. Itoh, H. Aoki, H. Yanagita, K. Okada, K. Yamanaka, Y. Matsuoka, H. Toratani, *Optical Fiber Conference, Victoria, Canada, PD1–1* (Optical Society of America, Washington, DC, 1997)
10. V. Krasteva, A. Yurkina, D. Machewirth, G. Sigel Jr., J. Non-Cryst. Solids **213–214**, 304 (1997)
11. D.A. Turnbull, S.Q. Gu, S.G. Bishop, J. Appl. Phys. **80**, 2436 (1996)
12. G. Edwards, R. Logan, M. Copeland, L. Reinisch, J. Davidson, B. Johnson, R. Maciunas, M. Mendenhall, R. Ossoff, J. Tribble, J. Wekhaven, D. O'Day, Tissue ablation by a free-electron laser tuned to the amide II band. Nature **371**, 416–418 (1994)
13. J. Hetch, *City of Lights* (Oxford University Press, New York, 1999)
14. J.A. Harrington, *Infrared Fibers and Their Applications* (SPIE Press, Bellingham, WA, 2004)
15. P.A. Thielen, Modeling of a mid-IR chalcogenide fiber Raman laser. Opt. Express **11**(24), 3284 (2003)
16. S.F. Carter, M.W. Moore, D. Szebesta, D. Ransom, P.W. France, Low loss fluoride fibre by reduced pressure casting. Electron. Lett. **26**, 2115–2117 (1990)

17. R. Nubling, J.A. Harrington, Optical properties of single-crystal sapphire fibers. Appl. Opt. **36**, 5934–5940 (1997)
18. J. Nishii, S. Morimoto, I. Inagawa, R. Iizuka, T. Yamashita, T. Yamagishi, Recent advances and trends in chalcogenide glass fiber technology: a review. J. Non-Cryst. Solids **140**, 199–208 (1992)
19. V. Artjushenko, V. Ionov, K.I. Kalaidjian, A.P. Kryukov, E.F. Kuzin, A.A. Lerman, A.S. Prokhorov, E.V. Stepanov, K. Bakhshpour, K.B. Moran, W. Neuberger, Infrared fibers: power delivery and medical applications. Proc. SPIE **2396**, 25–36 (1995)
20. Y. Matsuura, T. Abel, J.A. Harrington, Optical properties of small-bore hollow glass waveguides. Appl. Opt. **34**, 6842–6847 (1995)
21. J.A. Harrington, *"Infrared Fiber Optics" OSA Handbook*, vol IV (McGraw-Hill, USA, 2001), p. 14.2. ISBN 0-07-136456-0
22. J.S. Sanghera, L. Brandon Shaw, I.D. Aggarwal, Applications of chalcogenide glass optical fiber. C.R. Chimie **5**, 873–883 (2002)
23. J.S. Sanghera, I.D. Aggarwal, Active and passive chalcogenide glass optical fibers for IR applications: a review. J. Non-Cryst. Solids **256–257**, 6–16 (1996)
24. J.S. Sanghera, I.D. Aggarwal, *Infrared Fiber Optics* (CRC Press Inc., Boca Raton, FL, 1998)
25. Y. Kanamori, Y. Terunuma, T. Miyashita, Preparation of chalcogenide optical fiber. Rev. Electrical Comm. Lab **32**, 469–477 (1984)
26. L.E. Busse, J.A. Moon, J.S. Sanghera, I.D. Aggarwal, Laser Focus World **32**, 143 (1996)
27. L. Busse, J. Moon, J.S. Sanghera, I.D. Aggarwal, J. Harrington, K.K. Lum, *Proceedings of the 1995 IRIS Speciality Group on Materials* (Erim, Ann Arbor, MI, 1995), p. 237
28. I.D. Aggarwal, L.E. Busse, L.B. Shaw, B. Cole, J.S. Sanghera, in *Proceedings of the Diode Laser Technology Review*, Albuquerque, NM, 1998
29. J.S. Sanghera, I.D. Aggarwal, in *Proceedings of the 18th International Congress on Glass*, San Francisco, CA, 5–10 July 1998
30. P. Melling, *Commercial Literature*, Rempec Inc
31. J. Heo, M. Rodrigues, S. Saggese, G.H. Sigel Jr., Appl. Opt. **30**, 3944 (1991)
32. J.S. Sanghera, F.H. Kung, L.E. Busse, P.C. Pureza, I.D. Aggarwal, J. Am. Ceram. Soc. **78**, 2198 (1995)
33. X.H. Zhang, M.V. Duhamel, H.L. Ma, C. Blanchetiere, J. Lucas, J. Non-Cryst. Solids **161**, 547 (1993)
34. M. Druy, in *Infrared Fiber Optics*, ed. by J.S. Sanghera, I.D. Aggarwal (CRC, Boca Raton, FL, 1998). ch. 8
35. J.S. Sanghera, I.D. Aggarwal, L. Busse, P. Pureza, V. Nguyen, R. Miklos, F. Kung, R. Mossadegh, SPIE **2396**, 71 (1995)
36. M. Saito, Technology digest first workshop on optical fiber sensors, Jpn. Soc. Appl. Phys. 113 (1985)
37. G. Nau, F. Bucholtz, K.J. Ewing, S.T. Vohra, J.S. Sanghera, I.D. Aggarwal, SPIE **2883**, 682 (1996)
38. J.S. Sanghera, G. Nau, P.C. Pureza, I.D. Aggarwal, US Patent 5,525,800, 1996
39. B. Rigas, P.T.T. Wong, Cancer Res. **52**, 84 (1992)
40. T. Ueda, K. Yamad, T. Sugita, J. Eng. Ind. **114**, 317 (1992)
41. N.S. Kapany, R.J. Simms, Infrared Phys. **5**, 69 (1996)
42. M. Saito, M. Takizawa, S. Sakuragi, F. Tanei, Appl. Opt. **24**(9), 2304 (1985)
43. J. Nishii, T. Yamashita, T. Yamagishi, C. Tanaka, H. Stone, Appl. Phys. Lett. **59**, 2639 (1991)
44. H. Suto, Infrared Phys. Technol. **38**, 93 (1997)
45. M.K. Hong, S. Erramilli, P. Huie, G. James, A. Jeung, SPIE **2863**, 54 (1997)
46. D.T. Schaafsma, R. Mossadegh, J.S. Sanghera, I.D. Aggarwal, J.M. Gilligan, N.H. Tolk, M. Luce, R. Generosi, P. Perfetti, A. Cricenti, G. Margaritondo, Ultramicroscopy **77**, 77 (1999)
47. J.S. Sanghera, L.B. Shaw, L.E. Busse, D. Talley, I.D. Aggarwal, SPIE, Infrared transmitting fiber optics for biomedical applications. Proceedings Photonics West, San Diego, CA, SPIE **3596**, 178 (1999)

48. D.T. Schaafsma, J.A. Moon, J.S. Sanghera, I.D. Aggarwal, J. Lightwave Technol. **15**(12), 2242 (1997)
49. T. Schweizer, B.N. Samson, R.C. Moore, D.W. Hewak, D.N. Payne, Electron. Lett. **33**(5), 414 (1997)
50. A. Mori, Y. Ohishi, T. Kanamori, S. Sudo, Appl. Phys. Lett. **70**(10), 1230 (1997)
51. L.B. Shaw, B.J. Cole, J.S. Sanghera, I.D. Aggarwal, D.T. Schaafsma, *Optical Fiber Communications* (Paper WG8, San Jose, CA, 1998)
52. L.B. Shaw, D.T. Schaafsma, B.J. Cole, B. Harbison, J.S. Sanghera, I.D. Aggarwal, SPIE **3368**, 42 (1998)
53. M. Asobe, T. Ohara, I. Yokohama, T. Kaino, Electron. Lett. **32**, 1611 (1996)
54. M. Asobe, T. Kanamori, K. Kubodera, IEEE Photon. Technol. Lett. **4**, 362 (1992)
55. M.T. de Aruajo, M.V.D. Vermelho, A.S. Gouveia-Net, A.S.B. Sombra, J.A. Medeiros Neto, IEEE Photon. Technol. Lett. **8**, 821 (1996)
56. R. Reisfeld, A. Bornstein, Chem. Phys. Lett. **47**, 194 (1997)
57. A. Bornstein, R. Reisfeld, J. Non-Cryst. Solids **50**, 23 (1982)
58. R. Reisfeld, A. Bornstein, J. Non-Cryst. Solids **27**, 143 (1978)
59. R. Reisfeld, Ann. Chim. Fr. **7**, 147 (1982)
60. C.C. Ye, D.W. Hewak, M. Hempstead, B.N. Samson, D.N. Payne, J. Non-Cryst. Solids **208**, 56 (1996)
61. J. Moon, B.B. Harbison, J.S. Sanghera, I.D. Aggarwal, in *Proc. Photonics' 96*, Madras, India, 9–13 December 1996
62. Y.B. Shin, W.Y. Cho, J. Heo, J. Non-Cryst. Solids **208**, 29 (1996)
63. Y.B. Shin, J.N. Jang, J. Heo, Opt. Quant. Electron. **27**, 379 (1995)
64. L.B. Shaw, B.H. Harbison, B. Cole, J.S. Sanghera, I.D. Aggarwal, Opt. Express **1**, 87 (1997)
65. T. Schweizer, D.W. Hewak, B.N. Samson, D.N. Payne, J. Lumin. **72–74**, 419 (1997)
66. P.F. Moulton et al., Tm-doped fiber lasers: fundamentals and power scaling. IEEE J. Sel. Top. Quant. Electron. **15**(1), 85–92 (2009)
67. D. Faucher et al., 20 W passively cooled single-mode all-fiber laser at 2.8 m. Opt. Lett. **36**(7), 1104–1106 (2011)
68. D. Faucher et al., Erbium-doped all-fiber laser at 2.94 m. Opt. Lett. **34**(21), 3313–3315 (2009)
69. D. Hudson et al., 1 W diode-pumped tunable Ho^{3+}, Pr^{3+}-doped fluoride glass fibre laser. Electron. Lett. **47**(17), 985–986 (2011)
70. T. Hu, D.D. Hudson, S.D. Jackson, Actively Q-switched 2.9 m $Ho^{3+}Pr^{3+}$-doped fluoride fiber laser. Opt. Lett. **37**(11), 2145–2147 (2012)
71. C. Carbonnier, H. Többen, U.B. Unrau, Room temperature CW fibre laser at 3.22 m. Electron. Lett. **34**(9), 893–894 (1998)
72. H. Többen, Room temperature CW fibre laser at 3.5 m in Er3+-doped ZBLAN glass. Electron. Lett. **28**(14), 1361–1362 (1992)
73. J. Schneider, C. Carbonnier, U.B. Unrau, Characterization of a Ho3+-doped fluoride fiber laser with a 3.9 m emission wavelength. Appl. Opt. **36**(33), 8595–8600 (1997)
74. S.D. Jackson, Nat. Photon. **6**, 423 (2012)
75. S.D. Jackson, A. Lauto, Lasers Surg. Med. **30**, 184 (2002)
76. M. Yamada, M. Shimizu, Y. Ohishi, J. Temmyo, M. Wada, T. Kanamori, M. Horiguchi, S. Takahashi, IEEE Photon. Technol. Lett. **9**, 994 (1992)
77. Biolase, http://www.biolase.com
78. Kavo Dental, http://www.kavo.com
79. Convergent Laser Technologies, http://www.convergentlaser.com
80. IPG Photonics, www.ipgphotonics.com
81. Y. Yao, A.J. Hoffman, C.F. Gmachl, Mid-infrared quantum cascade lasers. Nat. Photon. **6**(7), 432–439 (2012)
82. Nufern, http://www.nufern.com
83. M. Bernier, V. Fortin, M. El-Amraoui, Y. Messaddeq, R. Vallée, 3.77 μm fiber laser based on cascaded Raman gain in a chalcogenide glass fiber. Opt. Lett. **39**(7), 2052–2055 (2014)

84. M. De Sario, L. Mescia, F. Prudenzano et al., Feasibility of Er3+-doped, $Ga_5Ge_{20}Sb_{10}S_{65}$ chalcogenide microstructured optical fiber amplifiers. Opt. Laser Technol. **41**(1), 99–106 (2009)
85. J.S. Sanghera, L.B. Shaw, C.M. Florea, P. Pureza, V.Q. Nguyen, F. Kung, D. Gibson, I.D. Aggarwal, in *Nonlinear Properties of Chalcogenide Glass Fibers, Frontiers in Guided Wave Optics and Optoelectronics* ed. by B. Pal (InTech, 2010) ISBN: 978-953-7619-82-4. http://www.intechopen.com/books/frontiers-in-guided-wave-optics-and-optoelectronics/nonlinear-properties-ofchalcogenide-glass-fibers
86. M. Asobe, T. Kanamori, K. Naganuma, H. Itoh, T. Kaino, Third order nonlinear spectroscopy in As2S3 chalcogenide glass fiber. J. Appl. Phys. **77**, 5518–5523 (1995)
87. J.S. Sanghera, L.B. Shaw, P. Pureza, V.Q. Nguyen, D. Gibson, I.D. Aggarwal, *Progress of Chalcogenide Glass Fibers* (OSA, Washington, DC, 2007). ISBN 1-55752-830-6
88. R.E. Slusher, J. Hodelin, J.S. Sanghera, L.B. Shaw, I.D. Aggarwal, JOSA-B **21**, 1146 (2004)
89. V. Moizan, V. Nazabal, J. Troles et al., Er3+-doped GeGaSbS glasses for mid-IR fibre laser application: synthesis and rare earth spectroscopy. Opt. Mater. **31**(1), 39–46 (2008)
90. F. Prudenzano, L. Mescia, L. Allegretti, V. Moizan, V. Nazabal, F. Smektala, Theoretical study of cascade laser in erbium-doped chalcogenide glass fibers. Opt. Mater. **33**(2), 241–245 (2010)
91. L. Mescia, F. Smektala, F. Prudenzano, Review article: new trends in amplifiers and sources via chalcogenide photonic crystal fibers. Int. J. Opt. **2012**, 575818 (2012). doi:10.1155/2012/575818. 8 pages
92. M. Bernier, M. El-Amraoui, J.F. Couillard, Y. Messaddeq, R. Vallée, Writing of Bragg gratings through the polymer jacket of low-loss As_2S_3 fibers using femtosecond pulses at 800 nm. Opt. Lett. **37**(18), 3900–3902 (2012)
93. https://en.wikipedia.org/wiki/Fiber_Bragg_grating
94. M. Bernier, K. Asatryan, R. Vallee, T. Galstian, S.A. Vasil'ev, O.I. Medvedkov, V.G. Plotnichenko, P.I. Gnusin, E.M. Dianov, Quant. Electron. **41**, 465 (2011)
95. C. Florea, J.S. Sanghera, B. Shaw, I.D. Aggarwal, Opt. Mater. **31**, 942 (2009)
96. S.J. Mihailov, D. Grobnic, C.W. Smelser, P. Lu, R.B. Walker, H. Ding, Laser Chem. **2008**, 416251 (2008)
97. S.J. Mihailov, D. Grobnic, C.W. Smelser, Electron. Lett. **43**, 442 (2007)
98. M. Bernier, S. Gagnon, R. Vallée, Opt. Mater. Express **1**, 832 (2011)
99. C. Florea, J. Sanghera, I. Aggarwal, Opt. Mater. **30**, 1603 (2008)
100. D. Grobnic, S.J. Mihailov, C.W. Smelser, R.B. Walker, Proc. SPIE **6796**, 67961K (2007)
101. V. Fortin, M. Bernier, D. Faucher, J. Carrier, R. Vallée, Opt. Express **20**, 19412 (2012)
102. S. Kasap, P. Capper (eds.), *Springer Handbook of Electronic and Photonic Materials* (Springer, New York, 2007), p. 1063
103. R.L. Sutherland, *Handbook of Nonlinear Optics*, 2nd edn. (Marcel Dekker, New York, 2003)
104. R.W. Boyd, *Nonlinear Optics*, 2nd edn. (Academic, San Diego, 1992)
105. K. Tanaka, Optical nonlinearity in photonic glasses. J. Mater. Sci.: Mater. Electron. **16**, 633–643 (2005)
106. T. Morioka, M. Saruwatari, IEEE J. Sel. Area Commun. **6**, 1186 (1988)
107. B.P. Nelson, K.J. Blow, P.D. Constantine, N.J. Noran, J.K. Lucek, I.W. Marshall, K. Smith, Electron. Lett. **27**, 704 (1991)
108. M.A. Newhouse, D.L. Weidman, D.W. Hall, Opt. Lett. **15**, 1185 (1990)
109. K. Petkov, P.J.S. Ewen, J. Non-Cryst, Solids **249**, 150–159 (1999)
110. K.A. Cerqua-Richardson, J.M. Mckinley, B. Lawrence, S. Joshi, A. Villeneuve, Opt. Mater. **10**, 155 (1998)
111. T. Cardinal, K.A. Richardson, H. Shim, A. Schulte, R. Beatty, K. Le Foulgoc, C. Meneghini, J.F. Viens, A. Villeneuve, J. Non-Cryst. Solids **256–257**, 353 (1999)
112. G. Lenz, J. Zimmerman, T. Katsufuji, M.E. Lines, H.Y. Hwang, S. Spalter, R.E. Slusher, S.W. Cheong, J.S. Sanghera, I.D. Aggarwal, Opt. Lett. **25**, 254 (2000)

113. G. Kaur, F. Wang, Y.M. Yiu, D.W. Shoesmith, M. Zinke-Allmang, T.-K. Sham, Z. Ding, Surface second harmonic generation of Se–Te–Sb films. J. Mater. Sci. Mater. Electron. **20**, S164–S169 (2009). doi:10.1007/s10854-007-9498-8
114. W. Wadsworth, A. Ortigosa-Blanch, J. Knight, T. Birks, T. Man, P. Russell, J. Opt. Soc. Am. B **19**, 2148 (2002)
115. J.M. Dudley, G. Gënty, S. Coen, Rev. Mod. Phys. **78**, 1135 (2006)
116. G. Gënty, S. Coen, J.M. Dudley, J. Opt. Soc. Am. B **24**, 1771 (2007)
117. S.V. Smirnov, J.D. Ania-Castanon, T.J. Ellingham, S.M. Kobtsev, S. Kukarin, S.K. Turitsyn, Opt. Fiber Technol. **12**, 122 (2006)
118. R.H.T. Udem, T.W. Hänsch, Nature **416**, 233 (2002)
119. I. Hartl, X.D. Li, C. Chudoba, R.K. Ghanta, T.H. Ko, J.G. Fujimoto, J.K. Ranka, R.S. Windeler, Opt. Lett. **26**, 608 (2001)
120. S.T. Sanders, Appl. Phys. B **75**, 799 (2002)
121. G.P. Agrawal, *Nonlinear Fiber Optics*, 3rd edn. (Academic, San Diego, 2001)
122. G.I. Stegeman, R.H. Stolen, J. Opt. Soc. Am. B **6**, 652 (1989)
123. A. Zheltikov, Appl. Phys. B: Lasers Opt. **77**, 143 (2003)
124. D.D. Hudson, S.A. Dekker, E.C. Mägi, A.C. Judge, S.D. Jackson, E. Li, J.S. Sanghera, L.B. Shaw, I.D. Aggarwal, B.J. Eggleton, Opt. Lett. **36**, 1122 (2011)
125. B.J. Eggleton, B. Luther-Davies, K. Richardson, Nat. Photon. **5**, 1749 (2011)
126. E.C. Mägi, L. Fu, H. Nguyen, M. Lamont, D. Yeom, B. Eggleton, Opt. Express **15**, 10324 (2007)
127. C. Baker, M. Rochette, IEEE Photon. J. **4**, 960 (2012)
128. D.I. Yeom, E.C. Mägi, M.R.E. Lamont, M.A.F. Roelens, L. Fu, B.J. Eggleton, Opt. Lett. **33**, 660 (2008)
129. M. Sheik-Bahae, D.C. Hutchings, D.J. Hagan, E.W. Van Stryland, IEEE J. Quant. Electron. **27**, 1296 (1991)
130. N.F. Mott, E.A. Davis, *Electronic Processes in Non-Crystalline Materials*, 2nd edn. (Oxford University Press, New York, 1979)
131. J. Troles, F. Smektala, G. Boudebs, A. Monteil, Opt. Mater. **22**, 335 (2003)
132. V. Mizrahi, K.W. DeLong, G.I. Stegeman, M.A. Saifi, M.J. Andrejco, Opt. Lett. **14**, 1140 (1989)
133. A. Tuniz, G. Brawley, D.J. Moss, B.J. Eggleton, Opt. Express **16**, 18524 (2008)
134. H. Steffensen, C. Agger, O. Bang, J. Opt. Soc. Am. B **29**, 484 (2012)
135. R. Ahmad, M. Rochette, Opt. Express **20**, 10095 (2012)
136. A. Al-kadry, C. Baker, M. El Amraoui, Y. Messaddeq, M. Rochette, Broadband supercontinuum generation in As_2Se_3 chalcogenide wires by avoiding the two-photon absorption effect. Opt. Lett. **38**(7), 1185 (2013)
137. J. Nishii, T. Yamashita, T. Yamagishi, Appl. Opt. **28**, 5122 (1988)
138. T. Kanamori, Y. Terunuma, S. Takahashi, T. Miyashita, J. Lightwave Technol. **2**, 607 (1984)
139. A.V. Vasil'ev, G.G. Devyatykh, E.M. Dianov, A.N. Gur'yanov, A.Y. Laptev, V.G. Plotnichenko, Y.N. Pyrkov, G.E. Snopatin, I.V. Skripachev, M.F. Churbanov, V.A. Shipunov, Quant. Electron. **23**, 89 (1993)
140. T.M. Monro, Y.D. West, D.W. Hewak, N.G.R. Broderick, D.J. Richardson, Chalcogenide holey fibres. Electron. Lett. **36**(24), 1998–2000 (2000)
141. F. Désévédavy, G. Renversez, L. Brilland, P. Houizot, J. Troles, Q. Coulombier, F. Smektala, N. Traynor, J.L. Adam, Small-core chalcogenide microstructured fibers for the infrared. Appl. Opt. **47**(32), 6014–6021 (2008)
142. F. Smektala, F. Désévédavy, L. Brilland, P. Houizot, J. Troles, N. Traynor, Advances in the elaboration of chalcogenide photonic crystal fibers for the mid infrared. SPIE **6588**, 58803 (2007)
143. X. Feng, A.K. Mairaj, D.W. Hewak, T.M. Monro, Non silica glasses for holey fibers. J. Lightwave Technol. **23**(6), 2046–2053 (2005)

144. M. El-Amraoui, G. Gadret, J. C. Jules, J. Fatome, C. Fortier, F. Désévédavy, I. Skripatchev, Y. Messaddeq, J. Troles, L. Brilland, W. Gao, T. Suzuki, Y. Ohishi, F. Smektala (2010) Microstructured chalcogenide optical fibers from As2S3 glass: towards new IR broadband sources. Opt. Express 18(25), 26655
145. M.F. Churbanov, High purity chalcogenide glasses as materials for fiber optics. J. Non-Cryst. Solids **184**, 25–29 (1995)
146. G.G. Devyatykh, M.F. Churbanov, I.V. Scripachev, G.E. Snopatin, E.M. Dianov, V.G. Plotnichenko, Recent developments in As-S glass fibres. J. Non-Cryst. Solids **256&257**, 318–322 (1999)
147. D.L. Wood, J. Tauc, Weak absorption tails in amorphous semiconductors. Phys. Rev. B **5**(8), 3144–3151 (1972)
148. G.E. Snopatin, M.F. Churbanov, A.A. Pushkin, V.V. Gerasimenko, E.M. Dianov, V.G. Plotnichenko, High purity arsenic-sulfide glasses and fibers with minimum attenuation of 12 dB/km. Optoelectron. Adv. Mater. Rapid Commun. **3**(7), 669–671 (2009)

Part III
Selenium

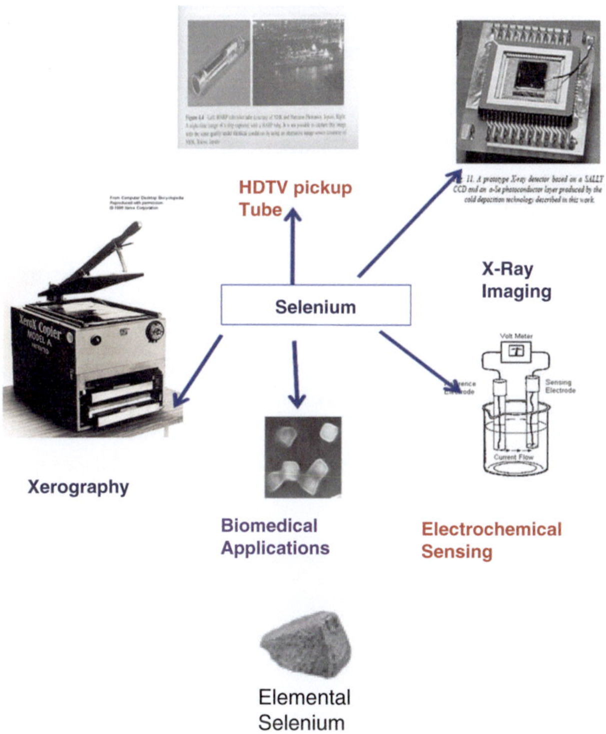

Selenium

The Element

Se—34—$1s^2\ 2s^2\ 2p^6\ 3s^2\ 3p^6\ 3d^{10}\ 4s^2\ 4p^4$

Atomic weight—78.96; m.p.—221 °C and has a B.P.—685 °C. Specific gravity—4.79 (vitreous); valence of −2, +4, or +6 under different conditions.

Selenium is a nonmetallic chemical element with symbol Se and atomic number 34. It was discovered in 1817 by Jons Jacob Berzelius who found it associated with tellurium meaning Earth. The term selenium originates from Latin silex, silicis, flint and the Greek word selene meaning moon. Selenium is found in a few rare minerals, such as crooksite and clausthalite. In years past, it has been obtained from flue dusts remaining from processing copper sulfide ores, but the anode muds from electrolyte copper refineries now provide the source of most of world's selenium. Selenium is recovered by roasting the muds with soda or sulfuric acid, or by smelting them with soda and niter. The color of amorphous selenium is either red, in powder form, or black, in vitreous form Fig. 1. Crystalline monoclinic selenium is a deep red, crystalline hexagonal selenium; the most stable variety is gray. Natural selenium contains six stable isotopes. Elemental selenium is nontoxic and is considered to be an essential trace element; however, hydrogen selenide and other selenium compounds are extremely toxic and resemble arsenic in their physiological reactions. Hydrogen selenide in a concentration of 1.5 ppm is intolerable to man.

(Atomic symbol: Se, atomic number: 34) is a Block P, Group 16, Period 4 element with an atomic radius of 78.96. The number of electrons in each of selenium's shells is 2, 8, 18, and 6 and its electron configuration is [Ar] $3d^{10}\ 4s^2\ 4p^4$. The selenium atom has a radius of 120 pm and a van der Waals radius of 190 pm.

Fig. 1 Black and red allotropes

Structure of trigonal selenium

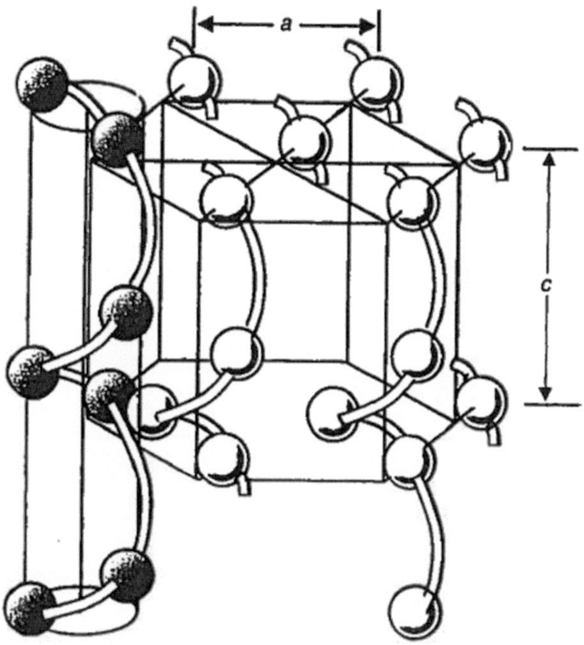

Selenium is a nonmetal with several allotropes: a black, vitreous form with an irregular crystal structure; three red-colored forms with monoclinic crystal structures; and a gray form with a hexagonal crystal structure, the most stable and dense form of the element. The two major crystalline allotropes are α-monoclinic (α-) and trigonal (γ-), of which the latter is the most stable at room temperature. One of the most common uses for selenium is in glass production; the red tint that it lends to glass neutralizes green or yellow tints from impurities in the glass materials.

There are two basic arrangements in crystalline Se. The hexagonal form is stable at room temperature; the atoms in this structure are arranged in long parallel chains with threefold spiral about the c-axis, with the three atoms in each unit of the spiral exhibiting D2 molecular symmetry. The bonding between atoms in a chain is covalent, and the bonding between chains is of van der Waals type. The α and β monoclinic forms of selenium have the same molecular unit, which consists of an eight-atom puckered ring with D4d molecular symmetry. Alternate atoms in the ring lie in two planes separated by about 1.3A, with the four atoms in each plane at the corners of a square. Bonding within the rings is covalent, and between the rings, van der Waals. The vitreous form is readily obtained on cooling the liquid, with the glass transition at about 40 °C.

Chapter 5
Imaging and Detection

Gurinder Kaur Ahluwalia

Aurora Picture taken with HARP-based camera. Courtesy: Dr. Kenkichi Tanioka, NHK, Japan

G.K. Ahluwalia, Ph.D (✉)
Materials and Nanotechnology Research Laboratory, Department of Physics College of The North Atlantic, Labrador West Campus, 1600 Nichols Adam Highway, Labrador City, A2V 0B8, NL, Canada
e-mail: gurinder.kaur@cna.nl.ca

G.K. Ahluwalia (ed.), *Applications of Chalcogenides: S, Se, and Te*,
DOI 10.1007/978-3-319-41190-3_5

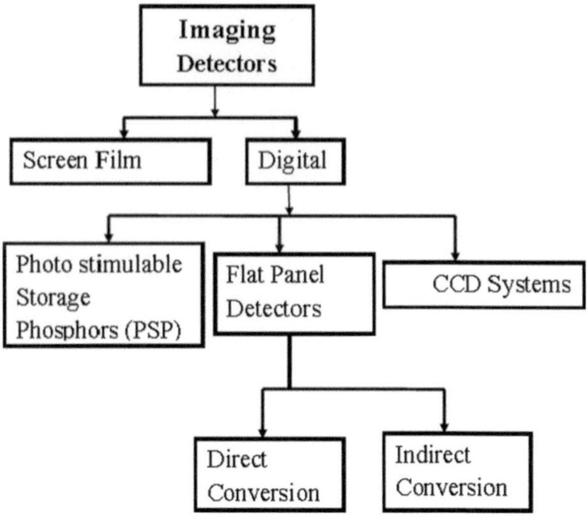

Fig. 5.1 Classification of imaging detectors

5.1 Introduction

Radiation-based diagnostic techniques such as X-ray imaging, computed tomography (CT), fluoroscopy, and molecular imaging have revolutionized the treatment of disorders in the field of medical sciences over the last century. Prototypic thin-film transistor-based systems with X-ray-sensitive elements can achieve detective quantum efficiencies of 65 % or more which greatly exceeds the performance of photostimulable storage phosphor systems, which have efficiencies of 20–35 % [1], and of traditional screen–film systems for chest radiography, which have nominal efficiencies of 25 % [2], and is comparable to the performance of a commercial selenium drum digital chest radiographic system [3]. Figure 5.1 gives a broad classification of the type of imaging detectors used for X-ray imaging techniques. Digital detectors are discussed in more detail in the subsequent sections in this chapter.

5.2 X-Ray Imaging for Medical Applications

When imaging with X-rays, an X-ray beam produced by an X-ray tube passes through the body. On its way through the body, parts of energy of the X-ray beam are absorbed depending on density and composition of the material it passes through and resulting in attenuation of the X-ray beam. Bone is very dense and absorbs or attenuates a great deal of the X-rays. The soft tissue around the bones is much less dense and attenuates or absorbs far less X-ray energy. It is these differences in absorption and the corresponding varying exposure level of the film that creates the images which can clearly show broken bones, clogged blood vessels, cancerous tissues, and other abnormalities. On the opposite side of the body, detectors or a film capture the attenuated X-rays, resulting in a clinical image. In conventional methods, a special phosphor coating inside the cassette glows and exposes the film. The resulting film is then developed much like a regular photograph. It is the special energy and wavelength of the X-rays which allow them to pass through the body part and create the image of the internal structures such as bones of the hand. Traditionally, X-ray images were exposed onto photographic film and needed much processing before they could be viewed. From organizational point of view, storage of the bulky exposed films for long durations also posed several challenges in hospitals and institutions. On the other hand, use of digital X-rays involves taking an electronic image which is processed by a computer, enabling it to be stored digitally and viewed on screen immediately without processing in addition to an improved image quality. Charge-coupled devices were the first direct-readout detectors to be introduced in radiology approximately 30 years ago. The most important inconvenience of CCD-based X-ray detectors is the thickness of the detector system needed because of the optical system which has to be integrated into the detector. As a result, these systems are direct readout digital X-ray detectors but not flat-panel detectors. Thin-film transistor (TFT)-based flat panel image detectors are thus far most convenient for imaging applications.

Digital radiography systems have been replacing traditional analog or screen–film (SF) systems over the last three decades. Self-scanning, direct-readout digital detectors have been in use since the introduction of the charged-coupled device (CCD). Recent advances in manufacturing technology have made possible a new generation of large-area, flat-panel X-ray imaging (FPXI) detectors with integrated, thin-film transistor readout mechanisms. Table 5.1 gives a timetable of development of digital technologies since the 1980s. The new technology is based on two factors—the promise of very rapid access to digital images wherever radiography with stationary X-ray equipment is performed and the anticipation of image quality that exceeds that of both screen–film receptors and photostimulable storage

Table 5.1 Timetable for development of digital technologies [5]

Year	Digital technology availability
1980	Computed radiography (CR), storage phosphors
1987	Amorphous selenium-based image plates
1990	Charge-coupled device (CCD) slot-scan direct radiography (DR)
1994	Selenium drum DR
1995	Amorphous silicon-cesium iodide (scintillator) flat-panel detector
	Selenium-based flat-panel detector
1997	Gadolinium-based (scintillator) flat-panel detector
2001	Gadolinium-based (scintillator) portable flat-panel detector
2001	Dynamic flat-panel detector fluoroscopy–digital subtraction angiography
2006	Digital tomosynthesis
2009	Wireless DR (flat-panel detector)

Permission from Springer

Table 5.2 Basic technology and readout mechanism for different types of imaging detectors

Detector technology		Capture element	Coupling element	Charge readout
Computed radiography (CR)		BaFBr: Eu^{2+}	Photostimulated luminescence (PSL) light-guide	Photomultiplier tube; signal digitization
Direct radiography (DR)	Direct conversion	a-Se	None	TFT array
	Indirect conversion	Csl or Gd_2-O_2S	Contact layer	a-Si photodiode/TFT array

TFT thin-film transistor liquid crystal display

phosphor computed radiographic systems because of improvements in X-ray detector technology. The transition from an SF environment to a new digital environment is a complex process. Technical factors concerning image acquisition, management of patient dose, and diagnostic image quality are some issues that could influence this process. Detectors using a scintillator layer and a light-sensitive TFT photodiode are called indirect-conversion TFT detectors. Those using an X-ray-sensitive photoconductor layer and a TFT charge collector are called direct-conversion TFT detectors [4]. Table 5.2 gives the basic elements used in computed radiography (CR) and digital radiography (DR) techniques.

5.3 Factors Affecting the Performance of Detectors

5.3.1 Detector Size

From practical applications viewpoint, physical dimensions of the detector should be large enough to capture the desired anatomic views and sufficiently compact for the desired applications. For example, detectors for standing radiographic examination should be at least 43 × 43 cm (i.e., 17 × 17 in.) in size in order to allow both vertical (35 × 43 cm) and horizontal (43 × 35 cm) imaging orientations without detector rotation. With an appropriate mechanical design, a rectangular 35 × 43 cm (i.e., 14 × 17 in.) detector can be easily rotated and may perform as well as a large detector [6].

5.3.2 Modulation Transfer Function (MTF)

MTF characterizes an additional aspect of the detector's performance. The MTF quantifies the loss of contrast in the image of a given object related to use of the detector [7–10, 13]. The MTF is conditioned by the physical properties of the detector, so that two different detectors may have different MTFs. Loss of contrast depends on the dimensions of the object X-rayed, i.e., the spatial frequencies contained in the image between 0 and the cutoff frequency of the detector. The MTF varies between 1 and 0, where 1 corresponds to complete transmission of the contrast of the object over the entire gray scale of the image and 0 corresponds to no transmission of this contrast, i.e., loss of visualization of this object. By definition, the MTF of an imaging system is equal to 1 for zero spatial frequency. MTF decreases as spatial frequencies in the image increase, but is independent of the pixel size.

5.3.3 Nyquist Frequency

The maximum spatial resolution that can be visualized in an image is defined by the Nyquist frequency (also called cutoff frequency) of the detector calculated from the pixel size, specifically the center-to-center spacing between pixels or pitch [7, 12].

According to Nyquist's sampling theorem, the cutoff frequency is equal to:

$$f_{\mathrm{N}}(\text{line pairs/mm}) = [1/2 \times \text{pixel size (mm)}]$$

In practice, this means that a detector with a pixel size of 100 μm has a cutoff frequency of 5 line pairs/mm. Objects smaller than 200 μm are therefore not detected. Visualization of smaller objects requires a detector with a smaller pixel size, i.e., with a higher cutoff frequency.

5.3.4 Spatial Resolution

Image spatial resolution, often expressed as the modulation transfer function (MTF), depends on physical detector characteristics and can vary substantially. The intrinsic spatial resolution of amorphous selenium (with direct conversion) and of structured cesium iodide (with indirect conversion) is higher than that of unstructured scintillators [14], the intrinsic resolution of selenium is much higher. Spatial resolution determines the dimensions of the smallest visible object. It is measured under special high contrast and low noise test conditions using a low kilovolt age and a high mAs value [7–10]. These measuring conditions are therefore very different from the exposure parameters used in the diagnostic setting. A higher spatial resolution improves image sharpness and allows better detection of small details in the image.

The maximum spatial resolution visualized in an image is equal to the Nyquist frequency, which is inversely proportional to the pixel size: a small pixel size corresponds to a high cutoff frequency and a large pixel size to a low cutoff frequency (Fig. 5.2). Objects or details of objects with a spatial frequency higher than the cutoff frequency are visualized in the form of artifacts superimposed on lower frequency objects. This phenomenon is called aliasing.

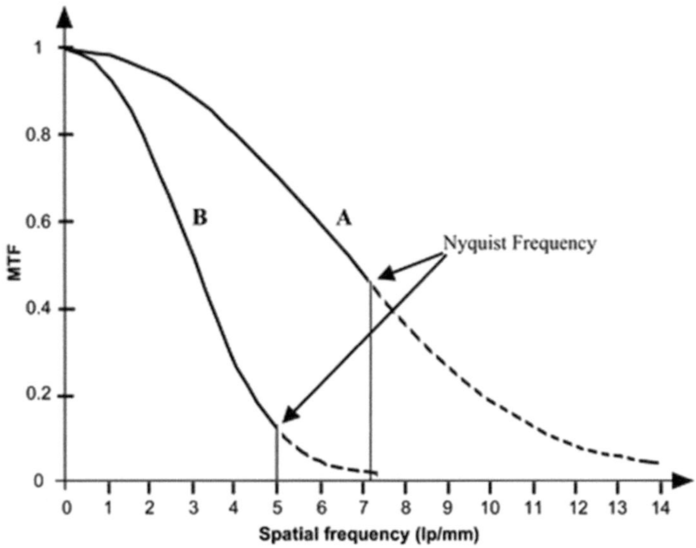

Fig. 5.2 MTF curves for two types of detectors A and B with different pixel sizes, 70 μm for A and 100 μm for B, respectively. Detector A displays a better MTF at all spatial frequencies than does detector B. Higher spatial resolution is provided by system A with higher Nyquist (cutoff) frequency (7.2 lp/mm). Lower spatial resolution is provided by system B with lower Nyquist frequency (5 lp/mm). Changing the pixel size in B from 100 to 70 μm would raise the cutoff frequency of the system from 5 to 7.2 lp/mm, the shape of the MTF curve being unchanged [14] (Reproduced with permission from Springer)

When comparing two detectors with different MTFs but the same pixel size, the aliasing effect is stronger for the detector with the highest MTF, i.e., the detector that provides a better contrast transmission at the cutoff frequency, but also non-negligible transmission beyond that frequency. In order to eliminate aliasing, the MTF of the ideal detector should be equal to zero beyond its Nyquist frequency (Fig. 5.3), which is obviously not possible. This means that for a real detector, the better the MTF, the higher the required Nyquist frequency, i.e., the higher the spatial resolution to be reached to limit the aliasing phenomenon.

5.3.5 *Detective Quantum Efficiency (DQE)*

DQE expresses the efficiency with which a detector uses the incident photons to form an image and is defined as the ratio of signal-to-noise ratio at the input and output stages of the detector:

$$\mathrm{DQE} = (\mathrm{SNR_{out}}/\mathrm{SNR_{in}})^2 \quad 0 \le \mathrm{DQE} \le 1$$

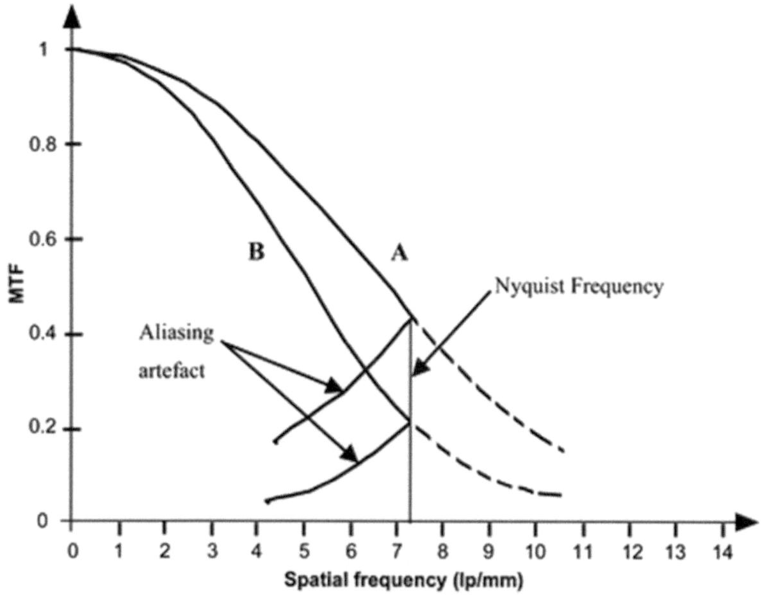

Fig. 5.3 MTF curves for two imaging systems with the same pixel size of 100 μm, hence the same Nyquist (cutoff) frequency (7.2 lp/mm). System A that exhibits a higher MTF at the Nyquist frequency than system B is prone to more aliasing artifacts [14] (Reproduced with permission from Springer)

DQE is influenced by the noise generated by the detector and by its spatial resolution [7–11, 12]. The DQE of an ideal detector would be equal to 1 when all X-ray photons reaching the surface of the detector would be used to form the image, and this detector would not generate any noise. The DQE of a real detector is less than 1 and reflects degradation of the signal-to-noise ratio (SNR) occurring between the detector input and output: Since image quality depends on the SNR, a reduction of noise improves the detection of low contrast objects in the image, i.e., small and/or low contrast masses. Noise is inversely proportional to the square root of the number of photons collected to form the image [7–9]. The SNR can therefore be improved either by increasing the number of incident photons (through a higher mA s value), together with an increase in the dose to the patient, or by using a larger portion of the photons reaching the detector, which increases the DQE.

Digital images can be processed to alter apparent image sharpness; however, excessive processing can lead to an increase in perceived noise. Thus, the inherent detector modulation transfer function, when expressed as a function of spatial frequency, is not as useful a measure of overall system performance as is the detective quantum efficiency (DQE). A better DQE improves image quality and remains unaffected by increasing or decreasing the dose to the patient. A thicker detection medium can be used for improving the DQE as this will increase the amount of X-rays absorbed. However, this may lead to loss of spatial resolution due to increase in scattering. With phosphor-based (i.e., indirect conversion) detectors, the emitted light spreads laterally in proportion to the scintillator thickness. Structured scintillators, which limit this effect, increase the DQE while maintaining spatial resolution. For a-Se direct conversion detectors, an electrical field channels the electronic charges, thus limiting their side scattering.

5.3.6 *Image Quality*

Detective quantum efficiency combines spatial resolution (i.e., modulation transfer function) and image noise (i.e., noise power spectrum) to provide a measure of the signal-to-noise ratio of the various frequency components of the image. Higher detective quantum efficiency values suggest greater image quality, and the results should be evaluated at all frequencies to estimate the ability of the image to depict both small and large image structures.

5.3.7 *Pixel Size and Matrix Size*

Pixel size and spacing are important parameters for defining the maximum spatial resolution of an image. However, more pixels do not always mean higher spatial resolution, because image blurring can result from X-ray scatter, light scatter, or both, in the detector. For chest radiography, some investigators (15, 16) have concluded that 0.2-mm pixel spacing (2.5 lp/mm, approximately 2000 pixels per row) is adequate for most diagnostic tasks, and in one commercial selenium-drum digital chest radiographic system, these 2K images have been reported to offer superior depiction of anatomic structures and an image appearance that is preferred over that of conventional screen–film radiographs. For mammography, smaller pixel sizes—probably in the range of 50–100 μm per pixel—are required. Although some detectors that offer substantially higher pixel pitch will soon be available, it should be remembered that the advantages that may be gained by using larger image matrices also have implications in practical implementation, including increased network traffic, increased archiving requirements, and the likely need for images reduction when images are displayed on video monitors.

5.3.8 *Monolithic Panels Versus Tiled Arrays*

Full-size thin-film transistor detector panels are technically challenging and can have relatively low yields. Manufacturing cost can be reduced by constructing detectors that consist of two or more smaller panels in a tiled configuration. In these detectors, digital image processing is used to “stitch together” the image sections to eliminate the appearance of the tile junctions. Although attractive in concept, this approach can be problematic in that it is difficult to seamlessly reconstruct a complete image without compromising image quality [17]. In addition, the effect of physical and thermal stress on the long-term structural integrity of tiled detectors has not yet been determined.

5.3.9 Bad Pixels and Rows

Perfect thin-section transistor arrays are difficult to make, so practical limits with regard to the number of defects allowed in these devices must be established by each manufacturer [6]. Defects may manifest as individual bad pixels or entire bad pixel columns or rows. Image corrections made by using system calibration masks and additional image processing can ameliorate some of these defects; however, as with detector tiling, one must exercise caution so that the image processing does not alter the image in ways that adversely affect the diagnosis.

5.3.10 Image Acquisition Time and Rate

Image quality with electronic detectors is affected by circuit design for charge collection, the charge measurement technique, analog-to-digital conversion rates during readout, and in some detectors, the device recharging between exposures. Generally, electronic noise is the predominant noise source in these detectors and the factor that limits image quality, particularly at the low-dose rates associated with fluoroscopy. For both fluoroscopy and static radiographic imaging, the precision of a pixel value (i.e., the number of gray levels) is related to the readout time, with longer times generally resulting in more precise image values. Since image readout involves the active flow of the current to and from each pixel element, an incomplete charge transfer can lead to inaccurate pixel values later if a subsequent exposure is used without taking the residual charge into account; the result is an electronic memory artifact. Similarly, rapid image acquisition can also lead to image artifacts owing to X-ray-induced charge trapping [18, 19]; however, compensation for this effect may be designed into the readout electronics of the detector [20, 21]. For these reasons, not all thin-film transistor system designs are practical for use where high exposure levels or high-speed image acquisition is required (e.g., in fluoroscopy); this factor should be considered when selecting a digital detector and how it will be used.

5.3.11 Dynamic Range

Digital radiography offers principal advantage due to decoupled image acquisition and image display which allows the detector to have a wide dynamic range and a linear response to X-ray exposure. For a given analog to digital converter (i.e., a given pixel depth), there can be a trade-off between the detector's sensitivity range (i.e., latitude) and contrast resolution (i.e., bits per pixel), and some manufacturers may choose to limit the dynamic range of the detector to maximize the number of gray values on the digital image [14]. If used with an automatic exposure control mechanism (i.e., phototimer), however, even detector systems with limited latitude (i.e., two to three decades) should have no operational problems in clinical use, because the photodetector will ensure that the exposure is appropriate for the available dynamic range. If no photo timer is incorporated into the system, it is recommended that the detector have a minimum of four decades of latitude, equivalent to the latitude of photostimulable storage phosphor systems, to avoid data loss owing to overexposure or underexposure.

5.4 Available Technologies

Historically, the first digital radiography system was the storage phosphor radiography system (SPR), developed by Fuji (Tokyo, Japan) and introduced to the market in the beginning of the 1980s. It was followed approximately 10 years later by a dedicated digital chest radiography unit, using a drum coated with a layer of amorphous selenium as detector (Thoravision, Philips, Hamburg Germany). Large-area flat-panel radiography detectors were introduced about a decade later.

5.4.1 Photostimulable Storage Phosphors (PSP)

This is a removable detector inserted as a cassette in a conventional X-ray imaging unit. Upon irradiation, electrons of the crystalline network of the phosphor are excited to a higher energy level. The number of excited electrons is proportional to the irradiation on each point on the screen. These excited electrons then spontaneously return to the ground state resulting in emission of radiation proportional to the number of excited electrons. With PSP plates [22, 23], the excited electrons are trapped into the crystalline structure where they remain stable for some time. After exposure, a laser beam of appropriate wavelength releases the trapped electrons, inducing the emission of light (i.e., photostimulation). The light emitted at each point of the PSP plate is obtained by scanning the screen with the laser beam. Light is collected, converted into an electrical signal, and amplified by a photomultiplier before digital conversion [24] as shown in Fig. 5.4.

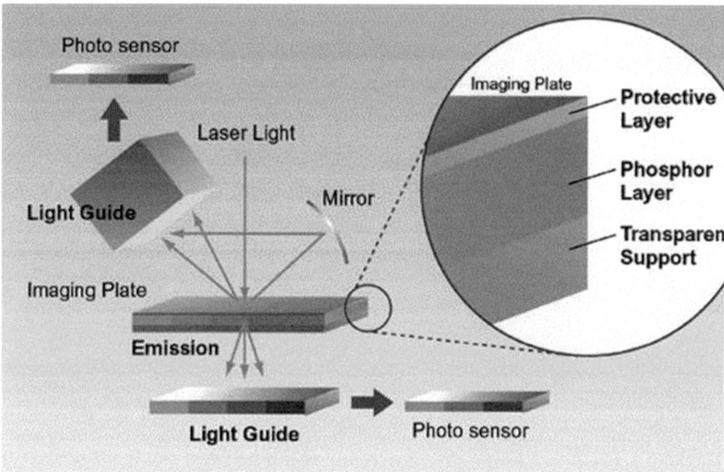

Fig. 5.4 Principle of simultaneous dual side reading of the photostimulable screen by means of a transparent support [14]. Courtesy of Fuji Medical Systems, France. Permission requested from Springer

In a digital system, on the other hand, image acquisition and display are two independent processes with the advantage of a generally linear response between the intensity of transmitted X-rays and the electronic signal recorded by the system (wide dynamic range). The image display can be optimized by adjustment of the display curve (linear or nonlinear, window settings) before printing the film or during visualization on the monitor.

5.4.2 Charge Coupled Devices (CCD)

Charge-coupled devices were the first direct-readout detectors to be introduced in radiology over 30 years ago. A CCD is an integrated circuit etched onto a silicon surface forming light sensitive elements called pixels. The CCD optical detector chip is composed of up to several million independent pixels (picture elements) where incoming photons induce charge. The pixels are represented by p-doped MOS capacitors. Figure 5.5 shows a CCD detector commercialized by Spectral Instruments.

These capacitors are biased above the threshold for inversion when image acquisition begins, allowing the conversion of incoming photons into electron charges at the semiconductor–oxide interface; the CCD is then used to read out these charges. When illuminated with X-rays, the array of pixels in the chip accumulates a charge pattern corresponding to the illumination pattern. This charge pattern can then be read out pixel by pixel, be digitized and transferred to computer memory. The X-ray quanta are transformed into visible light by an X-ray

Fig. 5.5 Commercialized CCD Detectors. Courtesy of Spectral Instruments

scintillator like gadolinium disulphide. The visible light is then transmitted to a CCD detector similar to those used in video cameras. The CCD transforms the visible light into a digital image. Since CCDs are physically much smaller (2–4 cm^2, integrated on a chip) than the image area, the light emitted by the scintillator screen must be transmitted to the CCD by some kind of optics reducing the size of the image. This optical system consists of lenses or optical fibers, and sometimes an image intensifier is used. Whereas CCDs are very sensitive detector systems, any of these optical systems reduces the number of photons that reaches the CCD, thus increasing noise of the system and degrading image quality. Geometric distortions and light scatter are another consequence of the use of optical reduction. With modern, cooled CCDs, thermal noise within the CCD itself can degrade the image quality.

5.4.3 *Flat Panel Detectors: Direct Versus Indirect Conversion*

Flat-panel X-ray detector (FPXD) technology (Fig. 5.6) developed in the last decade, fueled by improvements in large-area amorphous Silicon (a-Si) thin-film transistor (TFT) arrays, and innovations in deposition techniques for scintillators and photoconductors. *Direct-readout* electronic X-ray detectors use either a direct technique or an indirect technique for converting X-rays into an electric charge. Direct-conversion detectors (Fig. 5.7a) have an X-ray photoconductor, such as amorphous selenium, that converts X-ray photons into an electric charge directly, with no intermediate stage. *Indirect-conversion* devices (Fig. 5.7b) have a

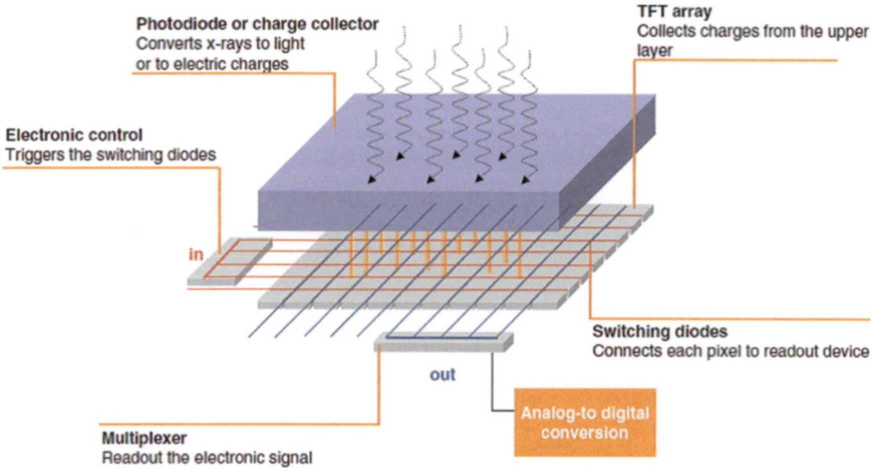

Fig. 5.6 Schematic diagram showing Flat panel detector. Reproduced with permission from [5]

Fig. 5.7 Schematic working of direct (**a**) and indirect (**b**) conversion detectors

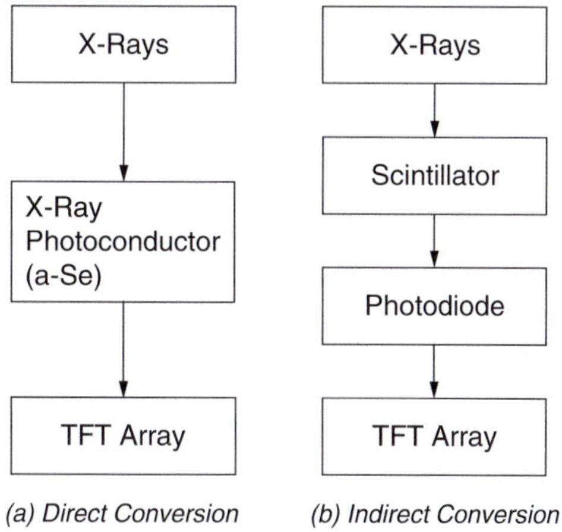

scintillator that first converts x rays into visible light. That light is then converted into an electric charge by using an amorphous silicon photodiode array or a CCD. Thin-film transistor (*TFT*) arrays (Fig. 5.8) may be used in both direct- and indirect-conversion detectors.

With indirect detection, the light generated by the X-ray interaction in the scintillator spreads in every direction and consequently degrades the spatial resolution. Needle-type scintillators can greatly help in guiding the light to a pixel, but a certain degree of light still will spread to neighboring pixels. The direct X-ray conversion method has the main benefit of having a higher modulation transfer function (MTF), leading to a high spatial resolution and highly contrasted images.

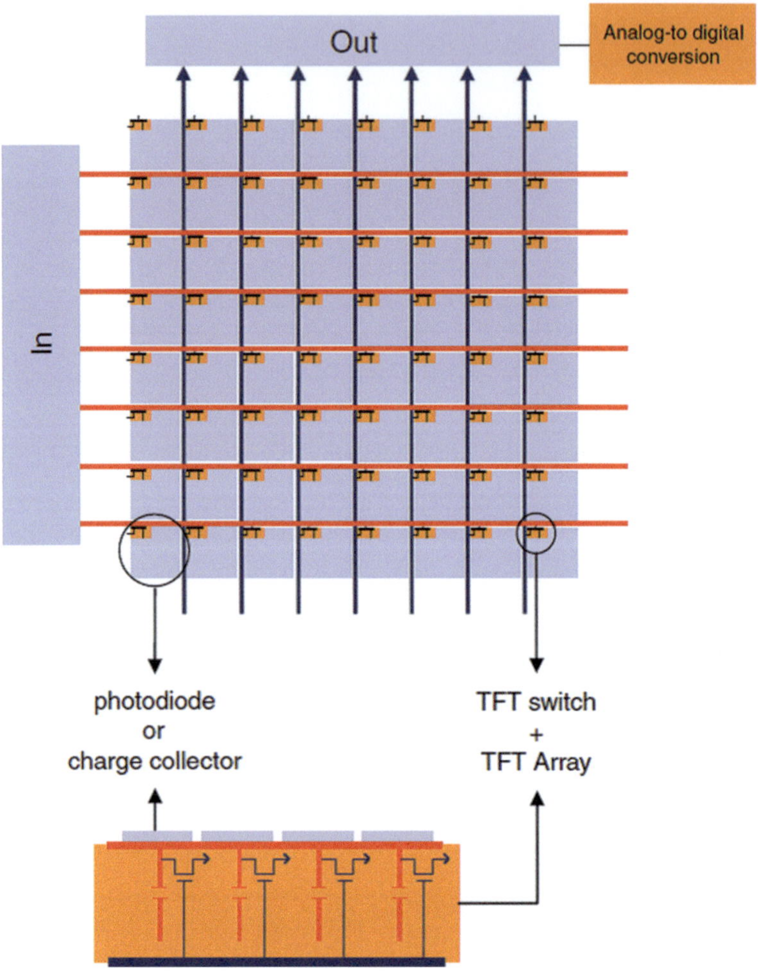

Fig. 5.8 TFT array: TFT thin-film transistor. Reproduced with permission from [5]

5.4.3.1 Direct Conversion Detectors

In direct detection, the X-rays are directly absorbed and converted to electrical charge in the detector. X-rays are absorbed in a photoconductor (such as a-Se) and generate charges that drift through the conductor and become collected on a storage capacitor *C* as shown in Fig. 5.9 [25].

The collected charge *Q* on the capacitor *C* then constitutes the signal. At an appropriate (access) time, the thin-film transistor (TFT) is activated and turns on to pass the charge *Q* on *C*, over an interconnect line, to a charge amplifier, which feeds the charge to the data line as a voltage signal. The a-Se photoconductor layer is called stabilized a-Se because it is a-Se alloyed with 0.2–0.5 % arsenic (As) and

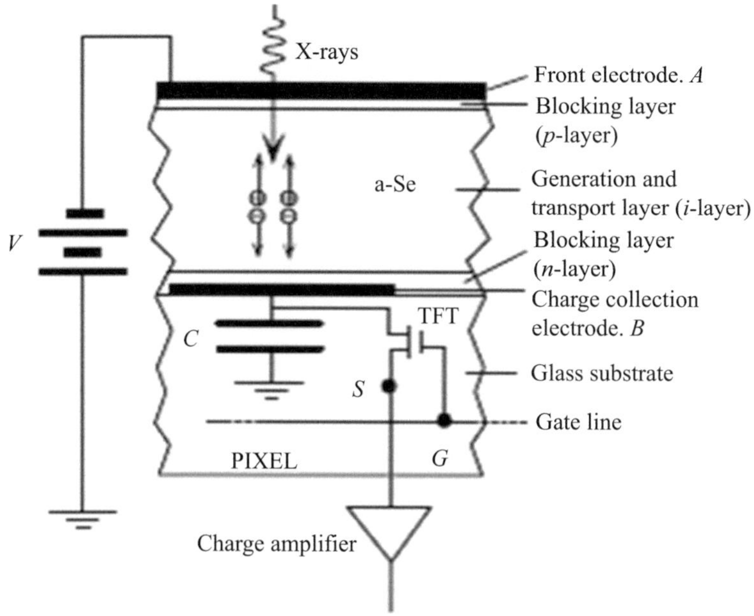

Fig. 5.9 Schematic diagram of one of the pixels in direct conversion flat panel image detectors. Reproduced with permission from [25]

doped with chlorine (Cl) in the ppm range. The arsenic alloying prevents the structure from crystallizing and Cl doping compensates for As introduced hole traps [25]. Although stabilized a-Se is now a well-established X-ray photoconductor for use in these direct conversion flat panel X-ray image detectors, there are many scientific aspects of the work that have not been resolved.

To obtain a reasonable sensitivity, the a-Se photoconductor needs an applied field of 5–10 V/μm and, at these fields; the dark current I_d in simple metal/a-Se/metal devices is higher than acceptable limits of 1–10 nA/cm^2 [26]. The dark current is reduced by either using a thin insulating dielectric layer between the a-Se layer and the positive electrode [27] or resorting to "p–i–n"-like multilayer structures [28]. In the latter, an intrinsic-like (i-layer) thick a-Se (with both hole and electron transport) is sandwiched between two thin layers of p-like (doped a-Se that has good hole transport but negligible electron transport) and n-like a-Se (doped material that has good electron transport but negligible hole transport) as illustrated in Fig. 5.10a.

The p- and n-like layers act as blocking layers to reduce dark current through the multilayer structure. The i-like layer transports both holes and electrons and therefore functions as an efficient X-ray photoconductor. The p-like and n-like definitions for a-Se do not follow conventional semiconductor physics definitions. They are based on the relative magnitudes of the mobility-lifetime product $\mu\tau$; e.g., $\mu_h\tau_h \gg \mu_e\tau_e$ implies a p-type like a-Se even though the exact position of the Fermi level is not well known. The thin n-layer consists of a-Se doped with an alkali element (typically sodium (Na)) and alloyed together with up to 10 % Arsenic (As) to stabilize a-Se (i.e., prevent crystallization). The n-like layer is the most important blocking layer in the multilayer structure because the dark current is primarily controlled by hole injection. Holes injected into the n-like layer become deeply trapped, and the resulting positive bulk space charge in this hole trapping layer reduces the actual field at the metal to n-type a-Se interface. The latter reduces the rate of hole injection from the contact as depicted in Fig. 5.10b. The term "carrier range" is used to represent the average distance a carrier drifts before it is deeply trapped per unit field, i.e., range is the mobility and lifetime product, $\mu\tau$. The *schubweg* is usually defined as the actual distance drifted $\mu\tau F$, where F is the field. Figure 5.11 shows two pixels of a direct conversion flat panel image detector [30].

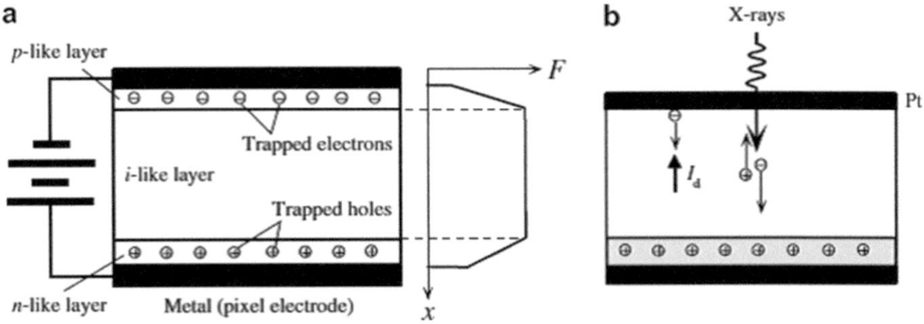

Fig. 5.10 (**a**) A simplified diagram of a-Se p-i-n detector structure. (**b**) double layer structure developed by Kasap et al. Reproduced with permission from [29]

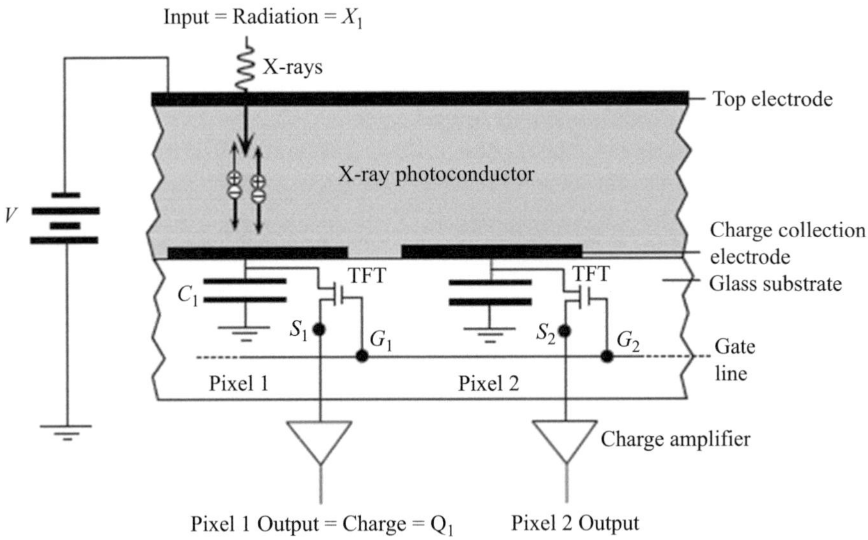

Fig. 5.11 Schematic diagram of the cross-sectional structure of two pixels of a direct conversion flat panel image detector. Reproduced with permission from [30]

Direct conversion X-ray imaging detectors based on a-Se have already been commercialized and shown to provide excellent X-ray images especially in mammographic applications. To obtain a reasonable sensitivity, the a-Se photoconductor needs an applied field ~5–10 V/μm. At these fields, the dark current in simple metal/a-Se/metal devices is higher than acceptable limits of 1–10 nA/cm^2. The dark current is reduced by either using a thin insulating dielectric layer reduced by either using a thin insulating dielectric layer between the a-Se layer and the positive electrode or resorting to p-i-n like multilayer structures [29]. Figure 5.12 shows a simple double layer detector structure based on a cold deposited n-layer on which an i-layer is grown. It was shown to have dark current densities lower than 10^{-10} A/cm^2 at a field of 10 V/μm, comparable to those commercialized p-i-n detectors in which the n-like layer is produced by doping a-Se with an alkali dopant.

Figure 5.13 shows a stabilized a-Se-based FPXI (AXS-2430) for mammography that has been developed and marketed by Anrad.

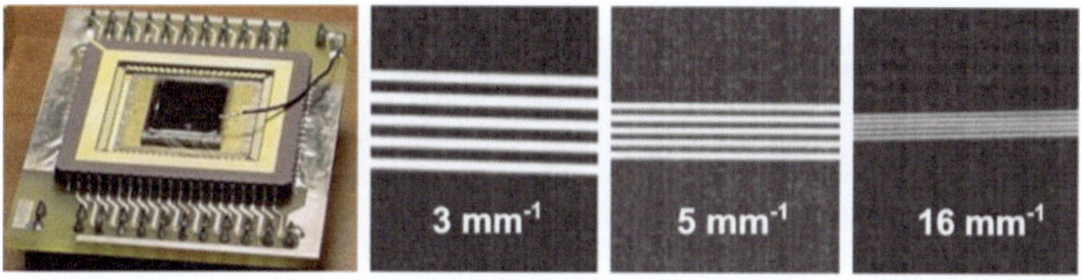

Fig. 5.12 A prototype X-ray detector based on a CCD readout circuit and a double layer (i–n) a-Se photoconductor layer produced by the cold deposition technology [29]. Differently scaled X-ray images of a resolution bar phantom that show the a-Se CCD X-ray detector can resolve bar patterns with resolution better than 16 lp/mm

Fig. 5.13 Anrad's mammographic FPXI AXS-2430 (previously LMAM) is used in the USA and European mammography markets. The field of view is 24 cm × 30 cm. These FPXIs have a pixel pitch of 85 μm, high DQE, high MTF, high contrast, high dynamic range, high patient throughput, and are capable of tomosynthesis. After [31] (Open access)

5.4.3.2 Indirect Detection

In indirect detectors, the X-rays are first converted into visible light by a phosphor layer such as CsI_2, which in turn is converted to electrical charge in the photodetector. The electrical charge is read out by means of an amorphous silicon (a-Si) active matrix thin-film transistor (TFT) array at each pixel. Indirect conversion detectors consist of a scintillator layer which converts X-ray photons into visible light. This is followed by an amorphous silicon photodiode layer, which transforms visible light into electric charges and, finally, the TFT-readout circuitry transforming the charge into digital values. One potential drawback of transforming X-rays into visible light is that lateral diffusion of the light reduces sharpness and spatial resolution of the image. To overcome this problem, some indirect conversion detectors use structured scintillators, consisting of cesium iodide crystals that are grown perpendicularly to the detector surface [4]. The single crystals have a diameter of approximately 5–10 μm and significantly reduce lateral diffusion of the scintillator light, much like optics. The reduction of light diffusion in turn allows to use thicker scintillator layers, thus increasing the detective quantum efficiency (DQE) of the detector system.

5.5 Material Selection for Imaging Applications

The overall performance of an imaging sensor is closely related to the photoconductor material [31]. In addition to a-Se, there have been a number of other photoconductors, such as polycrystalline layers of TlBr [32, 33], PbI_2 [34, 35], HgI_2 [36–42], CdZnTe [43, 44], and PbO [45, 46], that have been investigated and some of these, in particular HgI_2 and PbO, have shown potential for use in commercial FPXI applications. Most of these photoconductors to date either suffer from possessing too large a dark current or not having sufficient charge collection efficiency. In some cases, there are technological problems in manufacturing a uniform and homogenous layer over a large area. Nonetheless, with dedicated research, these problems are likely to be solved in the same way one had to solve similar problems in the development of commercial a-Se FPXIs. Some important parameters to be kept in mind for an appropriate material to be used as a photoconductor for direct conversion detectors [31] are given in the following sections [31].

5.5.1 Quantum Efficiency: A_Q

Nearly all the incident X-ray radiation should be absorbed within a practical photoconductor thickness to avoid unnecessary patient exposure. Over the energy range of interest, the linear attenuation coefficient α must be sufficiently large to allow the incident photons to be attenuated inside the photoconductor. Put differently, the X-ray attenuation depth δ, the reciprocal of α, must be substantially less than the photoconductor layer thickness L. The fraction of incident photons in the beam that are attenuated by the photoconductor depends on α of the photoconductor material and its thickness L; and is given by

$$A_Q = \text{Attenuated fraction} = [1 - \exp(-\alpha L)]$$

where $\alpha = \alpha(E_{ph}, Z, \rho)$ is a function of photon energy E_{ph}, atomic number Z, and density ρ of the material. A_Q is called the *quantum efficiency* (QE) because it describes the efficiency with which the medium attenuates photons. The attenuation depth δ is where the beam has been attenuated by 63 %.

The law of diminishing returns arising from exponential absorption in the above equation indicates that a doubling of L can only result in a further 63 % attenuation of the remaining beam $(1-0.63=37\ \%)$ for a total attenuation of $63+23=86\ \%$. The commercial mammographic detector in Fig. 5.10 has an a-Se layer that is nominally 200 μm thick. At 20 keV (within the mammographic range), the attenuation depth is 49 μm so that the photoconductor's A_Q is 98.3 %. At higher energies, for example at 60 keV, δ is 998 μm so that for an a-Se layer of thickness 1000 μm, A_Q is only 63.2 %. Although increasing the thickness would increase A_Q, it becomes practically very difficult to design sensors using thicker a-Se layers. Photoconductors with higher-Z components such as PbO, PbI_2, HgI_2, CdZnTe exhibit very good quantum efficiencies [31] in the high energy range that covers modalities such as chest radiography and angiography.

5.5.2 *Electron–Hole Pair Creation Energy* $W_\pm$

The photoconductor to be used for direct conversion imaging detectors should have as high intrinsic X-ray sensitivity as possible, i.e., it must be able to generate as many collectable (free) electron–hole pairs (EHPs) as possible per unit of absorbed radiation. The amount of radiation energy required, denoted as $W_\pm$, to create a single *free* electron and hole pair is called the *electron–hole pair creation energy* or the *ionization energy*; and it should be as low as possible because the free (or collectable) charge, ΔQ, generated from an incident and absorbed radiation of energy ΔE is simply $e\Delta E/W_\pm$, where e is the elementary charge. For many material systems $W_\pm$ is proportional to the bandgap E_g. For many crystalline semiconductors it is well known that $W_\pm \approx 3E_g$ according to the *Klein rule* [47, 48], which means that a lower $W_\pm$ requirement suggests a semiconductor with a narrower bandgap. Narrow bandgap semiconductors do not have sufficiently low thermal equilibrium concentrations of carriers to result in a low dark current. Amorphous Se as in the case of various other low-mobility solids, is an exception to the Klein rule in as much as $W_{\pm}$ decreases sharply with the applied electric field F and exhibits a behavior that follows [25, 49]

$$W_{\pm} = W_{\pm}{}^{\circ} + B/F$$

where $W_{\pm}{}^{\circ}$ is the intrinsic EHP creation energy at "infinite field," B is a constant that weakly depends on the X-ray photon energy, and F is the applied field. In the 20–40 keV range one can simply take $W_{\pm}{}^{\circ} \approx 6$ eV and $B \approx 4.4 \times 10^2$ eV·V/μm, which makes $W_{\pm}$ about 50 eV at $F = 10$ V/μm. $W_{\pm}$ for a-Se at typical operating fields (10–20 V/μm) is poorer than the competing large area X-ray photoconductors such as HgI_2 which has a $W_{\pm}$ of 5 eV. However, while lower $W_{\pm}$ is important in generating as much charge as possible from absorbed radiation, this charge still needs be collected. Although nearly all large area polycrystalline photoconductors have a lower $W_{\pm}$ than a-Se (Table 5.3) they do not currently possess sufficiently good charge collection efficiency for both electrons and holes at an operating field that results in an acceptably low dark current. Consequently, the sensitivity of these polycrystalline photoconductors as gauged by the actual charge collected per unit incident radiation may not necessarily be better. $W_{\pm}$ enters the calculation of the overall X-ray sensitivity and DQE of the photoconductor as well.

5.5.3 *Dark Current*

The dark current of the photoconductor under a bias voltage should be negligibly small. The dark current in most photoconductive semiconductors is normally attributed to one of two factors—the rate of injection of carriers from the contacts into the photoconductor and the rate of thermal generation of carriers. A small dark current implies that the contacts to the photoconductor should be non-injecting, and the rate of thermal generation of carriers from various defects or states in the bandgap should be negligibly small (i.e., dark conductivity is practically zero). Small dark conductivity generally requires a wide bandgap semiconductor that conflicts with the condition of smaller ionization energy. There have been several attempts to estimate what would constitute a negligible dark current density, J_d [50]; it is generally accepted that J_d should preferably not exceed 10 pA/mm^2 depending on the clinical application.

5.5.4 *Charge Collection Efficiency (CCE)*

Once the charges are generated by the absorption of X-rays, these charges have to be collected. The applied field drifts the electrons and holes in opposite directions towards their respective electrodes. During charge transport loss of carriers due to recombination or trapping should be minimum. If τ is the mean lifetime of a charge carrier, which could be due to recombination or deep trapping, and the drift mobility of the carrier is μ, then $\mu\tau F$ represents the mean distance drifted by the carrier before it is trapped (or disappears by recombination); this distance is called the *schubweg*. Since it is necessary to collect most of the charges, both electrons and holes, we need to ensure that the electron and hole schubwegs are both much longer than the thickness of the photoconductive layer, that is, $\mu\tau F \gg L$ for both electrons and holes. If the photoconductor layer is made thicker to capture more of the radiation (towards increasing A_Q), the $\mu\tau F \gg L$ condition would eventually be lost, and charge collection efficiency would start limiting the sensitivity. The drift mobility $\times$ lifetime product, $\mu\tau$, is normally called the *range* of the carrier, that is, its schubweg per unit field. Table 5.3 compares the carrier ranges among various large area X-ray photoconductors. One of the distinct advantages of a-Se is the fact that both electrons and holes possess reasonable ranges, which allows both electrons and holes to be collected upon their photogeneration.

In the case of polycrystalline semiconductors, the carrier ranges are highly dependent on the quality of the layer such as the grain size (polycrystallinity) and purity. For example, Schieber and Zuck quote maximum $\mu\tau$ values of 10^{-3} cm^2/V for electrons and 10^{-5} cm^2/V for holes in their best quality physical vapor deposited polycrystalline HgI_2 layers [51]. These values would render HgI_2 as one of the best polycrystalline X-ray photoconductors in terms of charge transport properties.

5.5.5 X-Ray Damage and Fatigue

As the photoconductor is exposed to X-rays, over time, its properties will deteriorate to some extent. During the irradiation, the photoconductor structure itself can become either temporarily or permanently *damaged* by the generation of various defects. One can expect X-ray induced new defects as well as an increase in the population of certain intrinsic defects. These defect populations may return to their thermal equilibrium concentrations but the rate of defect relaxation is generally thermally activated (e.g., diffusion controlled) and its time scale therefore highly dependent on the activation energy. In some cases, defects anneal out when the photoconductor is left unexposed for a few days and in some cases the defects take much longer to anneal and may appear permanent over the observation time.

Another possible effect of radiation damage could be the buildup of trapped charge carriers in the bulk of the photoconductor [52]. In the case of a-Se photoconductors, electrons are more likely to become trapped since they have shorter schubweg than holes [53]. Electron release times from deep traps may be of the order of hours. Accumulation of trapped electrons results in two undesirable effects:

1. Photogenerated holes can recombine with previously trapped electrons and are thereby prevented from reaching the collection electrode. The latter results in a

drop in the X-ray sensitivity of the exposed region of the X-ray sensor, and can lead to an effect called *ghosting* in X-ray imaging [54].

2. Net space charge due to trapped carriers (e.g., electrons) modifies the internal field and hence modifies the charge collection efficiency. In the case of a-Se, the modified field leads to changes in the photogenerated number of carriers since $W_{\pm}$ depends on the field. The modification of the internal field normally leads to a drop in the X-ray sensitivity [55]. In some X-ray photoconductors one species of carriers has a very limited range and becomes deeply trapped very quickly. The trapped carriers effectively polarize the sample, and modify the field and can reduce the charge collection efficiency [56].

5.5.6 Large Area Fabrication

An important requirement for a large area X-ray photoconductor is that it must be capable of being coated to the required thickness over a large area to cover the AMA, e.g., 24 cm × 30 cm for mammography. This rules out the use of X-ray sensitive crystalline semiconductors which are difficult to grow in such large areas, and would require process temperatures incompatible with the glass AMA substrate and it is a-Si:H electronics. Various polycrystalline semiconductors, such as $Cd_xZn_{1-x}Te$ (CZT), HgI_2, PbI_2, and PbO as summarized in Table 5.3 [31], have the feasibility to be prepared in large areas but their main drawback is the adverse effect of grain boundaries in limiting charge transport and, further, the high substrate and annealing temperatures required to optimize the semiconductor properties. High substrate temperature (in excess of 200 °C) are not compatible with a-Si:H TFT-AMA substrates onto which these polycrystalline semiconductors have to be coated. Nonetheless, much progress has been made in the last decade in preparing large-area polycrystalline photoconductors that show promise in FPXI applications. For example, screen printed polycrystalline HgI_2 does not need high substrate temperatures and has exhibited reasonable properties in terms of X-ray sensitivity [57].

Organic photoconductors currently dominate the xerographic photoreceptor industry (where they replaced a-Se and a-As_2Se_3) and can be cheaply prepared in large areas. However, they are of limited value for X-ray imaging because they do not absorb X-rays sufficiently well due to their poor attenuation coefficient arising from the very low Z elements comprising the organics (H, C, N, O). On the other hand, amorphous semiconductors such as a-Se, a-As_2Se_3, and a-Si:H are routinely prepared in large areas for such applications as xerographic photoreceptors and solar cells and are therefore well suited for flat panel X-ray detector applications. Amongst the three, a-Se is particularly well suited because it has a much greater X-ray absorption coefficient than a-Si:H, due to its greater atomic number, and good charge transport properties for both holes and electrons compared with a-As_2Se_3 in which electrons become trapped and the hole mobility is much smaller. Further, the dark current in a-Se is much smaller than that in a-As_2Se_3. Table 5.3

Table 5.3 Some typical or expected properties of selected X-ray photoconductors for large area applications

Photoconductor State preparation	δ at 20 keV δ at 60 keV	E (eV)	W (eV)	Electron, $\mu_e\tau_\varepsilon$ (cm^2/V)	Hole, $\mu_h\tau_h$ (cm^2/V)
Stabilized a-Se Amorphous	49 μm 998 μm	2.2	45 at 10 V/μm 20 at 30 V/μm	3×10^{-7}–10^{-5}	10^{-6}–6×10^{-5}
HgI_2 Polycrystalline	32 μm 252 μm	2.1	5	10^{-5}–10^{-3}	10^{-6}–10^{-5}
HgI_2 Polycrystalline	32 μm 252 μm	2.1	5	10^{-6}–10^{-5}	~10^{-7}
$Cd_{.95}Zn_{.05}Te$ polycrystalline Vacuum deposition (sublimination)	80 μm 250 μm	1.7	5	~2×10^{-4}	~3×10^{-6}
PbI_2, polycrystalline Normally PVD	28 μm 259 μm	2.3	5	7×10^{-8}	~2×10^{-6}
PbO, polycrystalline Vacuum deposition	12 μm 218 μm	1.9	8–20	5×10^{-7}	Small
T1Br Polycrystalline	18 μm 317 μm	2.7	6.5	Small	1.5–3×10^{-6}

δ is the attenuation depth at the shown photon energy, either 20 or 60 keV. From various references and combined selectively. $\mu\tau$ represents the carrier range (SP refers to screen printing, and PVD to physical vapor deposition) (As reported in [31])

provides a summary of typical preparation procedures for some selected X-ray photoconductor materials. Due to its commercial use as an electro photographic photoreceptor, a-Se is one of the most highly developed photoconductors [58]. It can be easily coated as thick films (e.g., 100–500 μm) onto suitable substrates by conventional vacuum deposition techniques and without the need to raise the substrate temperature beyond 60–70 °C. Its amorphous state maintains uniform characteristics to very fine scales over large area.

5.5.7 *Need for Speed*

There is a need to use the FPXIs in real time applications such as fluoroscopy. Once the electrons and holes are photogenerated by the absorption of X-rays, they should then drift within a reasonable time and become collected. The collection time for the X-ray generated charge is limited by the slowest carriers, that is, the carrier that has the lowest drift mobility. In the case of a-Se (assuming stabilized a-Se), electrons have a drift mobility that is roughly 2×10^{-3} cm^2 V^{-1} s^{-1} and much lower than that for holes. Taking a worst case calculation, the electron transit time across a 1000 μm layer (assuming, for example, a fluoroscopic application in which the a-Se layer has to be reasonably thick to absorb the X-rays) is typically 0.5 ms under an applied field of 10 V^{-1} μm^{-1}. This is much shorter than the shortest interframe time (33 ms for 30 fps) that one can expect to encounter in various common X-ray imaging applications. In the case of the mammographic sensor described above, with a thickness of 200 μm, the electron transit time is shorter than 0.1 ms. A flat panel digital radiographic system based on using a large area thin-film transistor (TFT) active matrix array (AMA) similar to the ones used in flat panels displays, with an electroded X-ray photoconductor has been reported in Ref [14].

5.6 Selenium: Band Structure Considerations

Selection and design of the photoconductor plays crucial role in the performance of a direct conversion X-ray imager. Important parameters which render a-Se an ideal candidate for the application are:

1. Ability to be easily coated over a large area as a thick layer (100–1000 μm).
2. Can be coated onto any suitable substrate by conventional vacuum deposition techniques.
3. Long-term stability of properties, i.e., minimal X-ray damage and fatigue.
4. Acceptable X-ray absorption coefficient.
5. Good transport properties for both holes and electrons.
6. Dark current is much smaller than many competing polycrystalline layers. This is due to its larger energy gap (E_g) and a smaller concentration of deep localized states in the mobility gap.
7. Much greater X-ray absorption coefficient due to its greater Z.
8. Better charge transport for both holes and electrons and much smaller dark current as compared to polycrystalline competing technologies.

5.7 Density of States (DOS) in a-Se

The density of states (DOS) diagram for a semiconductor is the key to understanding its electrical and optical properties. Even though a-Se has been extensively studied over many years [59–65], there are still various uncertainties and controversies in its DOS function. It has been reported [66] alloying pure a-Se with As (0.2–0.5 % As) prevents crystallization, but introduces deep hole traps and shortens the hole range. When the alloy is doped with 5–20 ppm of a halogen, the hole range is improved, but at the expense of electron range. DOS have been estimated [67] in pure and stabilized a-Se using various techniques such as electron time-of-flight (TOF) transient photocurrents as a function of electric field, and as a function of alloying with As and doping with Cl. The distribution of localized states *s* DOS *d* in stabilized *a*-Se has been investigated by comparing the measured and calculated transient photocurrents. Theoretical calculations of DOS in pure Se as well as Se doped with arsenic and chlorine have been discussed in Chap. 1. DOS in stabilized *a*-Se prepared by conventional vacuum deposition techniques have been compared with measured and calculated transient photocurrents by Safa et al. [67]. A tentative schematic diagram [30] which highlights the salient features is sketched in Fig. 5.14. A first peak at 30 eV below E_c with a relatively stable amplitude of ~10^{17} eV^{-1} cm^{-3}, which may be associated with such *intrinsic defects* as distortions of the normal dihedral angle. A second small peak s or shoulder d at 0.45–0.50 eV below E_c with an amplitude 10^{14}–10^{15} eV^{-1} cm^{-3}, which may be related to the existence of VAPs consisting of Se^{3+} and Se^{1-} defects. Deep states with an integral concentration of 10^{14}–10^{15} cm^{-3} lying below 0.65 eV, whose exact distribution could not be resolved over the time scale of the present experiments. The bandgap $E_c - E_v$ is about 2.1 eV. Over the time scale of typical transit times involved in a-Se X-ray photoconductor layers, the electron drift mobility μ_e is controlled primarily by the shallow traps at about 0.30–0.35 eV below E_c. There is a peak in the DOS at about 0.45–0.50 eV above E_v, and a distribution of deep hole traps at ~0.85–0.90 eV above E_v. The Fermi level is about 0.95-1 eV above E_v, slightly below mid gap (a-Se is *p-type*). The positions of valence alternation pair (VAP) type defects (Se^{3+} and Se^{1-}) are shown in the structure. The nature and the exact origin of the shallow traps and the deep traps at have not yet been fully resolved. Two different peaks have been shown in Fig. 5.14 for electron and hole traps above and below E_F based on the discharge of the saturated cycled-up xerographic potential [68] and cycled time-of-flight experiments [69]. (One "broad" peak, which straddles the Fermi level within the distribution, cannot be totally ruled out since states above E_F would be deep electron traps, and stated below E_F would be deep hole traps.)

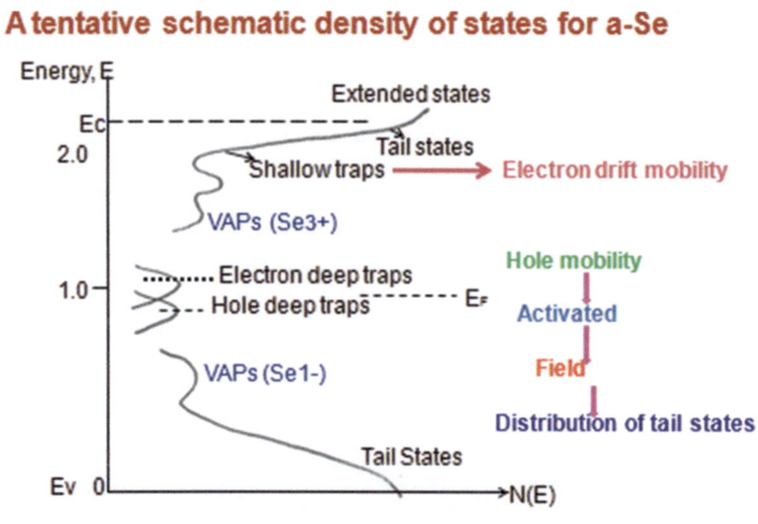

Fig. 5.14 Proposed DOS distribution in a-Se. After [30]

5.7.1 Avalanche Breakdown and HARP

One of the biggest surprises in the amorphous semiconductors field was the experimental observation that it is indeed possible to obtain impact ionization and avalanche multiplication in an amorphous semiconductor [70–76]. Even though the existence of avalanche multiplication was initially controversial and attributed to a probable photoconductive gain which is distinctly different from impact ionization, it soon became commercialized [30] in the HARP video tube—an ultra-high sensitive TV pick up tube uses avalanche multiplication in a-Se. to capture images at internally low light intensities.

Tanioka, working with coworkers at Japans television broadcast corporation (NHK), Research laboratories was able to develop a supersensitive photoconductor target called the HARP (High-gain Avalanche Rushing Amorphous Photoconductor). With its avalanche multiplication, the Harp was so sensitive that it was immediately used in a HDTV, cameras pick up tube: The tube incorporating the HARP target has been called the HARP video tube or a-Se avalanche tube. It was generally believed that avalanche would be noisy if it were to occur at all and so was not considered as a worthy project for investigation. However, while measuring what he thought to be the photoconductive gain of an a-Se. camera tube, Tanioka noticed that as he moved the object away from the camera, the image happened to move as well. This contradicted what was known about the relatively long time lag of photoconductive gain, the widely accepted source of gain in many photoconductors. This led to the idea that avalanche multiplication was the mechanism behind his observation. His group was then able to provide evidence for the occurrence of avalanche multiplication in the a-Se layer of the HARP tube for fields above 80 V/μm. Among their observations were the independence of the target capacitance on bias voltage, which includes the possibility that the charge multiplication is due to electron injection from the scanning beam, the apparently noise free amplification above 80 V/um and the variation of the signal current above 80 V/μm with target thickness.

Amorphous selenium HARP structures have been developed by NHK in Japan as photoconductive targets of broadcast video cameras and are now used routinely for electronic news gathering in HDTV, i.e., operation at low light conditions. Figure 5.15 shows an image from a real time movie of a rainbow under moonlight close to a waterfall taken by the NHK Camera using HARP tube shown in Figs. 5.16a, b respectively.

TV pick up tubes using such HARP targets have clearly demonstrated a far superior sensitivity than conventional TV pick-up tubes. In fact, these ultra-high sensitive HARP video tubes have been able to capture even star lit images. The need for high gain vidicons in HD TV is due to the use of small aperture lenses with much greater depth of field than in conventional TV that allow realistic large screen effects.

For use in optical imaging, a-Se HARP structure is deposited on a glass substrate covered with an ITO (indium tin oxide) coating, which serves as a transparent anode. The back of the a-Se HARP structure is free, that is, it has no physical

Fig. 5.15 Picture of a rainbow under moonlight at a waterfall taken using NHK camera using HARP tube. Courtesy Dr. Tanioka (NHK)

Fig. 5.16 (**a**) NHK Camera and (**b**) HARP TV pickup tube. Courtesy Dr. Tanioka (NHK)

electrode, which allows it to form a latent charge image. A scanning electron beam serves as a virtual cathode, biasing the free surface (see Fig. 5.17). Light photons incident on the front a-Se surface through a positively biased T electrode are a sorted and create EHPs. The freed holes drift to the free surface of the a-Se layer and if the electric field exceeds E_{th}, the drifting holes undergo avalanche multiplication. The holes accumulate as a latent charge image at the free surface in an amount proportional to the incident light intensity. An electron beam scans the free surface, completing the circuit, and enabling the accumulated positive charge to be sensed by the ITO electrode as a current. There have been many examples on the uses of Harpicons in low-light level applications in which they outperform all standard imaging chips; Figure 5.19 has images taken under low light conditions using an ordinary camera (left) and the one using HARP pick up tube (right).

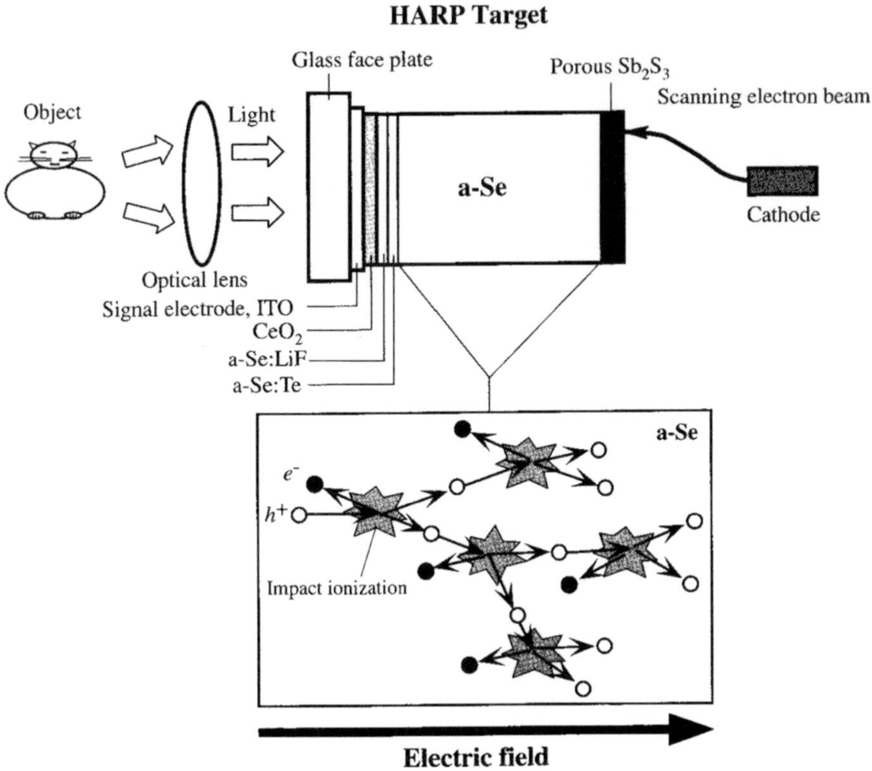

Fig. 5.17 Basic structure and principle of operation of HARP TV tube [30]. Reproduced with permission

5.7.1.1 HARP: Structure Details and Operation

1. The entire target is of the order of 1–35 μm thick [30].
2. The transparent signal electrode made of indium tin oxide (ITO) is biased positively with respect to the cathode.
3. The CeO_2 and a-Se. doped with Li F layers act as a hole blocking contact for hole injection.
4. The Sb_2S_3 layer reduces electron injection from the scanning electron beam and surpasses secondary electron emission.
5. The incident light from the object is absorbed mainly in the a-Se layer (or the photo generated layer a-Se: Te).
6. The e–h pairs photo generated in the a-Se layer are then separated by the applied electric field.
7. The electrons are neutralized quickly as they as close to the electrode, whereas holes have to drift across the bulk of the a-Se layer to reach the negative electrode.

8. These drifting holes constitute the signal current.
9. As the photo generated holes drift through the a-Se layer, as a result of the large applied electric field (greater than, 8×10^5 V/μm or 80 V/μm), they exhibit avalanche multiplication and hence yield an effective quantum gain greater than unity.
10. The quantum efficiency is the number of electron–hole pairs (EHPs) generated per absorbed photon and the quantum gain, or yield in the number of EHPs collected per absorbed photon.
11. The effective quantum gain resulting from avalanche multiplication depends on the field as well as the photo conductor thickness.
12. The thicker the a-Se layer, the greater the multiplication at a given field. In fact the multiplication factor increases exponentially with the thickness.
13. For example, in a 2 μm thick HARP target the quantum gain is about 10 at a field of 120 V/μm, whereas the gain is about 1000 in an a-Se target of thickness 24.8 μm at a field of 100 V/μm.
14. Avalanche multiplication is normally characterized in terms of an impact ionization coefficient, which is the probability per unit distance that a carrier, driven by an applied field, will generate an e-h pair by impact ionization.
15. The multiplication "M" therefore depends on the thickness "L"; $M = \exp(\alpha L)$, where α is the impact ionization coefficient.

The ionization coefficient α increases strangely with the field "F" in as much as a carrier can gain more energy from the field and thus has a higher probability of avalanching. The value of α depends on the details of impact ionization properties. The exact dependence of "α" on the field has been a subject of decades of discussion. The observed multiplication therefore increases sharply with the field. At $M = 100$ (at 96 V/μm) the dark current is still below 1 nA compared with the signal.

Tanioka and coworkers have successfully incorporated HARP layer, operating with avalanche gain, onto a CMOS image sensor to produce a prototype HARP-CMOS high sensitivity image sensor [77]. A conventional CMOS imaging array with electrodes at each pixel is coated with HARP that is biased to operate in the avalanche multiplication mode. When the HARP at a particular pixel receives signal photons as shown in Fig. 5.18, it photogenerates holes that multiply as they drift towards the negative electrode [30]. This multiplied photocurrent charges a capacitor that stores the signal as charge. The signal charge is later read out by appropriately addressing that particular pixel. Thus each pixel of the CMOS sensor receives an avalanche multiplied photocurrent signal and consequently the overall signal-to-noise ratio becomes substantially improved; the signal is multiplied at each pixel. The images from the HARP-CMOS sensor have proved to be superior to those from conventional imagers as seen in Fig. 5.19.

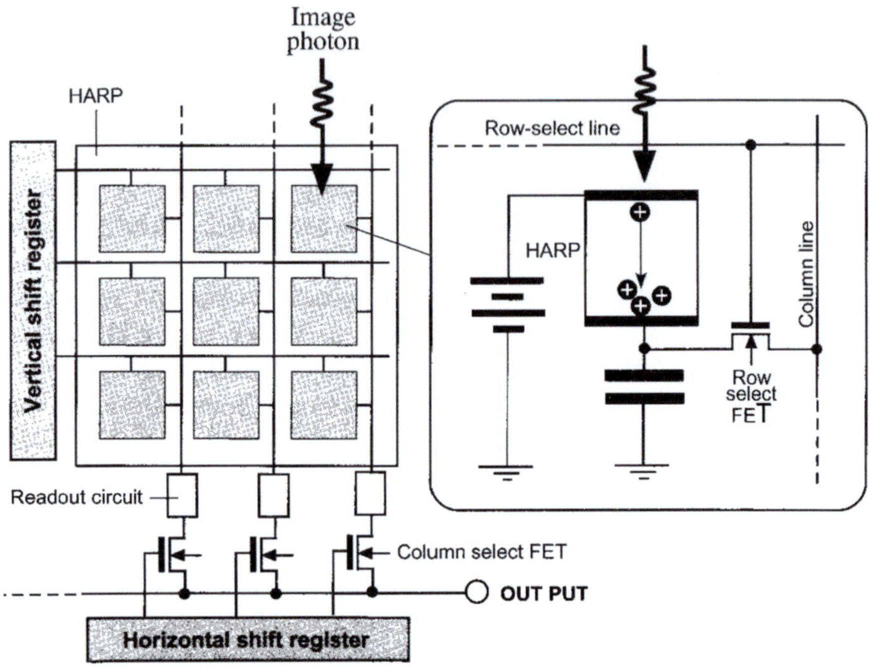

Fig. 5.18 Principle of the CMOS-HARP imager [30]. Reproduced with permission

Fig. 5.19 Images taken using standard camera (*left*) and using HARP-based NHK camera (*right*). Courtesy: Dr. Tanioka, NHK, Japan

5.8 Future Outlook and Summary

Electronically readable, large area X-ray detectors promise rapid access to the image for diagnosis, improved image quality relative to that of screen–film and storage phosphor-based radiography, reduced patient time, reduction of consumable agents, and possibility of reduced patient exposure. Amorphous selenium is one of the most highly developed photoconductors [78, 79]. It can be easily coated over large areas as a thick layer (~100–1000 μm). It can readily be coated onto any suitable substrate by conventional vacuum deposition techniques and without the need to raise the substrate temperature beyond 6–70 °C. Thus currently a-Se is one of the practical photoconductors for clinical medical imagers because it has an acceptable X-ray absorption coefficient, good transport properties for both holes and electrons, and dark current smaller than many polycrystalline layers.

References

1. J.T. Dobbins, D.L. Ergun, L. Rutz, D.A. Hinshaw, H. Blume, D.C. Clark, DQE(f) of four generations of computed radiography acquisition devices. Med. Phys. **22**, 1581–1593 (1995)
2. P.C. Bunch, Performance characteristics of asymmetric zero-crossover screen-film systems. Proc. SPIE **1992**, 46–65 (1653)
3. U. Neitzel, I. Maack, S. Gunther-Kohfahl, Image quality of a digital chest radiography system based on a selenium detector. Med. Phys. **21**, 509–516 (1994)
4. E. Kotter, M. Langer, Digital radiography with large-area flat-panel detectors. Eur. Radiol. **12**, 2562–2570 (2002)
5. L. Lanca, A. Silva, *Digital Radiography Detectors: A Technical Overview* (Springer Science + Bussiness Media, New York, 2013). Ch 2, Digital Imaging Systems for plain radiography
6. G. Harrell, C. James, T. Dobbins, C.E. Ravin, Principles of digital radiography with large-area, electronically readable detectors: a review of the basics. Radiology **210**, 595–599 (1999)
7. J.C. Dainty, R. Shaw, *Medical Imaging. Image Science* (Academic, London, 1974). ISBN 0-12 200850-2
8. J.T. Dobbins III, Effect of under sampling on the proper interpretation of modulation transfer function, noise power spectra and noise equivalent quanta of digital imaging systems. Med. Phys. **22**, 171–181 (1995)
9. M. Lundqvist, Silicon strip detectors for scanned multi-slit X-ray imaging, PhD Thesis, Kungl Tekniska Högskolan, Fysiska Institutionen, Stockholm, 2003. ISBN 91-7283-512-5, ISSN 0280-316X
10. B. Mikulec, Single photon detection with semiconductor pixel arrays for medical imaging applications, PhD Thesis, University of Vienna, Austria, 2000. CERN-THESIS 2000-021
11. W.R. Hendee, E.R. Ritenour, *Medical Imaging Physics*, 4th edn. (Wiley, New York, 2002), pp. F1–F88. ISBN 0-471-38226-4
12. J.T. Dobbins III, Determination of MTF, NPS and DQE in practice, in *Lectures Notes: Physics of Medical X-ray Imaging*, ed. by L.G. Mansson (European Commission, ERPET Course, Malmö, 1999), pp. 8–12
13. J.P. Moy, Signal-to-noise ratio and spatial resolution in X-ray electronic imagers: is the MTF a relevant parameter? Med. Phys. **27**, 86–93 (2000)
14. A. Noel, F. Thibault, Digital detectors for mammography: the technical challenges, *European Radiology*. (Springer-Verlag, 2004), 10.1007/s00330-004-2446-6
15. C. Floyd Jr., J.A. Baker, H.G. Chotas, D.M. Delong, C.E. Ravin, Selenium based digital radiography of the chest: radiologist's preference compared with film screen radiographs. AJR **165**, 1353–1358 (1995)
16. P.K. Woodard, R.M. Stone, D.S. Gierada, G.G. Reiker, T.K. Pilgram, R.G. Jost, Chest radiography; depiction of normal anatomy and pathologic structures with selenium based digital radiography versus conventional screen-film radiography. Radiology **203**, 197–201 (1997)

17. C.H. Slump, P.W. van Dijk, G. Laanstra et al., Real time diagnostic imaging with a novel X-ray detector with multiple screen CCD sensors. Proc. SPIE **3336**, 418–429 (1999)
18. H.G. Chotas, C. Floyd Jr., C.E. Ravin, Memory artifact related to selenium based digital radiography systems. Radiology **203**, 881–883 (1997)
19. S.O. Kasap, V. Aiyah, X-ray induced hole trapping in electroradiographic plates. J. Appl. Phys. **69**, 7087–7096 (1991)
20. W. Zhao, I. Blevis, S. Germann, J.A. Rowlands, D. Waetcher, Z. Huang, Digital radiology using active matrix readout of amorphous selenium; construction and evaluation of a prototype real time detector. Med. Phys. **24**, 1834–1843 (1997)
21. A. Tsukamoto, S. Yamada, T. Tomisaki et al., Development of a selenium based flat panel detector for real time radiography and fluoroscopy. Proc. SPIE **3336**, 388–395 (1999)
22. H. von Seggern, T. Voigt, W. Knopfer, G. Lange, Physical model of photostimulated luminescence of X-ray irradiated BaFBr:Eu2+. J. Appl. Phys. **64**, 1405–1412 (1998)
23. M.J. Yaffe, Digital mammography, in *Handbook of Medical Imaging*, ed. by J. Beuttel, H.L. Kundel, R.L. Van Metter (SPIE, Bellingham, 2000), pp. 329–372
24. E. Samei, J.A. Selbert, C.E. Willis, M.J. Flynn, E. Mah, K.L. Junck, Performance evaluation of computed radiography systems. Med. Phys. **28**, 361–371 (2000)
25. G. Belev, S.O. Kasap, Amorphous selenium as an X-ray photoconductor. J. Non-Cryst. Solids **345–346**, 484–488 (2004)
26. R.E. Johanson, S.O. Kasap, J.A. Rowlands, B. Polischuk, J. Non Cryst. Solids **227**, 1359 (1988)
27. S.O. Kasap, J.A. Rowlands, IEEE Proc. Circ. Dev. Syst. **149**, 83 (2002)
28. O. Tousignant, M. Choquette, Y. Demers, L. Laperriere, J. Leoeuf, M. Honda, M. Nishiki, A. Takahashi, A. Tsukamoto, SPIE Proc. **4682**, 503 (2002)
29. G. Belev, K. Safa, J.A. Rowlands, D. Hunter, M. Yaffe, Dependence of the electrical properties of stabilized a-Se on the preparation conditions and the development of a double layer X-ray detector structure. Curr. Appl. Phys. **8**, 383–387 (2008)
30. S. Kasap, J. Rowlands, K. Tanioka, A. Nathan, Applications of disordered semiconductors in modern electronics: selected examples, in *Charge Transport in Disordered Solids*, ed. by S. Baranovskii (Wiley, New York, 2006)
31. S. Kasap, J.B. Frey, G. Belev, O. Tousignant, H. Mani, J. Greenspan, L. Laperriere, O. Bubon, A. Reznik, G. DeCrescenzo, K.S. Karim, J.A. Rowlands, Amorphous and polycrystalline photoconductors for direct conversion flat panel X-ray image sensors. Sensors **11**, 5112–5157 (2011). doi:10.3390/s110505112
32. N. Destefano, M. Mulato, Influence of multi-depositions on the final properties of thermally evaporated TlBr films. Nucl. Instrum. Methods Phys. Res. A **624**, 114–117 (2010)
33. P.R. Bennett, K.S. Shah, L.J. Cirignano, M.B. Klugerman, L.P. Moy, M.R. Squillante, Characterization of polycrystalline TlBr films for radiographic detectors. IEEE Trans. Nucl. Sci. **46**, 689–693 (1999)
34. M. Yun, S. Cho, R. Lee, G. Jang, Y. Kim, W. Shin, S. Nam, Investigation of PbI_2 film fabricated by a new sedimentation method as an X-ray conversion material. Jpn. J. Appl. Phys. **49**, 041801–041805 (2010)
35. K.S. Shah, R.A. Street, Y. Dmitriyev, P. Bennett, L. Cirignano, M. Klugermaa, M.R. Squillante, G. Entine, X-ray imaging with PbI_2-based a-Si:H flat panel detectors. Nucl. Instrum. Methods Phys. Res. A **458**, 140–147 (2001)
36. Q. Zhao, L.E. Antonuk, Y. El-Mohri, Y. Wang, H. Du, A. Sawant, Z. Su, J. Yamamoto, Performance evaluation of polycrystalline HgI_2 photoconductors for radiation therapy imaging. Med. Phys. **37**, 2738–2748 (2010)
37. H. Du, L.E. Antonuk, Y. El-Mohri, Q. Zhao, Z. Su, J. Yamamoto, Y. Wang, Investigation of the signal behavior at diagnostic energies of prototype, direct detection, active matrix, flat-panel imagers incorporating polycrystalline HgI_2. Phys. Med. Biol. **53**, 1325–135 (2010)
38. G. Zentai, L. Partain, R. Pavlyuchkova, Dark current and DQE improvements of mercuric iodide medical imagers. *Proc. SPIE* 2007, 6510, doi:10.1117/12.713848

39. K. Kim, S. Kang, J. Park, S. Cho, B. Cha, J. Shin, S. Nam, J. Kim, Quantitative evaluation of mercuric iodide thick film for X-ray imaging device. *Proc. SPIE* 2007, 6142, doi:10.1117/12.653002
40. A. Zuck, M. Schieber, O. Khakhan, Z. Burshtein, Near single-crystal electrical properties of polycrystalline HgI_2 produced by physical vapor deposition. IEEE Trans. Nucl. Sci. **50**, 991–997 (2003)
41. J.S. Iwanczyk, B.E. Patt, C.R. Tull, L.R. MacDonald, N. Skinner, E.J. Hoffman, L. Fornaro, HgI_2 polycrystalline films for digital X-ray imagers. IEEE Trans. Nucl. Sci. **49**, 160–164 (2002)
42. R.A. Street, S.E. Ready, K. van Schuylenbergh, J. Ho, J.B. Boyec, P. Nylen, K. Shah, L. Melekhov, Comparison of PbI_2 and HgI_2 for direct detection active matrix X-ray image sensors. J. Appl. Phys. **91**, 3345 (2002)
43. S. Tokuda, H. Kishihara, S. Adachi, T. Sato, Improvement of temporal response and output uniformity of polycrystalline CdZnTe films for high-sensitivity X-ray imaging. Proc. SPIE **5030**, 861–870 (2003)
44. S. Tokuda, H. Kishihara, S. Adachi, T. Sato, Preparation and characterization of polycrystalline CdZnTe films for large-area, high-sensitivity X-ray detectors. J. Mater. Sci. Mater. Electron. **5**, 1–8 (2004)
45. M. Simon, R.A. Ford, A.R. Franklin, S.P. Grabowski, B. Mensor, G. Much, A. Nascetti, M. Overdick, M.J. Powell, D.U. Wiechert, PbO as direct conversion X-ray detector material. Proc. SPIE **5368**, 188–199 (2004)
46. M. Simon, R.A. Ford, A.R. Franklin, S.P. Grabowski, B. Mensor, G. Much, A. Nascetti, M. Overdick, M.J. Powell, D.U. Wiechert, Analysis of lead oxide (PbO) layers for direct conversion X-ray detection. IEEE Trans. Nucl. Sci. **52**, 2035–2040 (2005)
47. C.A. Klein, Bandgap dependence and related features of radiation ionization energies in semiconductors. J. Appl. Phys. **39**, 2029–2038 (1968)
48. R.C. Alig, S. Bloom, Electron-hole pair creation energies in semiconductors. Phys. Rev. Lett. **35**, 1522–1525 (1975)
49. C. Haugen, S.O. Kasap, J.A. Rowlands, Charge transport and electron-hole creation energy in stabilized a-Se X-ray photoconductors. J. Phys. D Appl. Phys. **32**, 200–207 (1999)
50. S.O. Kasap, J.A. Rowlands, Direct-conversion flat-panel X-ray image detectors. IEEE Proc. Circ. Dev. Syst. **149**, 85–96 (2002)
51. D. Tonchev, S.O. Kasap, Effect of ageing on glass transformation by temperature modulated DSC. Mater. Sci. Eng. A **328**, 1–5 (2002)
52. B. Fogal, R.E. Johanson, G. Belev, S. Leary, S. Kasap, X-ray induced effects in stabilized a-Se X-ray photoconductors. J. Non-Cryst. Solids **299**, 993–997 (2000)
53. S.O. Kasap, C. Juhasz, Charge transport in selenium based amorphous xerographic photoreceptors. Photograph Sci. Eng. SPSE **26**, 239–244 (1982)
54. W. Zhao, G. DeCrescenzo, S.O. Kasap, J.A. Rowlands, Ghosting caused by bulk charge trapping in direct conversion flat-panel detectors using amorphous selenium. Med. Phy. **32**, 488–500 (2005)
55. M.Z. Kabir, M. Yunus, S.O. Kasap, O. Tousignant, H. Mani, P. Gauthier, Sensitivity of stabilized a-Se based X-ray photoconductors. Curr. Appl. Phys. **6**, 393–398 (2006)
56. M.H. Chowdhury, M.Z. Kabir, Electrical properties of grain boundaries in polycrystalline materials under intrinsic or low doping. J. Phys. D Appl. Phys. **44**, 015102–015106 (2011)
57. M. Schieber, A. Zuck, Advances in physical vapor deposited polycrystalline-HgI_2 X-ray imaging detectors. J. Optoelectron. Adv. Mater. **5**, 1299–1303 (2003)
58. S. Kasap, J.B. Frey, G. Belev, O. Tousignant, H. Mani, L. Laperriere, A. Reznik, J.A. Rowlands, Amorphous Selenium and its Alloys from early xerography to high resolution x-ray image detectors and ultrasensitive imaging tubes. Phys. Status Sol. B Basic Sol. State Phys. **246**, 1794–1805 (2009). and references therein
59. M. Abkowitz, Philos. Mag. Letts. **58**, 53 (1998)
60. H.Z. Song, G.J. Adriaenssens, E.V. Emelianova, V.I. Arkhipov, Phys. Rev. B **59**, 10607 (1999)

61. S.O. Kasap, J.A. Rowlands, J. Mater. Sci. Mater. Electron. **11**, 179 (2000)
62. L. Benkhedir, M.S. Aida, G.J. Adriaenssens, J. Non-Cryst. Solids **344**, 193 (2004)
63. L. Benkhedir, M. Brinza, G.J. Adriaenssens, J. Phys. Condens. Matter **16**, S5253 (2004)
64. N. Qamhich, M.L. Benkhedir, M. Brinza, J. Willekens, G.J. Adriaenssens, J. Phys. Condens. Matter **16**, 3827 (2004)
65. K. Koughia, Z. Shakoor, S.O. Kasap, J.M. Marshall, J. Appl. Phys. **97**, 33706 (2005)
66. K.V. Koughia, B. Fogal, G. Belev, R.E. Johanson, S.O. Kasap, Density of states in the mobility gap of stabilized a-Se from electron time-of flight photocurrent analysis. JNCS **338–340**, 569–573 (2004)
67. K. Koughia, Z. Shakoor, S.O. Kasap, J.M. Marshall, Density of localized electronic states in a-Se from electron time of flight photocurrent measurements. J. Appl. Phys. **97**, 033706 (2005)
68. M. Abkowitz, E. Enck, Phys Rev. B **25**, 2567 (1982)
69. J. Veres, C. Juhasz, J. Non-Cryst. Solids **164–166**, 407 (1993)
70. K. Tanioka, J. Yamazaki, K. Shidara, K. Taketoshi, T. Hirai, Y. Takasaki, Adv. Electron. Electron. Phys. **74**, 379 (1988)
71. G. Juska, K. Arlauskas, Phys. Stat. Sol. **59**, 389 (1980)
72. G. Juska, K. Arlauskas, Phys. Stat. Sol. **77**, 387 (1983)
73. K. Tanioka, J. Yamazaki, K. Shidara, K. Taketoshi, T. Kawamura, S. Ishioka, Y. Takasaki, IEEE Electron. Dev. Lett. **EDL-8**(392) (1987)
74. K. Tsuji, Y. Takasaki, T. Hirai, K. Taketoshi, JNCS **14**, 94 (1989)
75. K. Tsuji, T. Ohshima, T. Hirai, N. Gotoh, K. Tanioka, K. Shidara, Mater. Res. Symp. Proc. **219**, 507 (1991)
76. K. Tsuji, Y. Takasaki, T. Hirai, J. Yamazaki, K. Tanioka, Optoelectron. Dev. Technol. (Jpn.) **9**, 367 (1994)
77. T. Watanabe, M. Goto, H. Ohtake, H. Maruyama, M. Abe, K. Tanioka, N. Egami, IEEE Trans. Electron. Dev. **50**, 63 (2003)
78. J. Mort, *The Anatomy of Xerography* (McFarland, London, 1989)
79. S.O. Kasap, in *The Handbook of Imaging Materials*, ed. by A. Diamond (Marcel Dekker, New York, 1991), ch. 8 and references therein

Chapter 6
Electrochemical Sensors

James J. Noël and Gurinder Kaur Ahluwalia

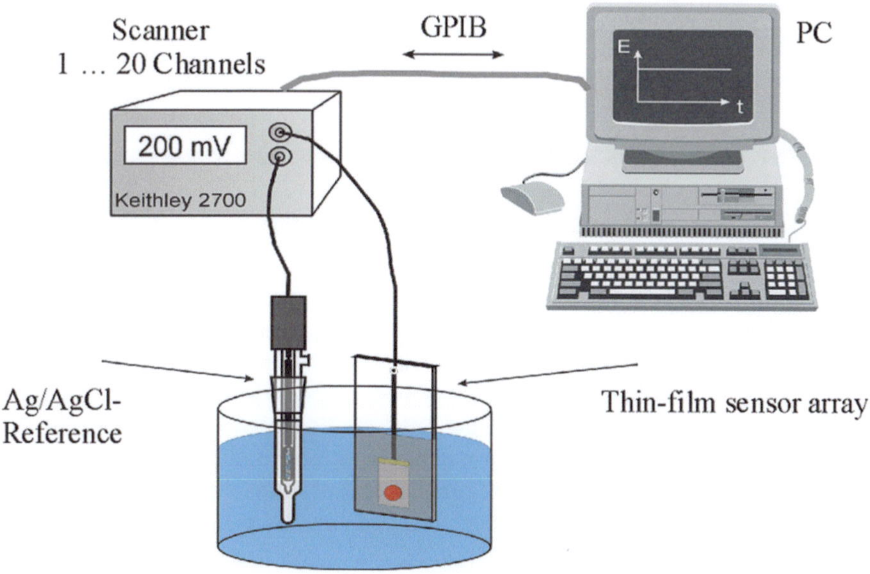

Measurement set-up for the electrochemical characterization of the chalcogenide glass-based thin-film sensor array

J.J. Noël, Ph.D. (✉)
Department of Chemistry, The University of Western Ontario, 1151 Richmond St., London, N6A 5B7, ON, Canada
e-mail: jjnoel@uwo.ca

G.K. Ahluwalia, Ph.D.
Materials and Nanotechnology Research Laboratory, Department of Physics College of The North Atlantic, Labrador West Campus, 1600 Nichols Adam Highway, Labrador City, A2V 0B8, NL, Canada
e-mail: gurinder.kaur@cna.nl.ca

G.K. Ahluwalia (ed.), *Applications of Chalcogenides: S, Se, and Te*,
DOI 10.1007/978-3-319-41190-3_6

6.1 Introduction

Potentiometric sensor electrodes constitute a class of chemical sensors that can indicate the presence and quantity of a species of interest (the analyte), based on the electrochemical response that the analyte stimulates on the sensor. Whereas some other types of electrochemical sensors rely on a change in the electrolytic current (amperometric sensors) or charge (coulometric sensors) to sense and quantify the analyte, potentiometric sensors generate a certain electrochemical potential difference in response to the presence of a specific analyte, and that potential difference is related quantitatively to the abundance of that species.

This chapter begins with a summary of the principles underlying the function of potentiometric sensors in general, and a description of analytical approaches employing them, before turning to focus on sensors employing chalcogen-containing compounds, especially ISE based on chalcogenide glass membranes, and briefly mentioning some future trends in the field.

6.2 ISE Principles and Theory

6.2.1 Sensor Response and the Phase Boundary Potential

A chemical sensor is a transducer that converts some chemical information into an electrical signal that can be detected by an appropriate instrument and reported to a user. In the case of potentiometric sensors, the chemical information transduced by the sensor is the concentration or, more precisely, the *activity* of a chemical species.

This transduction is effected via the influence of the chemical activity of the species of interest on the position of a phase boundary equilibrium involving that species, which is indicated by the electrochemical potential that develops between the sensor and a reference electrode included for the purpose of making the measurement (Fig. 6.1).

The potentiometric response of the sensor can be described quantitatively, according to equilibrium thermodynamic theory, beginning with the concept of the electrochemical potential, $\overline{\mu}$. For analyte species, A, in phase p,

$$\overline{\mu}_{A_p} = \mu_{A_p}{}^{\circ} + RT\ln a_{A_p} + z_A F \varphi_p \tag{6.1}$$

where $\overline{\mu}_{A_p}$ is the electrochemical potential of ion or molecule A in phase p, $\mu_{A_p}{}^{\circ}$ the standard chemical potential of ion or molecule A in phase p, R the universal gas constant, T the absolute temperature, a_{A_p} the activity of species A in phase p, F Faraday's constant, z_A the charge on each ion or molecule ($z_A = 0$) of species A, and φ_p the electrical potential within phase p [1].

When two phases are in equilibrium, the electrochemical potential of each species present must be the same in both phases. In the case of the sensor immersed

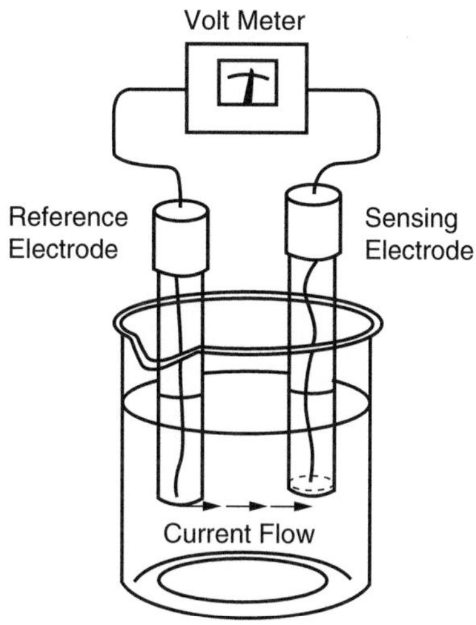

Fig. 6.1 An electrochemical potential difference develops between the sensor and a reference electrode

in a solution of the analyte species, when species A in the sensor surface (A_m) comes to equilibrium with species A in the solution (A_s),

$$A_s \rightleftarrows A_m \tag{6.2}$$

then,

$$\overline{\mu}_{A_s} = \overline{\mu}_{A_m} \tag{6.3}$$

where $\overline{\mu}_{A_s}$ and $\overline{\mu}_{A_m}$ represent the electrochemical potentials of species A in the sample solution and in the sensor surface, respectively.

Expanding the terms of Eq. (6.2) using Eq. (6.1), and rearranging, yields:

$$\varphi_m - \varphi_s = \frac{\left(\mu_{A_s}{}^{\circ} - \mu_{A_m}{}^{\circ}\right)}{z_A F} + \frac{RT}{z_A F}\ln\frac{a_{A_s}}{a_{A_m}} \tag{6.4}$$

If one notes that $\varphi_m - \varphi_s$ is the electrical potential difference, E_{ISE}, between the sensor and sample; that the difference in standard chemical potentials is the standard free energy of the interfacial transfer, $\Delta G = \mu_{A_s}{}^{\circ} - \mu_{A_m}{}^{\circ}$; and that the conversion between energy expressed in chemical units and in electrical units is $\Delta G = -zFE$, then one can rewrite Eq. (6.3) in electrochemical terms:

$$E_{ISE} = E^{\circ} + \frac{RT}{z_A F}\ln\frac{a_{A_s}}{a_{A_m}} \tag{6.5}$$

where E° is the standard potential [1]. Thus, measurement of the potential difference between the sample solution and the sensor, which is experimentally quite easy, yields a means to determine the activity of an analyte species.

The trick, then, in potentiometric sensor design, is to establish a phase boundary equilibrium system involving the species of interest, in a way that the equilibrium potential can be sensed by a pair of electrodes. Generally, this is accomplished by preparing a membrane that will develop a phase boundary equilibrium that specifically involves the analyte of interest, then exposing one side of that membrane to the solution containing the unknown to be determined, and measuring the resultant electrochemical potential difference across the membrane by means of two reference electrodes (one on each side of the membrane) or between one reference electrode and a second conductor in ohmic contact (as opposed to electrolytic contact) with the opposite side of the membrane. The concentration of the analyte species can then be determined, for example, by comparing the measured potential difference with a calibration curve developed by measuring the sensor's response to an appropriate series of standard solutions of the analyte species.

6.2.2 Selectivity

A key feature of ion-selective electrodes is that they are selective for a particular species of interest. That is, they are able to perform quantitative determinations of a particular chemical species even in the presence of other, perhaps similar, chemical species. Chemical species other than the species of interest that elicit a response from the ISE are called interferences. Each interference also generates a phase boundary potential according to its own version of Eq. (6.5). The degree to which the ISE responds to an interfering species is expressed in terms of a selectivity coefficient, K, the strength of the response to each interference being proportional to the product of the selectivity coefficient and the activity of the interference [2].

Since the phase boundary equilibria are responsible for the entire potentiometric response of a sensor electrode, the selectivity of an ISE is directly related to the types and extent of thermodynamic equilibria taking place at the sensor–solution interface and within the sensor membrane itself [3]. Selectivity coefficients can vary widely, depending on the chemistry of the sensor and that of the interfering species. For example, a certain NO_3^--selective electrode has selectivity coefficients of 6×10^{-5} for F^- ion and 1000 for ClO_4^- ion [2]. Many selectivity coefficients in the 10^{-12} range and some as low as 10^{-14} have been measured [4].

An impressive achievement of many years of studies of the influences of interfering species is that the influences of all interferences on E_{ISE} can be predicted quite precisely, even for mixtures containing multiple interferences with different ionic charges. The sensor response to a mixture of any number of mono-, di-, and tri-valent ions is given by [5],

$$E_{\mathrm{ISE}} = E_{\mathrm{A}}^{\circ} + \frac{RT}{F}\ln\left[\frac{1}{3}\sum_i\left(K_i^{\frac{1}{z_{\mathrm{A}}}}a_i\right) + \frac{\sqrt[3]{2}\,\theta}{3\sqrt[3]{\vartheta + \sqrt{-4\theta^3 + \vartheta^2}}} + \frac{\sqrt[3]{\vartheta + \sqrt{-4\theta^3 + \vartheta^2}}}{\sqrt[3]{2}}\right] \tag{6.6}$$

where

$$\theta = \left(\sum_i\left(K_i^{\frac{1}{z_{\mathrm{A}}}}a_i\right)\right)^2 + 3\sum_j\left(K_j^{\frac{2}{z_{\mathrm{A}}}}a_j\right) \tag{6.6b}$$

and

$$\vartheta = 2\left(\sum_i\left(K_i^{\frac{1}{z_{\mathrm{A}}}}a_i\right)\right)^3 + 9\left(\sum_i\left(K_i^{\frac{1}{z_{\mathrm{A}}}}a_i\right)\right)\left(\sum_j\left(K_j^{\frac{2}{z_{\mathrm{A}}}}a_j\right)\right) + 27\sum_k\left(K_k^{\frac{3}{z_{\mathrm{A}}}}a_k\right) \tag{6.6c}$$

and i, j, and k are indices representing all involved mono-, di-, and tri-valent ions, respectively, the contributions from ions of differing charges being summed separately. Note that the analyte species must be included within the valency-appropriate summation in Eq. (6.6), with selectivity coefficient $K_{\mathrm{A}} = 1$ and activity a_{A}.

In traditional, single-species ISE analyses it is therefore advantageous to employ a sensor with very small selectivity coefficients, and/or eliminate any interfering species from the test solution by some form of pre-purification. However, it is now possible in some cases to determine the selectivity coefficients with sufficient precision that, if the interference activities can be determined independently, the measured ISE response can be corrected for contributions from the interferences and an accurate, precise measurement of analyte concentration made even in their presence [3].

6.2.3 Sensitivity

Until the late 1990s, ISE were not useful for determinations of concentrations below about 10^{-6} M, and the best selectivity coefficients achievable were in the 10^{-5} range [3]. Vast improvements in both sensitivity and selectivity suddenly became possible when it was realized that ion fluxes from the ISE themselves were the limiting factors in both of these performance characteristics. High concentrations of the analyte species present in sensor membranes or in their internal reference solutions resulted in emissions of significant fluxes of analyte ions from the sensor into the solution being analyzed whenever the establishment of phase boundary equilibrium favoured such a flow. Consequently, for each ISE, a limiting minimum concentration of analyte species was always present in the solution near the sensor surface.

Although ion flux effects cannot be completely eliminated from ISE measurements, now that they are recognized and understood their influence can be decreased by a variety of means, including adjusting the composition of the inner reference solution, decreasing the rate of ionic diffusion within the sensor membrane, increasing the rate of ion transport on the analyte solution side of the sensor membrane, or compensating ion fluxes by applying an external electrical current. With such improvements in place, it is now possible to determine concentrations in the 10^{-8} to 10^{-11} M range (i.e., ppt range) and to obtain selectivity coefficients in the 10^{-10} to 10^{-15} range [3].

6.3 Analytical Methodologies

A number of analytical approaches can be employed in determinations based on ISE measurements. The most common are single-point calibrations, calibration curves, standard addition, and titrations. These techniques are, of course, not unique to ISE methods, but some aspects of their application are, so a brief description of each is given here.

6.3.1 Single-Point Calibration

If one describes analyte activity in terms of a "*p*-function" (pH, pOH, etc.) such that,

$$pA = -\log a_A \quad (6.7)$$

one can rearrange Eq. (6.5) to give,

$$pA = \frac{nF}{2.303RT}(L - E_{Meas}) \quad (6.8)$$

where E_{Meas} is the measured potential difference across the two terminals of the potentiometric sensor system and L a constant that conveniently incorporates other constant values that need not be evaluated individually, or may not even be knowable (e.g., those resulting from the constant activity of species A within the sensor surface or in the sensor's internal solution). The difference $L - E_{Meas}$ incorporates E_{ISE} and any other potential differences affecting the measurement, such as reference electrode potentials, junction potentials, ohmic drops, and asymmetry potentials.

The value of the constant L, needed to evaluate pA from E_{Meas}, can be obtained experimentally by determining E_{Meas} for a standard solution of known pA. To use this calibration for subsequently determining an unknown, one must assume that L does not change between calibration and measurement. In practice the evaluation of L by the single point calibration method may remain valid for up to a few hours, after which re-calibration is necessary [2].

Although this method of calibration is relatively quick and easy, it has the significant disadvantage that it is inherently erroneous, and the degree of the error is variable and uncertain. The problem arises because of the necessary assumption that L is the same in both the standard solution used for calibration and the solution to be analyzed. Differences in the composition of the two solutions, even if only the concentration of analyte, result in different junction potentials, thereby changing L to an extent that cannot be calculated a priori, but that could easily introduce a relative error of several percent, or worse [2].

6.3.2 *Calibration Curves*

The relationship between E_{Meas} and the analyte concentration can be directly determined empirically by using a series of calibration standards to develop a calibration curve (i.e., a plot of E_{Meas} vs. analyte concentration). One advantage of the calibration curve is that it relates E_{Meas} directly to analyte *concentration*, rather than activity, thereby accounting for changes in the activity coefficient intrinsically.

For the calibration curve method to work properly, the ionic strength of the solution to be determined should be the same as that of the standard solutions used to generate the calibration curve. When the concentration of electrolytes in the sample to be determined is not too great, ionic strengths can be made approximately equal by dosing both the sample and the calibration standards with a measured excess of inert electrolyte [2].

6.3.3 *Standard Addition*

The concentration of analyte in a sample can be quickly determined directly from E_{Meas}, without evaluation of constant L or preparation of a calibration curve, by using the standard addition method. In this approach, a potential measurement is made both before and after adding a small aliquot of a standard solution of the analyte to a known volume of the solution to be determined. As long as the addition results in minimal changes to the ionic strength of the sample, the activity coefficient of the analyte should remain unaffected. One must also assume that the junction potentials do not change either [2].

Equation (6.8) can be re-formulated as,

$$-\log a_A = -\log C_A \gamma_A = \frac{nF}{2.303RT}(L - E_1) \tag{6.9}$$

where C_A is the concentration of analyte A, γ_A the activity coefficient of species A, and E_1 the measured cell potential before addition.

After volume V_s of standard solution containing concentration C_s of species A is added to volume V_A of the solution to be determined, Eq. (6.9) becomes,

$$-\log\left[\left(\frac{C_A V_A + C_s V_s}{V_A + V_s}\right)\gamma_A\right] = \frac{nF}{2.303RT}(L - E_2) \tag{6.10}$$

where E_2 is the measured cell potential after the addition. By subtracting Eq. (6.9) from Eq. (6.10) one can obtain:

$$\log\left(\frac{C_A(V_A + V_s)}{C_A V_A + C_s V_s}\right) = \frac{nF}{2.303RT}(E_1 - E_2) \tag{6.11}$$

Solving for C_A yields:

$$C_A = \frac{C_s V_s}{(V_A + V_s)\exp\left(\frac{nF}{RT}(E_2 - E_1)\right) - V_A} \tag{6.12}$$

6.4 Types of ISE and Species Sensed

6.4.1 First-Order Indicator Electrodes

A first-order electrode measures the activity of its own cation [2]. For example, a silver sensor can be made by simply immersing a piece of silver metal into a solution containing Ag^+ ions. The redox equilibrium represented by the half-reaction

$$Ag^+_{(aq)} + e^- \rightleftarrows Ag_{(s)} \tag{6.13}$$

will be established quickly. In this case, no membrane is required, and the potential difference between the Ag metal electrode and a standard reference electrode can be measured directly and, for example, compared with a calibration curve to yield the activity of Ag^+ in the analyte solution. For a first-order indicator electrode to function, the solid electrode must be an electronic conductor, and the redox equilibrium involved must be reversible. These requirements can be satisfied for certain metals in contact with a solution containing their ions, such as Ag, Cu, Hg, Pb, and Cd, but preclude the development of first-order ISE for chalcogens.

6.4.2 Second-Order Indicator Electrodes

These ISE are denoted second-order because the anion in the salt layer on the metal surface is only indirectly involved in any redox process with the underlying metal [2]. This type of ISE can be prepared for metals whose cations form low solubility

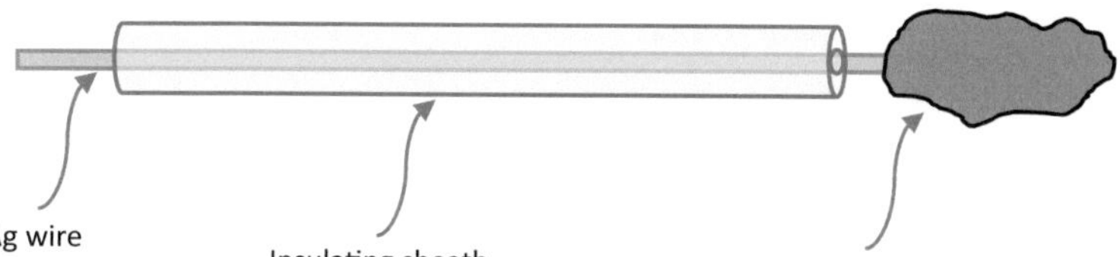

Fig. 6.2 Simple second-order ISE created by coating a Ag wire with a deposit of Ag_2S. This ISE could be used to quantify either Ag^+ or S^{2-} ions

salts with the anion of interest, simply by coating a wire with a layer of a salt composed of cations of that same metal and the anion of interest, as shown in Fig. 6.2. Coated wire electrodes of this sort are often prepared by anodizing the wire in a solution containing a reasonably high concentration of the anion of interest, by dipping the wire into a molten bath of the low solubility salt to deposit a layer, or by reaction of the metal with a precursor of the intended anion in an aggressive gas-phase environment.

For example, a silver sulphide-covered silver wire responds to sulphide activity via the phase boundary equilibrium represented by:

$$2Ag^{+}_{(aq)} + S^{2-}_{(aq)} \rightleftarrows Ag_2S_{(s)} \tag{6.14}$$

Here the Ag_2S layer on the Ag surface effectively acts as a sensor membrane. Note that this electrode would also function as a sensor for Ag^+ ions [6].

In order for this electrode to achieve phase boundary equilibrium and function properly as a sulphide sensor, the solution, at least in the vicinity of the electrode, should be saturated with the low-solubility salt to satisfy the solubility equilibrium given in reaction (6.14). This would result in an ion flux that would limit the sensitivity of this sensor.

The potential response of the silver wire to sulphide ions at 25 °C is,

$$E_{ISE} = -0.604\ V + 0.0284\,V \times pS \tag{6.15}$$

where E_{ISE} is given in V vs. the standard hydrogen electrode (i.e., V_{SHE}), and $pS = -\log a_{S^{2-}}$, where $a_{S^{2-}}$ is the activity of sulphide ion in the analyte solution [6].

In practice, due to the effects of a number of complicating factors, such as the reversibility of the reaction, electronic and ionic conductivity of the salt layer on the metal wire, temperature fluctuations, and slow kinetics, the measured potential of an ISE may not be identical to that predicted by Eq. (6.5). A plot of E_{ISE} vs. pS should still be linear for sulphide concentrations between about 10^{-6} and 10^{-1} M, but possibly with a different slope and intercept than indicated in Eq. (6.8). However, instead of evaluating the individual contributions of each of the non-idealities, it is much easier to overcome them by the use of one of the calibration methods described above. In practice, a valid and useful calibration curve is often obtained from second-order ISE over a concentration range of 10^{-7} to 10^{-1} M.

6.4.3 Membrane-Based ISE

The vast majority of ISE are based on sensor membranes of one sort or another. The possibilities for sensor membrane materials include glass (as in pH electrodes), solid oxides, salt layers (formed by precipitation, pressing, sintering, etc., and sometimes doped for increased conductivity), porous polymer films saturated with liquid ion exchange compound, composite materials, or even slices of biological tissues. The key is to find a material in which a strong phase boundary equilibrium with the intended analyte exists and that has suitable other properties (stability, sensitivity, selectivity, conductivity, robustness, cost, etc.).

Membrane-based ISE are prepared by positioning two reference electrodes on opposite sides of the active membrane to measure the electrochemical potential difference across the membrane, as shown in Fig. 6.3a. One side of the membrane is in contact with an electrolyte solution with some of the species of interest dissolved in it (the internal solution).

Usually this solution and the local (internal) reference electrode are encapsulated in a sealed compartment, the sensor membrane closing one end of the compartment. On the other side of the membrane, the second (external) reference electrode rests in the solution to be analyzed. A difference in analyte species activity between the solution in the internal compartment and the external solution to be analyzed results in a potential difference across the membrane, which can be detected as a potential difference between the two reference electrodes. The generic potentiometric cell that describes the sensor can be represented by the standard cell notation shown in Fig. 6.4.

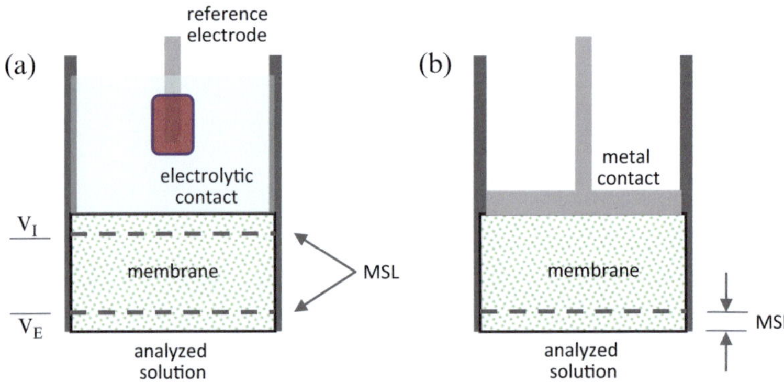

Fig. 6.3 Membrane-based ISE schematic; (**a**) ISE with electrolytic contact to internal reference electrode immersed in internal solution reservoir, and (**b**) all-solid-state ISE with ohmic contact to conductor on the back of the membrane. The membrane in panel (**a**) contacts electrolyte solution and develops a modified surface layer (MSL) on both sides. E_{ISE} is created by the difference in phase boundary equilibrium potentials V_I and V_E. Only one side of the membrane is in contact with solution in the all-solid-state electrode, and therefore only one MSL is created

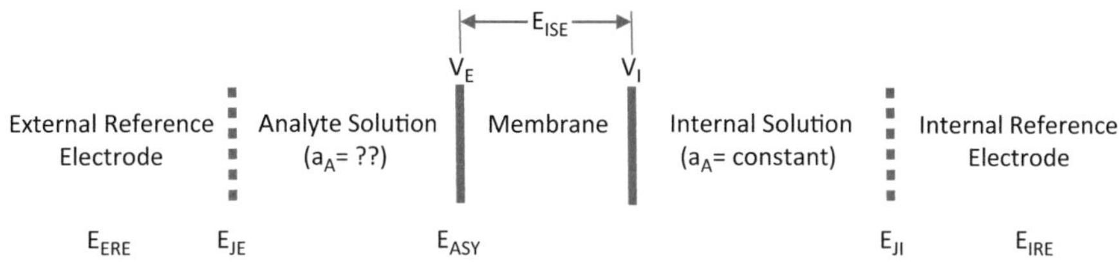

Fig. 6.4 Standard cell notation describing a membrane-based ISE. The membrane is effectively a salt bridge between the internal and external compartments

The measured potential of the cell will be,

$$E_{\text{Meas}} = E_{\text{IRE}} - E_{\text{ERE}} + E_{\text{JE}} + E_{\text{ASY}} + E_{\text{ISE}} + E_{\text{JI}} \tag{6.16}$$

where $E_{\text{IRE}} - E_{\text{ERE}}$ is the difference between reference electrode potential values (a known constant in all cases), E_{JE} and E_{JI} the junction potentials arising from differences in ionic mobility across the junction (porous frit, salt bridge, etc.) between the reference electrode and the solution in which it is immersed, E_{ASY} the "asymmetry potential" caused by factors that are not fully understood but that are related to the degree to which the two sides of the membrane do not respond identically to the same stimulus, and finally E_{ISE} the potential difference across the membrane caused by differences in the activities of chemical species on either side of the membrane, where $E_{\text{ISE}} = v_{\text{E}} - v_{\text{I}}$ and v_{E} and v_{I} the phase boundary potentials at the interfaces of the membrane with the external and internal solutions, respectively, established by interaction with the analyte. The values of v_{E} and v_{I} are determined by the activity of the analyte species, according to Eq. (6.5),

$$v_{\text{E}} = E^{\circ} + \frac{RT}{z_{\text{A}}F}\ln\frac{a_{\text{A}_\text{s}}}{a_{\text{A}_\text{m}}} \quad \text{and} \quad v_{\text{I}} = E^{\circ} + \frac{RT}{z_{\text{A}}F}\ln\frac{a_{\text{A}_\text{I}}}{a_{\text{A}_\text{m}}} \tag{6.17}$$

where a_{A_I} is the activity of the analyte species in the internal solution. Since the internal solution has a fixed composition, the values of a_{A_I}, and consequently v_{I}, are constants. The membrane composition is also fixed, so a_{A_m} is also constant. Therefore E_{ISE} changes only as a_{A_s} changes, and since E_{IRE}, E_{ERE}, E_{JE}, E_{JI}, and E_{ASY} are all constant over the duration of the measurement, E_{Meas} changes solely in response to changes in the activity of the analyte species in the solution being analyzed. Therefore the relationship between the analyte activity in the sample and the ISE response can be expressed as,

$$E_{\text{Meas}} = L + \frac{RT}{z_{\text{A}}F}\ln(a_{\text{A}_\text{s}}) \tag{6.18}$$

where L is a constant (obtained by combining all of the constants in Eqs. (6.16) and (6.17), including the standard potential, junction potentials, and asymmetry potential, as described previously).

As in the case of second-order electrodes (described above), it is not necessary (and, in fact, not possible) to evaluate all of the contributing terms in Eq. (6.16) separately or from theory. Again, the practical approach is to use a standard addition method, perform a single-point calibration, or prepare a calibration curve based on the electrode response to a series of standard solutions of the species to be analyzed and then use the calibration curve to translate the potential difference measured with the unknown solution present into the concentration of the analyte species.

A variation on the membrane ISE cell involves replacing the internal solution and reference electrode with a conductor (metal or conducting polymer) in direct ohmic (vs. electrolytic) contact with the membrane, as shown in Fig. 6.2b. This configuration is analogous to the second-order indicator electrodes described above, except that the membrane need not be a deposited salt film. Elimination of the internal solution and reference electrode can afford certain advantages for the all-solid-state ISE, including increased robustness and longevity, lower cost, easier fabrication and maintenance, and often better analytical performance [3, 7].

6.4.4 *Gas Sensing ISE*

ISE-based sensors for gaseous species can be prepared if a suitable gas-aqueous phase equilibrium involving the gas of interest exists. These devices might be more appropriately called gas sensing cells, since they involve a compartment containing two electrodes, between which the potential difference is measured. The approach, illustrated in Fig. 6.5, is to fabricate a compartment with the gas-permeable membrane incorporated into the vessel wall (the interior of the compartment

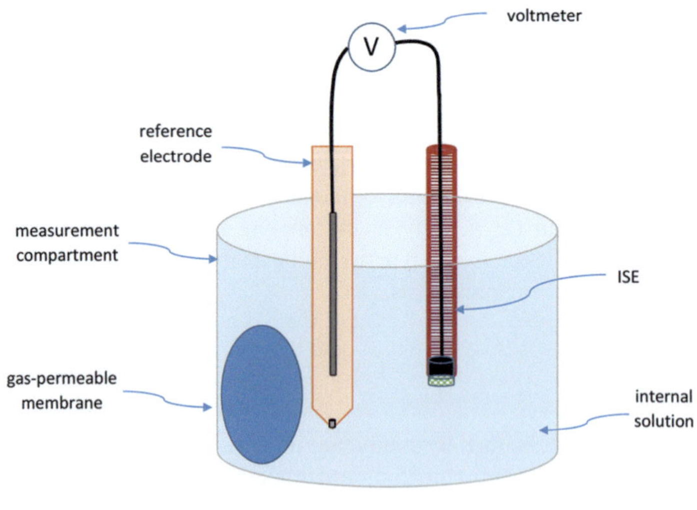

Fig. 6.5 Potentiometric gas-sensing cell

being filled with an appropriate solution) and expose the membrane exterior to the gas sample or atmosphere to be analyzed.

The gaseous species penetrating the membrane establishes an equilibrium with a species that is soluble in the internal solution, for example [8],

$$SO_{2(g)} + H_2O_{(l)} \rightleftarrows HSO^-_{3(aq)} + H^+_{(aq)} \tag{6.19}$$

and the detection is actually carried out by an ISE that is selective for one of the products of the equilibrium reaction, in this example a $HSO^-_{3(aq)}$-sensitive ISE, working in conjunction with a reference electrode, in the internal solution. Such gas-sensing electrodes have also been developed for a variety of other gases, including CO_2 (using a pH electrode), NO_2 (using a nitrate-selective electrode), H_2S, HF, HCN, and NH_3 [2].

6.4.5 *Non-specific Multisensors*

It is possible to make simultaneous potentiometric determinations of all the components of a mixture by using an array of poorly selective sensors having different selectivity coefficients [9]. Since the response of a potentiometric sensor to a multicomponent mixture can be fully predicted (based on Eq. (6.6)), once the selectivity coefficients of all of the active species are known, by using a sufficient number of sensors that respond to multiple components of the mixture one can generate a system of independent equations sufficient to yield the concentrations of all of the analytes, if the mathematical solution can be determined. Using currently available computing power it may be possible to deconvolute the linked set of electrode responses by multivariate regression, statistical approaches, or for very complicated mixtures, pattern recognition and artificial neural networks [7, 9, 10]. This approach can be used to produce what is commonly called an "electronic tongue" that in one "taste" of a mixture can determine its composition or at least distinguish it from another very similar mixture. For example, such a "tongue" has been used to distinguish various samples of wines [11].

6.5 Chalcogenide Membrane-Based Potentiometric Sensor Electrodes

As this is a volume devoted to chalcogen chemistry, this chapter must now turn its focus toward chalcogen-based sensors. In that regard, ISE based on chalcogenide glass (ChG) membranes have become quite important and widely used, as a consequence of the variety of species that can be sensed and the many favourable characteristics (stability, robustness, low solubility, sensitivity, etc.) that they possess.

6.5.1 Membrane Fabrication

The propensity of chalcogens to react with a very wide variety of main group elements and transition metals to yield low-solubility salts (in many cases these could be considered nearly covalent compounds) makes them likely candidates for construction of ISE sensor membranes. Because chalcogenide compounds that have been solidified in a glassy state tend to be more durable in acidic and redox-active environments than their polycrystalline counterparts, and because electrodes fabricated from ChG often exhibit a longer service life, better selectivity, and higher reproducibility than do crystalline chalcogenides, they merit significant attention as sensor membrane materials [7, 12].

To prepare bulk ChG suitable for ISE membranes, high-purity starting materials are reacted under carefully controlled conditions at temperatures above the melting point of the resultant chalcogenide, homogenized, and then rapidly quenched to yield the solid chalcogenide compound in the glassy state, which can be verified by X-ray diffraction [7]. Membranes can then be prepared by slicing the bulk ChG ingots thinly, or by pressing powdered ChG into sheets. The bulk ChG membrane can then be mechanically shaped, polished, and assembled into a sensor element.

If the ISE is to have electrolytic contact to an inner reference electrode (as in Fig. 6.3a), then it is usually sealed into a glass or polyethylene tube using epoxy glue, and the tube filled with the internal solution and reference electrode. To make an all-solid-state probe with ohmic contact, the inner membrane surface is coated with a metal film (e.g. Cu or Ag) then soldered to a wire lead, or directly attached to the wire with conducting epoxy [7].

The advantages of bulk ChG electrodes are that they are long-lived and their proper function can easily be restored by mechanical polishing of the membrane surface, followed by conditioning in a solution of the analyte ion (10^{-3} to 10^{-2} M). On the other hand, bulk membrane sensors tend to have a relatively slow response to stimulus, a high potential drift rate (3–4 mV/h), which requires frequent (hourly) recalibration, and they can be fragile [7].

To overcome the latter problem, composite membranes can be prepared by dispersing ChG powder in an inert matrix. Unfortunately, though more mechanically robust, these composite membranes generally have slower response and lower potential-activity slope than the corresponding bulk ChG membranes. Better performance can be achieved by all-solid-state composite sensors made by covering Ag/AgCl coated-wire type reference electrodes with a layer of (~1 %) ChG powder-impregnated polymer [7].

Alternatively, ChG membranes can be prepared on a wide variety of suitable substrate materials by thin-film deposition techniques, such as vacuum thermal evaporation [7], cathode sputtering [7], spin coating [13], or pulsed laser deposition from a bulk ChG target source [10]. These thin-film deposition techniques are especially important since they are compatible with other microelectronics fabrication methods, and facilitate the production of miniaturized sensors. In addition, ChG thin films offer advantages of higher conductivity and faster response times compared to the bulk materials.

The incorporation of halides into the ChG network [14] increases its defectiveness, thereby enhancing ion transport within the MSL and ion exchange with the solution [7], both of which improve the performance of the ISE. The membrane composition also influences the ISE sensitivity and selectivity via effects on the solubility of the metal chalcogenides present at the surface. For example, the selectivity of the sulphide glass electrode increases in the order $Ni^{2+} < Zn^{2+} < Cd^{2+} < Pb^{2+} < Cu^{2+}$ [7].

6.5.2 Membrane Composition and Structure

Chalcogens, on their own or as chalcogenide compounds with elements of Groups 13–15, readily form amorphous (glassy) solids, even at relatively low rates of cooling from the melt. The glassy state can be retained even when these are doped with up to 30 at.% alkali metals, halogens, transition metals, and rare-earth metals [12]. In particular, As_2S_3, As_2Se_3, Sb_2Se_3, and even GeS_2 and $GeSe_2$ commonly form the glassy matrix [7]. Doping helps to increase conductivity and provides the basis for potential-determining equilibria, since the ChG membrane responds reversibly to ions that are included in its composition. Some examples of glass compositions and the relevant analyte species for each are given in Table 6.1 [7].

Chalcogenide glasses have shown higher stability against dissolution in the electrolyte and chemical attack in corrosive media and better selectivity in the presence of some interfering ions [12, 58, 59]. They have therefore been successfully applied as membrane materials for detection of metal ions such as Ag^+ [18], Fe^{3+} [28], Pb^{2+} [29, 37], Cd^{2+} [29], Cu^{2+} [7, 60], Cr^{2+} [61], Br^- [29], and Hg^{2+} [35, 62].

When a doped ChG membrane is conditioned by exposure to a solution of the analyte species (10^{-3} to 10^{-2} M), dissolution and degradation of the ChG surface take place, leading to the development of a MSL. Formation of the MSL is accompanied by the creation of a large number of vacancies within the glassy network (the atomic density of the MSL is less than half that of the bulk ChG). This provides increased conductivity and ionic mobility within the membrane. The MSL formation also generates active exchange centres, which provide absorption/desorption sites for specific ions; an example of a mechanism whereby charged defect sites can act as ion exchange centres is shown in Fig. 6.6 [7].

6.5.3 Mechanism of Ion Exchange for Zn and Sn ISE

6.5.3.1 Zn ISE

The mechanism of Zn ISE based on As_2Se_3–Sb_2Se_3–ZnSe (ASZ) and $GeSe_2$–ZnSe–ZnTe (GZZ) glasses and the ionic sensitivity of chalcogenide glass sensors

Table 6.1 ChG systems investigated as active membrane materials [7]

Determined ion	Glass composition	Reference
Ag^+	Ag–As–S	[15]
	Ag–As–Se	[16]
	$Ag_{1.5–50.1}S_{24.0–33.5}Fe_{0.3–1.9}+As$	[17]
	$Ag_{15}As_{42.5}Se_{42.5-x}Te_x$, $x=0–40$	[16, 18]
Na^+	$(GeSe)_{20–85}(Ga_3N_2)_{5–35}(NaCl)_{5–30}+GeS$	[19]
	GeS_2–Ga_2S_3–NaCl	[14]
Zn^{2+}	$(GeSe_2)_{72–54}(Sb_2Se_3)_{36–16}(ZnSe)_{10,20}$	[20]
	$(As_2Se_3)_{90–63}(Sb_2Se_3)_{0–27}(ZnSe)_{5,10}$	[20]
	$(GeSe_2)_{94}(ZnSe)_4(ZnTe)_2$	[21]
Cu^{2+}	$Se_{60}Ge_{28}Sb_{12}+2\,\%$ Fe	[22]
	$Cu_x(As_2S_3)_{100-x}$, $x=5–25$	[23]
	$Cu_x(As_2Se_3)_{100-x}$, $x=0–25$; $Cu_{25}(As_2Se_3)_{75}$	[24–26]
	Cu–As–Se, [Cu] = 40–90	[27]
	Cu–As–Se–Te; $Cu_{25}(As_2(Se_{0.5}Te_{0.5})_3)_{75}$	[24–26]
	$Cu_xAg_{25-x}As_{37.5}$, $x=0–25$	[12, 28, 29]
	$Cu_xAg_{25-x}As_{37.5}Se_{37.5}$, $x=0–25$	[12, 28, 29]
	$Cu_{23}Ag_2As_{37.5}Se_{37.5}$; $Cu_{23}Ag_2Se_{38}As_{17}S_{20}$	[30]
	Cu_2Se–As_2Se_3	[25]
	$Se_{60}Ge_{28}Sb_{12}+20.8\,\%$ Cu	[31]
Cu^{2+} microsensors	Cu–Ag–As–Se–Te	[32, 33]
	$Cu_xAg_y(As_2S_3)_{100-(x+y)}$, $x=15–30$, $y=10–20$	[34]
	$Cu_xAg_y(As_2Se_3)_{100-(x+y)}$, $x=15–30$, $y=10–20$	[34]
Cd^{2+}	$(Ge_{20}Te_{30}Se_{50})_{100-x}(CdS)_x$, $x\leq 0.5$	[25]
	CdS–Ag_2S–As_2S_3; CdI_2–Ag_2S–As_2S_3	[29]
Cd^{2+} microsensors	$(CdS)_x(AgI–Sb_2S_3)_{1-x}$, $x=0–5$	[35]
Pb^{2+}	$(AgAsS_2)_{60}(PbI_2)_{40}$	[36]
	$Pb_{20}(As_2Se_3)_{80}$; $Pb_{30}(As_2Se_3)_{70}$	[24]
	$Pb_{20}Cu_5(As_2Se_3)_{75}$	[24]
	PbS–Ag_2S–As_2S_3; PbI_2–Ag_2S–As_2S_3	[29, 37–40]
	Ag–As–S	[32]
Pb^{2+} microsensors	Pb–Ag–I–As–S	[41]
Hg^{2+}	$Ge_{20}Te_{30}Se_{50}$–HgS	[25]
	The composition is not given	[17, 42]
	AgBr–Ag_2S–As_2S_3–HgI_2	[43, 44]
Hg^{2+} microsensors	As_2S_3–Ag_2S–AgBr	[45, 46]
Fe^{3+}	Ge–Se–Sb + Fe; $Fe_x(Se_{60}Ge_{28}Sb_{12})_{98-x}$, $x=1–10$	[22, 47–50]
	$Se_{80}As_{10}S_5Fe_5$	[51]
	$(GeSe_3)_x(FeSe_2)_y(Ag_2Se)_z$, $x=76–85$; $y=9–14$; $z=5–15$	[20]
Cr^{6+}	As-based ChG, the composition is not given	[13, 17, 52]
Br^-	$(AgBr)_{55–65}(Ag_2S)_{15–30}(As_2S_3)_{10–30}$	[53, 54]
Multisensor systems	Ag–As–S	[33]
	$(AgI)_x(Sb_2S_3)_{100-x}$, $x=30, 40, 50, 60$	[55]
	Doped As_2S_3 and As_2Se_3 films	[56]
	Pb–Ag–As–I–S, Cd–Ag–As–I–S, Cu–Ag–As–Se	[57]

Fig. 6.6 Active ion adsorption centres in As_2Se_3–Sb_2Se_3–ZnSe type ChG; (**a**) before conditioning, (**b**) after conditioning and under working conditions, (**c**) at the next measurement. Reprinted with permission from [7]

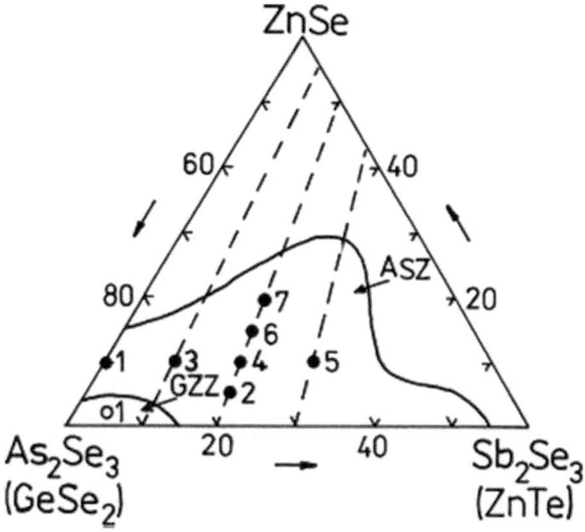

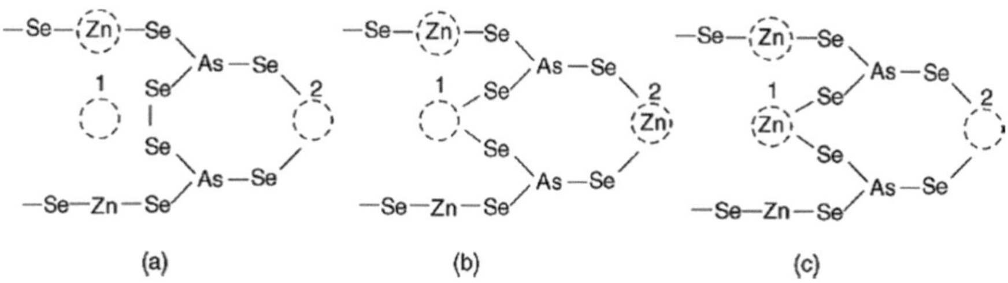

Fig. 6.7 Glass-forming regions in the As_2Se_3–Sb_2Se_3–ZnSe and $GeSe_2$–ZnSe–ZnTe system and the compositions studied

has been reported in literature [21]. Basic analytical characteristics such as stability, linearity, slope of the electrode function, limits of detection, effect of pH on the electrode potential and response time were studied with varying glass compositions. A structural mechanism for explanation of the dependencies obtained is suggested. Appropriate compositions for ISE were selected on the basis of the glass forming regions of ASZ and GZZ as shown in Fig. 6.7. Eight groups of ISE (with three samples of each group) were studied. The membranes (1–3 mm thick) were prepared by cutting, burnishing and polishing of the glassy samples obtained by a conventional melt quenching technique. They were constructed with an inner liquid junction. The sensor element was fixed to the edge of a PVC tube (the electrode body) with epoxy resin. The contact between the membrane and the inner reference electrode (Ag/AgCl) was achieved through an electrolyte containing 10^{-3} M KCl. The electrode function was measured using an electrochemical cell of the type:

$$Hg|Hg_2Cl_2|KCl(\text{saturated})||\text{Analyte Solution}|\text{ISE Membrane}|KCl|AgCl|Ag$$

A comparison of the results of the studied ISE shows that compositions with higher ZnSe content (maximal values of z) and with higher $GeSe_2$ content (minimal values of m) are convenient for practical applications. In particular, 1 ASZ and 7 ASZ exhibit the best parameters: for 1 ASZ, the slope of the calibration curves, $S = 38$ mV dec^{-1}, linear range, $L = 10^{-5}$–10^{-1} M Zn(II), limit of detection, $LD = 4.0 \times 10^{-6}$ M Zn(II); for 7 ASZ, $S = 41$ mV dec^{-1}, $L = 10^{-5}$–10^{-2} M Zn (II), $LD = 2.0 \times 10^{-6}$ M Zn(II). In both cases, $(pH)_{min}$ 3–4, $(pH)_{max}$ 6–7, time to reach stable values of the electrode function, $\tau_{95} = 2$–30 s in the concentration range 10^{-6} to 10^{-1} M Zn(II).

Structural changes of the defective centre in the working regime of the ISE are illustrated in Fig. 6.6.

Before conditioning, two free positions, (1) and (2), exist (case (a)). After conditioning, Zn^{2+} ions coming from the solution occupy position (2) in the defect centre, which is accompanied by bond breaking between the two bridged Se atoms (case (b)). This results in creation of two new bonds. During the measurement cycle, the Zn^{2+} ions occupy position (1), while Zn^{2+} ions in position (2) pass into solution (case (c)). Again, two free bonds are formed. This consecutive process of occupation and delivery of the active positions (1) and (2) in the defect centre repeats continuously, ensuring a permanent working regime of the ISE. The relative value of S decreases during the working regime and the stabilization of the electrode function to values typical for a divalent ion (for more details the reader is referred to [21]) can be explained by the assumption of creation of doubly ionized defects (Fig. 6.6b, c). External influences (dirt, sediment, dissolution, oxidation, etc.) lead to blockage or exhaustion of the membrane surface, which can be restored by polishing and so the sensor continues to work by repeated conditioning, following the above mechanism.

6.5.3.2 Tin (Sn) ISE

All-solid-state Sn(II) ISE of the "coated-wire" type were created [63, 64] on the base of novel multicomponent chalcogenide glasses from the $GeSe_2$–Sb_2Se_3–SnSe system. The ISE construction consists of a standard Ag/AgCl electrode coated with a composite mixture of binding polymer filled with Sn-containing chalcogenide glassy powder as an ion-selective active membrane component. The proposed Sn (II) ISE guarantees reliable analytical behaviour with high reproducibility of the investigated electrode characteristics. They were found to possess good selectivity and reversibility to Sn^{2+} ions in the presence of Pb^{2+}, Ni^{2+}, Zn^{2+}, Cd^{2+}, Cu^{2+}, Al^{3+}, CH_3COO and Cl^- ions.

Glass formation has been observed in the binary $GeSe_2$–Sb_2Se_3system in the concentration range from 0 to 70 mol% Sb_2Se_3, as can be seen from the Gibbs' diagram presented in Fig. 6.8. No glasses were obtained in the binary $GeSe_2$–SnSe system. The maximal solubility of SnSe in the glassy alloys has been found to be up to 57 mol% on the compositional tie-line $(GeSe_2)_3Sb_2Se_3$–SnSe (Fig. 6.8).

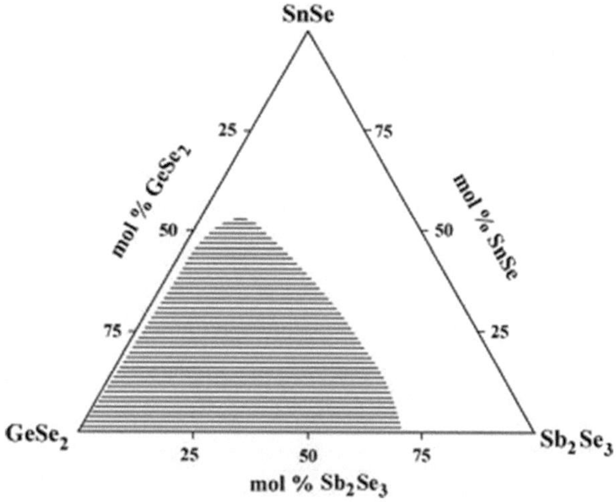

Fig. 6.8 Glass-forming region in the $GeSe_2$–Sb_2Se_3–SnSe system. Reprinted with permission from [64]

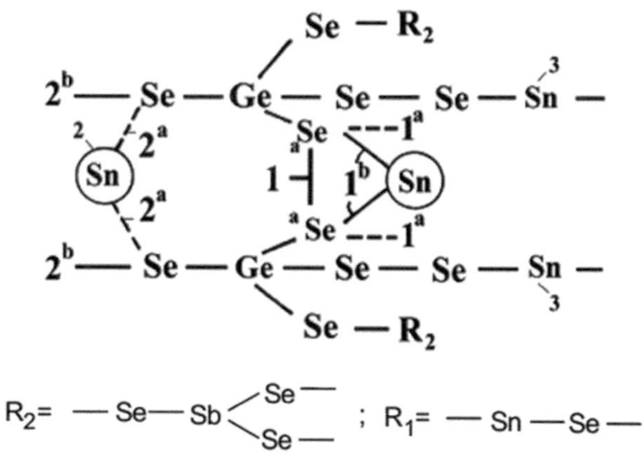

Fig. 6.9 Structure of divalent negatively charged defect centres in the ChGs from the $GeSe_2$–Sb_2Se_3–SnSe system [63]

During the ChG synthesis, Sn atoms occupy position 3 as shown in Fig. 6.9. During the initial measuring stage with the Sn (II) ISE, the Sn^{2+} ions from the negatively charged [Se ($GeSe_{3/2}R_1R_2$)] defects take part in the ion-exchange process at the membrane–solution interface. As a result, the electrode function has a slope typical for single valence anions, i.e., reversible to the slope of the single valence cation (~59 mV dec^{-1}). Sn atoms from position 3 participate in the exchange process between the ChG membrane and the measured solution at the beginning of the working stage of the Sn (II) ISE and immediately after membrane surface polishing. After conditioning or continuous measurements, Sn^{2+} ions from the solution occupy position 2 in the defect centre (Fig. 6.9), which causes simultaneously the bond breaking between the two bridged Se atoms. As a result, two unsaturated bonds are created (position 1a). During the next measuring stage with the Sn (II) ISE, Sn^{2+} ions coming from the solution occupy position 1 and connect the two free bonds 1a. Meanwhile, the Sn^{2+} ions from position 2 pass into the solution, forming again two unsaturated bonds 2b. In this way, the permanent working regime of the electrodes is ensured by the continuous repetition of the consecutive occupation and liberation of the active positions 1 and 2 in the defect centres from the Sn^{2+} ions. This model of the potential-generating mechanism, which assumes ion exchange in the charged defect centres created in the glassy network, explains satisfyingly a common peculiarity observed in the analytical behaviour of ISE based on multicomponent ChGs with a $GeSe_2$ glass-former.

6.6 Trends and Development

In addition to the ongoing synthesis, selection, and amelioration of ChG and chalcohalide glass membrane material compositions for improved sensitivity and selectivity, and for analysis of more chemical species, several particular trends toward areas of future development in ChG-based potentiometry are currently in evidence.

6.6.1 Multi-sensor Arrays and the Electronic Tongue

ChG membranes are well suited to the creation of electrodes with wide cross-sensitivity, since their glassy networks accept high levels of doping from multiple components, and they respond reversibly to each ion included in the composition. This means that they can be made to have simultaneous sensitivity to numerous solution species. In traditional single-species analysis by ISE this behaviour is opposite to what one would seek in a good potentiometric sensor, but now researchers are turning it into a feature that can be used to advantage in the analysis of complex mixtures.

If one assembles an array of sensor electrodes that each respond to a number of the components of an unknown solution, and the cross-sensitivity is overlapping (i.e., the different ISE respond to at least some of the same ions, even if/especially if the strength of the response to the same ion differs from electrode to electrode), then each solution of different composition should generate a unique set of responses from the sensor array. The full calibration of such an array would be difficult and very demanding, and for a complicated unknown solution the measurement output of such an array would be quite complicated. However, rather than seeking a complete analytical deconvolution of the electrode array output into a set of precise solution compositions, researchers are beginning to treat the analysis using pattern recognition and neural network-type software. In this way they are contributing to the development of a so-called "electronic tongue" that can qualify and distinguish complicated samples (mineral water, wine, etc.) from other very similar samples (e.g., water from a different well, wine from a neighbouring vineyard).

Challenges in this area include development and training of the pattern recognition software and fabrication of sensors with stable, reproducible responses to decrease the need for repeated training of the pattern recognition programs.

6.6.2 Metallic Doping of chalcogenides

The redox chemistry of ChG is such that metallic state atoms can be accommodated within the ChG matrix. ChG can play host to as much as 30 at.% metal ions, while still retaining the glassy structure and potentiometric sensitivity that makes them suitable for use in ISE. The presence of neutral metal atoms within the glass can increase its conductivity and ion transport properties by orders of magnitude, which enhances the performance as potentiometric sensors.

The main challenge in this development is to introduce metallic atoms into the ChG structure. Photoreduction of metal cations, especially Ag, is possible, but only within a few nanometres of the material surface because of the limited penetration of visible light into the ChG. However, the injection of Ag atoms enhances their own mobility within the matrix, resulting in the transport of the Ag atoms deeper into the material and making the outer surface available for the photoreduction and injection of more Ag atoms. Thus, a layer by layer injection of Ag into the ChG is possible by patient repetition. Other means of introducing metallic-state atoms into ChG include radiolytic photoreduction of incorporated metal cations by X-rays, gamma rays, or high-energy electron beams, and incorporation of metal nanoparticles or metal-generating precursor compounds during ChG synthesis [7]. Advances in metal atom doping of ChG and chalco-halide glass sensor membranes will also depend on gaining better control over the location and concentration of metal atoms in the membrane.

6.6.3 Sensor Miniaturization

In keeping with the rest of modern electronic technology, the movement toward smaller devices is also a trend in potentiometric sensor development. ChG-based ISE are well suited to incorporation into micro sensors and micro sensor arrays, due in part to their fine properties as potentiometric sensors, but in larger part because they can be fabricated by thin-film deposition techniques (such as thermal evaporation, cathode sputtering, spin coating, and laser ablation), similar to those already in use for producing other microelectronic devices.

6.6.3.1 Fabrication of Chalcogenide Glass-Based Thin-Film Sensor Arrays

A first step towards a microfabricated potentiometric thin-film sensor in the direction of an "electronic tongue" array for the simultaneous detection of Pb^{2+}, Cd^{2+}, and Cu^{2+} [57] using sensitive layers based on chalcogenide glass materials has been demonstrated. Thin-film chalcogenide glass materials consisting of mixtures of Pb-Ag-As-I-S, Cd-Ag-As-I-S, or Cu-Ag-As-Se were prepared by a pulsed laser deposition technique. For the fabrication of the sensor arrays, a p-doped (Bor) 4″ single-crystal silicon wafer with ⟨100⟩ orientation, specific resistance of >1000 Ω cm and thickness of 381 ± 25 μm was used. The complete fabrication steps to realize the thin-film sensor array are summarized in Fig. 6.10.

The first fabrication step is the generation of a SiO_2 layer with a thickness of about 500 nm as insulating layer between the substrate and the electrodes as well as the electrodes (fabrication step I). This SiO_2 layer was grown by means of thermal oxidation. To create the respective thin-film sensors as an array on the silicon wafer, photolithographic patterning has been performed (fabrication steps II + III). Therefore, a positive photoresist was spin-coated onto the silicon substrate.

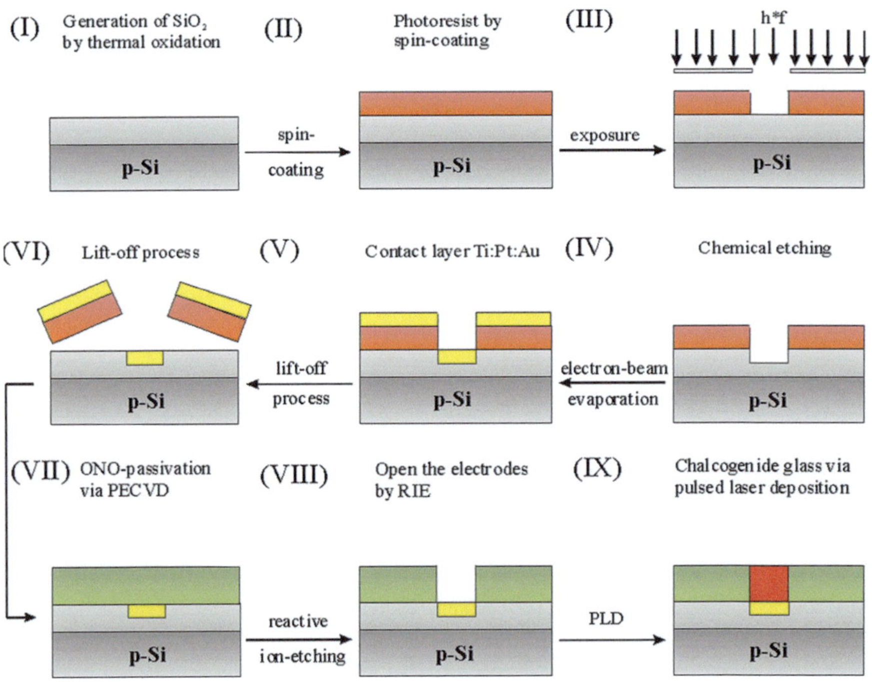

Fig. 6.10 Survey on the fabrication steps of the chalcogenide glass-based thin-film sensor array (I: thermal oxidation, II: spin-coating, III: exposure, IV: chemical etching, V: electron beam evaporation, VI: lift-off process, VII: ONO passivation, VIII: RIE, IX: PLD). Reproduced from [21]. Open access

Hexamethyldisiloxane was used to improve the adhesion between the photoresist and the silicon substrate. For the exposure of the photoresist through a mask, which consists of the structure of the sensor array, a mask aligner was used. After developing the photoresist, the structure of the sensor electrodes was chemically wet-etched by hydrofluoric solution as shown in fabrication step IV. In this way, trenches have been etched into the SiO_2 layer, in order to embed the contact layers for the sensor structures. Fabrication step V depicts the deposition of the contact layers by means of electron beam evaporation. The contact layers are built up of a layer system of Ti: Pt: Au with a thickness ratio of 30:175:300 nm.

The deposition of the contact layers was completed by a lift-off process in acetone that removes all photoresist-coated metallic areas (see fabrication step VI). To prevent the later corrosion of the silicon substrate and the contact layers, an oxide-nitride-oxide (ONO) passivation layer system has been deposited by means of a plasma-enhanced chemical vapour deposition process [65]. This ONO layer system was made of SiO_2:Si_3N_4:SiO_2 with a thickness ratio of 130:530:130 nm (see fabrication step VII). By means of a reactive ion etching process, only the areas which serve as contact for the sensor layers, as well as the contact pads, have been opened, as shown in fabrication step VIII. Subsequent to the described process steps, the heavy metal-selective chalcogenide glass materials such as Pb-Ag-As-I-S, Cd-Ag-As-I-S and Cu-Ag-As-Se were deposited by means of an "off-axis" pulsed laser deposition (PLD) process (step IX). A KrF excimer laser with a wavelength of 248 nm and a repetition rate of 10 s^{-1} was used to deposit the individual materials of the respective chalcogenide glass target on the Si/SiO_2 substrate, including the contact layers. A laser energy density of 5 J cm^{-2} was applied to the chalcogenide glass target at room temperature. The PLD process took place in a N_2 atmosphere at a pressure of 2×10^{-1} mbar to prevent any oxidation of the deposited materials. A process time of approximately 10 min was used to get thin films with a thickness of about 500 nm. To ensure the patterning of the different thin-film sensors on the silicon chip, a home-made Al mask was used. After cutting the wafer into single sensor chips of 10 mm $\times$ 10 mm, each sensor chip was glued on a printed circuit board (PCB), wire-bonded, and encapsulated by an epoxy resin. Figure 6.11 shows the cross-sectional view of a single sensor chip including the chalcogenide glass-based thin-film sensor array, after connection to the PCB and encapsulation.

The complete thin-film sensor array after wire bonding and encapsulation is presented in Figs. 6.12a, b. The sensor array from Fig. 6.12 was then attached to the PCB, bonded and encapsulated with epoxy resin. In the left part of the figure, the conduction tracks and the contact pins to the measurement device can be seen.

The chalcogenide glass-based thin-film sensor array was found to exhibit a nearly Nernstian response to the respective primary ions, such as Pb, Cu, and Cd in the concentration range 10^{-6} to 10^{-2} mol/L for Pb^{2+} and Cd^{2+} ions and 10^{-7} to 10^{-3} mol/L for Cu^{2+} ions. The results obtained in terms of cross-sensitivities, i.e., selectivities, are comparable to those measurements achieved with conventional bulk electrodes.

Table 6.2 gives the electrochemical parameters of the chalcogenide glass microsensors as reported in [66]. The selectivity (K_{sel}^{*}) was determined by the fixed interference method.

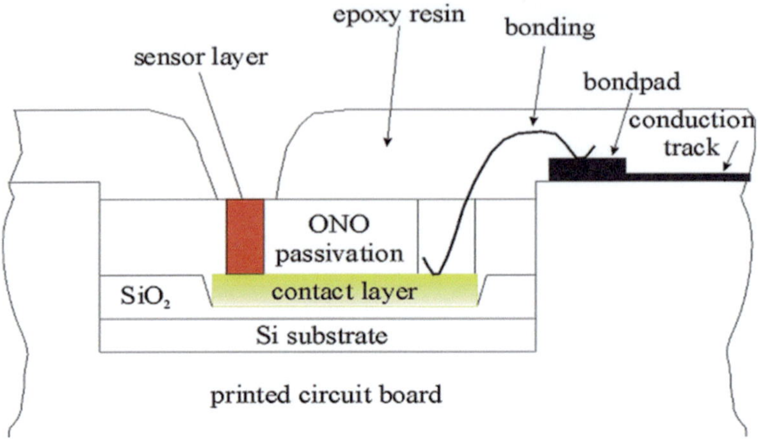

Fig. 6.11 Schematic cross section of the chalcogenide glass-based sensor chip after bonding and encapsulation. Reproduced from [21]

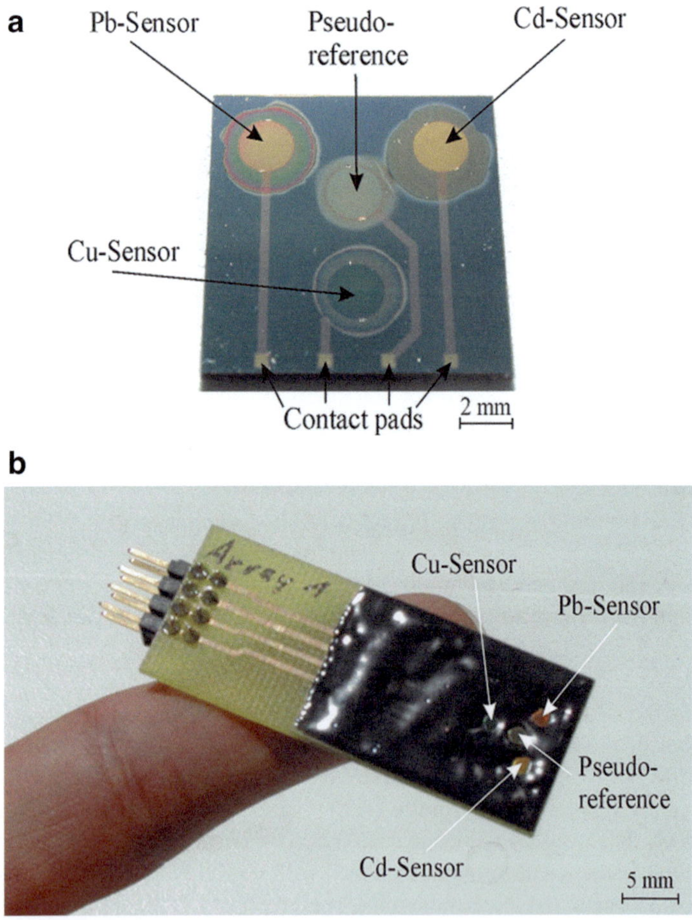

Fig. 6.12 (**a**) Micrograph of the thin-film sensor array after the PLD process. The different chalcogenide glasses on the electrodes are shown (**b**) Photograph of the complete thin-film sensor array after wire bonding and encapsulation. Reproduced from [21]

Table 6.2 Electrochemical parameters of the chalcogenide glass microsensors [66]

Material	$PbS-AgI-As_2S_3$	$CdS-AgI-As_2S_3$	Cu-Ag-As-Se-Te	T1-Ag-As-I-S
Ion	Pb^{2+}	Cd^{2+}	Cu^{2+}	$T1^{+}$
Sensitivity	26–29 mV/pPb	25–28 mV/pCd	27-30 mV/pCu	54–60 mV/pT1
Selectivity: (K_{sel})	$Cd^{2+} \rightarrow 0.05$	$Pb^{2+} \rightarrow 33$	$Pb^{2+} \rightarrow 5 \times 10^{-5}$	$Pb^{2+} \rightarrow 1 \times 10^{-2}$
	$Zn^{2+} \rightarrow 2 \times 10^{-5}$	$Zn^{2+} \rightarrow 2 \times 10^{-4}$	$Cd^{2+} \rightarrow 4 \times 10^{-5}$	$Cd^{2+} \rightarrow 1 \times 10^{-2}$
	$Ni^{2+} \rightarrow 1 \times 10^{-5}$	$Ni^{2+} \rightarrow 1 \times 10^{-3}$	$Zn^{2+} \rightarrow 3 \times 10^{-6}$	$Cu^{2+} \rightarrow 0.5$
	$Co^{2+} \rightarrow 2 \times 10^{-5}$	$Co^{2+} \rightarrow 1 \times 10^{-3}$	$Ni^{2+} \rightarrow 6 \times 10^{-6}$	$Zn^{2+} \rightarrow 5 \times 10^{-4}$
	$Fe^{2+} \rightarrow 1 \times 10^{-4}$	$Fe^{2+} \rightarrow 2 \times 10^{-3}$	$Co^{2+} \rightarrow 5 \times 10^{-6}$	$Ca^{2+} \rightarrow 2 \times 10^{-5}$
	$Ca^{2+} \rightarrow 5 \times 10^{-5}$	$Ca^{2+} \rightarrow 3 \times 10^{-3}$	$Ca^{2+} \rightarrow 1 \times 10^{-5}$	$Mg^{2+} \rightarrow 2 \times 10^{-5}$
Stability	>250 d	>230 d	>230 d	>160 d
pH range	2–7	2.5–8	2–6	2–5
Detection limit (mol/l)	1×10^{-7}	4×10^{-7}	1×10^{-7}	3×10^{-5}

6.7 Summary

Ion-selective electrodes are the chemical sensors with the longest history and probably still the largest number of applications. Chalcogenide glass materials as membranes for potentiometric sensors for chemical analysis in solutions, e.g., ion-selective electrodes, have been studied for more than three decades. Already in the 1970s, the first papers dealing with the ion-sensing properties of chalcogenide glasses were published, followed by systematic investigations of their solid-state chemistry, the sensing mechanism, and their analytical characteristics. Combining the analytical performance of chalcogenide glass membranes with techniques originally invented for semiconductor processing, a new generation of silicon-based thin-film sensors with chalcogenide glass materials for heavy metal detection has been created. In the beginning, co-sputtering of Cu and As_2Se_3 resulted in a Cu^{2+}-selective field-effect sensor, while in recent years, a variety of chalcogenide glass materials have been stoichiometrically deposited onto ISFETs (ion-sensitive field-effect transistors), EIS (electrolyte–insulator–semiconductor) devices, LAPS (light-addressable potentiometric sensors), and ISE structures, respectively, to serve as miniaturized chemical sensors [67].

References

1. P.W. Atkins, *Physical Chemistry*, 2nd ed. (W.H. Freeman and Co., New York, 1982). pp. 258–264, 331–341
2. D.A. Skoog, D.M. West, *Fundamentals of Analytical Chemistry*, 4th edn. (Saunders College Publishing, Toronto, 1982), pp. 396–424
3. E. Pretsch, Trends Anal. Chem. **26**(1), 46 (2007)
4. E. Bakker, E. Pretsch, Trends Anal. Chem. **20**(1), 11 (2001)

5. E. Bakker, P. Bhlmann, E. Pretsch, Talanta **63**, 3 (2004)
6. A. Eftekhari, Anal. Lett. **34**(7), 1087 (2001)
7. V.S. Vassilev, S.V. Boycheva, Talanta **67**, 20 (2005)
8. M.D. Mowery, R.S. Hutchins, P. Molina, M. Alajarín, A. Vidal, L.G. Bachas, Anal. Chem. **71**, 201 (1999)
9. B.R. Kowalski, Anal. Chem. **52**(5), 112R (1980)
10. Y. Mourzina, J. Schubert, W. Zander, A. Legin, Y.G. Vlasov, H. Lüth, M.J. Schöning, Electrochim. Acta **47**, 251 (2001)
11. A. Legin, A. Rudnitskaya, Y. Vlasov, C. Di Natale, E. Mazzone, A. D'Amico, Sens. Actuator B **65**, 232 (2000)
12. Y.G. Vlasov, E.A. Bychkov, Ion Selective Electrode Rev. **9**, 5 (1987)
13. A.V. Legin, E.A. Bychkov, Y.G. Vlasov, Sens. Actuator B **15–16**, 184 (1993)
14. Y.G. Vlasov, E.A. Bychkov, E.A. Kazakova, Z.U. Borisova, Zh. Anal. Khim. **39**, 452 (1984)
15. Y.G. Vlasov, E.A. Bychkov, B.L. Seleznev, Solid State Ion. **24**, 179 (1987)
16. Y.G. Vlasov, E.A. Buchkov, A.V. Legin, Talanta **6**, 1059 (1994)
17. Y.G. Vlasov, E.A. Bychkov, B.L. Seleznev, Sens. Actuator B **2**, 23 (1990)
18. Y.G. Vlasov, E.A. Bychkov, USSR Patent A.C. 1,402,913, MKI G 0127/30, LGU-1469958/31-25, 1986.
19. Y.G. Vlasov, E.A. Bychkov, Anal. Lett. **22**(5), 1125 (1989)
20. S. Boycheva, V. Vassilev, I. Markova, in *Proc. 16th Cong. Chemists and Technologists of Macedonia*, Skopje, vol. 2, 28–30 October 1999, p. 379
21. S. Boycheva, V. Vassilev, Z.G. Ivanova, J. Appl. Electrochem. **32**(3), 281 (2002)
22. C.T. Baker, I. Trachtenberg, J. Electrochem. Soc. **118**, 571 (1971)
23. R. Jasinski, I. Trachtenberg, G. Rice, J. Electrochem. Soc. **121**(3), 363 (1974)
24. A.E. Owen, J. Non-Cryst. Solids **35–36**, 999 (1980)
25. N. Tohge, M. Tanaka, J. Non-Cryst. Solids **80**, 550 (1986)
26. M.I. Frazer, A.E. Owen, J. Non-Cryst. Solids **59–60**, 1031 (1983)
27. E.A. Bychkov, M. Bruns, H. Klewe-Nebenius, G. Pfennig, W. Hoffmann, H. Ache, Sens. Actuator B **25**, 733 (1995)
28. Y.G. Vlasov, E.A. Bychkov, A.M. Medvedev, Anal. Chim. Acta **185**, 137 (1986)
29. Y.G. Vlasov, E.A. Bychkov, A.M. Medvedev, Zh. Anal. Khim. **40**, 438 (1985)
30. S. Hadjinikolova, S. Boycheva, V. Vassilev, L. Ilcheva, L. Nikolova, in *Proc. 13th Conference on Glass and Ceramics*, Varna, vol. 1, 29, September to 1 October 1999 (Publishing House "Science Invest", Sofia), p. 304
31. S.P. Awasthi, N.S. Shenoy, T.P. Radhakrishnan, Analyst **119**, 1361 (1994)
32. J. Schubert, M.J. Schöning, C. Schmidt, M. Siegert, S. Mesters, W. Zander, P. Kordos, H. Lüth, A. Legin, Y. Mourzina, B. Seleznev, Y.G. Vlasov, Appl. Phys. A: Mater. Sci. Process. **69**(7), 803 (1999)
33. M.J. Schöning, C. Schmidt, J. Schubert, W. Zander, S. Mesters, P. Kordos, H. Lüth, A. Legin, B. Seleznev, Y.G. Vlasov, Sens. Actuator B **68**, 254 (2000)
34. R. Tomova, G. Spasov, R. Stoycheva-Topalova, A. Buroff, J. Non-Cryst. Solids **266–269**, 985 (2000)
35. A. Guessous, J. Sarradin, P. Papet, K. Elkacemi, S. Belcadi, A. Pradel, M. Ribes, Sens. Actuator B **53**, 13 (1998)
36. C. Brohnke, J.P. Malugani, A. Saida, G. Robert, Electrochim. Acta **26**, 1137 (1981)
37. Y.G. Vlasov, E.A. Bychkov, A.V. Legin, Electrokhimia **21**, 1444 (1989)
38. Y.G. Vlasov, E.A. Buchkov, A.V. Legin, E. Pungor, Ion-Selective Electrode **4**, 657 (1984)
39. Y.G. Vlasov, E.A. Bychkov, A.V. Legin, Elektrokhimia **22**, 1472 (1986)
40. A.V. Legin, E.A. Bychkov, Y.G. Vlasov, Phys. Chem. Glasses **13**, 302 (1987)
41. Y. Mourzina, M.J. Schöning, J. Schubert, W. Zander, A. Legin, Y.G. Vlasov, P. Kordos, H. Lüth, Sens. Actuator B **71**, 13 (2000)
42. R. De-Marco, J. Shacleton, Talanta **49**, 385 (1999)
43. R. De-Marco, B. Pejcic, K. Prince, A. Riessen, Analyst **128**(6), 742 (2003)

44. B. Pejcic, R. De-Marco, Appl. Surf. Sci. **228**, 378 (2004)
45. A. Pradel, F. Genty, M. Miloshova, V. Tsegelnik, E. Bychkov, P. Papet, J. Sarradin, M. Ribes, Electrochem. Soc. Proc. **22–95**, 76 (1995)
46. M. Miloshovam, E. Buchkov, V. Tsegelnik, V. Strykanov, H. KleweNebenius, M. Bruns, W. Hoffmann, P. Papet, J. Sarradin, A. Pardel, M. Ribes, Sens. Actuator B **57**, 171 (1999)
47. R. Jasinski, I. Trachtenberg, J. Electrochem. Soc. **120**(7), 1169 (1973)
48. S. Hadjinikolova, S. Boycheva, V. Vassilev, Z.G. Ivanova, in *Proc. Int. Semiconductor Conf., 22nd*, Ed., Sinaia, Romania, vol. 2, 5–9 October 1999, CAS'99 Proceedings, p. 507
49. S. Hadjinikolova, S. Ecova, L. Ilcheva, Mater. Forum **62–64**, 291 (1990)
50. R. De Marco, D.J. Mackey, Mar. Chem. **68**(4), 28 (2000)
51. S. Hadjinikolova, S. Boycheva, V. Vassilev, L. Ilcheva, Ann. Univ. Chem. Technol. Metall. **36**, 4 (1999)
52. J.H. Christensen, P. Clemmensen, G.H. Hansen, K. Liltorp, J. Mortensen, Sens. Actuator B **45**, 239 (1997)
53. Y.G. Vlasov, D.V. Golikov, E.A. Bychkov, L.N. Moskvin, Zh. Anal. Khim. **41**, 1635 (1986)
54. Y.G. Vlasov, L.N. Moskvin, E.A. Bychkov, D.V. Golikov, Analyst **114**, 185 (1989)
55. Y.G. Vlasov, A.V. Legin, A.M. Rudnitskaya, Zh. Anal. Khim. **52**(8), 837 (1997)
56. R. Tomova, R. Stoycheva-Topalova, A. Buroff, J. Mater. Sci. Mater. Electron. **14**, 843 (2003)
57. J.P. Kloock, Y. Mourzina, J. Schubert, M.J. Schöning, Sensors **2**, 356 (2002)
58. Y.G. Vlassov, E.A. Bychkov, Sens. Actuators **12**, 275 (1987)
59. C.E. Koenig, E.W. Grabner, Electroanalysis **7**(11), 1090 (1995)
60. R. Tomova, R. Stoycheva-Topalova, A. Buroff, J. Optoelectron. Adv. Mater. **7**(3), 1399 (2005)
61. V. Vassilev, V. Parvanova, L. Aljihmani, B. Pejova, S. Hadjinikolova, P. Petkov, in *XVIII Cong. Chemists and Technologists of Macedonia*, 23–25 September, 2004, Ohrid, Macedonia, Extended Abstract, ICT-19, p. 58
62. A. Guessous, P. Papet, J. Sarradin, M. Ribes, Sens. Actuator B **24–25**, 296 (1995)
63. V. Vassilev, K. Tomova, S. Boycheva, V. Parvanova, J. Non-Cryst. Solids **355**, 1566 (2009)
64. V. Vassilev, K. Tomova, V. Parvanova, S. Boycheva, J. Alloys Compd. doi:10.1016/j.jallcom.2009.06.02
65. F. Fassbender, G. Schmitt, M.J. Schöning, H. Lüth, G. Buß, J.W. Schultze, Sens. Actuator B **68**, 128 (2000)
66. J.P. Klook, Y. Mourzina, Y. Ermolenko, T. Doll, J. Schubert, M.J. Schöning, Sensors **4**, 156 (2004)
67. M.J. Schöning, J.P. Kloock, Electroanalysis **19**(19-20), 2029 (2007)

Chapter 7
Biomedical Applications

Mandeep Singh Bakshi and Gurinder Kaur Ahluwalia

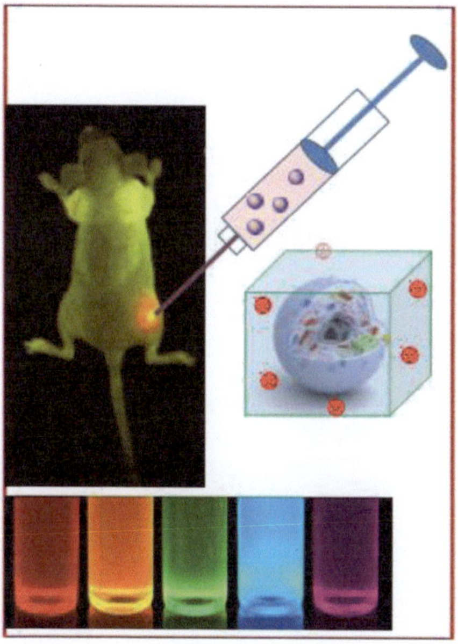

M.S. Bakshi, Ph.D. (✉)
Department of Natural and Applied Sciences, University of Wisconsin - Green Bay, 2420 Nicolet Drive, Green Bay, WI 54311–7001, USA
e-mail: ms_bakshi@yahoo.com

G.K. Ahluwalia, Ph.D
Materials and Nanotechnology Research Laboratory, Department of Physics College of The North Atlantic, Labrador West Campus, 1600 Nichols Adam Highway, Labrador City, A2V 0B8, NL, Canada
e-mail: gurinder.kaur@cna.nl.ca

G.K. Ahluwalia (ed.), *Applications of Chalcogenides: S, Se, and Te*,
DOI 10.1007/978-3-319-41190-3_7

7.1 Introduction

Chalcogenides an amazing class of materials with astonishingly superb applications in electronic industry have made our life easy on this planet are not limited to this field; they are increasingly playing important roles in tissue engineering, biomarkers, and cytotoxicity. This chapter deals with some of the important biomedical applications of chalcogenide materials in early detection as well as fast cure of some of the critical illnesses. We discussed in the previous chapter about the importance of the shape and size of the chalcogenide nanomaterials. It is not only from the photophysical aspects of these materials but also from the "size disparity" in comparison to animal cell. A biomedical application requires the incorporation of nanoparticles (NPs) in deep tissue, uptake in the cell, or in the systemic circulation. In all these cases, smaller size and monodisperse shape is the important issues to be addressed before considering these materials to use in these applications. For instant, an animal cell is of several micrometers in size in comparison to nm size of NPs. For a single cell target to deliver drug or a cytotoxic analysis, several hundreds of NPs are available to perform their specific action which becomes possible only due to the "size disparity" (Fig. 7.1). That is why surgical

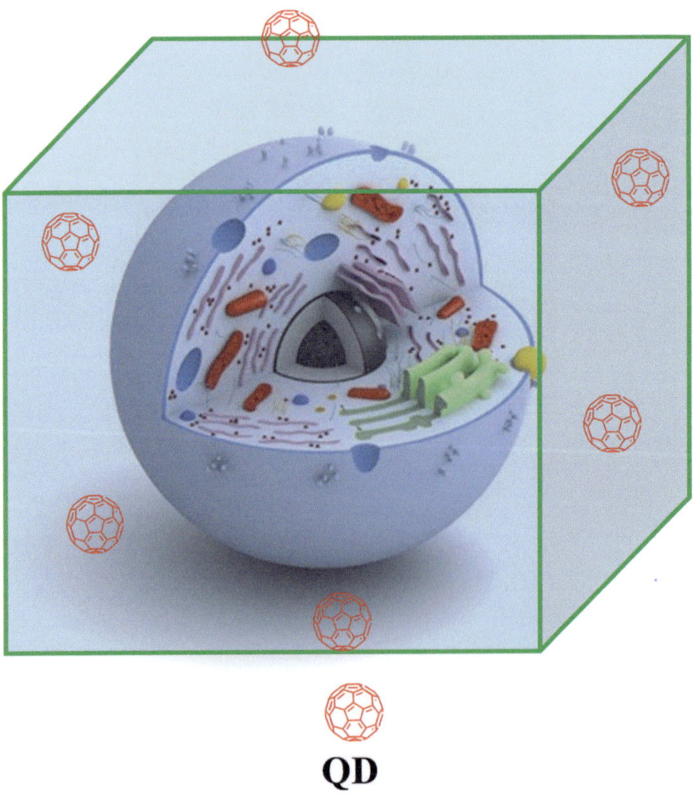

Fig. 7.1 Size disparity between the animal cell and quantum dot (drawing is not to scale)

removal of the cancerous tumor is not the final cure but a target drug delivery up to a single cell or removal of each and every cancerous cell in terms of cytotoxicity is probably the final solution. In addition, NPs can reach deep inside the desired marked sites where other contrast agents or drugs cannot usually reach and can cross some biological barriers such as the gastrointestinal barrier and blood–brain barrier. In terms of NPs toxicity, coated NPs are usually nontoxic because the free NP surface which bears a tremendous potential to interact with the cell wall is passivated by a suitable coating. However, NPs are absorbed and cleared in cells by endocytosis and macrophages, while the endocytic pathway is dependent on the size and shape of the NPs because endocytosis of spherical NPs is easier and faster than that of rod-shaped or fiber like NPs. Hence, the shape and size of the nanostructured chalcogenides discussed in the previous chapter plays a key role in biological applications. Bio-sensing and bio-imaging are the two most important biomedical applications of chalcogenide materials which have been well exploited recently in several molecular imaging techniques such as magnetic resonance imaging, X-ray computed tomography, positron emission tomography, and optical imaging.

7.2 Toxicity of Nanomaterials

Before starting with the useful biological applications, it is equally important to understand the toxicity of nanomaterials, because they have to be incorporated in the biological systems for their appropriate use [1]. The safety of designed nanomaterials is in focus since countless materials have been synthesized and characterized for various applications [2]. Already numerous consumer products are using such materials and they have eventually grown exponentially within a decade [3], and hence, their toxicological screening is important in terms of their applications in whole-animal [4, 5] living systems. Thus, more robust models for in vitro toxicological testing such as high-throughput screening (HTS) are required in terms of a comparison between in vitro and in vivo conditions. The multidimensional data set generated by HTS analysis includes 4032 data points obtained from (2 cell lines $\times$ 4 cytotoxicity responses $\times$ 8 time points $\times$ 7 NPs $\times$ 9 doses) and it is presented in toxicological response characteristics through the use of self-organizing maps (SOMs) onto a two dimensional display where in the spatial distribution of the different NPs at given concentrations provide a qualitative indicator of the degree of similarities or differences between materials that generate one or more of the cellular response outcomes (Fig. 7.2). Four clusters of NPs are identified based on the similarities between their lethal response outcomes (C1–C4). Cluster C1 consists of toxic QD1 (CdSe/ZnS in toluene) and ZnO at indicated doses in brackets. Cluster C2 consists of lower doses (0.3–50 μg/mL) of ZnO and high doses (50–100 μg/mL) of Ag NPs, while cluster 3 includes moderately toxic NPs (Pt and SiO_2). Cluster 4 comprises non-cytotoxic NPs (Au, Al_2O_3) or ineffective doses of potentially cytotoxic NPs. SOM units that are closer to one another are

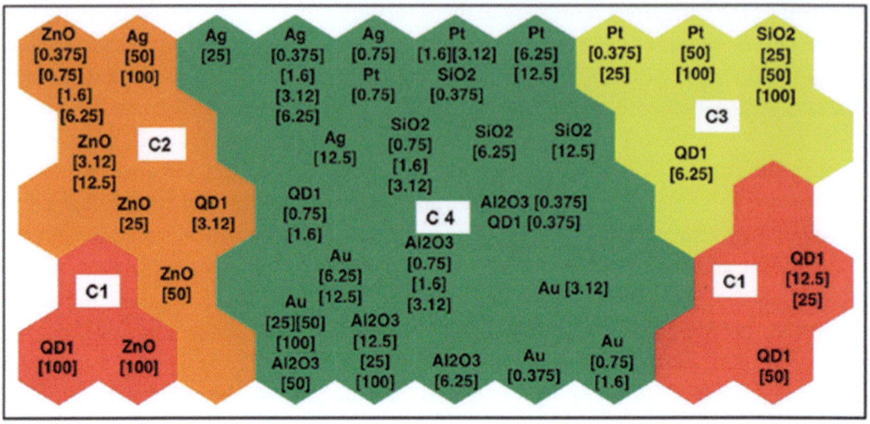

Fig. 7.2 Use of self organizing maps (SOMs) to identify similarity patterns for NPs and cellular responses. SOM defined by the distribution of NPs on a 2D grid that considers all response characteristics in both cell types. It is generated by the integration of the entire data set to distribute NPs based on the biological response measurements. NPs showing similarity in their cytotoxicity responses cluster together. Noteworthy is the clustering of highly toxic NPs (ZnO and QD1) at moderate to high concentrations in clusters C1 and C2, while cluster C3 comprises Pt and SiO_2, with cluster four incorporating most noncytotoxic NPs (Reproduced with permission from [1])

more similar with respect to toxicity profiles, as clusters 1 and 2 consist of highly cytotoxic ZnO and QD1, while cluster C4 is more heterogeneous and includes Ag, Pt, SiO_2, Au, and Al_2O_3.

These results have been compared with in vivo outcome of zebrafish (*Danio rerio*), as a model organism to screen for chemical and NP toxicity [6–8]. Zebrafish embryos exposed to NP suspensions at 4 h post-fertilization (hpf) are assessed for toxicity every 24 h, in a period of 120 h observation (Fig. 7.3). Toxicity end points involve scoring of hatching rates, mortality rates, cardiac rate, and the appearance of abnormal morphological features (e.g., pericardial/yolk sac edema, large/small yolk, short tail, body length, bent spine, etc.). This analysis shows that QD1, ZnO, and Ag exert morphological effects, while Ag and Pt influenced the cardiac rate. Nano-Ag-treated embryos develop bent spines, short body sizes, pericardial edema, submandibular edema, and bent tails by 120 h, while the majority of ZnO-treated embryos remained unhatched and embryos treated with QD1 died early and underwent complete dissolution within 24 h. ZnO consistently inhibited embryo hatching at all concentrations tested, while Ag and Pt NPs exerted effects on hatching at 15 and 25 μg/mL, respectively (Fig. 7.3b). While toluene containing QD1 exerted high lethality at concentrations >5 μg/mL, ZnO and Pt induced sublethal responses that became lethal at 25 μg/mL (Fig. 7.3c). A summary of in vivo toxicological profile is shown in Table 7.1.

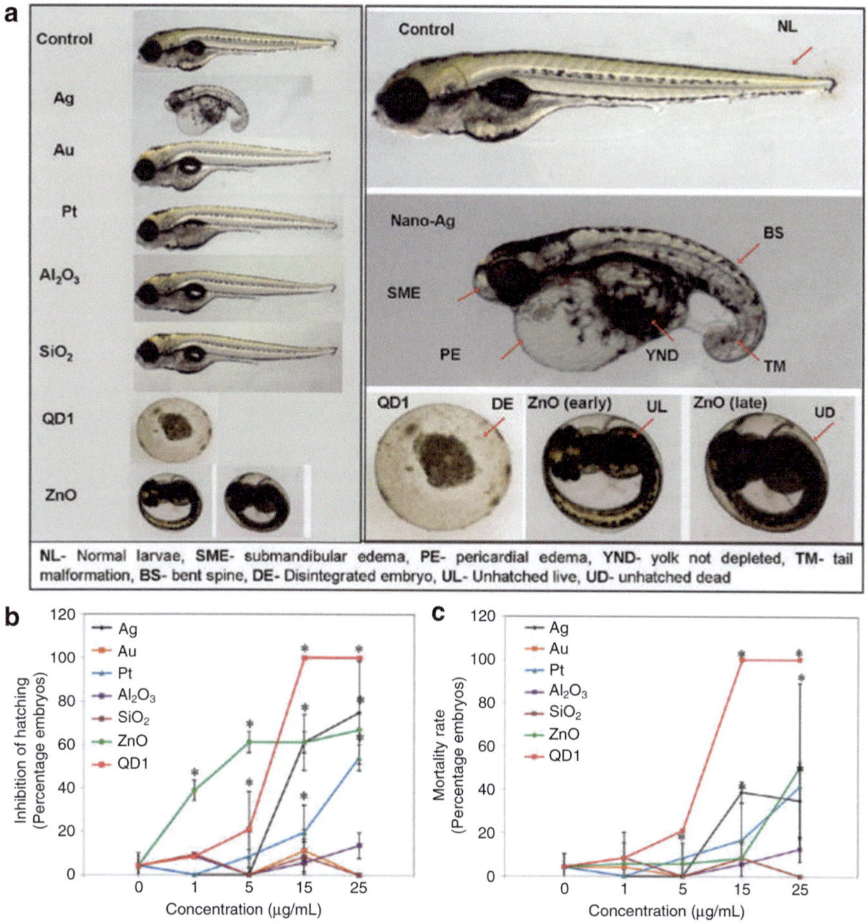

Fig. 7.3 NP toxicity in zebrafish embryos. (**a**) Images of the zebrafish embryos exposed to Holtfreter's medium with and without the NPs added at 15 μg/mL for 72 hpf. Note the gross morphological defects seen in nano-Ag-treated embryos. The QD1 treatment resulted in the disintegration of embryo, while ZnO treatment inhibited embryo hatching. Zebrafish embryos exposed to increasing doses of NPs were also assessed for (**b**) the hatching rate at 72 hpf and (c) the mortality rate at 120 hpf. The average value was calculated for a total of 36 embryos from three experiments with 12 embryos in each group. *Values statistically significant from control (PE 0.05). Error bars represent standard deviation from average value. (Reproduced with permission from [1])

7.3 QDs as Fluorescent Biomarker

QDs have been successfully used in vivo whole animal fluorescence imaging. Fluorescent $CuInSe_xS_{2-x}$ nanocrystals encapsulated in poly (lactic-co-glycolic) acid microparticles are used in vivo microparticle transport and localization [9] (Fig. 7.4). *Y. pseudotuberculosis*-derived protein invasion targets intestinal M cells

Table 7.1 Ranking of NPs based on the toxicity end points in zebrafish embryos

Ranking	Attributes	NPs	Morphological defects	Physiological defects
0	No morphological or physiological defects	Au, Al_2O_3, SiO_2	No observable morphological defects	No physiological defects
1	Single morphological/ physiological defect			
2	Multiple morphological and physiological defects	Pt, QD2	Pericardial edema	Low heart beat, mortality
3	Severe multiple morphological and physiological defects	Ag, QD3	Bent spine, opaque yolk, pericardial edema, yolk not depleted, short body, submandibular edema, tail malformation	Mortality and reduced hatching rate and low heart beat
4	Embryos do not survive	ZnO, QD1	Unhatched embryo, disintegrated embryo	Embryos do not survive, embryos fail to hatch

QD2 water-soluble core/shell CdSe/ZnS, *QD3* core-only CdSe particles with mercaptoundecanoic acid (MUA) surface modification (Reproduced with permission from [1])

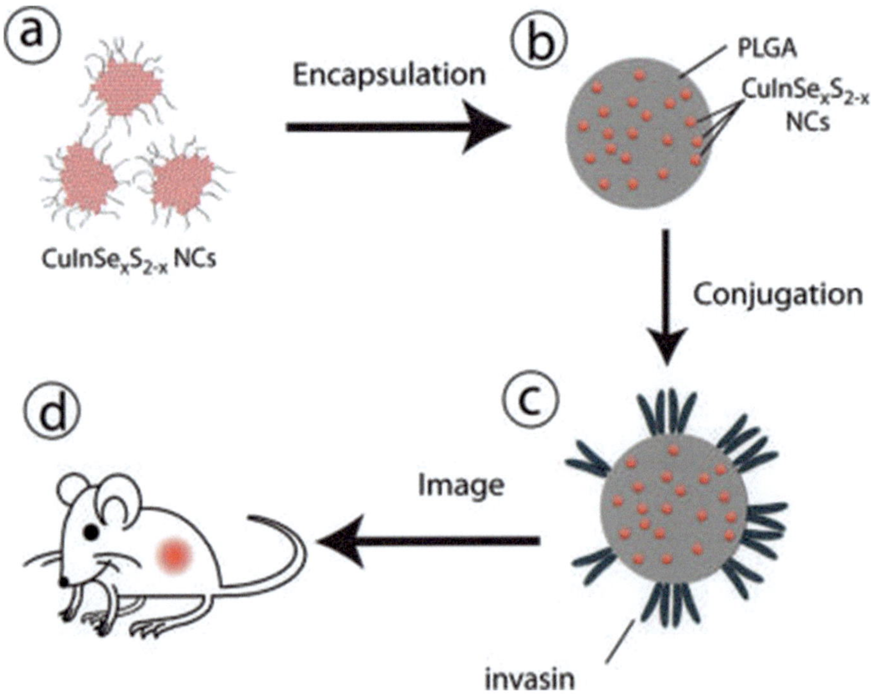

Fig. 7.4 Illustration of the experimental procedure. (**a**) $CuInSe_xS_{2-x}$/ZnS quantum dots (NCs) dispersed in dichloromethane are (**b**) encapsulated in PLGA microspheres, prior to (**c**) conjugation with wild-type invasin. (**d**) After intragastric gavage, in vivo fluorescence is monitored (Reproduced with permission from [9])

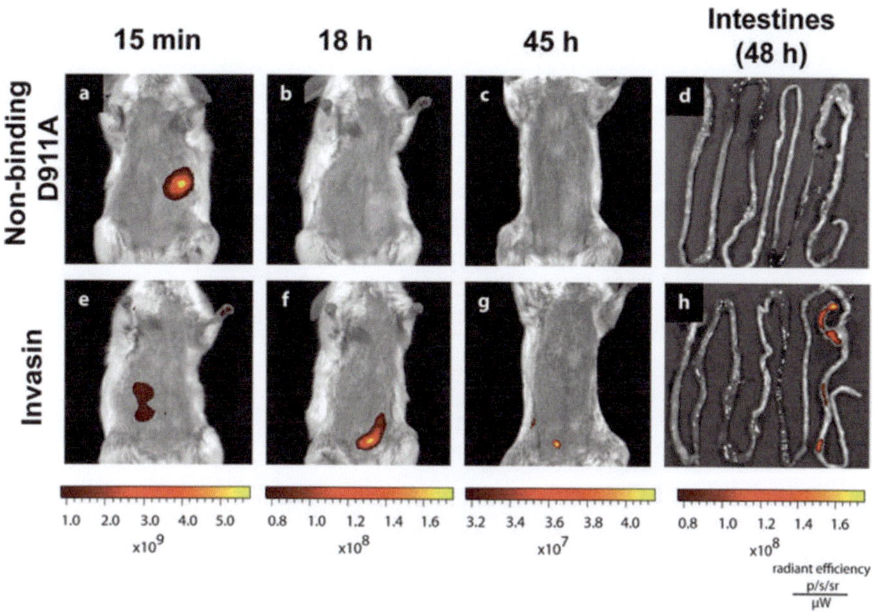

Fig. 7.5 Representative in vivo fluorescence images tracking the same subjects at 15 min (**a**, **e**), 18 h (**b**, **f**), and 45 h (**c**, **g**) after oral gavage and the intestines after dissection at 48 h (**d**, **h**). Mice ($n = 3$) were administered 15 mg $CuInSe_xS_{2-x}$ microparticles conjugated with (**e**–**h**) wild type invasion (Reproduced with permission from [9])

which are present in the gastrointestinal tract [10] and represent the first line of intestinal immune responses to combat bacterial or viral pathogens [11, 12]. Since access to lymphocytes underlying M cells is a major limitation in oral delivery of subunit vaccines, targeting M cells has emerged as a general strategy for oral vaccine delivery [13–17]. Quantum dot fluorescence is clearly visible in the whole animal images and identified the location of the microparticles in vivo even in the stomach [18]. The fluorescence signal in Fig. 7.5a, e is located in stomach and restricted to the cecum and the colon where the invasin-conjugated particles are retained. M cells are found along the gastrointestinal tract in Peyer's patches in the distal small intestine, isolated lymphoid follicles, cecal patches, and colonic patches [19, 20]. $CuInSe_xS_{2-x}$ nanocrystals loaded into invasion-functionalized poly(lactic-co-glycolic) acid microparticles are observed in deep tissue within the gastrointestinal tract by in vivo whole animal fluorescence imaging. The $CuInSe_xS_{2-x}$ nanocrystals are promising new biocompatible diagnostic agents.

Glutathiose (GSH) stabilized Ag_2S quantum dots Ag_2S-GSH-SNO are used for cell imaging and small animal imaging (Fig. 7.6) to demonstrate the fluorescence imaging ability of such nanoparticles [21]. The presence of "NO" functional group has its own very important functions in biological systems, is generated in various tissues from the amino acid L-arginine by different forms of nitric oxide synthase [22]. Thus, incorporation of fluorophores that are highly effective for in vivo

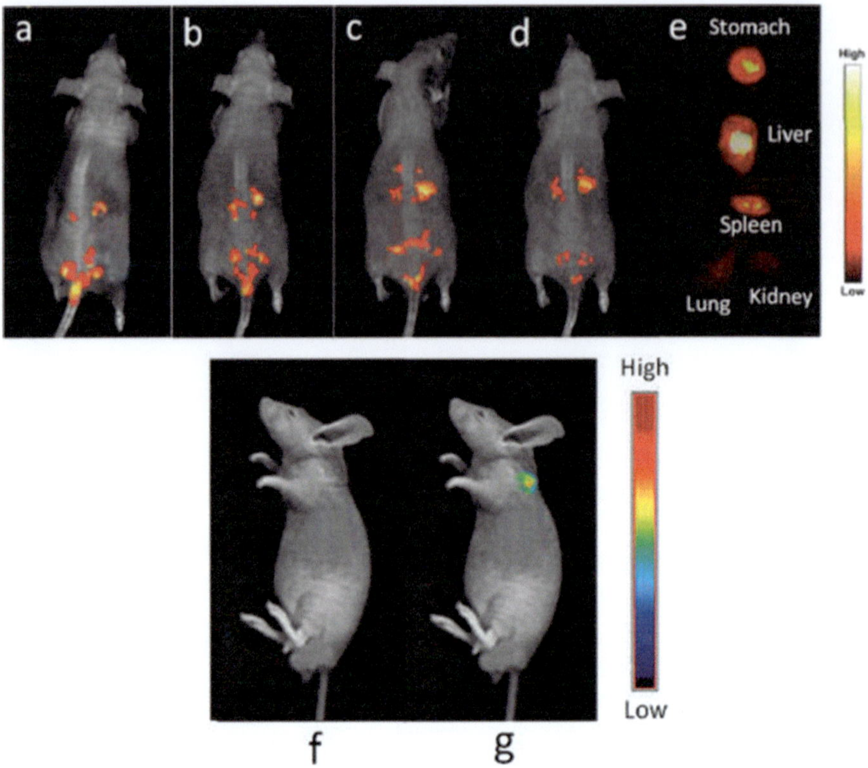

Fig. 7.6 In vivo NIR images of nude mice at 1 h (**a**), 6 h (**b**), 12 h (**c**) and 24 h (d) after tail intravenous injection of Ag_2S-GSH-SNO nanoparticles. (**e**) Ex vivo NIR images of viscera from a sacrificed nude mouse at 12 h post-injection. (**f**) Bright-field image and (**g**) fluorescence image of a nude mouse at 5 min after subcutaneous injection of CuFL and Ag_2S-GSH-SNO nanoparticles under irradiation of 488 nm laser (Reproduced with permission from [21])

fluorescence imaging is urgent need for therapeutic applications of NO donors and a controlled delivery of NO from fluorescent complexes with low cytotoxicity, and ability of efficient cell uptake is still challenging. In terms of therapeutic applications of the Ag_2S-GSH-SNO nanoparticles, a 5 μg dose of the nanoparticles was administered into anesthetized mice by tail vein injection. At different time points post-injection, the mice were imaged under 808 nm excitation. The NIR fluorescence signals derived from the nanoparticles in the mice body are visualized at 1 h post-injection. At longer time post-injection, the nanoparticles are distributed in several regions of the mice body via blood circulation. The good contrast of the images due to deep tissue penetration of the NIR fluorescence facilitates diagnosis and treatment monitoring. The readily detectable signals further indicated the strong tissue penetrability of the NIR fluorescence emitted by the nanoparticles. The NIR signals were clearly detected in the liver, stomach, and spleen, and the fluorescence intensities of the lung and kidney are poor. The Ag_2S-GSH-SNO

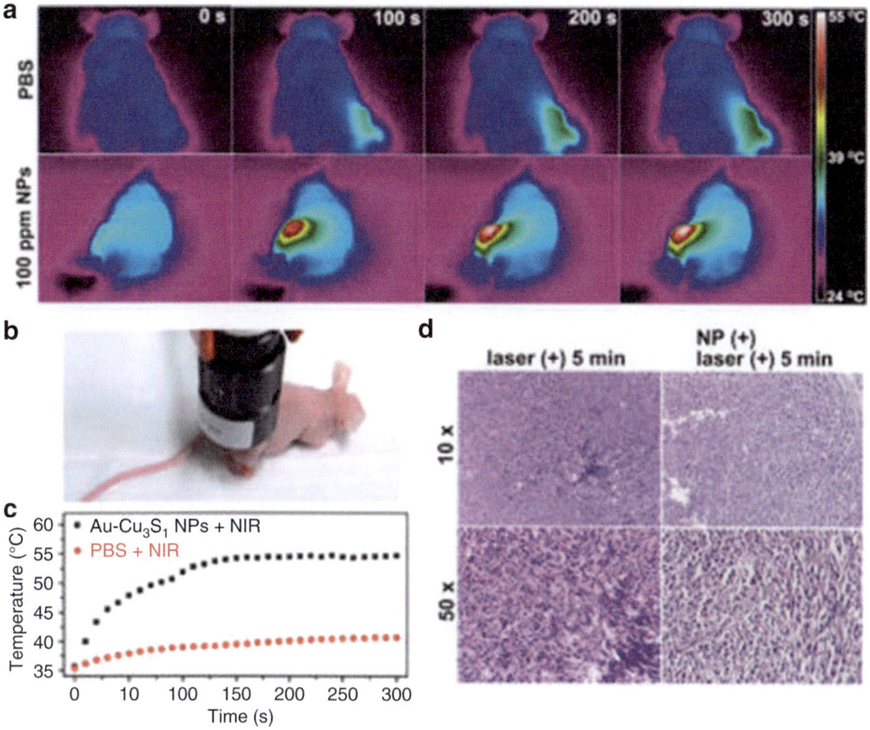

Fig. 7.7 (**a**) Representative thermal images of tumor-bearing mice exposed to 1064 nm laser (0.6 W/cm^2, 5 min) after intratumorally injected with 100 μL of PBS (control) and Au–Cu_9S_5 NPs (100 μg/mL). (**b**) A photograph illustrating the experimental setup for in vivo photothermal treatment. (**c**) The corresponding temperature rise profiles at the tumor site over 5 min laser irradiation with NPs or PBS only. (**d**) H&E staining of tumor tissue sections at different magnifications after photothermal treatments with and without Au–Cu_9S_5 NPs (Reproduced with permission from [23])

nanoparticles tend to be enriched in the liver, stomach, and spleen. The injection region of a mouse emits fluorescence under 488 nm irradiation at 5 min post-injection indicated the in vivo NO release from the nanoparticles.

The NIR-II window can offer larger tissue penetration depth relative to the commonly researched NIR-I window [23], due to lower absorption and scattering by tissues in this spectral range. The deep tissue photothermal ability in NIR-II window using the synthesized Au–Cu_9S_5 NPs under 1064 nm laser irradiation has been explored by using pig epithelial tissues of varying thickness (2, 5, 7, 10, 15, and 20 mm) (Fig. 7.7) [24]. It is further explored in vivo by taking the cancer cells of nude mice bearing CT26 tumor and injected intratumorally with 100 μL of aqueous dispersion of 100 μg/mL of Au–Cu_9S_5 hybrid NPs and the tumors are then exposed to 1064 nm laser at a power density of 0.6 W/cm^2 for 5 min. The spatial temperature distribution at the tumor surface was monitored real-time by a thermal imaging camera. The temperature of the tumor regions with Au–Cu_9S_5 NPs

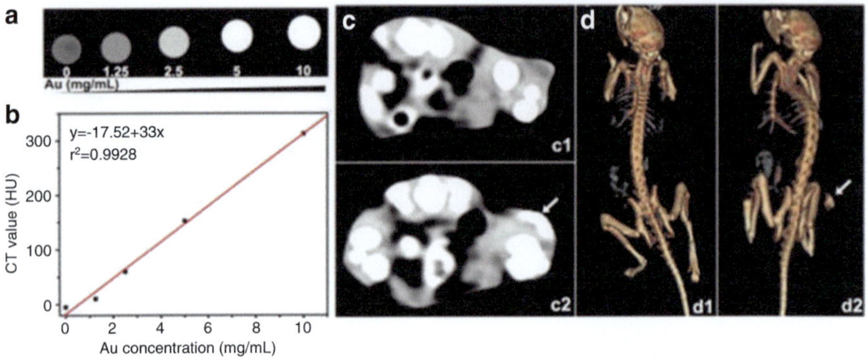

Fig. 7.8 (**a**) CT images and (**b**) CT value (HU) of Au–Cu_9S_5 NPs dispersion as a function of the particle concentrations. Cross-sectional CT images of tumor before (**c1**) and after injection of Au–Cu_9S_5 NPs (**c2**), and their corresponding 3D reconstructed CT images before (**d1**) and after injection (**d2**) of Au–Cu_9S_5 NPs. The arrow heads point to the tumor sites (Reproduced with permission from [23])

increases rapidly from 35 to 54 °C in 2 min. The effect of NIR laser treatment on tumor with or without Au–Cu_9S_5 is evaluated by histology studies where Au–Cu_9S_5 NPs + NIR treated tumor tissues show necrosis, such as severe degree of cell shrinkage to lost contact, nuclear condensation, and the corruption of the tumor extracellular matrix.

Au–Cu_9S_5 NPs are also investigated for contrast enhancement in CT imaging [23], which is a widely used medical imaging technique for diagnosis due to its deep tissue penetration and high resolution characteristics. High atomic number element such as Au has shown excellent X-ray attenuation ability, which triggered research on using biocompatible Au NPs as contrast enhancing agent for CT imaging [25, 26]. The feasibility of the synthesized hybrid NPs to be a CT contrast enhancing agent was first examined in phantom (Fig. 7.8a, b). CT contrast enhancing effect of the NPs is also evaluated on nude tumor-bearing mice with a clinical scanner. After being anesthetized, mice are intratumorally injected with 100 μL of Au–Cu_9S_5 NPs at a dosage of 0.5 mg (Au element), CT images are acquired before and after injection (Fig. 7.8c, d). High image contrast at the tumor sites could be observed after NP injection, as the tumor CT image became much brighter. The reconstructed three-dimensional (3D) CT image of the mouse clearly displayed enhanced contrast in the tumor region after intratumorally administered NPs, revealing that Au–Cu_9S_5 NPs can be simultaneously used as a contrast enhancing agent for tumor imaging.

Surface coating of quantum dots is a critical parameter in the development of new diagnostic or therapeutic agents. CdTe/ZnS QDs, with a core diameter of ~10–15 nm coated with methoxy-PEG chains of different lengths have been used to slow down the speed of their first pass extraction toward the liver and their bio-distribution is assessed using noninvasive fluorescence imaging in a mouse [27]. Immediately after intravenous injection, fluorescence from all QDs is easily seen in the superficial vasculature (Fig. 7.9). Subsequently, QDs are seen to be

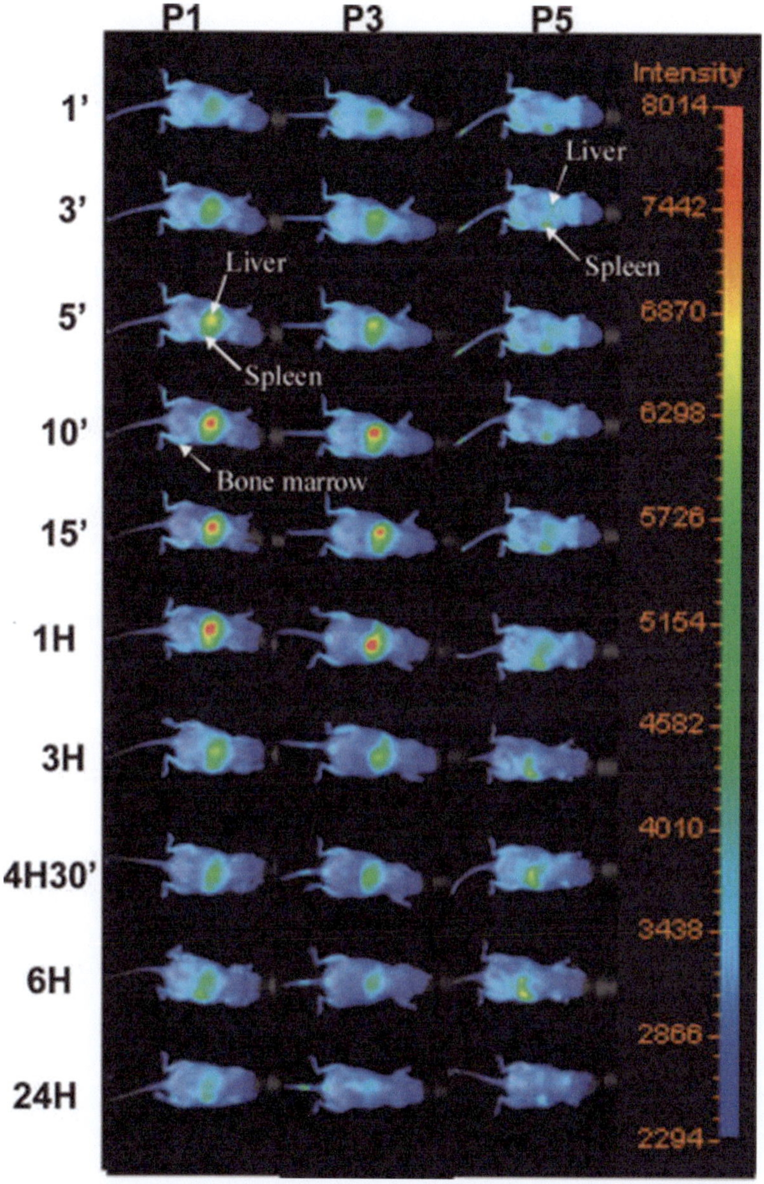

Fig. 7.9 Time-dependent imaging of QD-PEG*Y*-OCH3 (P1: *Y*) 2750; P3: *Y*) 7000; and P5: *Y*) 22,000) in live animals over a 24 h period. Fluorescence images (in color scale, *right* part of the figure) are superimposed to visible light images (in *white* and *black*). For fluorescence images, the integration time is 500 ms and the contrast has been set between 2294 and 8014 (Reproduced with permission from [27])

deposited in the liver, spleen, skin, and bone marrow, in a PEG molecular weight-dependent manner. Important fluorescence signal is displayed by the liver and spleen, as well as the stomach and skin. Increasing the PEG length leads to a slower uptake of the particles. The PEG coating is particularly efficient when exceeding 12,000 Da molecular weight. This slower accumulation in the liver should go along with a longer blood circulation half-life of the QDs. Therefore, the increase of the PEG chain length seems rather beneficial for decreasing the uptake of the particles, and thus, targeting a tumor area is easier with long circulating neutral particles coated with long PEG chains [28]. The optimal nanoparticle size for taking full advantage of the enhanced permeability and retention effect [29] at the tumor site is around 50-nm diameter [30], and QDs coated with the 22,000 Da PEG chains satisfies this. Smaller particles would not be retained efficiently in the tissue or bind with sufficient affinity to the targeted cells.

Dendron-coated InP/ZnS core/shell QDs are a type of promising NIR fluorescent probes for biomedical applications [31, 32]. The dendrimer and dendron molecules exhibit unique physicochemical and biological properties, which have great potential for use in a variety of applications, including drug delivery and surface engineering [33, 34]. The robust and small dendron molecules not only stabilize the QDs but also minimize the size increase of coated QDs in aqueous solution [35–37]. QD710-Dendron demonstrates near-infrared (NIR) emission, high stability in biological media, suitable size with possible renal clearance, and ability of extravasation; while QD710-Dendron-RGD2 is highly specific in targeting tumor cells resulted in high tumor uptake and long retention of the nanoprobe at tumor sites [38]. The fluorescent signals derived from both QD710-Dendron-RGD2 and QD710-Dendron appeared in tumors are readily distinguished from surrounding tissues after 4 h in both groups (Fig. 7.10). The tumor uptake of QD710-Dendron

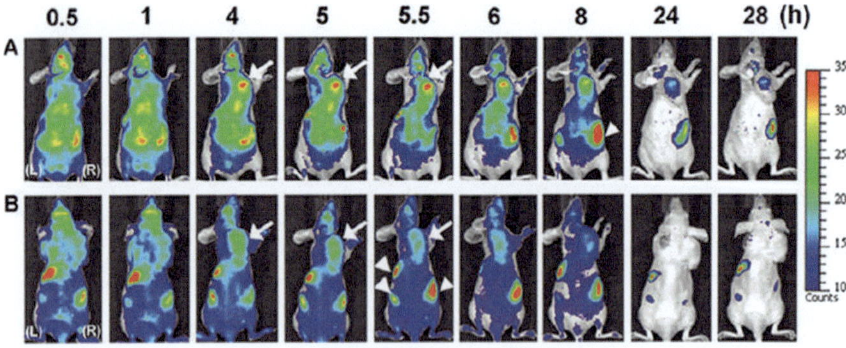

Fig. 7.10 In vivo NIR fluorescence imaging. The dorsal images of SKOV3 tumor-bearing (*arrows*) mice (*L*, left side; *R*, right side) injected with (**a**) QD710-Dendorn-RGD2 (200 pmol) and (**b**) QD710-Dendron (200 pmol) at 0.5, 1, 4, 5, 5.5, 6, 8, 24, and 28 h, respectively. The incidental high fluorescent signals in other body parts (*arrowheads*) might have originated from regular rodent food in stomach and feces in intestine (Reproduced with permission from [38])

with visible contrast from surrounding tissues indicated the possibility of enhanced permeability and retention effect for the suitably small-sized QD710-Dendron, but the tumor fluorescent intensity dramatically decreased over time (Fig. 7.10b). By contrast, in the mice injected with QD710-Dendron-RGD2, the tumor contrast is still apparent even after 24 h (Fig. 7.10a), indicating that the highly specific targeting of QD710-Dendron-RGD2 to integrin αvβ3-positive SKOV3 tumor induced the long term retention of QDs in the tumor site [39, 40].

A uniform ternary CuInS2 (CIS) and core/shell CIS/ZnS QDs coated with O-carboxymethyl chitosan (OCMCS) are used for cytotoxicity with HeLa and OECM-1 cell lines. *C. elegans* is selected as a simple organic model to characterize the toxicity and chemical stability of QDs during digestion progress (Fig. 7.11) [41]. In exposed experiments, CIS/OCMCS and CIS/ZnS/OCMCS QDs are chosen with emission at near-infrared region and average size is 200 nm. The QDs are

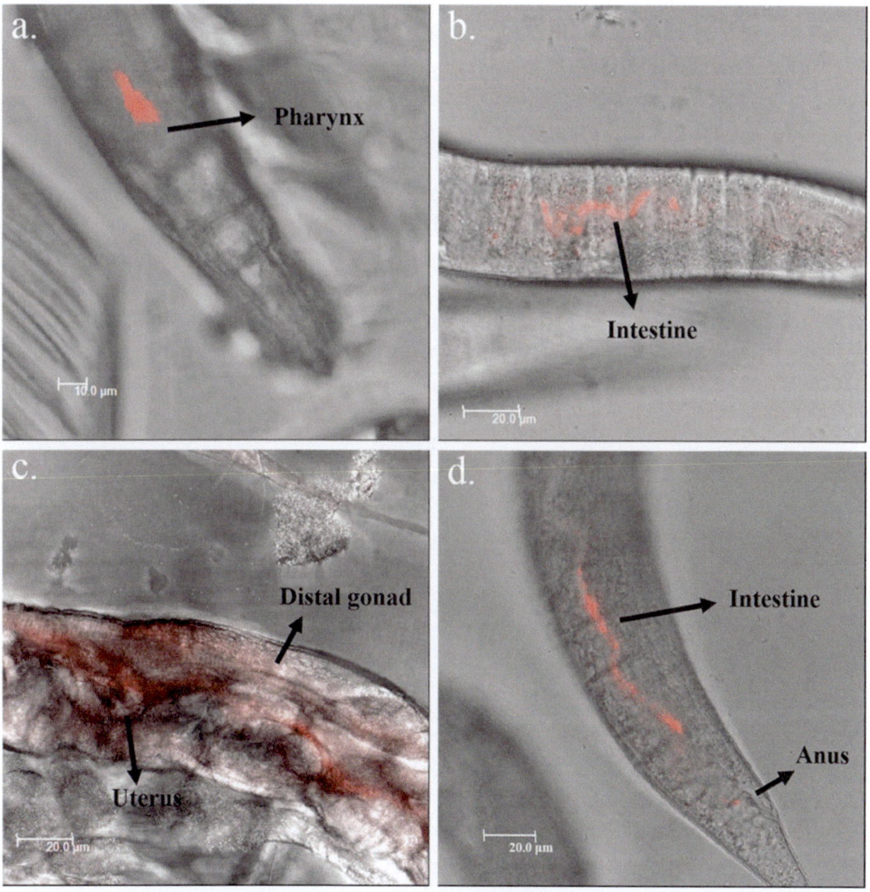

Fig. 7.11 Confocal images of *C. elegans* treated with CIS/ZnS/OCMCS QDs for (**a**) 12, (**b**) 48, (**c**) 72, and (**d**) 96 h (Reproduced with permission from [41])

exposed with the food source, *E. coli* OP50 to *C. elegans* to digest. They are found located at the head and pharynx after exposed for 12 h. The pumping by the pharynx muscle is the major cause for the digestion that transfers QDs from the pharynx to the intestine. At this exposed time, most of QDs remained in the pharynx. For 48 h, the QDs localized before the intestine, and after 72 h, the QDs localized after the intestine and some of them penetrate from the alimentary system to the reproductive system and eggs, and some are found localized at the distal gonad and uterus. After exposing for 96 h, the QDs transfer to the anus and they still emit fluorescence and are thus still stable inside the *C. elegans*. The chemical stability of QDs in *C. elegans* from shortest to longest exposure times is determined by measuring copper and zinc K-edge absorption spectra. X-ray absorption near edge structure results show that both CIS and CIS/ZnS QDs sustain highly chemically stable in *C. elegans* for the shortest and longest exposure times. It confirms that CIS QDs are very stable in long biological digestion and can be considered for better applications in bioimaging and drug delivery.

CdTe QDs have found numerous applications ranging from electronics but their use for biological applications is still very limited because the CdTe QDs are more toxic than the core/shell CdSe/ZnS, CdSe/CdS/ZnS, and InP/ZnS QDs [42–44]. A long-term contact of the functionalized CdTe QDs with biological fluids can result in partial desorption of the hydrophilic moieties, thus exposing the bare nanocrystal surface to the biological system, with high chances of releasing toxic cadmium ions to the surrounding. As a result, CdTe QDs are not widely used for in vitro and in vivo imaging. However, CdTe QDs can be safely utilized as imaging contrast agents when CdTe QDs are covered with a shell of a wide-bandgap semiconductor that protects the core from degradation. It provides long-term colloidal and optical stability, and surface passivation with biocompatible molecules. Therefore, CdTe/ZnSe core/shell QDs have been used in in vitro imaging of chicken tissue and embryo [45]. The fertilized eggs are incubated for 19 days in the incubator and after the incubation, the embryos are removed from the shell and then different CdTe/ZnSe QDs are injected under the skin by injection. The fluorescence properties of the QDs in the embryo body are detected and the fluorescence intensity of QDs is analyzed by Carestream Molecular Imaging Software. The image of chicken embryo with all the CdTe/ZnSe QDs is shown in Fig. 7.12. All the QDs50-120 were applied into the same chicken embryo and analyzed. Considerable differences between the QDs were observed. It is observed that the CdTe/ZnSe QDs of all temperature can be used for the in vitro experiments.

To demonstrate the capability of CdTe/ZnTe QDs for in vivo targeting, they are injected to the mouse through subcutaneous and tail vein injections [46]. The high-luminescence signal at different spots of treated mouse can be easily visualized above the background autofluorescence signal. This indicates that the CdTe/ZnTe QDs can serve as excellent ultrasensitive probes for in vivo multiplex imaging (Fig. 7.13). For the distribution and clearance of the functionalized QDs in vivo, mice are sacrificed, surgically opened, and imaged and are shown in Fig. 7.13a and b, respectively. No luminescence signal from the QDs is observed in the major organs because the QDs are excreted from the animal body. Histological analysis is

Fig. 7.12 All QDs (CdTe/ZnSe °C) were applied into the chicken embryo (25 lL under the skin): a—QDs50, b—QDs60, c—QDs70, d—QDs80, e—QDs90, f—QDs100, g—QDs110, and h—QDs120. Detected by Carestream Molecular Imaging Software (Rochester, NY, USA). The fluorescence and the X-ray images were overlaid by PhotoFiltre Studio X (Reproduced with permission from [45])

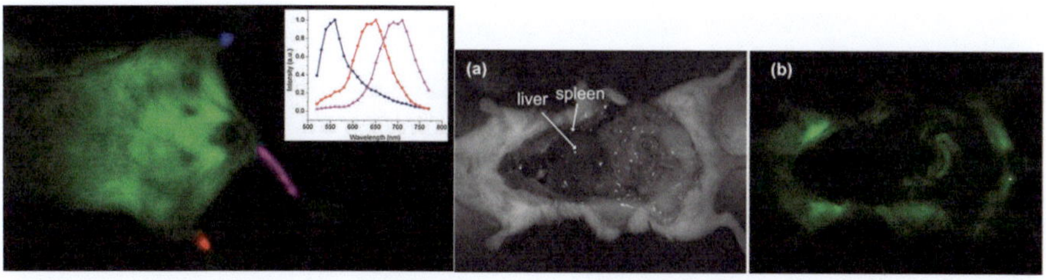

Fig. 7.13 Fluorescence images of mouse injected with three different sizes emitting CdTe/ZnTe QDs. The corresponding spectrum profiles are shown in the *inset*. (**a**) Image and (**b**) fluorescence image of a surgically opened mouse following treatment with CdTe/ZnTe QDs for 2 weeks (Reproduced with permission from [46])

performed on the tissues obtained from the major organs such as liver, spleen, kidney, lung, and heart to investigate for any sign of acute toxicity. No signs of toxicity are observed in the tissues from the QDs.

A suitably coated CdTe QDs is an excellent tool for the high-content screening (HCS) technology which is a recent advance in the integration and automation of quantitative fluorescence microscopy and image analysis [47, 48]. A comprehensive investigation of CdTe QD induced cytotoxicity in NG108-15 murine neuroblastoma cells using various HCS assays has been reported [49]. The cellular response of the NG108-15 murine neuroblastoma cells with two types of CdTe QDs; thioglycolic acid (TGA)-capped CdTe QD (TGA-QD), and TGA-capped CdTe QD produced in the presence of gelatin (Gelatin-QD), has been studied. Both of these QDs have a diameter of 3 nm and emit in the green-yellow region of the spectrum (~560 nm). NG108-15 murine neuroblastoma cells possess neuron-like properties and can be differentiated relatively easily and quickly.

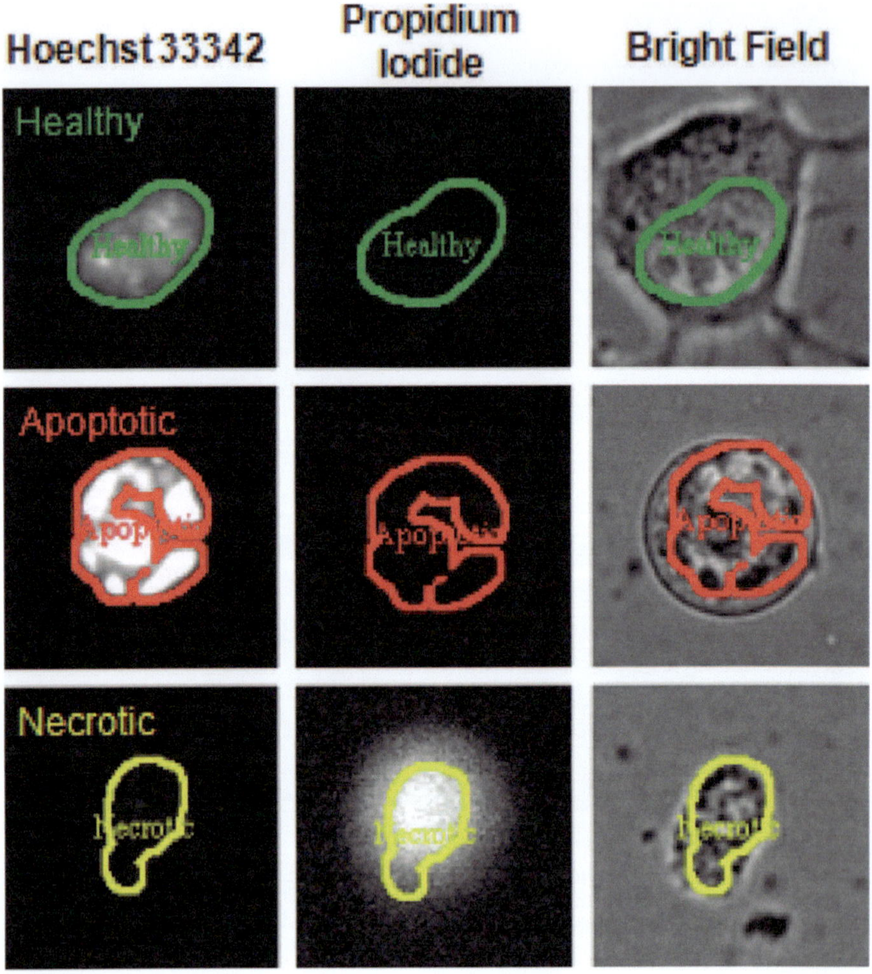

Fig. 7.14 Representative fluorescence and bright field images of a healthy (*green* outline), an apoptotic (*red* outline), and a necrotic (*yellow* outline) cell. Cells were stained simultaneously with Hoechst 33342 (*blue* channel, *first column*) and propidium iodide (*red* channel, *middle column*). Outline and classification of cells were generated by the IN Cell Investigator image analysis software using the supervised classification capability (Reproduced with permission from [49])

Figure 7.14 illustrates representative images of healthy, apoptotic, and necrotic cells. The plasma membranes of necrotic cells are permeable to propidium iodide, so necrotic cells were identified by their intense red fluorescence. For both undifferentiated and differentiated cells, treatment with QDs induced predominantly an increase in apoptotic cell population, as the number of necrotic cells remained relatively constant regardless of QD dosage or treatment time [50]. For undifferentiated NG108-15 cells, the number of apoptotic cells remained constant between 1.5 and 6 h of incubation with QDs and increased sharply after 24 h, while for differentiated cells, an incremental increase in apoptotic population is observed between the three time points, indicating differentiated cells to be more sensitive to QD treatment. Although TGA-QDs and Gelatin-QDs produced similar cellular response in NG108-15 cells, the most obvious difference between the two QDs is seen in the apoptotic population of the undifferentiated cells. It increases with TGA-QD and remains unchanged with Gelatin-QD.

7.4 Instrumentation for Bioimaging

A combination of analytical transmission electron microscopy and serial block face scanning electron microscopy can provide a comprehensive description of the internalization of nanoparticles by endocytically active cells [51]. The use of quantum dot as biomarkers is becoming well established and these applications demand an appropriate understanding of the cellular uptake, quantification and the fate of internalized particles. Serial block face scanning electron microscopy (SBF SEM) allows the collection of 3D [51]. Electron microscopy of a cellular sample embedded in resin provides a fixed time point on which to conduct analysis. Although this may be perceived as a disadvantage when monitoring a dynamic biological process such as particle uptake, it does provide the opportunity for extensive examination at a resolution where individual nanoparticles can be identified and counted.

Both the 3D SBF SEM reconstruction and TEM imaging of thin sections of cells indicates that quantum dot uptake is still taking place at the surface of the human osteosarcoma U-2 OS cells after 1 h exposure. Near the cell membrane it is possible that quantum dots that appear to be internalized may actually be in the early stages of endocytosis, and are part of larger agglomerates still connected to the cell surface [52]. Quantum dots are present in a variety of locations, some are outside the cell, presumably in the process of being taken up into the cell at the time of sample processing, whereas other nearby groups appear to be in membrane bound vesicles. TEM imaging is limited by a lack of knowledge of what is present in the thin sections cut immediately prior to and following the section being examined. Reconstruction of the whole cell shows that quantum dot agglomerates are present in more than one microtome slice. There are slices where the agglomerate appears to be internalized in the cell and other slices where it is apparently just attached to the surface [52]. Thus 3D imaging and reconstruction shows which quantum dots are completely and which are incompletely internalized. SBF SEM is particularly useful in these nanomedicinal studies because of the capacity to collect 3D image sets across whole cells, although not operating at a resolution to identify individual quantum dots.

7.5 Unique Properties of Selenium: Anticarcinogenic Effect

Although we discussed several biological applications of chalcogenide nanomaterials, elemental selenium is one of the most important components for the biological system with several useful applications [53–55]. It is biologically active element which complexes with enzymes or proteins. The toxicity of the Se [56, 57] is related to its nonspecific complexation with the protein which renders biological functionality of the protein. Nonspecific interactions of Se with a protein are considered to generate toxic effect. There are several selenoproteins which perform many important biological functions. Selenite supplementation significantly reduces oxidative DNA damage. Se deficiency increases the viral infection. Thyroid gland contains more selenium per gram of tissue than any other organ. Thyroid problems have been linked to selenium deficiency. That is why, the thyroid hormone and selenium-containing deiodinase enzymes are critically important for brain development during pregnancy, and hence, it strengthens the immune system. The natural Se containing antioxidant enzymes have the potential to reduce the effect of free radicals on the aging process while Se deficiency is associated with increased cardiovascular disease. Selenomethionine is the major Se compound available in cereals, legumes, and soybeans. Meats and fish can be good sources of Se while garlic, onions, and broccoli contain enough Se in the form of methyl selenocysteine. Usually, the Se contents of foods depend mainly on the Se contents of the soil from they are produced.

Selenium has well known *anticarcinogenic* effects. It demonstrates inhibitory effects in various tumor cell lines and transplanted tumors [58, 59]. Se reduces side effects of chemotherapeutic agents during the cancer treatments and improves the immune system of cancer patients. A low serum Se is considered to be the main contributory factor for the gastrointestinal, prostatic, and thyroid cancer and is a pre-diagnostic indicator of higher cancer risk [60, 61]. Apart from this, a number of other elements such as Cd, As, Zn, Hg, and Pb present in foods, drinking water, and atmosphere complexes with Se to form unreactive selenides or nonspecific protein complexes which reduces the required amount of Se intake. It also stimulates the oxygen radical production, inhibits DNA-repair enzymes, exhibits cocarcinogenic activity and hence stimulates the growth of tumor cells. Excess of Zn intake from different kinds of red meats allows Zn to form ZnSe and reduces the required amount of Se needed for its anticarcinogenic effects. At the cellular level, adequate supply of Se reduces the damaging effects of endogenously or exogenously generated oxygen radicals which may cause mutations and the oxidative activation of chemical carcinogens. Therefore, a daily extradietary supplement of 200 mg of Se reduces cancer risk.

References

1. S. George, T. Xia, R. Rallo, Y. Zhao, Z. Ji, S. Lin, X. Wang, H. Zhang, B. France, D. Schoenfeld, R. Damoiseaux, R. Liu, S. Lin, K.A. Bradley, Y. Cohen, A.E. Nel, Use of a high-throughput screening approach coupled with in vivo zebrafish embryo screening to develop hazard ranking for engineered nanomaterials. ACS Nano **5**, 1805–1817 (2011)
2. A. Nel, T. Xia, L. Madler, N. Li, Toxic potential of materials at the nanolevel. Science **311**, 622–627 (2006)
3. Nanotechnology Consumer Product Inventory, Project on emerging nanotechnology, Woodrow Wilson International Center for Scholars, Washington, DC. http://www.nanotechproject.org/inventories/consumer/
4. R.F. Service, Nanotechnology; can high-speed tests sort out which nanomaterials are safe? Science **321**, 1036–1037 (2008)
5. T. Hartung, Toxicology for the twenty-first century. Nature **460**, 208–212 (2009)
6. X. Zhu, J. Wang, X. Zhang, Y. Chang, Y. Chen, The impact of ZnO nanoparticle aggregates on the embryonic development of Zebrafish (Danio rerio). Nanotechnology **20**, 195103 (2009)
7. T.C. King-Heiden, P.N. Wiecinski, A.N. Mangham, K.M. Metz, D. Nesbit, J.A. Pedersen, R.J. Hamers, W. Heideman, R.E. Peterson, Quantum dot nanotoxicity assessment using the Zebrafish embryo. Environ. Sci. Technol. **43**, 1605–1611 (2009)
8. V.E. Fako, D.Y. Furgeson, Zebrafish as a correlative and predictive model for assessing biomaterial nanotoxicity. Adv. Drug Deliv. Rev. **61**, 478–486 (2009)
9. M.G. Panthani, T.A. Khan, D.K. Reid, D.J. Hellebusch, M.R. Rasch, J.A. Maynard, B.A. Korgel, In vivo whole animal fluorescence imaging of a microparticle-based oral vaccine containing ($CuInSe_xS_{2-x}$)/ZnS core/shell quantum dots. Nano Lett **13**, 4294–4298 (2013)
10. P.C. Tyrer, A. Ruth Foxwell, J.M. Kyd, D.C. Otczyk, A.W. Cripps, Vaccine **25**, 3204–3209 (2007)
11. H. McDaniel, N. Fuke, J.M. Pietryga, V.I.J. Klimov, Phys. Chem. Lett. **4**, 355–361 (2013)
12. H. Zhong, Y. Zhou, M. Ye, Y. He, J. Ye, C. He, C. Yang, Y. Li, Chem. Mater. **20**, 6434–6443 (2008)
13. R.N. Palumbo, C. Wang, Curr. Drug Deliv. **3**, 47–53 (2006)
14. A. Shukla, O.P. Katare, B. Singh, S.P. Vyas, Int. J. Pharm. **385**, 47–52 (2010)
15. T. Nochi, Y. Yuki, A. Matsumura, M. Mejima, K. Terahara, D.-Y. Kim, S. Fukuyama, K. Iwatsuki-Horimoto, Y. Kawaoka, T. Kohda, S. Kozaki, O. Igarashi, H.J. Kiyono, Exp. Med. **204**, 2789–2796 (2007)
16. R. KuoLee, W. Chen, Expert Opin. Drug Deliv. **5**, 693–702 (2008)
17. V. Fievez, L. Plapied, A. des Rieux, V. Pourcelle, H. Freichels, V. Wascotte, M.-L. Vanderhaeghen, C. Jerôme, A. Vanderplasschen, J. Marchand-Brynaert, Y.-J. Schneider, V. Préat, Eur. J. Pharm. Biopharm. **73**, 16–24 (2009)
18. R. Schwarz, A. Kaspar, J. Seelig, B. Kunnecke, Magn. Reson. Med. **48**, 255–261 (2002)
19. K. Hase, K. Kawano, T. Nochi, G.S. Pontes, S. Fukuda, M. Ebisawa, K. Kadokura, T. Tobe, Y. Fujimura, S. Kawano, A. Yabashi, S. Waguri, G. Nakato, S. Kimura, T. Murakami, M. Iimura, K. Hamura, S.I. Fukuoka, A.W. Lowe, K. Itoh, H. Kiyono, H. Ohno, Nature **462**, 226–U101 (2009)
20. E.L. McConnell, A.W. Basit, S.J. Murdan, Pharm. Pharmacol. **60**, 63–70 (2008)
21. L. Tan, A. Wan, H. Li, Conjugating S-nitrosothiols with glutathiose stabilized silver sulfide quantum dots for controlled nitric oxide release and near-infrared fluorescence imaging. ACS Appl. Mater. Interfaces **5**, 11163–11171 (2013)
22. A.R. Butler, D.L.H. Williams, Chem. Soc. Rev. **22**, 233–241 (1993)
23. X. Ding, C.H. Liow, M. Zhang, R. Huang, C. Li, S. He, M. Liu, Z. Yu, N. Gao, Z. Zhang, Y. Li, Q. Wang, S. Li, J. Jiang, Surface plasmon resonance enhanced light absorption and photothermal therapy in the second near-infrared window. J. Am. Chem. Soc. **136**, 15684–15693 (2014)

24. A.N. Bashkatov, E.A. Genina, V.I. Kochubey, V.V. Tuchin, J. Phys. D Appl. Phys. **38**, 2543 (2005)
25. C. Xu, G.A. Tung, S. Sun, Chem. Mater. **20**, 4167 (2008)
26. D. Xi, S. Dong, X. Meng, Q. Lu, L. Meng, J. Ye, RSC Adv. **2**, 12515 (2012)
27. T. Jean Daou, L. Liang, P. Reiss, V. Josserand, I. Texier, Effect of poly(ethylene glycol) length on the in vivo behavior of coated quantum dots. Langmuir **25**, 3040–3044 (2009)
28. X. Gao, Y. Cui, R.M. Levenson, L.W. Chung, S. Nie, Nat. Biotechnol. **22**, 969–976 (2004)
29. H. Maeda, J. Wu, T. Sawa, Y. Matsumura, K. Hori, J. Control. Release **65**, 271–284 (2000)
30. P. Decuzzi, R. Pasqualini, W. Arap, M. Ferrari, Pharm. Res. **26**, 235–243 (2009)
31. R.G. Xie, D. Battaglia, X.G. Peng, J. Am. Chem. Soc. **129**, 15432–15433 (2007)
32. K.T. Yong, H. Ding, I. Roy, W.C. Law, E.J. Bergey, A. Maitra, P.N. Prasad, ACS Nano **3**, 502–510 (2009)
33. J.M. Oliveira, A.J. Salgado, N. Sousa, J.F. Mano, R.L. Reis, Prog. Polym. Sci. **35**, 1163–1194 (2010)
34. D. Astruc, E. Boisselier, C. Ornelas, Chem. Rev. **110**, 1857–1959 (2010)
35. Y.A. Wang, J.J. Li, H.Y. Chen, X.G. Peng, J. Am. Chem. Soc. **124**, 2293–2298 (2002)
36. W.H. Guo, J.J. Li, Y.A. Wang, X.G. Peng, J. Am. Chem. Soc. **125**, 3901–3909 (2003)
37. W.Z. Guo, J.J. Li, Y.A. Wang, X.G. Peng, Chem. Mater. **15**, 3125–3133 (2003)
38. J. Gao, K. Chen, R. Luong, D.M. Bouley, H. Mao, T. Qiao, S.S. Gambhir, Z. Cheng, A novel clinically translatable fluorescent nanoparticle for targeted molecular imaging of tumors in living subjects. Nano Lett **12**, 281–286 (2012)
39. J.P. Xiong, T. Stehle, R.G. Zhang, A. Joachimiak, M. Frech, S.L. Goodman, M.A. Aranout, Science **296**, 151–155 (2002)
40. R. Weissleder, K. Kelly, E.Y. Sun, T. Shtatland, L. Josephson, Nat. Biotechnol. **23**, 1418–1423 (2005)
41. C.-W. Chen, D.-Y. Wu, Y.-C. Chan, C.C. Lin, P.-H. Chung, M. Hsiao, R.-S. Liu, Evaluations of the chemical stability and cytotoxicity of $CuInS_2$ and $CuInS_2$/ZnS core/shell quantum dots. J. Phys. Chem. C **119**, 2852–2860 (2015)
42. S.J. Cho, D. Maysinger, M. Jain, B. Roder, S. Hackbarth, F.M. Winnik, Langmuir **23**, 1974 (2007)
43. J. Ma, J.Y. Chen, Y. Zhang, P.N. Wang, J. Guo, W.L. Yang, C.C. Wang, J. Phys. Chem. B **111**, 12012 (2007)
44. Y. Zhang, L. Mi, P.-N. Wang, S.-J. Lu, J.-Y. Chen, J. Guo, W.-L. Yang, C.-C. Wang, Small **4**, 777 (2008)
45. A. Moulick, I. Blazkova, V. Milosavljevic, Z. Fohlerova, J. Hubalek, P. Kopel, M. Vaculovicova, V. Adam, R. Kizek, Application of CdTe/ZnSe quantum dots in in vitro imaging of chicken tissue and embryo. Photochem. Photobiol. **91**, 417–423 (2015)
46. W.-C. Law, K.-T. Yong, I. Roy, H. Ding, R. Hu, W. Zhao, P.N. Prasad, Aqueous-phase synthesis of highly luminescent CdTe/ZnTe Core/shell quantum dots optimized for targeted bioimaging. Small **5**(11), 1302–1310 (2009)
47. K.A. Giuliano, J.R. Haskins, D.L. Taylor, Advances in high content screening for drug discovery. Assay Drug Dev. Technol. **1**, 565–577 (2003)
48. J. Comley, High content screening—emerging importance of novel reagents/probes and pathway analysis. Drug Discov. World **6**, 31–53 (2005)
49. E. Jan, S.J. Byrne, M. Cuddihy, A.M. Davies, Y. Volkov, Y.K. Gun'ko, N.A. Kotov, High-content screening as a universal tool for fingerprinting of cytotoxicity of nanoparticles. ACS Nano **2**(5), 928–938 (2008)
50. W.H. Chan, N.H. Shiao, P.Z. Lu, CdSe quantum dots induce apoptosis in human neuroblastoma cells via mitochondrial-dependent pathways and inhibition of survival signals. Toxicol. Lett. **167**, 191–200 (2006)
51. W. Denk, H. Horstmann, Serial block-face scanning electron microscopy to reconstruct three-dimensional tissue nanostructure. PLoS Biol. **2**, e329 (2004)

52. N.I.C.O. Hondow, M.R. Brown, T.O.B.I. Starborg, A.G. Monteith, R.I.K. Brydson, H.D. Summers, P.A.U.L. Rees, A.N.D.Y. Brown, Quantifying the cellular uptake of semiconductor quantum dot nanoparticles by analytical electron microscopy. J. Microsc. **261**, 167–176 (2016). doi:10.1111/jmi.12239
53. A.P. Fernandes, V. Gandin, Selenium compounds as therapeutic agents in cancer. Biochim. Biophys. Acta **1850**, 1642–1660 (2015)
54. J. Fernindez-Lodeiro, M.F. Pinatto-Botelho, A.N.A. Soares-Paulino, A.C. Gonalves, B.A. Sousa, C. Princival, A.A. Dos Santos, Synthesis and biological properties of selenium- and tellurium-containing dyes. Dyes Pigm. **110**, 28–48 (2014)
55. S. Roth, S. Zhang, J. Chiu, E.K. Wirth, U. Schweizer, Development of a serum-free supplement for primary neuron culture reveals the interplay of selenium and vitamin E in neuronal survival. J. Trace Elem. Med. Biol. **24**, 130–137 (2010)
56. E.P. Painter, The chemistry and toxicity of selenium compounds, with special reference to the selenium problem. Chem. Rev. **28**, 179–213 (1941)
57. A.M.H. deBruyn, P.M. Chapman, Selenium toxicity to invertebrates: will proposed thresholds for toxicity to fish and birds also protect their prey? Environ. Sci. Technol. **41**, 1766–1770 (2007)
58. C. Ip, Selenium inhibition of chemical carcinogenesis. Fed. Proc. **44**, 2573–2578 (1985)
59. J.A. Milner, Effect of selenium on virally induced and transplantable tumor models. Fed. Proc. **44**, 2568–2572 (1985)
60. Aaseth J., Glattre E., Frey H., Norheim G., Ringstad J., Thomassen, Y., Selenium in the human thyroid gland, in *Proc. 6th Internatl. Trace Elem. Symp*. ed. by M. Anke, W. Baumann, H. Bräunlich, C. Brückner, B. Groppel, M. Grün, vol 3 (VEB Krongreß-u. Werbedruck, Jena, 1989) pp. 991–914
61. E.M. Glattre, Y. Thomassen, S.Q. Thorensen, T. Haldorsen, P.G. Lund-Larsen, L. Theodosen et al., Prediagnostic serum selenium in a case-control study of thyroid cancer. Int. J. Epidemiol. **18**, 45–49 (1989)

Chapter 8
Selenide Glass Fibers for Biochemical Infrared Sensing

Pierre Lucas and Bruno Bureau

P. Lucas, Ph.D. (✉)
University of Arizona, Tucson, 85721, AZ, USA
e-mail: pierre@email.arizona.edu

B. Bureau, Ph.D.
Glass and Ceramics, ISCR UMR-CNRS 6226, University of Rennes 1, Campus de Beaulieu, Avenue du General Leclerc, Rennes, E35042, France

G.K. Ahluwalia (ed.), *Applications of Chalcogenides: S, Se, and Te*,
DOI 10.1007/978-3-319-41190-3_8

8.1 Introduction

Vibrational spectroscopy is established as one of the most powerful analytical technique for chemical and biological analysis due the highly specific nature of molecular vibration signatures. Every molecule possesses a very unique set of vibrational modes which can be probed by absorption or transmission spectroscopy to yield a selective spectral fingerprint. This spectral fingerprint is located in the region between 400 and 4000 cm^{-1} (2–20 μm) which encompasses the mid- and far-infrared domain. Optical probes in that spectral range are therefore of great interest and have led to the widespread development of Fourier Transform Infrared (FTIR). This technique possesses many advantages including the nondestructive and noninvasive character of its probing mechanism as well as its rapidity of detection which can enable quasi-real time monitoring of some chemical and biochemical processes [1, 2]. FTIR is consequently a prevailing technique for analytical chemistry but also finds increasing popularity in the biomedical field. Indeed, biological molecules exhibit a rich and complex spectral signature in the "fingerprint" region between 5 and 12 μm which can be exploited to identify the strain of microorganisms [3–5] or to distinguish between healthy and diseased tissues [6, 7]. The range of application of infrared spectroscopy techniques is therefore very broad and extends over many fields including environmental monitoring of toxicants [8, 9], quantification of metabolic processes [10, 11], or bio-fluid and tissue analysis [12, 13]. In addition, the intrinsic selectivity of this technique can be further improved through the use of advanced spectral analysis such as principal component analysis (PCA) [14, 15], partial least square (PLS) regressions [16, 17], or neural network analysis [18, 19].

However, transmission spectroscopy measurements are limited to samples exhibiting fairly good transparency in the infrared domain and do not lend themselves to in situ analysis. Consequently, various alternative collection techniques such as attenuated total reflectance (ATR) have been developed to extend the potential of infrared spectroscopy [20]. In that respect, one of the most versatile techniques that have been developed in the last decades involves the use of infrared transparent glass fibers for the delivery and collection of infrared signals. The signal collection mechanism is similar to that of ATR but permits remote and in situ analysis using extended fiber cables [21, 22]. This technique is called fiber evanescent wave spectroscopy (FEWS) and has been successfully used for chemical and biological sensing. Its applications include monitoring of chemical reactions [23–25], detection of pollutants in water supplies [26, 27], cancer diagnosis [28], analysis of tumorous tissues [29], metabolic monitoring of live cells and bacterial cultures [2, 30], or human serum analysis [31].

While standard infrared optical elements such as ATR are easily produced from crystalline solids (Ge, ZnSe CaF_2), these materials are inadequate for the production of optical fibers. Indeed, the production of complex elements such as fibers requires materials that possess appropriate rheological properties for molding or drawing. Moreover these materials must retain optimal infrared transparency during the shaping process and must be chemically resistant to standard chemical and biological environments. In that respect, selenide glasses exhibit an ideal combination of rheological properties [32], chemical resistance [33], and infrared transparency [34]. They are therefore the material of choice for the synthesis of infrared optical fibers for FEWS applications.

8.2 Methods

8.2.1 Fiber Evanescent Wave Spectroscopy

8.2.1.1 Infrared Absorption

When a beam of infrared light interacts with a target sample, photons of specific energy are absorbed as they excite the fundamental vibrational modes specific to the target molecules. A plot of the transmitted photon intensity as a function of wavelength constitutes the resulting absorption/transmission spectrum as depicted on Fig. 8.1 for a culture of human lung cells. In practice, the target samples is coated or supported on an optically clear substrate (window, ATR, fibers) during the collection process. In order to remove any signal from the substrate, a background signal I_B of the substrate alone is first collected without the sample. Then the substrate is put into contact with the samples and the total signal I_S is recorded. The absorption spectrum A is then obtained from Eq. (8.1):

$$A = -\log\left(\frac{I_S}{I_B}\right) \tag{8.1}$$

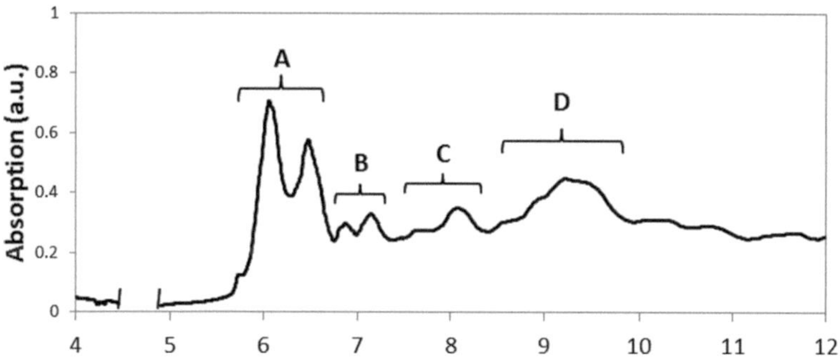

Fig. 8.1 Absorption vibrational spectrum of a human lung cell collected by fiber evanescent wave spectroscopy. The vibrational modes associated with each cell component can be identify including *A*: Amide I and II; *B*: fatty acids and amino acids; *C*: phospholipids and nucleic acids; *D*: polysaccharides

This is the standard process for an ATR experiments and it similarly applies to FEWS experiments where the signal of the fiber in air constitutes the background signal I_B. The human lung cell absorption spectrum shown in Fig. 8.1 was collected this way by FEWS using a selenide glass fiber.

8.2.1.2 Remote Spectroscopy

In a conventional FTIR experiment, the sample is placed between the light source and the detector and the optical signal is directly recorded in transmission mode (Fig. 8.2a). This setup implies that the sample has sufficient infrared transparency to yield a transmission spectrum and requires that the samples be removed from its environment and placed inside the spectrometer. This is not an adequate approach for biological samples that requires in situ or in vivo measurements, or for samples exhibiting high absorptions. Similarly this approach would not be possible for the analysis of coatings or cell cultures deposited on opaque substrates. Continuous monitoring of natural environment such as water supplies would also be precluded. For these cases, a remote spectroscopy method that brings the light to the samples and collects the signal back is desirable.

FEWS provide such a method through the use of infrared transparent optical fibers as depicted in Fig. 8.2b. The light source from a conventional FTIR spectrometer or a quantum cascade laser (QCL) is coupled into one end of the fiber and guided towards the target sample. The infrared signal then interacts with the sample through total internal reflection along a tapered sensing zone (Fig. 8.2b). The resulting absorption signal is subsequently collected on the other end of the fiber and analyzed on the detector. The fiber cable is flexible and permits to conveniently reach in situ environments.

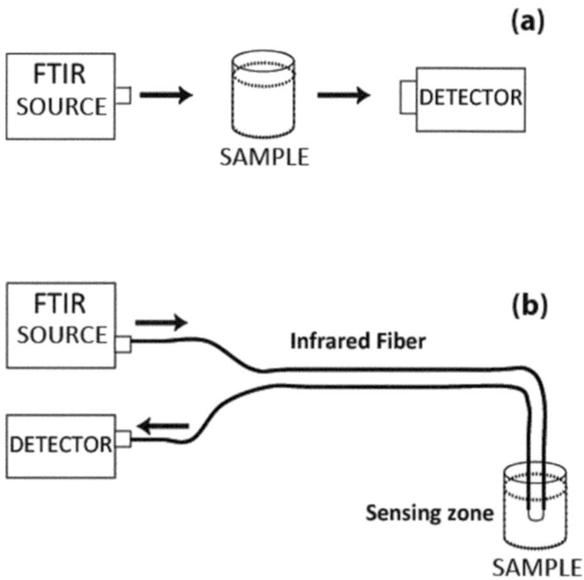

Fig. 8.2 (**a**) Conventional FTIR spectroscopy in transmission mode. (**b**) Setup for fiber evanescent wave spectroscopy allowing remote vibrational signal collection using an infrared transparent selenide glass fiber

8.2.1.3 Evanescent Waves, Total Internal Reflection

When light is coupled into an optical fiber, the beam is trapped by total internal reflection provided the refractive index of the fiber material is greater than he surrounding. The light then propagates through the fiber until it exists at the other end (Fig. 8.3a). However, the electric field of the light wave is not fully confined within the fiber but rather extend above the surface by up to 1 μm. This is termed the penetration depth d_p of the evanescent wave (Fig. 8.3a). If the fiber is exposed to air, the only optical attenuation observed are that of the material itself. However, if an absorbing medium is put in contact with the fiber, the interaction of the evanescent wave with molecular vibrations results in energy transfer and optical absorption (Fig. 8.3b). If a broadband infrared source is coupled into one end of the fiber, the signal collected on the other end results in an absorption spectrum identical to that of ATR. These absorption spectra are essentially identical to that of a transmission spectrum except for a minor wavelength dependence of the intensity [20]. However, the peak positions are not affected, and conventional spectral analysis can be performed to interpret the spectra.

The intensity of the evanescent wave decays exponentially with distance z above the fiber surface according to Eq. (8.2) where γ is the attenuation coefficient:

$$E = E_o \exp(-z\gamma) \tag{8.2}$$

The penetration depth d_p of the evanescent wave is defined as the distance z at which E has decayed to $1/e$ of its original value. Basic principles of internal reflection spectroscopy indicates that d_p is a function of the refractive index of the glass as well as that of the surrounding [20]. In most FEWS experiments, the surrounding is mainly composed of an aqueous environment and d_p can be estimated from Eq. (8.3):

$$d_p = \frac{\lambda}{2\pi\left(n_2^2 \sin^2\theta_i - n_1^2\right)^{\frac{1}{2}}} \tag{8.3}$$

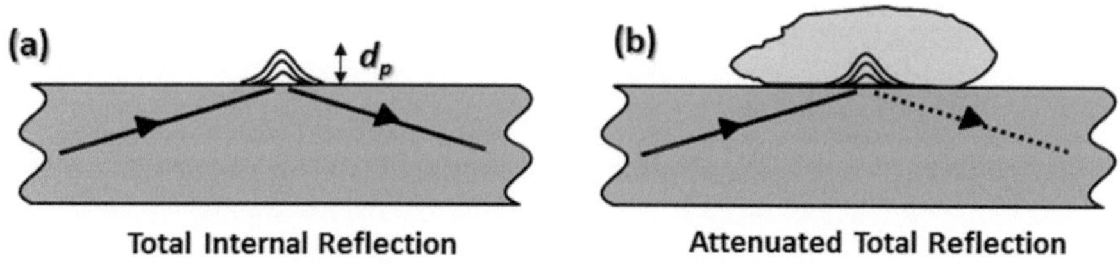

Fig. 8.3 (**a**) Schematics of light propagation through total internal reflection within a waveguide. The electric field of the light extends above the fiber surface to form an evanescent wave with penetration depth d_p. (**b**) Schematics of evanescent wave absorption occurring when an absorbing medium couples with the evanescent wave and results in energy transfer through activation of vibrational modes

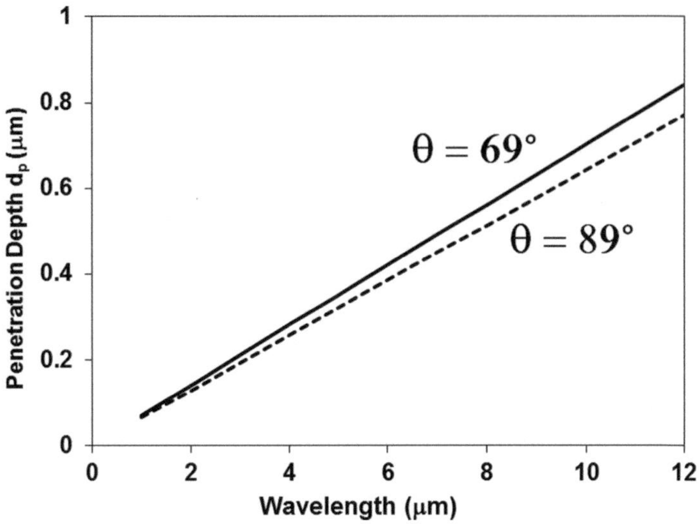

Fig. 8.4 Wavelength dependence of the penetration depth d_p during a typical FEWS experiment using a selenide glass fiber in an aqueous environment. The range depends on the incident light coupling angle

where λ is the light wavelength, θ_i is the angle of incidence on the fiber surface, n_2 is the refractive index of the glass, and n_1 is that of the surrounding. In a standard FEWS experiment, the refractive index of the selenide glasses is $n_2 \sim 2.8$ while that of the surrounding is $n_1 \sim 1.3$. The ratio of refractive index also determines the numerical aperture of the fiber and dictates the maximum coupling angle according to Snell's law [35]. The maximum coupling angle obtained in these conditions is $\theta_c = 21°$ so that the minimum angle of incidence on the fiber surface is $\theta_i = 90° - \theta_c = 69°$. The range of possible incident angle is therefore $90° \leq \theta_i \leq 69°$. The average penetration depth d_p can then be estimated from Eq. (8.3) as a function of wavelength. Figure 8.4 indicates that d_p reaches almost down to 1 μm in the long wavelength range. These results imply that the majority of the signal originate from a thin layer above the fiber and indicate that FEWS is a local probe which requires intimate contact with the target analyte.

8.2.1.4 Sensing Zone

Equation (8.3) clearly indicates that d_p is strongly inversely dependent on the incident angle θ_i. Hence, any strategy that permits to reduce θ_i should exponentially increase the local electric field and provide large enhancement in detection limits. A local reduction in the fiber diameter provides such as strategy. Figure 8.5 qualitatively shows that tapering the fiber results in lower θ_i. This results in increasing penetration depth and consequently great enhancement in the detection signal intensity.

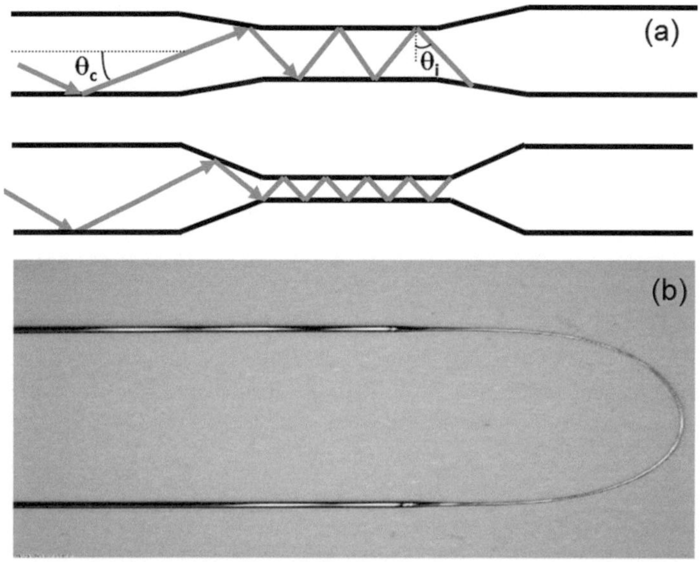

Fig. 8.5 (a) Effect of fiber tapering on the incident angle θ_i and the density of internal reflection event. (b) Example of a tapered selenide fiber with a sensing zone diameter of ~100 μm

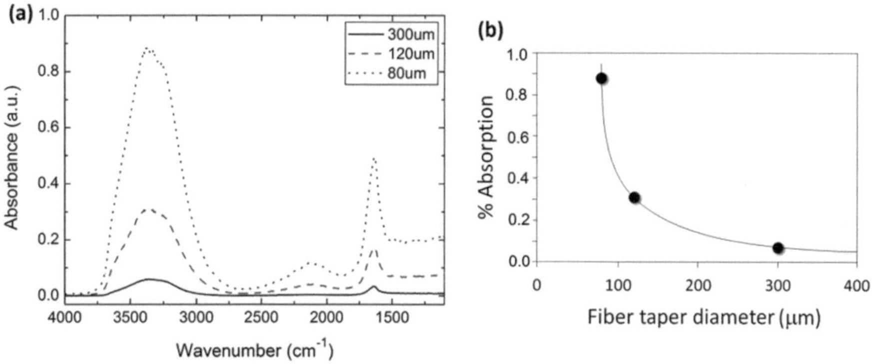

Fig. 8.6 (a) Absorption spectrum of a water sample collected by FEWS using decreasing sensing zone diameters. (b) Evolution of the water peak at 3300 cm^{-1} with decreasing fiber taper diameter

The effect of fiber tapering on the signal intensity is considerable as shown in Fig. 8.6. The detection intensity of a water sample collected with a 10 mm tapered sensing zone is shown to increase exponentially with decreasing fiber diameter as depicted in Fig. 8.6b. A decrease in fiber diameter from 300 to 80 μm results in signal intensity increase of more than one order of magnitude. Fiber tapering is therefore an important requirement for effective FEWS experiments. Selenide fibers lend themselves well to tapering using various methods including chemical, mechanical, or optical [36, 37].

The enhancement in detection signal intensity can also be explained in terms of geometrical optics and total internal reflection. Indeed, Fig. 8.5a clearly indicates that a reduction in diameter leads to an increased spatial density of internal reflection events and in turn leads to enhanced coupling into the target sample. The number of reflections per unit distance L can be computed based on simple geometrical considerations according to Eq. (8.4):

$$N(\theta, d, L) = L \times \frac{\tan(90 - \theta_c)}{d} \tag{8.4}$$

where d is the fiber diameter and θ_c is the coupling angle.

It should also be noted that the sharpness of the drop in diameter at the sensing zone additionally affects the reflection geometry and the intensity of the evanescent wave.

8.2.1.5 Advanced Spectral Analysis

As mentioned earlier, the absorption spectra collected using the FEWS technique are very similar to conventional FTIR spectra collected in transmission mode. Hence, advanced spectral analysis techniques such as PCA and PLS regression are applicable to the interpretation of FEWS spectra. While the infrared spectra of simple molecules are relatively simple and permit highly selective molecular detection based on straightforward analysis of peak positions and intensities, the spectra of biomolecules is extremely complex and requires advanced statistical analysis. For example the spectral differences between certain proteins or strains of bacteria are minute and not recognizable through simple inspection. However, systematic statistical analysis provides an effective means of differentiating and identifying biomolecules with high certainty [31, 38–40].

Most biomolecules contain components with similar functional groups such as amides and amino acids, and hence, their spectral analysis is not based on the identification of specific vibrational modes but instead relies on unsupervised statistical analysis of an ensemble of spectra. This analysis is usually not performed over the whole spectra but rather over a restricted wavelength range which exhibit large inter-sample variance. In the CPA technique, each spectrum is decomposed into a linear combination of eigenvectors weighted by a coefficient called a "score." Multiple spectra can then be compared along two or three dimensional plots where the axis represents the most significant eigenvectors and each spectrum is represented as a point which coordinate correspond to its "score"[22]. Spectra of individual species then tend to cluster in specific area of the CPA map. This enables the identification of any new spectrum. An example of this analysis is illustrated in Fig. 8.7. The serum of eight obese mice and eight normal control mice was extracted and analyzed using FEWS. The absorption spectra of the two mice types are shown in Fig. 8.7a and show minimal differences. However, when PCA is performed on the 1100–1000 cm^{-1} region corresponding to the sugar ring

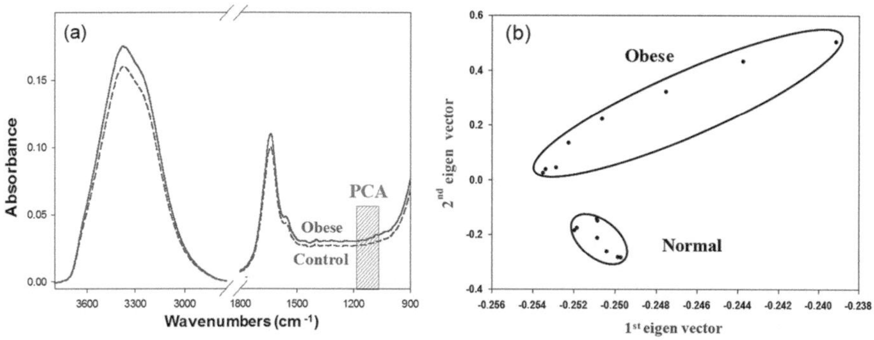

Fig. 8.7 (**a**) Infrared spectra of serum extracted from obese mice and normal control mice. PCA is performed on the spectral band shown near 1100 cm^{-1}. (**b**) PCA map of the 16 mice serum showing normal mice spectra and obese mice spectra cluster in different areas

vibrational modes, the two types of mice spectra can be differentiated. The PCA map of the 16 spectra shown in Fig. 8.7b indicates that the two types of mice can be distinguished from their position on the 2D map. This type of FEWS analysis requires only 10 μl of serum.

Logistic PLS is another highly effective spectral analysis method based on a supervised classification of spectra using a regressor model [40]. As in PCA the PLS method is applied on a restricted spectral region and requires a collection of spectra. A selection of the most discriminant frequencies is performed as part of the regressor model. This method was shown to be more effective that PCA in identifying the bacterial strain involved in the contamination of food products such as meat, cheese, or milk [40, 41]. PLS regression was also successfully used to analyze human serum and identify metabolic dysfunctions associated with cirrhosis, hemochromatosis, and other pathologies [31]. This method is very promising and is currently being developed for commercialization.

The quality of the infrared spectrum is quintessential to the accuracy of the PCA or PLS results, and hence, signal collection must be optimal. For FEWS experiments this requires the use of highly transparent infrared fibers to perform collection of high intensity signal. Chalcogenide glasses and particularly selenide glasses are ideally suited for this purpose.

8.2.2 Chalcogenide Glasses

8.2.2.1 Optical Transmission

Glasses are the materials of choice for the fabrication of optical elements due to their exceptional transparency and easy formability. In contrast, crystalline materials only exhibit adequate transparency in the single crystal form and cannot be

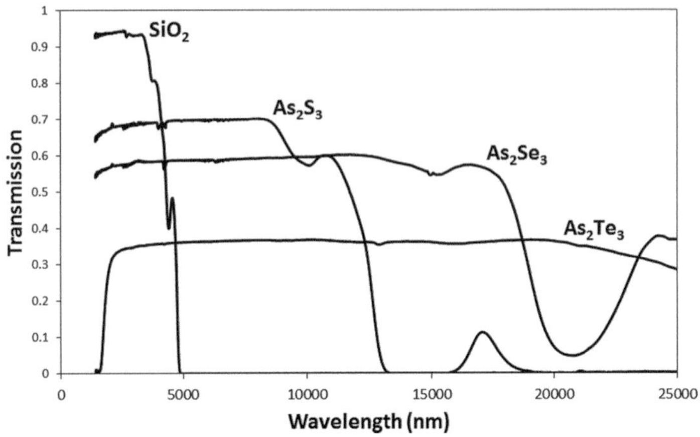

Fig. 8.8 Comparison of the transmission window of silica with that of chalcogenide glasses based on sulfide, selenide, and tellurides. Heavier component elements result in extended transmission in the infrared domain. The transmission was measured on 1–2 mm thick samples. The decreasing maximum in transmission is due to Fresnel reflection

used to produce complex optical elements such as fibers. This is why the vast majority of optical elements are made of glass. However, most conventional glasses are based on oxide materials such as silicates which exhibit excellent transparency in the visible and near-infrared range but are totally opaque in the mid-infrared range (Fig. 8.8). Hence, standard silicate glasses cannot be used for the design of infrared optics. In this case, glasses with heavier components are required to extend the transparency window through the mid-infrared. Chalcogenide glasses are the only family of amorphous material that can satisfy that requirement.

The transparency window of an optical material is limited on the short wavelength side by electronic transition across the band-gap and on the long wavelength side by vibrational transition of the component elements called phonons. The phonon frequency is largely controlled by the reduced mass m^* of the network oscillator which can be approximated for a diatomic oscillator using Eqs. (8.5) and (8.6).

$$E_{\mathrm{n}} = \left(n + \tfrac{1}{2}\right) h\nu \tag{8.5}$$

$$\nu = \frac{1}{2\pi}\sqrt{\frac{k}{m^*}} \tag{8.6}$$

Here, E_{n} is the energy required for a vibrational transition and ν is the vibrational frequency of the oscillator. Equation (8.6) indicates that elements of heavier mass will lead to lower frequency absorptions (longer wavelength) and consequently extend the transparency window further in the infrared. The absorption edge on the long wavelength side is therefore due to a broad density of phonons originating from the amorphous network and it is consequently termed the "multiphonon cutoff." The effect of increasing elemental mass is clearly visible on Fig. 8.8

where the lighter oxide cuts off near 4 μm, while the heavier sulfide cuts off near 10 μm, and the even heavier selenide and tellurides cut off near 18 μm and 25 μm respectively. It should also be noted that while Eq. (8.6) defines the vibrational frequency of the fundamental vibrations, the optical window is actually limited by overtones of these vibrations [42].

Among the different chalcogenide glass families shown in Fig. 8.8, selenide glasses emerge as the best candidate for biochemical sensing. Indeed, sulfide glasses cut off near 10 μm and may not capture important features at long wavelength, in particular in the "signature" region of biomolecules. Telluride glasses have exceptional transmission windows but they are not as good glass formers as selenides and may lead to nucleation and scattering losses during fiber drawing [43]. Selenides on the other hand combine both wide infrared transmission and good glass-formation ability which makes them ideal candidates for the production of infrared optical fibers and for bio-sensing.

8.2.2.2 Fiber Versus Window Transmission

One of the main motivations for using chalcogenide glasses is their unique formability and the ability to produce optical fibers. In that respect it must be considered that light transmission through fibers involves a long optical path which results in strong losses even for weak absorption phenomena such as high order overtones or impurities. Hence, the effective transparency window of a fiber can be significantly reduced in comparison to that of a disc. This effect is illustrated in Fig. 8.9 where the transmission of a 2 mm thick disc is compared to the attenuation of a fiber made of the same $Ge_{20}As_{20}Te_{45}Se_{15}$ glass. The transmission shows a weak overtone due to Se-vibration near 12 μm which only slightly alters the transmission of the disc but results in a strong absorption cut-off in the fiber effectively limiting its optical window. This should be taken into account when designing fibers for the long infrared domain such as application in the second atmospheric window between 8 and 12 μm.

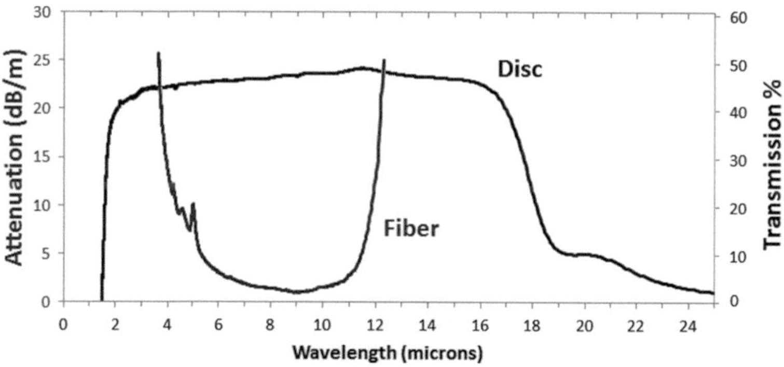

Fig. 8.9 Comparison of the optical transmission measured on a glass window (1 mm thick) and on a fiber made of the same $Ge_{20}As_{20}Te_{45}Se_{15}$ glass

8.2.2.3 Synthesis of Chalcogenide Glasses

Chalcogenide glasses are composed of the three chalcogen elements, S, Se, Te combined with neighboring elements such as Ge, As, Ga, Sb and others. They can form glass over wide compositional range which permits to tune their optical properties for specific application. For FEWS sensing, the selenide-based composition $Te_2As_3Se_5$ has been commonly used due to its outstanding glass stability and its wide infrared window [31, 32, 37, 40].

One of the most important aspects of chalcogenide glass synthesis is to avoid any contact with ambient oxygen or moisture during the preparation. Indeed light impurities such as oxygen and hydrogen produce high frequency absorption within the optical window which rapidly deteriorates the optical quality of the glass. Figure 8.10 illustrates the effect of oxygen and water impurities on the optical window of a $Te_2As_3Se_5$ glass. The purified sample shows a broad and flat optical window extending up to 18 μm while the contaminated sample exhibit strong absorption bands due to Se-H, Se-O and As-O impurity vibrations. Even small amount of impurities lead to large absorption and severely reduce the optical window as shown in Fig. 8.10. This would be particularly detrimental for fibers which would become largely opaque as a result of these impurity absorptions. The synthesis of chalcogenide glasses is therefore performed with extreme care to remove the maximum amount of impurities.

Chalcogenide glass syntheses are performed under high vacuum (10^{-6} Torr) in order to remove any trace of ambient oxygen and water as depicted in Fig. 8.11. First, the synthesis is performed using high purity starting elements ranging from 99.999 to 99.99999 %. However, even high purity elements tend to develop a thin

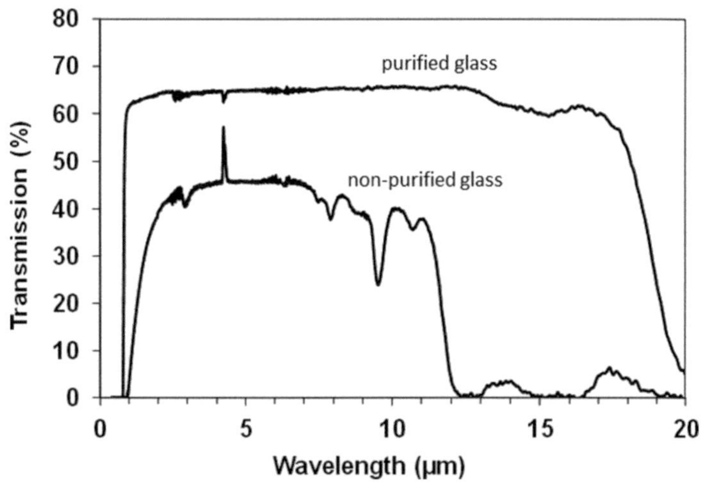

Fig. 8.10 Comparison of the optical window of a high purity $Te_2As_3Se_5$ glass and the same glass contaminated with oxygen (The *line* at 4.2 μm corresponds to CO_2 absorption from the spectrometer chamber)

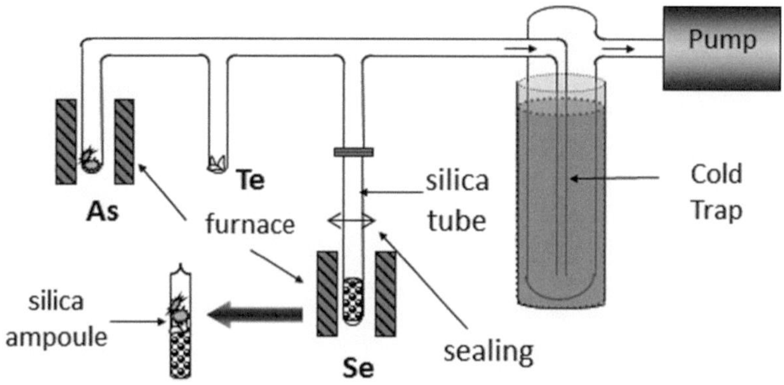

Fig. 8.11 High vacuum setup used for the preparation and purification of chalcogenide glasses

oxide layer which must be removed in situ by evaporation of high vapor pressure oxides. Each element is then heated separately to remove surface oxides which tend to have higher vapor pressures for As and Se and therefore evaporate and condense away from the reaction tube or in the cold trap (Fig. 8.11).

Purified elements are then combined in the main reaction tube by tilting the assembly shown in Fig. 8.11 without breaking the vacuum and the silica ampoule containing all the elements is sealed with a blow torch. The silica tube is then introduced in a rocking furnace and heated up to 800–900 °C for 10–20 h in order to obtain a perfectly homogeneous melt. The ampoule is then quenched in water to produce the glass and annealed near the glass transition temperature to release internal strains and prevent fractures. The resulting glass rod is carefully removed from the silica tube and used as a preform for drawing optical fibers.

This in situ purification process permits to remove the high vapor pressure contaminants such as water and oxide which are most detrimental to the optical window. However, more advanced methods using distillation steps are also used to remove low vapor pressure impurities such as carbon. These distillation steps also permit to use oxygen getters such as $TiCl_4$, Al, or Mg which form refractory oxides that can be removed when the glass is distilled out [34, 44].

These multiple purification steps permit to produce fiber with attenuations as low as 12 dB/km [45].

8.2.2.4 Fiber Production

Chalcogenide fibers are produced from a solid glass preform using a low temperature drawing tower to adapt to the low glass transitions of these materials, typically between 150 and 250 °C. The tip of the preform is heated in a thin ring furnace under a flow of helium to prevent any oxidation. The fiber is then wrapped on the drum for collection as depicted in Fig. 8.12.

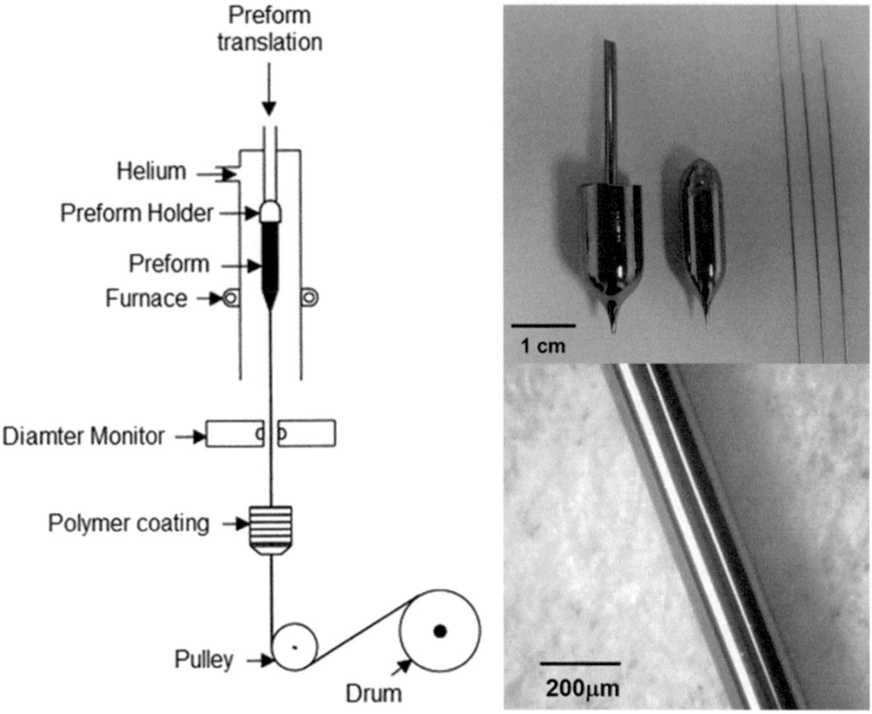

Fig. 8.12 Schematics of an optical fiber drawing tower for the production of chalcogenide glass fibers using single or double index preforms

The fiber diameter can then be controlled by tuning the preform translation speed into the furnace, drum rotation speed, and pulley tension. Typical fiber diameter for chalcogenide fibers is between 200 and 300 μm as shown in Fig. 8.12. The preform diameter is typically 1 cm or less in order to prevent any thermal gradient and avoid nucleation during the drawing process. Both, single index and double index fibers can be produce this way. In order to produce a double index core/clad fiber, a glass rod of lower index is drilled to produce a cladding tube and a narrow glass rod of higher index is inserted in the cladding to produce the core as depicted in Fig. 8.12. The double index preform is then drawn to produce core/clad fibers. Another common approach for drawing core/clad fiber involves the double crucible method [46]. In this method, the core and cladding glasses are melted in individual crucibles connected through a concentric double drawing nozzle. Pressure is applied on each crucible using helium gas and a double layer glass fiber forms at the output of the nozzle and can be drawn by pulling. This method is used commercially to produce single mode selenide fibers.

As mentioned above, optical fibers used for FEWS experiments must be tapered in order to produce a sensing zone and enhance the signal collection efficiency. The tapering process can be performed mechanically during the drawing process by imposing a short excess tensile stress on the fiber. This is the simplest and fastest

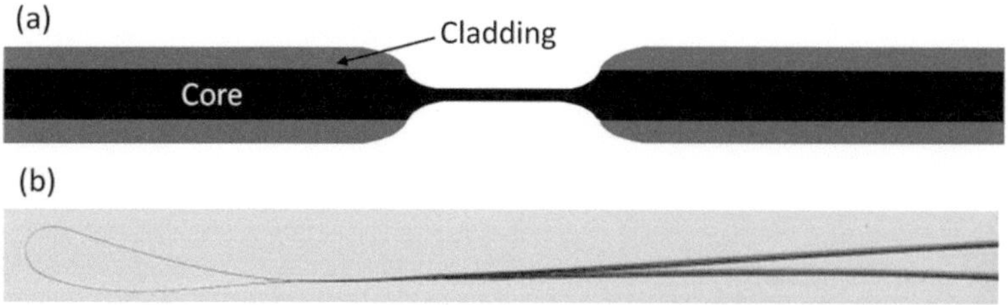

Fig. 8.13 (**a**) Schematics a sensing zone etch chemically from a double index fiber. (**b**) Image of a sensing zone produced on a $Te_2As_3Se_5$ fiber

method; however, it does not provide very good control over the geometry and length of the taper. Another common method involves the use of an etching solution to congruently dissolve the glass along the sensing zone [37]. This method has the advantage of providing good control over the taper length and yields excellent surface quality. But more importantly this method permits to produce sensing zones from core/clad fibers by selectively etching the cladding layer [47] as depicted in Fig. 8.13a.

An alternative method using laser microfabrication has also been demonstrated to produce fine tapers [36]. This approach takes advantage of the photo-fluidity effect exhibited by chalcogenide glass during irradiation with sub-bandgap light [48]. Upon irradiation with a high intensity laser, the fiber reaches a fluid state in the region of illumination and can be tapered mechanically with high accuracy. This process was shown to have no detrimental effect on the quality of the fiber transmission [36].

8.2.2.5 Biocompatibility and Chemical Stability

The presence of hazardous elements such as arsenic in the composition of many chalcogenide glasses raises concerns regarding the toxicity of the glass itself. Indeed soluble arsenate species such as $AsO_4{}^{3-}$ and $AsO_3{}^{3-}$ exhibit high toxicity towards living organisms. However, arsenic atoms present in chalcogenide glasses are covalently bonded in a trivalent configuration which is highly stable against chemical damage or dissolution [33, 49]. As a result, fresh chalcogenide glasses show no evidence of toxicity toward live organisms [2, 50]. Indeed, both rat alveolar macrophage cells and human lung cells were shown to form strong attachment on the surface of $Te_2As_3Se_5$ fibers and showed no evidence of toxic effects [50]. The culture of human lung cell depicted on Fig. 8.14 indicates strong adhesion and formation of an almost confluent cell layer on the fiber surface. The fluorescence assay shown in Fig. 8.14b further confirms that the cells exhibit healthy metabolism. Nevertheless it should also be pointed out that chalcogenide glasses that have sustained significant oxidation develop a surface oxide layer which can subsequently dissolve in aqueous environment and release small amounts of arsenic

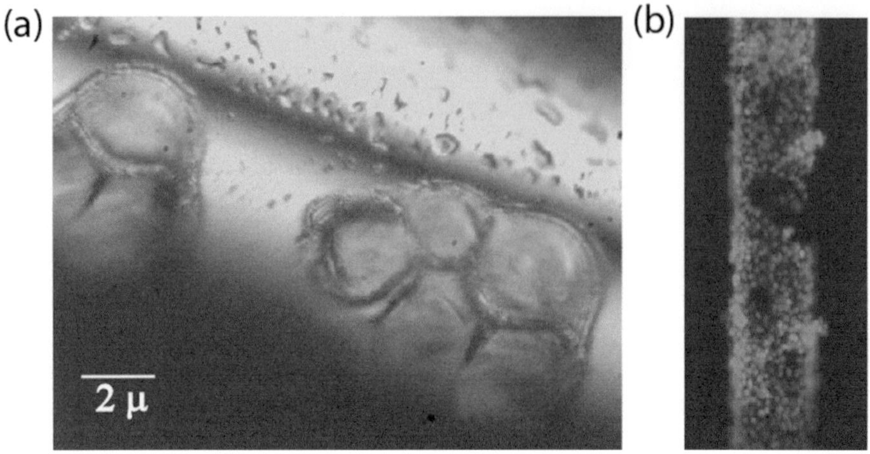

Fig. 8.14 (**a**) Optical micrograph of A549 type II human lung cells forming strong attachment on the surface of a $Te_2As_3Se_5$ fiber. (**b**) Fluorescence micrograph of a confluent layer of human lung cell coated at the surface of a $Te_2As_3Se_5$ fiber. The cells have been dyed with a fluorescence tag which indicates viability

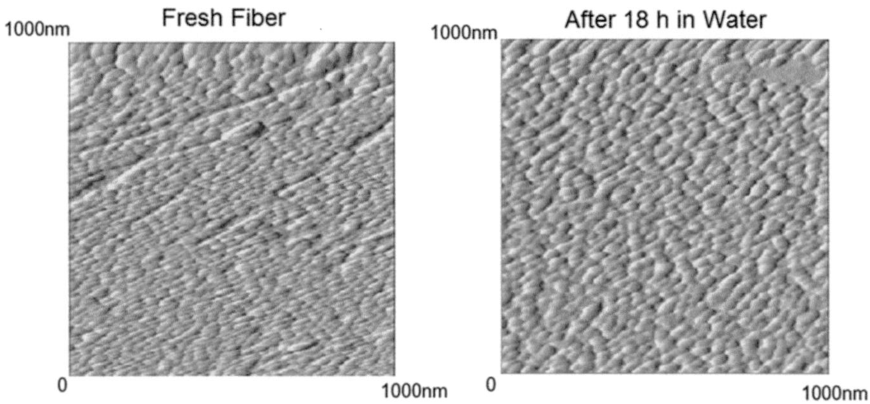

Fig. 8.15 Atomic force microscopy image of a $Te_2As_3Se_5$ fiber surface before and after exposure to water for up to 18 h

[50]. If this is the case care should be taken to rinse the fiber before use in order to remove the oxide layer. But overall, the use of fresh glass fibers for bio-sensing experiment in situ or in vivo does not pose direct issues of toxicity.

The absence of toxicity is the result of the excellent chemical stability of these glasses in particular towards the aqueous and biological environments encountered during typical FEWS bio-sensing measurements. For example $Te_2As_3Se_5$ fibers show outstanding stability against water exposure. The atomic force microscopy images of the fiber surface shown in Fig. 8.15 show no observable damage from water corrosion even after 18 h of immersion. Similar results were obtained after 5 days of immersion in water for other chalcogenide glasses [49].

While selenide glass exhibit good chemical stability in normal aqueous environment at room temperature it should also be noted that significant degradation start to occur at elevated temperature or at high pH values when selenium tends to react with water to form colloidal particles [51].

8.2.2.6 Hydrophobicity

Another advantageous property of chalcogenide fibers for bio-sensing is the hydrophobic character of their surface. Most biological and chemical sensing experiments involve the detection of organic molecules in aqueous environment. Enrichment of target organic molecules in the vicinity of the fiber surface is therefore desirable as it would increase the detection signal. Indeed the evanescent wave strength decreases exponentially above the fiber, and hence, the signal is most intense at the fiber surface. Hydrophobic surfaces have a tendency to repeal water molecules while favoring interactions with less polar organic molecules thereby enriching the surface with target molecules. A comparison of the FEWS and transmission spectra of the same ethanol solution indicate that this process indeed occurs at the surface of chalcogenide fibers (Fig. 8.16a). The main ethanol peak at 1050 cm^{-1} has a significantly higher intensity that the main water peak at 1650 cm^{-1} when collected by FEWS while the opposite occurs in transmission as shown in Fig. 8.16a. This effect is not unique to ethanol but is shown to be systematically observed for all organic molecules with an increasing effect for nonpolar molecules [52]. In fact a linear relationship is observed between the dielectric constant of the target molecule and its spectral enhancement relative to water when collected by FEWS as depicted in Fig. 8.16b. This correlation clearly establishes that the enhancement is associated with the hydrophobic nature of selenide fibers. Indeed contact angle measurements

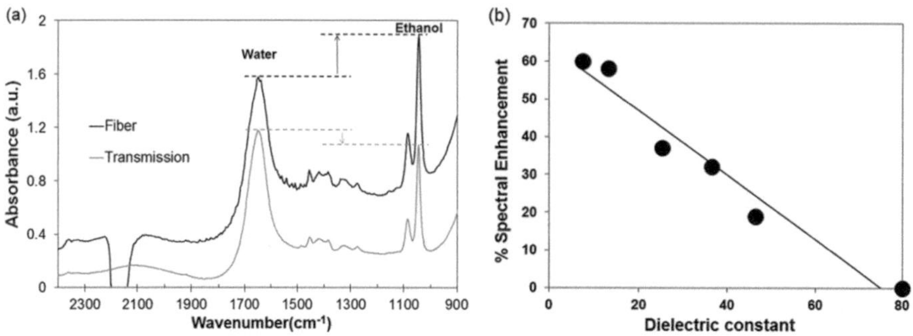

Fig. 8.16 (**a**) Comparison of the relative intensity of the main absorption peaks of water and ethanol collected in transmission mode and by FEWS. (**b**) Linear correlation between the dielectric constant of an organic target molecule and its spectral enhancement relative to water when collected by FEWS

of various organic solvents performed on chalcogenide glass discs of the same composition as the fiber indicate strong interactions (low contact angle) for nonpolar solvent and repulsion (high contact angle) for polar solvents [52].

8.2.2.7 Mechanical Properties

One of the main weaknesses of chalcogenide glasses is admittedly their moderate mechanical resistance relative to the conventional oxide glasses [53]. Chalcogenide glasses have a Young's modulus near ~20 GPa roughly four times lower than silica (~80 GPa) [54]. Similarly the Vickers Hardness of selenide glasses is near ~1.5 GPa, roughly six times lower than silica (8.9 GPa) [53, 55]. This moderate mechanical resistance is reflected in the properties of the fibers as well. For example, $Te_2As_3Se_5$ fibers exhibit modest tensile strength near ~110 MPa which further decrease with aging due to relaxation processes as well as the formation of surface defects [56, 57]. Fortunately, these mechanical properties can be greatly improved with the use of polymer coatings [58, 59].

Bare selenide fibers also exhibit moderate bending strength especially if the diameter exceed a few hundred micrometers [56]. However, the bending flexibility is greatly enhanced by tapering as shown in Fig. 8.17b. This is important for in situ sensing application where direct contact has to be applied locally. The robustness of the sensing zone can also be improve by designing looped sensing zones as depicted in Fig. 8.17a. This looped geometry yields notable improvement in robustness which provide much ease of handling for practical and commercial applications [60].

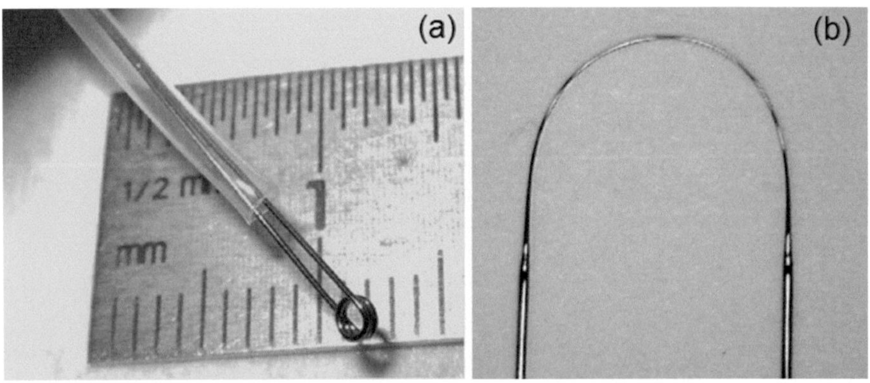

Fig. 8.17 (**a**) Looped fiber sensor made from $Te_2SAs_3Se_5$ glass. (**b**) Selenide fiber tapered down to 100 μm showing a bending radius of less than 1 cm

8.3 Application to Chemical and Biomedical Sensing

8.3.1 Food Safety

The regular occurrence of bacterial contamination in food supplies is a growing concern as it has recently lead to multiple deaths in the USA and Europe. A rapid and noninvasive means of detecting the presence of bacterial contamination in food products is therefore highly desirable for monitoring the safety of our food supply at a reasonable cost. However, bacterial pathogens are made of the same major constituents including amides, proteins, nucleic acids, polysaccharides, and phospholipids which complicates a highly selective identification between different strains based on simple chemical analysis. Biological assays involving sampling and growth are highly selective, but they would be too slow and costly for rapid screening. Nevertheless it was recently demonstrated that FEWS analysis of dairy and meat-based product could lead to selective identification of bacterial contamination within minutes [40]. The spectral information was collected through direct contact of a 1 cm long sensing zone with the food-product followed by PCA analysis. The results shown in Fig. 8.18 for the two meat products (sausages and minced meat) provide a means to differentiate between the harmless endogenous flora and dangerous pathogens such as Listeria, Salmonella, and Staphylococcus. Each spectrum was collected on a food sample incubated for 18 h with a pathogen or containing only the endogenous flora.

An identical experiment performed on dairy products (fresh milk, raw cheese) shows even better selectivity as depicted on Fig. 8.19. The series of spectra corresponding to sample contaminated with each pathogen are clustered in a specific region of the 3D PCA map, thereby allowing selective identification. The improved selectivity of the method with dairy product as compared to meat products is likely the result of the improved contact with the fiber probe in the case of the dairy samples.

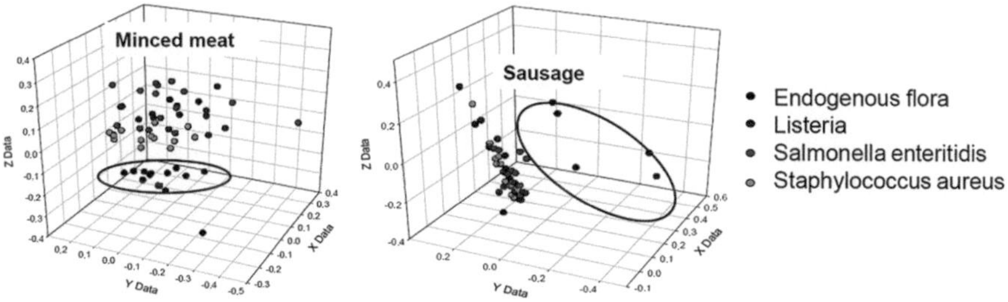

Fig. 8.18 Three-dimensional PCA map of FEWS spectra collected from meat-based product contaminated with pathogens including Listeria, Salmonella, and Staphylococcus as well as the same sample containing only endogenous flora. The points corresponding to the spectra of endogenous flora are *circled*

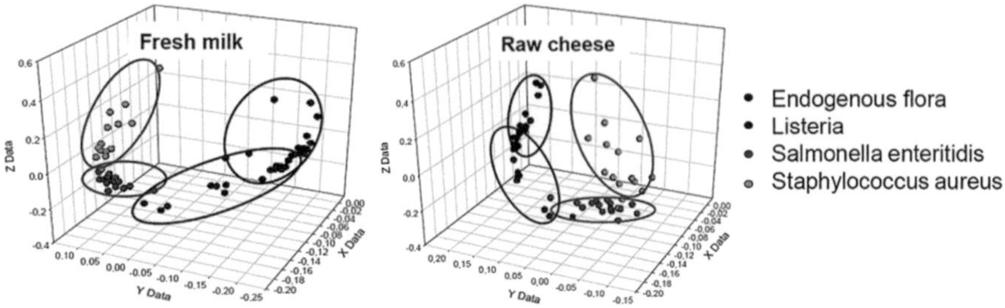

Fig. 8.19 Three-dimensional PCA map of FEWS spectra collected from dairy product contaminated with pathogens including Listeria, Salmonella, and Staphylococcus as well as the same sample containing only endogenous flora

For all the above samples, the PCA was performed on the 1000–1800 cm^{-1} and 2800–3000 cm^{-1} spectral regions which provided the greatest variance between dissimilar samples. The 2800–3000 cm^{-1} spectral region correspond to lipid hydrocarbon vibration while the 1000–1800 cm^{-1} is the rich "fingerprint" region of biomolecules.

But while PCA provides an effective means of visually assessing the differentiation between pathogens, it does not provide a definitive determination of the pathogen identity. In contrast, it can be shown that supervised classification method such as logistic PLS regression provide such a determination with high success rate [40]. The PLS analysis was performed on the same set of spectra as that used for the PCA above. The lowest error level is achieved for a selection of about 100 regressors (absorptions) dispersed over the same spectral regions as the PCA. These regressors are selected statistically based on their ability to discriminate between samples, but do not offer a biochemical interpretation to the origin of the differences between the samples. Nevertheless, the confusion matrix shown in Table 8.1 indicate that the PLS method is highly discriminative and classify the sample correctly in most cases. Overall, the PLS method permits to identify pathogen strains with a classification error lower than 3.5 % [40].

The combination of FEWS and statistical spectral analysis therefore provide an effective means of performing routine screening of food supply to selectively identify the presence of dangerous pathogen while avoiding false-positive from endogenous microorganisms. The method is rapid, noninvasive, and relatively low-cost compared to standard biological assays. Nevertheless larger scale studies are still necessary to validate these results and extend them over a wider range of pathogens. This would be of much interest as these types of sensing strategies can directly apply to the field of health care and can be easily implemented to biomedical sensing as shown in the next section.

Table 8.1 Confusion matrix for minced mint, raw cheese, and sausage obtained by logistic PLS regression using FEWS spectra

From	To			
	End	Lis	Sal	Sta
Minced meat				
End	15	0	0	0
Lis	0	15	0	0
Sal	0	0	15	0
Sta	0	0	0	15
Raw cheese				
End	14	1	0	0
Lis	0	15	0	0
Sal	0	0	15	0
Sta	0	0	0	15
Sausage				
End	13	1	0	1
Lis	0	13	1	1
Sal	0	1	13	1
Sta	1	1	0	13

8.3.2 Biomedical Diagnostics

8.3.2.1 Metabolic Profiling in Human

Early diagnostic of life threatening diseases is a growing imperative in the health care industry as it permits to potentially tackle a dangerous condition before it is fully developed. However, large-scale screening of patients using standard biological and imaging techniques can be prohibitively expensive. Novel diagnostic methods allowing rapid screening of specific diseases at low cost are therefore in demand. Clinical studies have shown that FEWS analysis of human serum is a very promising technique for early diagnostics [31, 38]. Only 10 μl of serum is necessary for a measurement and collection is completed in 30 s, thereby allowing fast screening. The technique is combined with statistical spectral analysis such as PCA or PLS to provide high specificity on the identification of pathology. Preliminary clinical studies of hepatic pathologies performed on groups of patient with metabolic abnormalities versus normal volunteers have shown that PLS analysis of sera spectra permits to identify the presence of cirrhosis and dysmetabolic hepatosiderosis with a specificity of 84 % and 87 % respectively [31]. Similar studies were also performed to discriminate between patients with and without cirrhosis using PCA [38]. The resulting three-dimensional PCA map shown in Fig. 8.20 show that it is possible to discriminate sera of healthy patients from those exhibiting metabolic disorder associated with cirrhosis [38]. The PCA is performed on a small spectral region between 1257 and 1308 cm^{-1} which offers the highest variance between samples.

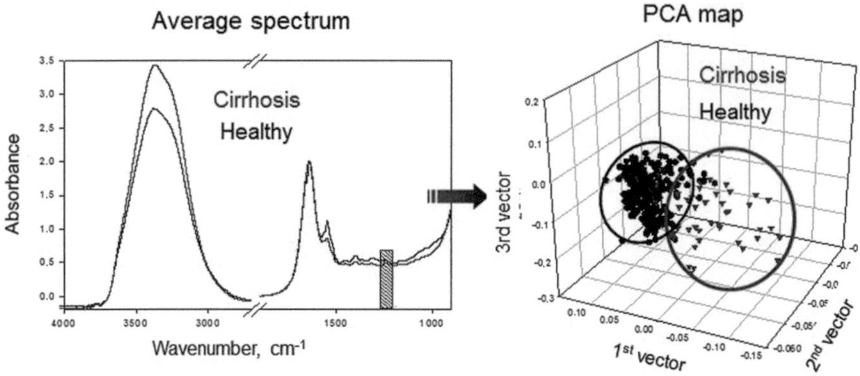

Fig. 8.20 PCA analysis of human serum from patients exhibiting metabolic disorder associated with cirrhosis compared to that of a healthy control group PCA is performed over the 1257–1308 cm^{-1} spectral region

8.3.2.2 Metabolic Profiling in Animals

Pharmaceutical studies and biological studies are widely performed on laboratory animals before transitioning to human trials. Hence, the ability to monitor metabolic processes in animals with fast and low cost techniques could fasten and reduce the cost of pharmaceutical research. This is another area where FEWS has shown some promising potentials [29, 61, 62]. FEWS analysis with and without statistical analysis can be performed on tissues as well as serum in order to monitor metabolic processes in laboratory animals such as mice.

Slices of tissues obtained by cryogenic section can be directly applied to the FEWS sensing zone for spectral collection as depicted in Fig. 8.21.

This method has been successfully used for the analysis of mice liver tissues [29, 62]. Studies on starved and normal mice showed increase in lipid and triglyceride bands in the tissues of starved mice. The accumulation of triglycerides in the liver tissues was indeed confirmed by histological studies performed by staining [62]. Additional studies on tumoral and non-tumoral liver tissues indicated similar accumulation of triglyceride in tumoral hepatocytes as indicated by a notable increase in the C=O vibration at 1746 cm^{-1} which is characteristic of triglycerides [29]. The spectral difference between the tumoral and non-tumoral tissues is illustrated in Fig. 8.22. The spectral changes are sufficiently intense and manifest that no statistical spectral analysis is required in this case.

Performing FEWS directly on tissue requires heavy procedures such as biopsy or cryogenic sectioning with often results in the animal's death. This is not optimal for long-term studies requiring continuous metabolic monitoring. Instead metabolic profiling can be effectively performed using small quantities (10 μl) of serum [61]. In this case the analysis of sera requires PCA treatment of the spectra to interpret the results with good selectivity. This method was tested to identify obese mice prone to hyperphagia and type II diabetes in comparison to healthy mice.

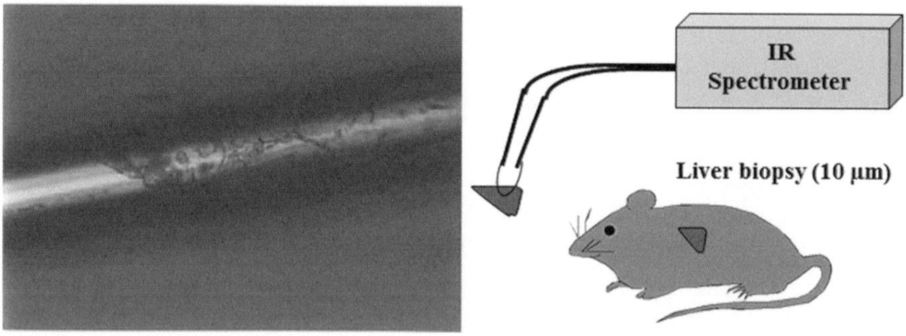

Fig. 8.21 FEWS experiments performed on animal tissues. Slices of tissues obtained from cryogenic section following biopsy are directly applied on the infrared fiber sensing zone

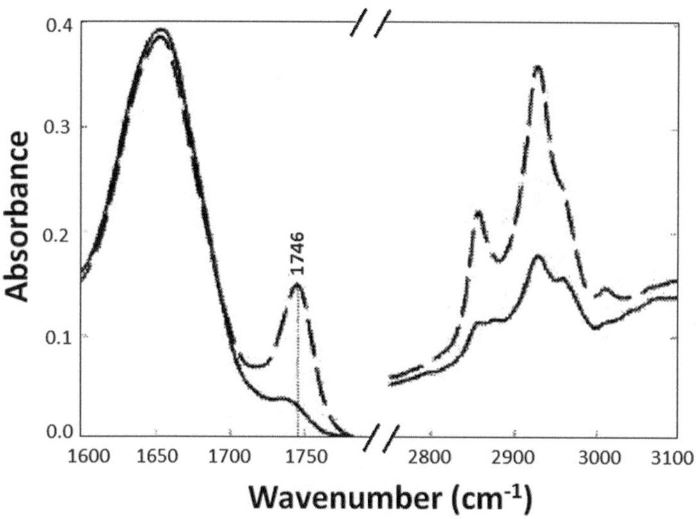

Fig. 8.22 FEWS spectra of tumoral and non-tumoral liver tissues. A strong intensity change at 1746 cm^{-1} is indicative of triglyceride accumulation in tumoral hepatocytes

The two-dimensional PCA map previously shown in Fig. 8.7 indeed indicate that the spectra of diseased mice can be discriminated using this method. Small quantities of serum can then be sampled at regular interval to monitor the metabolism of laboratory animal in a noninvasive way.

8.3.2.3 Metabolic Monitoring of Microorganisms

Biological studies of fundamental cellular processes often involve culturing cell lines and/or microorganisms to investigate the mechanism of metabolic changes.

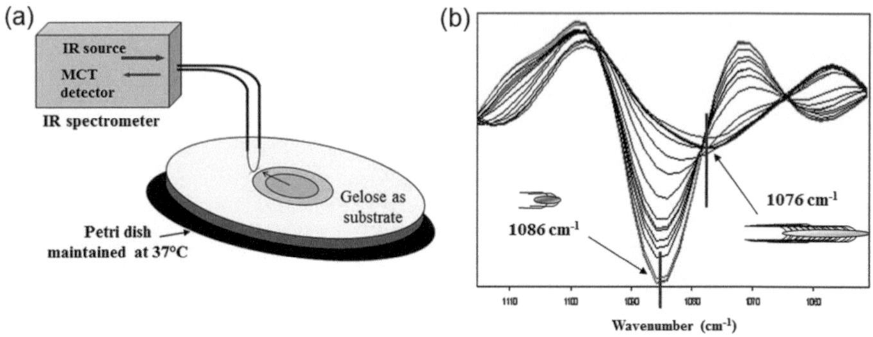

Fig. 8.23 (**a**) Experimental setup for monitoring the dynamic of bacterial biofilms using FEWS. (**b**) Time evolution of the second derivative spectra in the range of polysaccharide vibrations. The band at 1086 cm^{-1} corresponds to vegetative cells while the band at 1076 cm^{-1} corresponds to swarming cells

The ability to monitor these metabolic changes continuously over time is often key to elucidate their origin and to understand their mechanism. Hence, sensing methods that permits extended monitoring of biological environments are highly desirable. Collecting infrared spectra using FEWS only takes seconds, and hence, quasi-real time monitoring of biological processes can be performed to continuously probe metabolic changes in microorganisms [2, 30, 61].

An example of continuous monitoring of metabolic process in bacterial cultures is depicted in Fig. 8.23. A fiber put in contact with a Petri dish colonized by Proteus mirabilis bacteria is used to monitor the dynamics of the biofilm [30, 62]. The center of the dish was initially spotted with a bacterial culture (inoculum) and the dish was maintained at 37 °C while FEWS spectra were continuously recorded for 5 h. Bacteria are known to colonize large area in a matter of hours through the formation of biofilms. The front of the bacterial film is composed of swarmer cells that secrete exopolysaccharides as lubricant. The inner area of the biofilm is instead composed of dormant cells of the vegetative phenotype. Spectral monitoring indeed shows that the initial signal is only composed of the water signal from the gelose but as the bacterial front reaches the fiber, the signal first convert to that of the swarmer cells and finally to that of the dormant cells. The second derivative spectra of the polysaccharide region shown in Fig. 8.23b indicate a distinct spectral change with time which reflects the bacterial cell change. These results indicate that the infrared signature collected by FEWS permits to identify changes in bacterium phenotype. The changes shown here are clearly established from the second derivative spectra but more advanced statistical analysis can also be used to provide even further selectivity [30].

Another example of dynamic FEWS measurements involves monitoring the metabolic response of live cells to different toxins [2, 63]. In this case, cultures of human lung cells are directly deposited on the sensing zone of the optical fiber as depicted on Fig. 8.14. The spectra collected this way contains the rich "signature" region of biomolecules where the vibrational mode associated with each cell

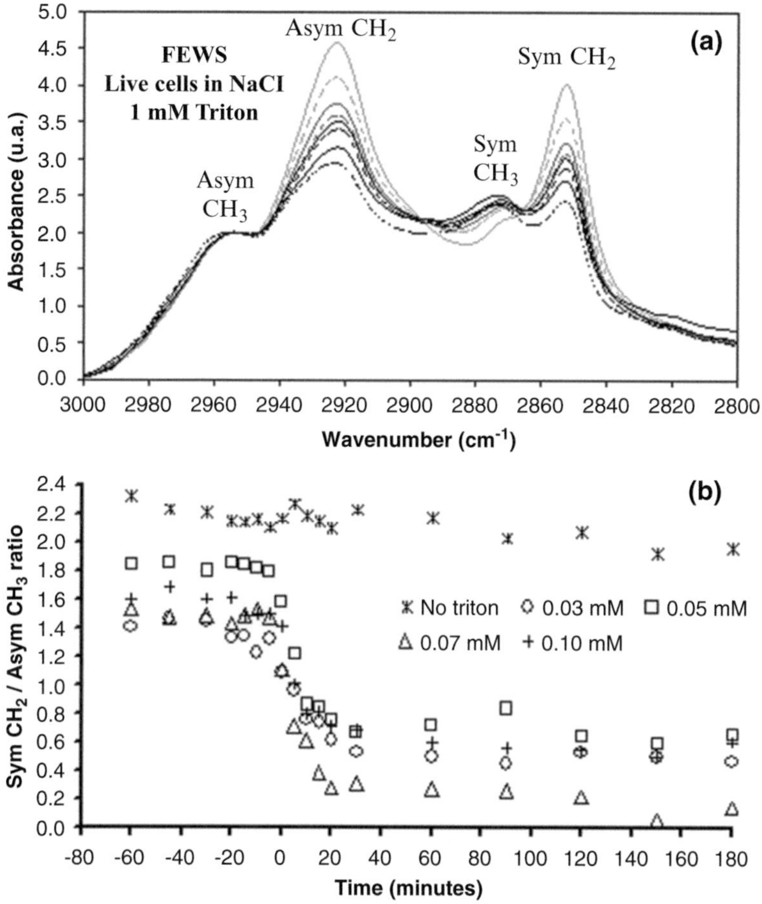

Fig. 8.24 (**a**) Spectral response collected by FEWS for live human lung cells exposed to 1 mM of surfactant Triton X-100. The spectral range corresponds to the hydrocarbon vibration of the membrane phospholipid bilayer. (**b**) Time dependence of spectral changes upon addition of surfactant to the cell environment. The Triton X-100 was added at time $t = 0$ and the cell were monitored before and after

components can be identified as depicted in Fig. 8.1. The response of the live cells to various toxicants can therefore be detected through changes to specific spectral features. This permits not only to detect the presence of a toxicant but also to learn about the pathways used by the toxicants to induce metabolic changes in the cell. For example the response of human lung cells to a surfactant (Triton X-100) is depicted in Fig. 8.24. The FEWS spectra permit to detect a rapid change in the membrane phospholipid signal between 2800 and 3000 cm^{-1} (Fig. 8.24a). The response is complete within 20 min for exposure concentration ranging from 1 mM to 30 μM. In contrast the signal remains stable if no Triton is added. This change in the phospholipid bilayer signal is consistent with the mode of metabolic disruption expected from a surfactant. Indeed Triton X-100 is a model surfactant that causes rapid cell death by dissolving the cell membrane. The cell response to only μM concentrations is clearly detected. This level of concentration is too low for direct FTIR detection, and hence, cell cultures can serve as a sensitizer for the detection of biologically active substances even in minute concentrations.

The results presented above demonstrate that the FEWS technique combine a unique set of attributes for bio-sensing, including high selectivity, rapidity, and in situ capability. The probing mechanism is also noninvasive which makes this method ideal for long term monitoring of metabolic processes.

8.3.2.4 Monitoring of Industrial Processes

Many industrial processes involve slow chemical reactions which can be affected by temperature, humidity and other environmental factors. Yet ensuring that the reaction is complete is a crucial requirement to guarantee consistency and quality of the final product. This applies to many industrial fields ranging from food processing to materials production.

For example, fermentation of apple juice to produce cider is a complex process that involves multiple types of sugars such as fructose, saccharose, and glucose. The kinetic of reaction that produces alcohol may vary between sugar types and the proportion of each sugar varies between apple types. Hence, an in situ method for monitoring the fermentation process over time is desirable to control the level of alcohol in the fermentation bins. As mentioned previously, infrared fiber cables based on $Te_2As_3Se_5$ are bio-compatible and highly resistant against degradation in aqueous environments, and hence, they are well suited for safe monitoring of food products. Indeed, it was shown that the sugar to alcohol content in fermented apple juice could be accurately monitored over time using FEWS analysis [23]. The sugar to alcohol ratio predicted by FEWS was corroborated using a gravimetric methods and shown to be in close agreement (Fig. 8.25b). The FEWS method provides the added advantage of allowing the selective identification of each sugar type as shown in Fig. 8.25a.

Monitoring of reaction media at room temperature can be effectively achieved with $Te_2Se_3As_5$ fibers; however, their glass transition temperature is only ~132 °C [64], and hence, they are not adequate for monitoring industrial processes at higher temperatures. Nevertheless, chalcogenide glasses can be produced over a wide range of compositions that allows for tuning their physical properties including their glass transition temperature. For example it was shown that the curing of epoxy resin could be monitored at temperatures up to 200 °C using $Ga_5Ge_{25}Sb_{10}Se_{65}$ glass fibers exhibiting a glass transition temperature $T_g = 283$ °C [65]. In this experiment, the fiber was introduced directly into a hot mold and then covered with the epoxy resin. FEWS signal were then collected every 5 min and analyzed by PCA. The results shown in Fig. 8.26 indicate that the isothermal curing process can be continuously monitored over time. The end of the curing process can

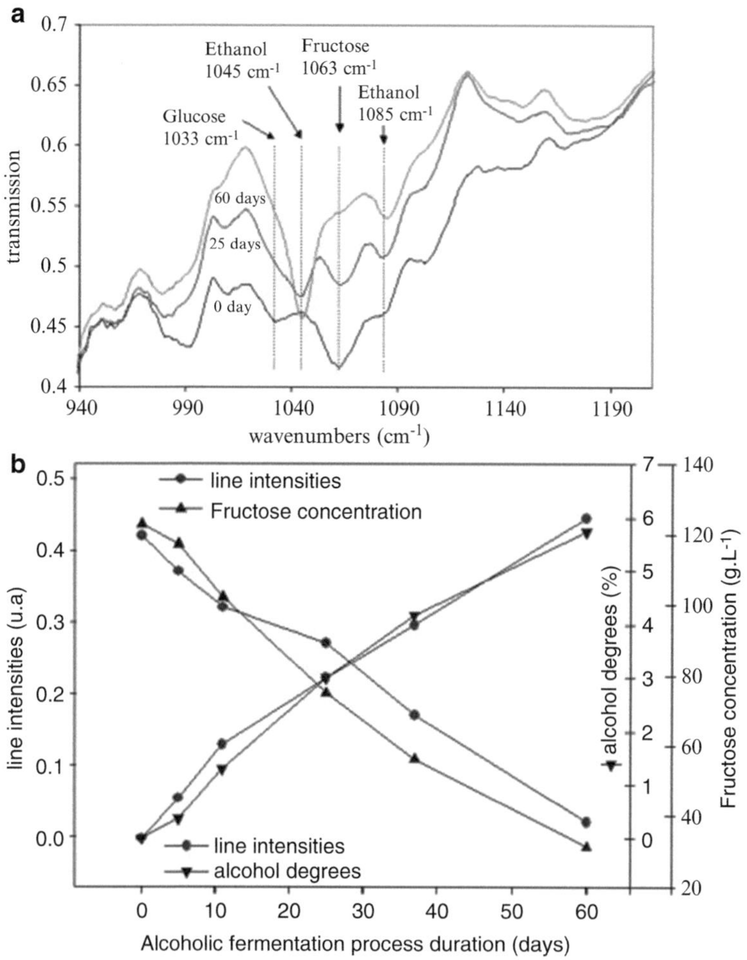

Fig. 8.25 (**a**) FEWS spectra of apple juice throughout the fermentation process. (**b**) Correlation between the sugar–alcohol ratio predicted by FEWS analysis and gravimetric measurements as a function of fermentation time

be clearly identified as the time at which the points on the PCA map remain at the same position. The PCA map was based on the spectral range between 970 and 1130 cm^{-1} which correspond to the range of vibrations for C–O bonds created during the curing process.

This example illustrates the potential for designing various selenide glass compositions with properties tailored for the specific conditions of a FEWS experiment. Further compositional engineering permits to raise the T_g even higher but a good balance must be found between thermal resistance and the optimal glass stability which enables the fabrication of high quality optical fibers. Overall, selenide fibers show reasonable resistance to reactive environments and it was also shown that they

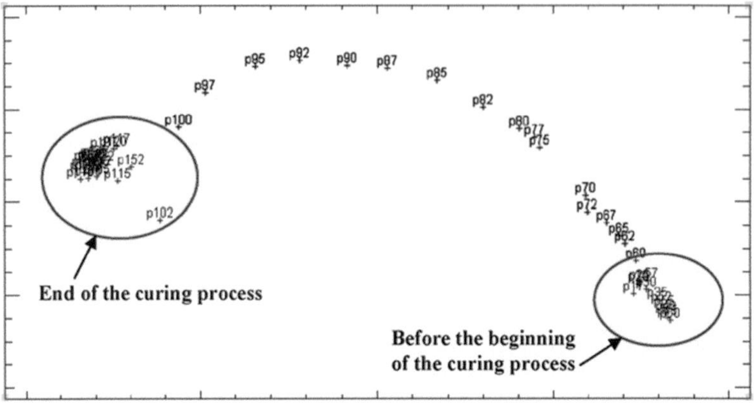

Fig. 8.26 Two-dimensional PCA map representing the spectral evolution of epoxy resin during curing at 180 °C for 150 min. The PCA was performed over the 970–1130 cm^{-1} spectral range

could be used to monitor microwave assisted chemical reactions without loss of optical quality [24].

8.3.2.5 Environmental Monitoring

With freshwater supplies quickly dwindling across the world, monitoring the quality and safety of drinking water reserves is becoming increasingly relevant. Organic pollutants released by various industries can easily find their way into aquifers and contaminate our drinking water supplies, and hence, sensing methods that permit to continuously monitor the presence of organic compounds in water are of much interest. Typically, chemical analysis of water supplies is performed by chromatography which is time consuming and cannot be performed continuously. In contrast, FEWS permits a quasi-real time detection of a wide range of organic pollutant at minimal cost. Indeed field tests in natural aquifer have shown that FEWS is an effective method to detect organic pollutants such as chlorinated hydrocarbons down to 1 ppm concentrations [66]. The field test involved introducing dichloroethylene and tetrachloroethylene in the aquifer inlet and testing the water quality in real time using a FEWS probe through a monitoring well. Periodic water sampling was also performed to quantify pollutant concentration by chromatography for comparison with the FEWS results. The results shown in Fig. 8.27 indicate an excellent correlation between the FEWS and chromatography results.

This experiment demonstrates the ability of FEWS to continuously probe water supplies for more than 10 h and detect ppm levels of pollutant. This constitute a serious advantage over discrete sampling techniques currently use for chemical testing of water supplies.

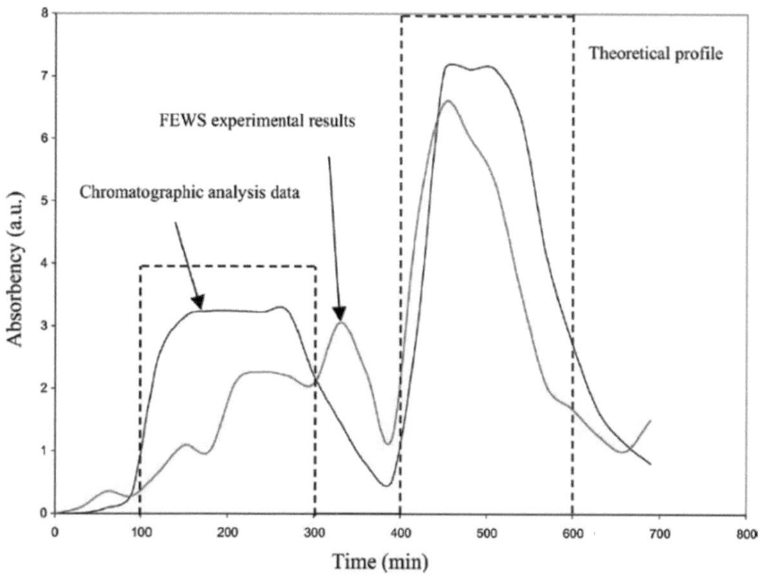

Fig. 8.27 Time evolution of the dichloromethylene concentration detected by FEWS and chromatography during field tests in a natural aquifer

8.4 Prospects

The field of infrared glass optics has been rapidly growing in the last decade, in particular with the development of molded lenses for the automotive industry [67, 68]. Similarly the field of infrared fibers has been rapidly developing and commercial source of chalcogenide fibers are now available around the world including Canada (Coractive), USA (IRFlex), UK (Fiber Photonics), Germany (Art Photonics), and China. While the defense industry is a major driver of the development of infrared fibers, particularly for the delivery of high power laser used in missile countermeasure applications, biomedical sensing is a growing application with large potentials. Fiber based biomedical sensors are likely to become one of the major consumer of infrared fibers in quantity since the fiber probes would be used as consumables. For example, selenide fiber probes are currently being commercialized for the early diagnostic of multiple pathology such as cirrhosis, leukemia, breast cancer etc. using noninvasive serum analysis [69]. Only a few microliters of serum is necessary for the diagnostic as shown in Fig. 8.28. This is a promising technique as it may avoid invasive procedures such as biopsies.

Selenide glasses are also increasingly used for the development of multi-material fibers involving a combination of chalcogenide glass and high T_g polymers [58, 59, 70–72]. An appropriate selection of inorganic and organic materials with similar thermal properties permits to co-draw complex composite fiber structures. The polymer structure can be designed to simply provide superior mechanical

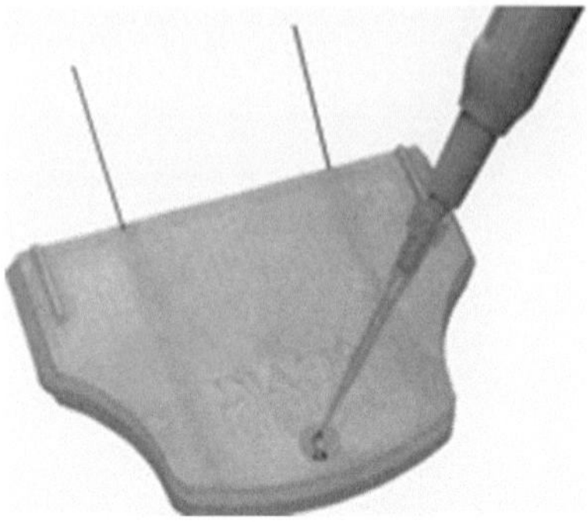

Fig. 8.28 Commercial infrared fiber probe based on selenide glass for the early diagnostic of pathologies using noninvasive serum analysis

resistance to the fiber cable in comparison to conventional fibers [58, 59]. But more complex structure based on alternating submicrometer-thick concentric layers of glass and polymer permit to produce hollow-core fibers where infrared light is guided through Bragg reflections [70]. These types of hollow fibers have the advantage of propagating light in air and consequently minimize absorptions or nonlinear effects that become limiting factors at high power density. In particular these fibers can deliver high power CO_2 laser light at 10.6 μm which is of much interest for laser surgery. Indeed these fibers are now broadly commercialized by OmniGuide and used in a variety of surgical tools [73]. Moreover, the multilayer structure enables the integration of additional functional elements such as thermal sensors for self-monitored optical transport [71, 72]. The field of multifunctional fibers is rapidly advancing with the development of additional functions for applications in chemical sensing and biotechnology [74, 75].

Finally selenide glasses have also been largely used for the production of integrated optics and sensors [76–81]. High quality waveguides and resonating structures can be produced by lithography [80] or by embossing [82] to produce complex circuits for sensing or telecommunication applications [76]. Such integrated circuits have several advantages including their small scale, robustness and large scale production through standard microfabrication techniques. More recently thin chalcogenide glass photonic circuits have been patterned over flexible plastic substrates [83]. The circuits produced this way retain excellent optical performance while exhibiting high flexibility with bending radii below one millimeter. These devices have potential for direct in situ bio-sensing applications such as skin patches and other wearable sensors.

8.5 Conclusion

For many years, the field of infrared optics has been largely dominated by crystalline solids such as Ge, ZnSe, and CaF_2. However, the production of optical elements from these materials requires the synthesis of single crystals followed by expensive forming processes such as diamond turning. Moreover, they do not permit the production of complex optical elements such as fiber waveguides. In contrast, chalcogenide glasses enable the low-cost molding of complex optics such as diffractive lenses and the production of fiber waveguides of high optical quality. Selenide glasses are particularly well suited for these applications due to their wide infrared transparency and superior glass-forming ability. Optical fibers drawn from selenide glasses were shown to be ideally suited for FEWS sensing and their use was demonstrated for a variety of applications ranging from medical diagnostics, pollutant detection, food safety, and monitoring of industrial processes. The field of infrared optics has recently undergone a broad resurgence following the development of new infrared technologies including high power sources such quantum cascade laser and low cost detectors such as bolometers. As a consequence, advanced infrared optics are in increasing demand and the commercial production of infrared fibers is rapidly growing across the world. Yet the first commercial device based on selenide fibers is just now reaching the market, and hence, there is still much potential for new technological development in the field of chalcogenide glass optics.

References

1. D.A.C. Compton, S.L. Hill, N.A. Wright, M.A. Druy, J. Piche, W.A. Stevenson, D.W. Vidrine, In situ FT-IR analysis of a composite curing reaction using a mid-infrared transmitting optical fiber. Appl. Spectrosc. **42**, 972 (1988)
2. P. Lucas, D. LeCoq, C. Juncker, J. Collier, D.E. Boesewetter, C. Boussard-Pledel, B. Bureau, M.R. Riley, Evaluation of toxic agent effects on lung cells by fiber evanescent wave spectroscopy. Appl. Spectrosc. **59**, 1–9 (2005)
3. D. Naumann, Microbiological characterizations by FT-IR spectroscopy. Nature **351**, 81 (1991)
4. M. Diem, S. Boydston-White, L. Chiriboga, Infrared spectroscopy of cells and tissues: shinning light onto a novel subject. Appl. Spectrosc. **53**, 148A (1999)
5. D. Naumann, Infrared spectroscopy in microbiology, in *Encyclopedia of Analytical Chemistry*, ed. by R.A. Meyers (John Wiley & Sons Ltd, Chichester, 2000), p. 102
6. M. Diem, K. Papamarkakis, J. Schubert, B. Bird, M.J. Romeo, M. Miljkovic, The infrared spectral signatures of disease: extracting the distinguishing spectral features between normal and diseased states. Appl. Spectrosc. **63**, 307A–318A (2009)
7. M. Diem, N. Laver, K. Bedrossian, J. Schubert, K. Papamarkakis, B. Bird, M. Miljkovic, Detection of viral infection in epithelial cells by infrared spectral cytopathology, in *Handbook of Biophotonics*, ed. by J. Popp (Wiley-VCH Verlag GmbH & Co. KGaA, Weinheim, 2011), pp. 251–258
8. M. Karlowatz, M. Kraft, B. Mizaikoff, Simultaneous quantitative determination of benzene, toluene, and xylenes in water using mid-infrared evanescent field. Anal. Chem. **76**, 2643 (2004)
9. R. Krska, E. Rosenber, K. Taga, R. Kellner, A. Messica, A. Katzir, Polymer coated silver halide infrared fibers as sensing devices for chlorinated hydrocarbons in water. Appl. Phys. Lett. **61**, 1778 (1992)
10. M. Diem, L. Chiriboga, P. Lasch, A. Pacifico, IR spectra and IR spectral maps of individual normal and cancerous cells. Biopolymers **61**, 349–353 (2002)

11. A. Pacifico, L.A. Chiriboga, P. Lasch, M. Diem, Infrared spectroscopy of cultured cells II. Spectra of exponentially growing, serum-deprived and confluent cells. Vib. Spectrosc. **32**, 107 (2003)
12. B. Rigas, S. Morgello, I.S. Goldman, P.T.T. Wong, Human colorectal cancers display abnormal FTIR spectra. Proc. Natl. Acad. Sci. U. S. A. **87**, 8140 (1990)
13. H. Fabian, P. Lasch, D. Naumann, Analysis of biofluids in aqueous environment based on mid-infrared spectroscopy. J. Biomed. Opt. **10**, 031103 (2005)
14. R. Nomen, J. Sempere, K. Aviles, Detection and characterization of water alcohol hydrates by on-line FTIR using multivariate data analysis. Chem. Eng. Sci. **56**, 6577 (2001)
15. T. Hasegawa, J. Nishijo, T. Imae, Q. Huo, R.M. Leblanc, Selective observation of boundary water near a solid/water interface by variable-angle polarization specific attenuated total reflection infrared spectroscopy and principal component analysis. J. Phys. Chem. B **105**, 12056 (2001)
16. F.A. Inon, J.M. Garrigues, S. Garrigues, A. Molina, M.D.L. Guardia, Selection of calibration set samples in determination of olive oil acidity by partial least squares-attenuated total reflectance-Fourier transform infrared spectroscopy. Anal. Chim. Acta **489**, 59 (2003)
17. J. Tewardi, J. Irudayaraj, Quantification of saccharides in multiple floral honeys using fourier transform infrared microattenuated total reflectance spectroscopy. J. Agric. Food Chem. **52**, 3237 (2004)
18. T. Udelhoven, D. Naumann, J. Schmitt, Development of a hierarchical classification system with artificial neural networks and FT-IR spectra for the identification of bacteria. Appl. Spectrosc. **54**, 1471 (2000)
19. O. Eytan, B.-A. Sela, A. Katzir, Fiber-optic evanescent-wave spectroscopy and neural networks: application to chemical blood analysis. Appl. Opt. **39**, 3357–3360 (2000)
20. N.J. Harrick, *Internal Reflection Spectroscopy* (Interscience Publishers, New York, 1967)
21. P. Lucas, M.R. Riley, C. Boussard-Pledel, B. Bureau, Advances in chalcogenide fiber evanescent wave biochemical sensing. Anal. Biochem. **351**, 1–10 (2006)
22. P. Lucas, B. Bureau, Advanced infrared glasses for biochemical sensing, in *Biointerface Characterization by Advanced IR Spectroscopy*, ed. by C.M. Pradier, Y.J. Chabal (Elsevier, Amsterdam, 2011), pp. 217–243
23. D. Le Coq, K. Michel, J. Keirsse, C. Boussard-Pledel, G. Fonteneau, B. Bureau, J.-M. Le Quere, O. Sire, J. Lucas, Infrared glass fibers for in-situ sensing, chemical and biochemical reactions. C. R. Chim. **5**, 907–913 (2002)
24. C. Boussard-Pledel, S. Hocde, G. Fonteneau, H.L. Ma, X.H. Zhang, K. Lefoulgoc, J. Lucas, Infrared glass fibers for evanescent wave spectroscopy. Proc. SPIE **3596**, 91 (1999)
25. M.A. Druy, P.J. Glatkowski, W.A. Stevenson, Mid-IR tapered chalcogenide fiber optic attenuated total attenuated reflectance (ATR) sensors for monitoring epoxy resin chemistry. Proc. SPIE **2069**, 113 (1993)
26. K. Michel, B. Bureau, C. Pouvreau, J.C. Sangleboeuf, C. Boussard-Plédel, T. Jouan, T. Rouxel, J.-L. Adam, K. Staubmann, H. Steinner, T. Baumann, A. Katzir, J. Bayona, W. Konz, Development of a chalcogenide glass fiber device for in situ pollutant detection. J. Non-Cryst. Solids **326**, 434 (2003)
27. H. Steiner, M. Jakusch, M. Kraft, M. Karlowatz, T. Baumann, R. Niessner, W. Konz, A. Brandenburg, K. Michel, C. Boussard-Pledel, B. Bureau, J. Lucas, Y. Reichlin, A. Katzir, N. Fleischmann, K. Staubmann, R. Allabashi, J.M. Bayona, B. Mizaikoff, In situ sensing of volatile organic compounds in groundwater: first field tests of a mid-infrared fiber-optic sensing system. Appl. Spectrosc. **57**, 607–613 (2003)
28. N. Afasnasyeva, R. Bruch, A. Katzir, Infrared fiberoptic evanescent wave spectroscopy: application in biology and medicine. Proc. SPIE **3596**, 152 (1999)
29. S. Hocde, O. Loreal, O. Sire, C. Boussard-Pledel, B. Bureau, B. Turlin, J. Keirsse, P. Leroyer, J. Lucas, Metabolic imaging of tissues by infrared fiber-optic spectroscopy: an efficient tool for medical diagnosis. J. Biomed. Opt. **9**, 404–407 (2004)

30. J. Keirsse, E. Lahaye, A. Bouter, V. Dupont, C. Boussard-Pledel, B. Bureau, J.-L. Adam, V. Monbet, O. Sire, Mapping bacterial surface population physiology in real-time: infrared spectroscopy of Proteus mirabilis swarm colonies. Appl. Spectrosc. **60**, 584–591 (2006)
31. M.L. Anne, C. Le Lan, V. Monbet, C. Boussard-Pledel, M. Ropert, O. Sire, M. Pouchard, C. Jard, J. Lucas, J.L. Adam, P. Brissot, B. Bureau, O. Loreal, Fiber evanescent wave spectroscopy using the mid-infrared provides useful fingerprints for metabolic profiling in humans. J. Biomed. Opt. **14**, 054033 (2009)
32. S. Hocde, C. Boussard-Pledel, G. Fonteneau, J. Lucas, Chalcogens based glasses for IR fiber chemical sensors. Solid State Sci. **3**, 279–284 (2001)
33. P. Lucas, A.A. Wilhelm, M. Videa, C. Boussard-Pledel, B. Bureau, Chemical stability of chalcogenide infrared glass fibers. Corros. Sci. **50**, 2047–2052 (2008)
34. V.S. Shiryaev, J.L. Adam, X.H. Zhang, C. Boussard-Pledel, J. Lucas, M.F. Churbanov, Infrared fibers based on Te-As-Se glass system with low optical losses. J. Non-Cryst. Solids **336**, 113–119 (2004)
35. E. Hecht, *Optics*, 2nd edn. (Addison-Wesley, Reading, MA, 1987)
36. E. Lepine, Z. Yang, Y. Gueguen, J. Troles, X.-H. Zhang, B. Bureau, C. Boussard-Pledel, J.-C. Sangleboeuf, P. Lucas, Optical microfabrication of tapers in low-loss chalcogenide fibers. J. Opt. Soc. Am. B **27**, 966–971 (2010)
37. D. Lecoq, K. Michel, G. Fonteneau, S. Hocde, C. Boussard-Pledel, J. Lucas, Infrared chalcogen glasses: chemical polishing and fiber remote spectroscopy. Int. J. Inorg. Mater. **3**, 233–239 (2001)
38. S. Cui, R. Chahal, C. Boussard-Pledel, V. Nazabal, J.-L. Doualan, J. Troles, J. Lucas, B. Bureau, From selenium- to tellurium-based glass optical fibers for infrared spectroscopies. Molecules **18**, 5373–5388 (2013)
39. Z. Yang, M.K. Fah, K.A. Reynolds, J.D. Sexton, M.R. Riley, M.-L. Anne, B. Bureau, P. Lucas, Opto-electrophoretic detection of bio-molecules using conducting chalcogenide glass sensors. Opt. Express **18**, 26754–26759 (2010)
40. M.L. Brandily, V. Monbet, B. Bureau, C. Boussard-Pledel, O. Loreal, J.L. Adam, O. Sire, Identification of foodborne pathogens within food matrices by IR spectroscopy. Sens. Actuators B **160**, 202–206 (2011)
41. B. Bureau, C. Boussard, S. Cui, R. Chahal, M.L. Anne, V. Nazabal, O. Sire, O. Loreal, P. Lucas, V. Monbet, J.-L. Doualan, P. Camy, H. Tariel, F. Charpentier, L. Quetel, J.-L. Adam, J. Lucas, Chalcogenide optical fibers for mid-infrared sensing. Opt. Eng. **53**(2014), 027101 (2014)
42. P. Lucas, G.J. Coleman, S. Jiang, T. Luo, Z. Yang, Chalcogenide glass fibers: optical window tailoring and suitability for bio-chemical sensing. Opt. Mater. **47**, 530–536 (2015)
43. Z. Yang, T. Luo, S. Jiang, J. Geng, P. Lucas, Single-mode low-loss optical fibers for long-wave infrared transmission. Opt. Lett. **35**, 3360–3362 (2010)
44. J. Troles, V. Shiryaev, M. Churbanov, P. Houizot, L. Brilland, F. Desevedavy, F. Charpentier, T. Pain, G. Snopatin, J.L. Adam, $GeSe_4$ glass fibres with low optical losses in the mid-IR. Opt. Mater. **32**, 212–215 (2009)
45. G.E. Snopatin, M.F. Churbanov, A. Apushkin, V.V. Gerasimenko, E.M. Dianov, V.G. Plotnichenko, High purity arsenic-sulfide glasses and fibers with minimum attenuation of 12 dB/km. Optoelectron. Adv. Mater. Rapid Commun. **3**, 669–671 (2009)
46. G.E. Snopatin, V.S. Shiryaev, V.G. Plotnichenko, E.M. Dianov, M.F. Churbanov, High-purity chalcogenide glasses for fiber optics. Inorg. Mater. **45**, 1439–1460 (2009)
47. S. Hocde, C. Boussard-Pledel, G. Fonteneau, D. Lecoq, H.L. Ma, J. Lucas, Recent developments in chemical sensing using infrared glass fibers. J. Non-Cryst. Solids **274**, 17 (2000)
48. Y. Gueguen, J.C. Sangleboeuf, V. Keryvin, E. Lepine, Z. Yang, T. Rouxel, C. Point, B. Bureau, X.-H. Zhang, P. Lucas, Photoinduced fluidity in chalcogenide glasses at low and high intensities: a model accounting for photon efficiency. Phys. Rev. B: Condens. Matter **82**, 134114 (2010)

49. Z. Yang, A.A. Wilhelm, P. Lucas, High-conductivity tellurium-based infrared transmitting glasses and their suitability for bio-optical detection. J. Am. Ceram. Soc. **93**, 1941–1944 (2010)
50. A.A. Wilhelm, P. Lucas, D.L. DeRosa, M.R. Riley, Biocompatibility of Te-As-Se glass fibers for cell-based bio-optic infrared sensors. J. Mater. Res. **22**, 1098–1104 (2007)
51. Y.-F. Niu, J.-P. Guin, T. Rouxel, A. Abdelouas, J. Troles, F. Smektala, Aqueous corrosion of the $GeSe_4$ chalcogenide glass: surface properties and corrosion mechanism. J. Am. Ceram. Soc. **92**, 1779–1787 (2009)
52. P. Lucas, M.A. Solis, C.D. Le, C. Juncker, M.R. Riley, J. Collier, D.E. Boesewetter, C. - Boussard-Pledel, B. Bureau, Infrared biosensors using hydrophobic chalcogenide fibers sensitized with live cells. Sens. Actuators B **B119**, 355–362 (2006)
53. J.P. Guin, T. Rouxel, J.C. Sangleboeuf, I. Melscoet, J. Lucas, Hardness, toughness, and scratchability of germanium-selenium chalcogenide glasses. J. Am. Ceram. Soc. **85**, 1545 (2002)
54. T. Rouxel, Elastic properties and short-to medium-range order in glasses. J. Am. Ceram. Soc. **90**, 3019–3039 (2007)
55. E. Lebourhis, P. Gadaud, J.P. Guin, N. Tournerie, X.H. Zhang, J. Lucas, T. Rouxel, Temperature dependence of the mechanical behaviour of a GeAsSe glass. Scr. Mater. **45**, 317 (2001)
56. G. Delaizir, J.-C. Sangleboeuf, E.A. King, Y. Gueguen, X.-H. Zhang, C. Boussard-Pledel, B. Bureau, P. Lucas, Influence of ageing conditions on the mechanical properties of Te-As-Se fibres. J. Phys. D Appl. Phys. **42**, 095405 (2009)
57. G. Yang, H. Chen, C. Boussard-Pledel, J.-C. Sangleboeuf, B. Bureau, Effect of physical aging on fracture behavior of Te2As3Se5 glass fibers. Ceram. Int. **41**, 4487–4491 (2015)
58. G. Tao, S. Shabahang, H. Ren, F. Khalilzadeh-Rezaie, R.E. Peale, Z. Yang, X. Wang, A.F. Abouraddy, Robust multimaterial tellurium-based chalcogenide glass fibers for mid-wave and long-wave infrared transmission. Opt. Lett. **39**, 4009–4012 (2014)
59. G. Tao, S. Shabahang, E.-H. Banaei, J.J. Kaufman, A.F. Abouraddy, Multimaterial preform coextrusion for robust chalcogenide optical fibers and tapers. Opt. Lett. **37**, 2751–2753 (2012)
60. P. Houizot, M.-L. Anne, C. Boussard-Pledel, O. Loreal, H. Tariel, J. Lucas, B. Bureau, Shaping of looped miniaturized chalcogenide fiber sensing heads for mid-infrared sensing. Sensors **14**, 17905–17914 (2014)
61. J. Keirsse, B. Bureau, C. Boussard-Pledel, P. Leroyer, M. Ropert, V. Dupont, M.L. Anne, C. Ribault, O. Sire, O. Loreal, J.L. Adam, Chalcogenide glass fibers for in-situ infrared spectroscopy in biology and medicine. Proc. SPIE-Int. Soc. Opt. Eng. **5459**, 61–68 (2004)
62. J. Keirsse, C. Boussard-Pledel, O. Loreal, O. Sire, B. Bureau, P. Leroyer, B. Turlin, J. Lucas, IR optical fiber sensor for biomedical applications. Vib. Spectrosc. **32**, 23–32 (2003)
63. M.R. Riley, P. Lucas, C.D. Le, C. Juncker, D.E. Boesewetter, J.L. Collier, D.M. DeRosa, M.E. Katterman, C. Boussard-Pledel, B. Bureau, Lung cell fiber evanescent wave spectroscopic biosensing of inhalation health hazards. Biotechnol. Bioeng. **95**, 599–612 (2006)
64. P. Lucas, E.A. King, Y. Gueguen, J.-C. Sangleboeuf, V. Keryvin, R.G. Erdmann, G. Delaizir, C. Boussard-Pledel, B. Bureau, X.-H. Zhang, T. Rouxel, Correlation between thermal and mechanical relaxation in chalcogenide glass fibers. J. Am. Ceram. Soc. **92**, 1986–1992 (2009)
65. M.L. Anne, E. Le Gal La Salle, B. Bureau, J. Tristant, F. Brochot, C. Boussard-Pledel, H.L. Ma, X.H. Zhang, J.L. Adam, Polymerisation of an industrial resin monitored by infrared fiber evanescent wave spectroscopy. Sens. Actuators B **137**, 687–691 (2009)
66. K. Michel, B. Bureau, C. Boussard-Plédel, T. Jouan, J.L. Adama, K. Staubmann, T. Baumannc, Monitoring of pollutant in waste water by infrared spectroscopy using chalcogenide glass optical fibers. Sens. Actuators B **101**, 252–259 (2004)
67. J. Franks, K. Rogers, Y. Guimond, Optical and thermo mechanical properties of infrared glasses. Proc. SPIE **6940**, 69400P/69401–69400P/69408 (2008)
68. Y. Guimond, Y. Bellec, K. Rogers, A new moldable infrared glass for thermal imaging and low cost sensing. Proc. SPIE-Int. Soc. Opt. Eng. **6542**, 654225/654221–654225/654226 (2007)
69. DIAFIR, http://www.diafir.com

70. B. Temelkuran, S.D. Hart, G. Benoit, J.D. Joannopoulos, Y. Fink, Wavelength-scalable hollow optical fibres with large photonic bandgaps for CO_2 laser transmission. Nature **420**, 650–653 (2002)
71. M. Bayindir, O. Shapira, D. Saygin-Hinczewski, J. Viens, A.F. Abouraddy, J.D. Joannopoulos, Y. Fink, Integrated fibres for self-monitored optical transport. Nat. Mater. **4**, 820–825 (2005)
72. A.F. Abouraddy, M. Bayindir, G. Benoit, S.D. Hart, K. Kuriki, N. Orf, O. Shapira, F. Sorin, B. Temelkuran, Y. Fink, Towards multimaterial multifunctional fibres that see, hear, sense and communicate. Nat. Mater. **6**, 336–347 (2007)
73. Omniguide, http://www.omni-guide.com/
74. A. Gumennik, A.M. Stolyarov, B.R. Schell, C. Hou, G. Lestoquoy, F. Sorin, W. McDaniel, A. Rose, J.D. Joannopoulos, Y. Fink, All-in-fiber chemical sensing. Adv. Mater. **24**, 6005–6009 (2012)
75. A. Canales, X. Jia, U.P. Froriep, R.A. Koppes, C.M. Tringides, J. Selvidge, C. Lu, C. Hou, L. Wei, Y. Fink, P. Anikeeva, Multifunctional fibers for simultaneous optical, electrical and chemical interrogation of neural circuits in vivo. Nat. Biotechnol. **33**, 277–284 (2015)
76. B.J. Eggleton, B. Luther-Davies, K. Richardson, Chalcogenide photonics. Nat. Photonics **5**, 141–148 (2011)
77. C. Monat, M. Spurny, C. Grillet, L. O'Faolain, T.F. Krauss, B.J. Eggleton, D. Bulla, S. Madden, B. Luther-Davies, Third-harmonic generation in slow-light chalcogenide glass photonic crystal waveguides. Opt. Lett. **36**, 2818–2820 (2011)
78. M.W. Lee, C. Grillet, C.L.C. Smith, D.J. Moss, B.J. Eggleton, D. Freeman, B. Luther-Davies, S. Madden, A. Rode, Y. Ruan, Y.-h. Lee, Photosensitive post tuning of chalcogenide photonic crystal waveguides. Opt. Express **15**, 1277–1285 (2007)
79. M.W. Lee, C. Grillet, C. Monat, E. Magi, S. Tomljenovic-Hanic, X. Gai, S. Madden, D.-Y. Choi, D. Bulla, B. Luther-Davies, B.J. Eggleton, Photosensitive and thermal nonlinear effects in chalcogenide photonic crystal cavities. Opt. Express **18**, 26695–26703 (2010)
80. J. Hu, N. Carlie, L. Petit, A. Agarwal, K. Richardson, L. Kimerling, Demonstration of chalcogenide glass racetrack microresonators. Opt. Lett. **33**, 761–763 (2008)
81. Y. Zou, D. Zhang, H. Lin, L. Li, L. Moreel, J. Zhou, Q. Du, O. Ogbuu, S. Danto, J.D. Musgraves, K. Richardson, K.D. Dobson, R. Birkmire, J. Hu, High-performance, high-index-contrast chalcogenide glass photonics on silicon and unconventional non-planar substrates. Adv. Opt. Mater. **2**, 478–486 (2014)
82. T. Han, S. Madden, D. Bulla, B. Luther-Davies, Low loss chalcogenide glass waveguides by thermal nano-imprint lithography. Opt. Express **18**, 19286–19291 (2010)
83. L. Li, H. Lin, S. Qiao, Y. Zou, S. Danto, K. Richardson, J.D. Musgraves, N. Lu, J. Hu, Integrated flexible chalcogenide glass photonic devices. Nat. Photonics **8**, 643–649 (2014)

Part IV
Tellurium

Tellurium: The Element

Te 52 $4d^{10}$ $5s^2$ $5p^4$

Tellurium is a metallic element with symbol Te and atomic number 52; at.wt. 127.60; m.p. 449.51 °C; b. p. 988 °C; sp.gr. 6.24 (20 °C); valence −2, 4 or 6. The term tellurium in Latin means *Goddess of the Earth* [1]. It was discovered by Muller von Reichenstein in 1782, and named by Klaproth who isolated it in 1798. Tellurium is occasionally found native, but is more often found as the telluride of gold (calaverite), and combined with other metals. It is recovered commercially from the anode muds produced during the electrolytic refining of blister copper. Crystalline tellurium has a silvery white appearance, and when pure exhibits a metallic lustre. It is brittle and easily pulverized. Amorphous tellurium is formed by precipitating Te from a solution of telluric or tellurous acid. Te is a p-type semiconductor, and shows greater conductivity in certain directions, depending on the alignment of atoms. Its conductivity increases slightly with exposure to light. Naturally occurring tellurium has eight isotopes. Five of those isotopes, ^{122}Te, ^{123}Te, ^{124}Te, ^{125}Te, and ^{126}Te, are stable [2]. The other three, ^{120}Te, ^{128}Te, and ^{130}Te, have been observed to be radioactive [3, 4, and 5]. The stable isotopes make up only 33.2 % of the naturally occurring tellurium; this is possibly due to the long half-lives of the unstable isotopes. They are in the range from 10^{13} to 2.2×10^{24} years (for ^{128}Te). This makes ^{128}Te the isotope with the longest half-life among all radionuclides [6] which is approximately 160 trillion (10^{12}) times the age of the known universe. There are 38 known nuclear isomers of tellurium with atomic masses that range from 105 to 142. Tellurium is among the lightest elements known to undergo alpha decay, with isotopes ^{106}Te to ^{110}Te being able to undergo this mode of decay [3]. The atomic mass of tellurium (127.60 g/mol) exceeds that of the following element iodine (126.90 g/mol) [7]. Natural tellurium consists of eight isotopes, two of which are radioactive with very long half-lives. Tellurium improves the machinability of copper and stainless steel, and its addition to lead decreases the corrosive action of sulfuric acid on lead and improves its strength and hardness. Gold telluride minerals are the most notable natural gold compounds. However, they are not a commercially significant source of tellurium itself, which is normally extracted as a by-product of copper and lead production [2]. Commercially, the primary use of tellurium is in alloys, foremost in steel and copper to improve machinability. Tellurium is also used in electronic and semiconductor devices. Bismuth telluride has been used in thermoelectric devices. Tellurium is also an important component of the material used for optical memories and radiation and photodetectors. Applications in solar panels and as a semiconductor material consume a considerable fraction of tellurium production.

Tellurium exhibits a narrow band gap and **phase change behavior** which makes its compound suitable for the following:

- Optical storage media (commercialized) (Ge-Sb-Te)
- Strong contenders to replace nonvolatile flash memories
- Compounds used in photovoltaic devices such as solar cells (PbSe/PbS/PbTe)
- Infrared detectors (PbSe/PbS/PbTe and Hg-Cd-Te)

Chapter 9
Data Storage Devices

Gurinder Kaur Ahluwalia

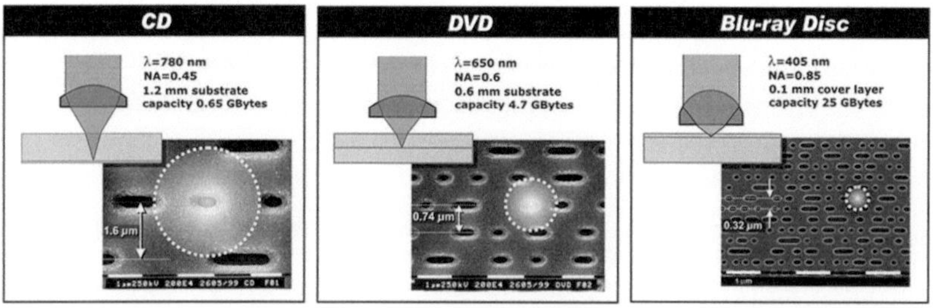

Three generations of optical disks, Courtesy, Philips Research Laboratories, Netherlands

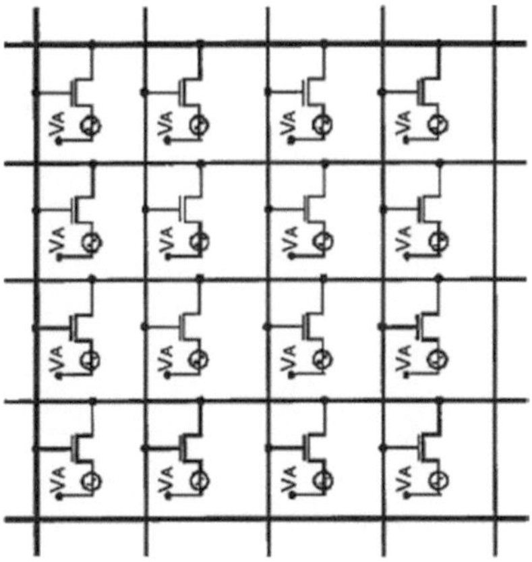

OUM

G.K. Ahluwalia, Ph.D (✉)
Materials and Nanotechnology Research Laboratory, Department of Physics College of The North Atlantic, Labrador West Campus, 1600 Nichols Adam Highway, Labrador City, A2V 0B8, NL, Canada
e-mail: gurinder.kaur@cna.nl.ca

G.K. Ahluwalia (ed.), *Applications of Chalcogenides: S, Se, and Te*,
DOI 10.1007/978-3-319-41190-3_9

(continued)

9.1 Introduction

Recent advances in multimedia would have been impossible without computer performance getting faster and the stored information denser [1]. Amorphous chalcogenides have been extensively studied over the last half century due to their applications in rewritable optical data storage and in non-volatile phase change memory devices. Tellurium based alloys (Ge-Sb-Te and AgInSbTe) have been commercialized in Optical Storage media and are strong contenders to replace non-volatile flash Memories. The term "Memory" is used to denote a device or system wherein the operations Write–Read–Erase of information can be accomplished in any desired permutation. The ability of media to be used repeatedly, that is, to be rewritable, is also an important requirement. Conventional methods such as magnetic recording, is limited in further increasing the recording density because it requires huge write field for high anisotropic media.

Phase Change Materials (PCM) based Solid-State Memory devices and Digital-Versatile Disk-Rewritable (DVD-RW)/Blu-ray disk (BD) are a billion dollar industry in the present multimedia market. Current multimedia (Compact disk (CD), Digital Versatile Disk-Random Access Memory (DVD-RAM)) strongly relies on rewritable phase change optical memories using $Ge_2Sb_2Te_5$ (GST) as the phase change material [2, 3]. GST is also so far considered to be a desirable material for solid-state memories—Phase Change Random Access Memories: PC-RAMs.

The idea to use an amorphous-to-crystalline phase transition for information storage dates back to the 1960s when S.R. Ovshinsky suggested a memory switch based on changes in the properties of amorphous and crystalline phases of multicomponent chalcogenides [4, 5]. A landmark was achieved in the late 1980s by Matsushita/Panasonic who developed phase change optical disk technology that remained stable over a million use cycles [6]. This technology became the mainstream in optical disk production and in the late 1990s resulted in the commercialization of 4.7 GB Digital Versatile Disk-Random Access Memory (DVD-RAM). During the development process, various materials were examined, and the best performing in terms of speed and stability were found to be $Ge_2Sb_2Te_5$ (GST) used in DVD-RAM and an AgInSbTe alloy used in CD-RW [7, 8].

Alternatives to the current flash memory technology, which is relatively slow and lacks random access capability (wherein data can be randomly read and written at every byte), are being actively pursued. Thus, search for a universal memory storage device that combines rapid read and write speeds, high storage density and non-volatility is driving the exploration of new materials in nanostructured form. Phase Change Random Access Memory (PCRAM) is one of the most promising candidates for next generation nonvolatile memory. Phase change materials, which can be switched between amorphous and crystalline states, are promising in this respect, but top down processing of these materials into nanostructures often damages their useful properties. Self-assembled nanowires-based phase change material memory devices offer an attractive solution owing to their sub-lithographic sizes and unique geometry, coupled with the facile etch-free processes with which they can be fabricated.

9.2 Phase Change Memory

9.2.1 Mechanism of Data Storage Using Phase Change Materials

Information is said to have been written in a chosen medium if its physical or chemical state is altered under the influence of the signal (or writing beam) energy. The altered state in a storage medium is brought about by making proper use of one or more of its characteristics properties (i.e., thermal, electrical, magnetic, elastic, optical etc.) Reading of the stored information is accomplished by sensing the altered state using an appropriate technique, and erasing of information is achieved using such means as electrical, magnetic, or optical either alone or in some suitable combination.

During the storage of digital data, 1 and 0 s are recorded as either glassy (high resistive and low reflective) or crystalline (low resistive and high reflective) structures (Fig. 9.1).

Optical memory disks use laser light to convert small portions of a thin chalcogenide film between the high and low reflective states. On the other hand, non-volatile phase change random access memory (PCRAM) uses a voltage bias to convert the material between the high and low resistive states. It stores data in a smaller area and with higher speeds for both read and write processes than the optical memory disks. Due to its capability of random access memory, it is sometimes referred to as unified memory. Applications explored for this technology span from wireless, embedded systems to solid-state storage automotive and space applications.

In case of rewritable media such as optical disks, an initially amorphous as-deposited GST layer is crystallized by exposure to a laser beam of intensity sufficient to heat the material to a temperature slightly above the glass-transition temperature. A subsequent exposure to an intense and short laser pulse melts the PCM that is then converted into the amorphous state on quenching. A recorded bit is an amorphized mark against the crystalline background as shown in Fig. 9.2. The reversibility of the crystallization–amorphization process allows fabrication of rewritable memory.

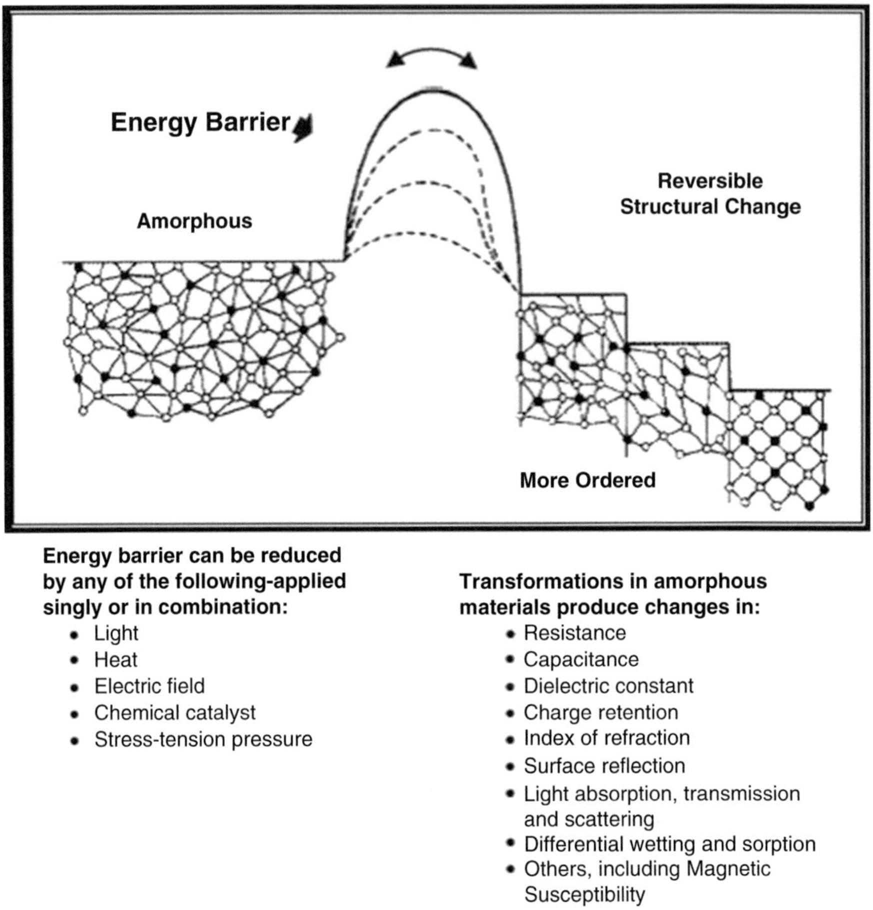

Fig. 9.1 Mechanism for information storage/retrieval and display by structural transformation (Reprinted with permission from IOP [5])

9.2.2 Phase Change Memories: A Materials Overview

Various kinds of materials including organic, inorganic, synthetic compounds, natural biomaterials, single component as well as hybrid systems have been extensively explored for application in data storage devices. Although several binary, ternary, and quaternary chalcogenide alloys can be realized, only some of these can exhibit amorphous → crystalline → amorphous transition and good read–write cyclability (no phase separation during reversible phase transitions) for applications in phase change memory devices.

Two classes of tellurium based materials (1) GeTe-Sb_2Te_3 (GST) and (2) $Ag_{3.4}In_{3.7}Sb_{76.4}Te_{16.5}$ (AIST) [9, 10] are now practically used as the memory

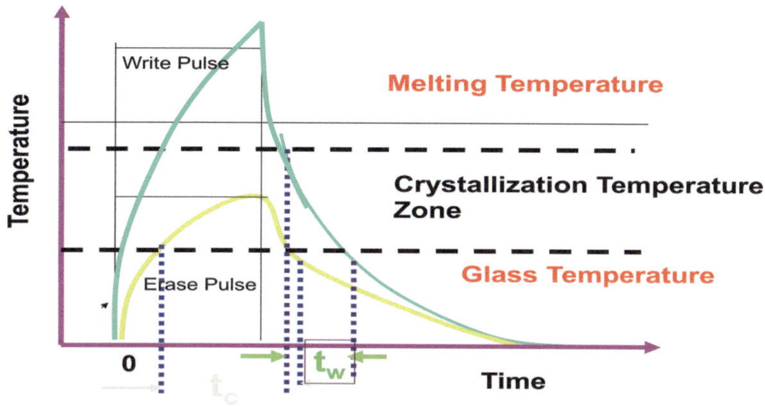

Fig. 9.2 Laser/electric pulse and temperature variation with time during phase change recording process

layers in phase change optical disk media, as well as in the memory cells of solid-state electrical memories. Recording can be achieved using laser irradiation or ohmic heating to cause reversible phase changes between amorphous and crystalline phases in these materials.

The basic requirements of a desirable material for PCM devices are: (a) high speed of crystallization; minimum erase time less than 1 ns. (b) good stability at room temperature; (c) lifetime of amorphous written spots longer than 10 years—data retention. (d) Good reversibility; (e) write–erase cycle larger than 10^9 times, and (f) cyclability endurance.

In order to meet these requirements, a desirable PC material must satisfy the following requirements: (1) A very highly symmetric, cubic crystalline phase with large random atomic distributions such as simple cubic and rock-salt structure. (2) Their crystalline phases should have very large thermal vibrations, i.e., large atomic displacements, particularly at high temperature. Consequently, their amorphous structure will reflect an isotropic atomic configuration in the liquid state. Hence, the resemblance of the atomic distribution between the amorphous and crystalline phases at high temperature will reduce the barrier between the two states that will shorten the time for atomic rearrangement for crystallization.

Other Criteria for material selection include

1. Large atomic mobility in amorphous and super cooled states.
2. Short atomic diffusion distance from the atomic location in amorphous state to the lattice sites of the crystalline states.
3. Melting point: 500–1000 °C; T_g ~2/3 T_m; T_g is the glass transition temperature and T_m is the melting point.
4. Thermal Stability: Long lifetime at room temperature.
5. Optical constants: Optical contrast between the amorphous and crystalline phases.

9.2.3 Why Chalcogenide Semiconductors for Memory Devices?

Amorphous chalcogenides are less toxic, easy to handle and can be fabricated over large areas. Extra functionality for optical fabrication and performance are provided by their unique set of photo-induced effects [10]. Taking advantage of the possibility of modifying the optical band gap and the index of refraction by illumination, optical elements for applications in communications, interconnections, and data storage have been patterned in chalcogenide glasses [11].

Several combinations of chalcogenide materials have the ability to repeatedly transform between glassy (disordered) and crystalline (ordered) atomic structures. The atomic configurations of conventional semiconductor materials Silicon (Si) or Germanium (Ge) is quite different from that in the chalcogens Tellurium (Te) or Selenium (Se). The former exhibits the rigid diamond type structure and each atom forms tight covalent bonds with four surrounding atoms. This results in high melting temperatures, T_m of 952 °C for Ge and 1414 °C for Si, and high viscosity in the molten state. These properties cause some difficulties for their practical application as phase change materials. It means that crystalline (c-) Si or c-Ge are difficult to amorphize due to their high T_m and simultaneously amorphous (a-) Si or a-Ge are difficult to crystallize due to their high viscosity (i.e., slow atomic diffusion). On the contrary, Te and Se are essentially twofold coordinated forming coil like structures; Te atoms form strong covalent bonds with two adjacent Te atoms resulting in a -Te-Te-Te coil like structure. Every -Te-Te-Te- coil is bound by weak van der Waals force to another to form a hexagonal crystal. This is the reason why Te has a very low T_m (=450 °C). Te has been reported [12], to have partially threefold coordination in molten form which explains its ability to easily form amorphous phase as follows: Once Te is heated up above T_m, it easily melts by cutting van der Waals bonds between -Te-Te-Te- coils. These coils form a random network structure since Te has partially three bonds in the molten removing the cross-linking bonds. Thus stable amorphous materials can be obtained by adding some cross-linking element to Te. The cross-linking elements should have threefold or fourfold coordination with Te such as P, As, and Sb (threefold) and Si, Ge, and Sn (fourfold) state. By rapid quenching, the network structure is frozen to form the amorphous solid. Since the threefold coordinated a-Te is not stable at room temperature, it automatically crystallizes by removing the cross-linking bonds [47].

The crystal structure of a photo-recording material has a strong influence on the optical characteristics, cyclability and record–erase characteristics [13]. Thus identifying a compound's crystal structure is a prerequisite for improving its performance as a recording medium.

Two important categories of chalcogen based phase change materials are (1) Ge-Te based, e.g., GeSbTe (GST) and (2) Sb-Te based such as AgInSbTe (AIST). The most common materials are GeTe–Sb_2Te_3 pseudo binary compounds [14] (group 1 in Fig. 9.3a) and Sb–Te binary compounds with small amounts of In,

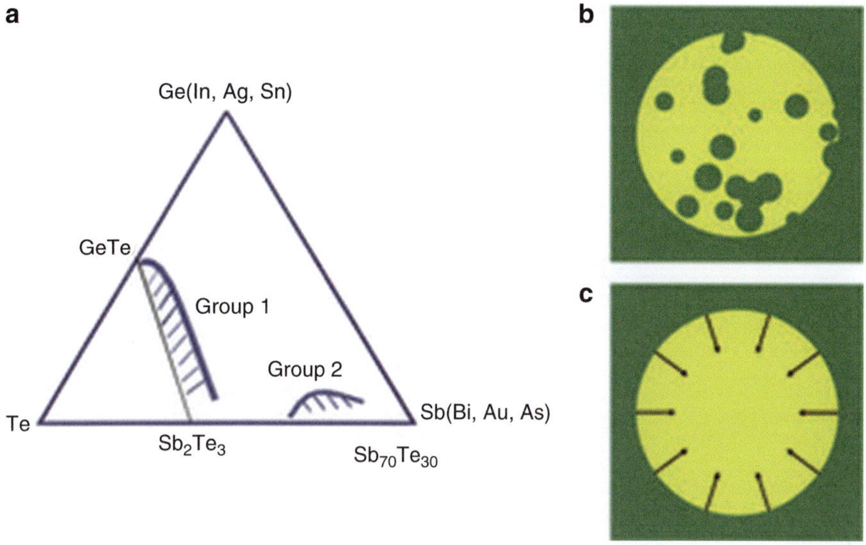

Fig. 9.3 (**a**) The most commonly used materials for optical recording are in groups 1 and 2. (**b**) Nucleation-dominated recrystallization (as in GST) [15]. (**c**) Growth-dominated recrystallization (AIST) [10]. Reprinted by permission from Macmillan Publishers Ltd: Nature Materials [16], copyright (2011)

Ag and/or Ge, for example $Ag_{3.5}In_{3.8}Sb_{75.0}Te_{17.7}$ (AIST), shown as group 2 in Fig. 9.3a [15]. It is well known that recrystallization in the two groups is strikingly different: in group 1 it proceeds mainly by way of nucleation inside the marks (Fig. 9.3b), in group 2 by way of crystal growth from the rim (Fig. 9.3c) [10].

By adding dopants to the phase-change material, variations in optical contrast, crystallization rate and amorphous stability can be obtained [17, 18]. In phase change recording the term dopant is generally used when atoms are added in concentrations of typically 2–10 %. Alloying antimony with certain elements enhances the amorphous phase stability. For Sb_2Te_3 a crystallization temperature of about 103 °C was measured and for $Ge_{15}Sb_{85}$ the crystallization temperature of the as-deposited amorphous phase is about 250 °C. Table 9.1 [19] summarizes the crystallization temperatures of the as-deposited sputtered amorphous phase of many fast growth (FG) phase-change materials in column 1d. All dopants in Table 9.1 enhance the thermal stability of the sputtered amorphous phase. Especially addition of Ge or Ga to an Sb-rich composition leads to a large increase in the crystallization temperature T_c. An estimate of the amorphous stability can be made a priori by calculating the glass transition temperature [20] which can be considered as the lower limit of the crystallization temperature. The glass transition temperature, and hence the crystallization temperature, increases by adding elements that increase the average coordination number and/or the average bond enthalpy. This explains the high amorphous stability of $Ge_{15}Sb_{85}$; Ge has a high bonding strength and a relatively large coordination number of 4. Tables 9.2 and 9.3 give minimum laser-

Table 9.1 Overview of crystallization temperatures (for as deposited amorphous layers and written amorphous marks), activation energies for crystallization of amorphous marks and extrapolated archival life stabilities at 50 °C of various Fast Growth (FG) materials [19]

Phase-change composition	T_c (as deposited) (°C)	T_c (written) (°C)	E_{act} (kJ/mol)	Archival life @ 50 °C
Sb_2Te	103			
$Ge_8Sb_{72}Te_{20}$	222	192		
$Ag_8Sb_{72}Te_{20}$	175	120	196	100 days
$In_8Sb_{72}Te_{20}$	200	93	121	1 h
$Ga_8Sb_{72}Te_{20}$	223	88	184	2 h
$Ga_8Sb_{77}Te_{15}$	230	84	155	0.5 h
$Sn_8Sb_{72}Te_{20}$	153			<Few days at RT
$Ge_6In_2Sb_{72}Te_{20}$	208	176	465	4×10^{13} years
$Ge_2In_7Sb_{72}Te_{20}$	196	133	290	4×10^{2} years
$Ge_{15}Sb_{85}$	250	208	237	1×10^{6} years
$Ge_{22}Sb_{78}$	271	246	346	3×10^{14} years
$In_{15}Sb_{85}$	186	117	265	2–30 years
$Ga_{10}Sb_{90}$	210	165	320	4×10^{6} years
$Ga_{15}Sb_{85}$	233	210	323	3×10^{10} years

Reprinted with permission from AIP publishing

Table 9.2 Overview of the minimum laser-induced crystallization times of an amorphous mark of 125 nm radius, as measured on a static tester setup, of various FG materials [19]

Phase-change composition	Crystallization time (mark radius of 125 nm) (ns)
$In_5Sb_{71}Te_{24}$	49
$In_5Sb_{74}Te_{21}$	36
$In_5Sb_{77}Te_{18}$	25
$In_5Sb_{80}Te_{15}$	18
$Ag_8Sb_{72}Te_{20}$	42
$Ge_8Sb_{72}Te_{20}$	33
$In_8Sb_{72}Te_{20}$	29
$Ga_5Sb_{72}Te_{20}$	20
$Sn_8Sb_{72}Te_{20}$	17
$Ge_{12}Sb_{11}$	13
$Ge_{15}Sb_{15}$	15
$Ge_{22}Sb_{72}$	23
$In_{15}Sb_{15}$	6
$Ga_{15}Sb_{15}$	31

Reprinted with permission from AIP publishing

induced crystallization times of an amorphous mark of 125 nm radius and nucleation times of various FG materials respectively.

Table 9.3 Overview of nucleation times of various FG materials [19]

Phase-change composition	Nucleation time (μs)
$Ag_8Sb_{72}Te_{20}$	15
$Ge_8Sb_{72}Te_{20}$	200
$In_8Sb_{72}Te_{20}$	4
$Ga_8Sb_{72}Te_{20}$	1.5
$Sn_8Sb_{72}Te_{20}$	0.6
$Ge_{12}Sb_{11}$	16
$Ge_{15}Sb_{15}$	49
$Ge_{22}Sb_{72}$	400
$In_{15}Sb_{15}$	1
$Ga_{15}Sb_{15}$	4.5

Reprinted with permission from AIP publishing

9.2.4 GeSbTe (GST)

Chalcogen based phase change material containing Germanium, antimony and tellurium (Ge-Sb-Te) has been widely used as optical rewritable (CD-RW) data storage media. In particular, $(GeTe)_{1-x}(Sb_2Te_3)_x$ pseudo binary compounds are widely used as memory layer in rewritable phase-change optical disks, making use of their differing optical properties such as reflectivity and transmissivity between the amorphous and crystalline phases. The characteristics of these compounds essential for use in optical disks include nanosecond scale rewritable speeds, extremely good reversibility between the amorphous and crystalline phases over 10^5 cycles, and long thermal stability over scores of years [10]. A typical composition for current digital versatile disk random access memory (DVD-RAM) is $x = 1/3$. This composition exhibits suitable crystallization and melting temperatures for laser-induced phase changes and rapid phase transition time [10, 21–23]. On the other hand, GeTe-rich compositions such as $x = 1/9$ ($Ge_8Sb_2Te_{11}$) are adopted for the Blu-ray disk on which as much as 50 gigabytes of information can be recorded using a blue-violet laser [23]. Despite these pseudo binary compounds being functional and commercially available over a wide composition range, systematic composition-dependent investigations are being intensively carried out and the physical origin of the structural changes accompanying the fast amorphous crystalline phase transition still remain unknown. Several theories to explain the phenomena have been put forth by researchers.

The stable crystal structure of GST is hexagonal [24] but the structure of a thin laser-crystallized layer is believed to be different [25]. On the basis of data obtained using EXAFS, XANES and XRD, Kolobov et al. [8] suggested that $Ge_2Sb_2Te_5$ does not possess the rock salt structure but more likely consists of well defined rigid building blocks that are randomly oriented in space consistent with cubic symmetry. Laser induced amorphization results in drastic shortening of covalent bonds and a decrease in the mean square relative displacement, demonstrating a substantial increase in the degree of short range ordering, resulting in sharp contrast to the amorphization of typical covalently bonded solids. It is proposed that this novel

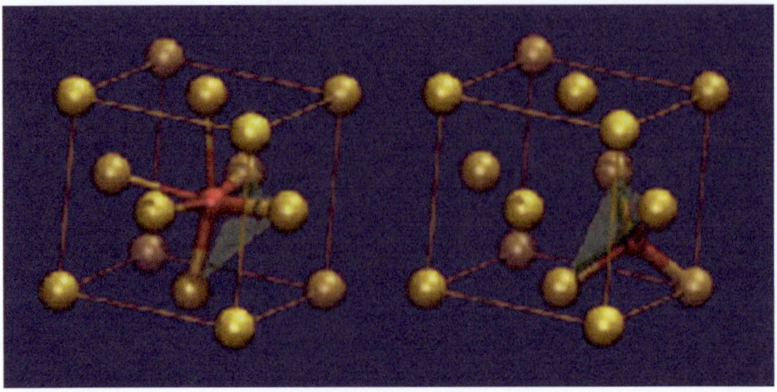

Fig. 9.4 Fragments of the local structure of GST around Ge atoms in the crystalline (*left*) and amorphous (*right*) states. Stronger covalent bonds are shown as *thicker lines* whereas weak interblock bonds are shown as *thinner lines*. The stronger covalent bonds remain intact on the umbrella-flip structural transformation rendering the Ge sublattice random. It is this nature of the structural change that makes the medium fast and stable. Reprinted by permission from Macmillan Publishers Ltd: [Nature Materials [8], copyright (2004)]

Fig. 9.5 Crystal structure of laser-amorphized GST: a schematic two-dimensional image of the lattice distortion of the rock salt structure due to charge redistribution between the constituent elements; atoms that form the building block of the GST structure are shown using *thick lines*. The *arrows* indicate displacements of atoms from the ideal rock salt positions. Reprinted by permission from Macmillan Publishers Ltd: [Nature Materials [8], Copyright (2004)]

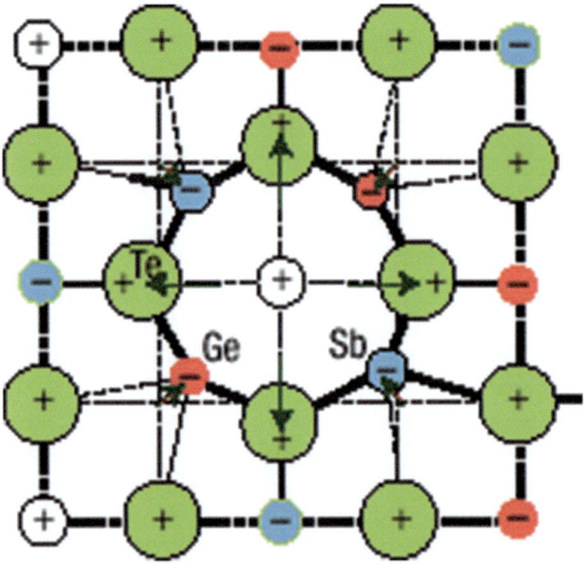

order-disorder transition is due to an umbrella flip of Ge atoms from an octahedral position into a tetrahedral position without rupture of strong covalent bonds as shown in Fig. 9.4. This unique two state nature of the transformation ensures fast DVD performance and repeatable switching over ten million cycles.

In the ideal rocksalt structure, in order to form six saturated covalent bonds with two electrons per bond, it is necessary for Ge, Sb, and Te to possess six valence electrons each. In particular, Ge thus requires two additional electrons whereas each

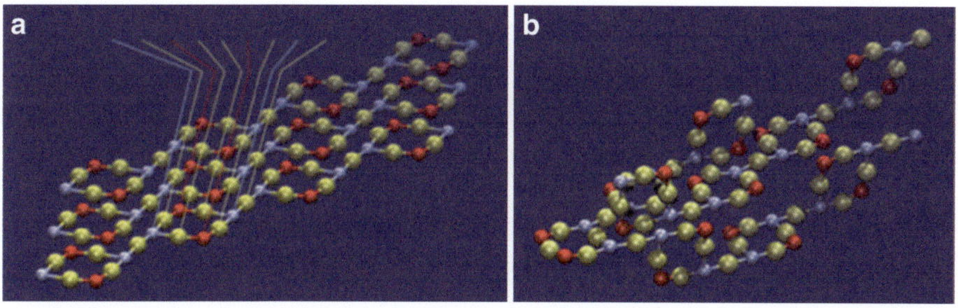

Fig. 9.6 (**a**) A schematic two-dimensional image of the lattice distortion of the rocksalt structure due to charge redistribution between the constituent elements; atoms that form the building block of the GST structure are shown using *thick lines*. The *arrows* indicate displacements of atoms from the ideal rocksalt positions. (**b**) The crystal structure of laser-amorphized GST. The ideal structure of GST, constructed from the above building blocks. Reprinted by permission from Macmillan Publishers Ltd: [Nature Materials [8], (2004)]

Sb requires one additional electron. These missing electrons must be donated to Ge/Sb and can be exactly provided by orbitals on Te pointing into vacancies (Fig. 9.5). Although this is a simplified picture, it strongly suggests that vacancies are an intrinsic part of the structure.

The ideal structure has been viewed as a "layered" structure of nine like-atom layers with the stacking sequence -Sb-Te-Ge-Te-Te-Ge-Te-Sb-Te-, as marked in Fig. 9.6a, b. This sequence is identical to that determined [25] for the stable hexagonal structure of GST. They believe that the formation of the ordered chain structure is a prelude for the phase transformation from the rocksalt-like to hexagonal structure [26].

Shortcomings of Ge-Sb-Te material: The material possesses limitations imposed by regularities in edge of bit marks as well as by jitter which is largely determined by the combined optical, thermal and crystallization properties of the medium and by the optical quality and recording strategy of the laser beam. The Ge-Sb-Te alloys have a major shortcoming in having a low crystallization temperature (~150 °C) which adversely affects the stability of the amorphous phase making the devices not suitable for application when high temperature regime is required. Hence the search for improved materials and technology is still ongoing.

9.2.5 AgInSbTe (AIST)

An equally important class of phase change materials is the Sb–Te-based alloys applied in DVD-RW disks. Matsunaga et al. [16] have performed extensive experimental X-ray diffraction (XRD), extended X-ray absorption fine structure (EXAFS) and hard X-ray photoelectron spectroscopy (HXPS) experiments with density functional simulations to determine the crystalline and amorphous structures of $Ag_{3:5}In_{3:8}Sb_{75:0}Te_{17:7}$ (AIST) and how they differ from GST.

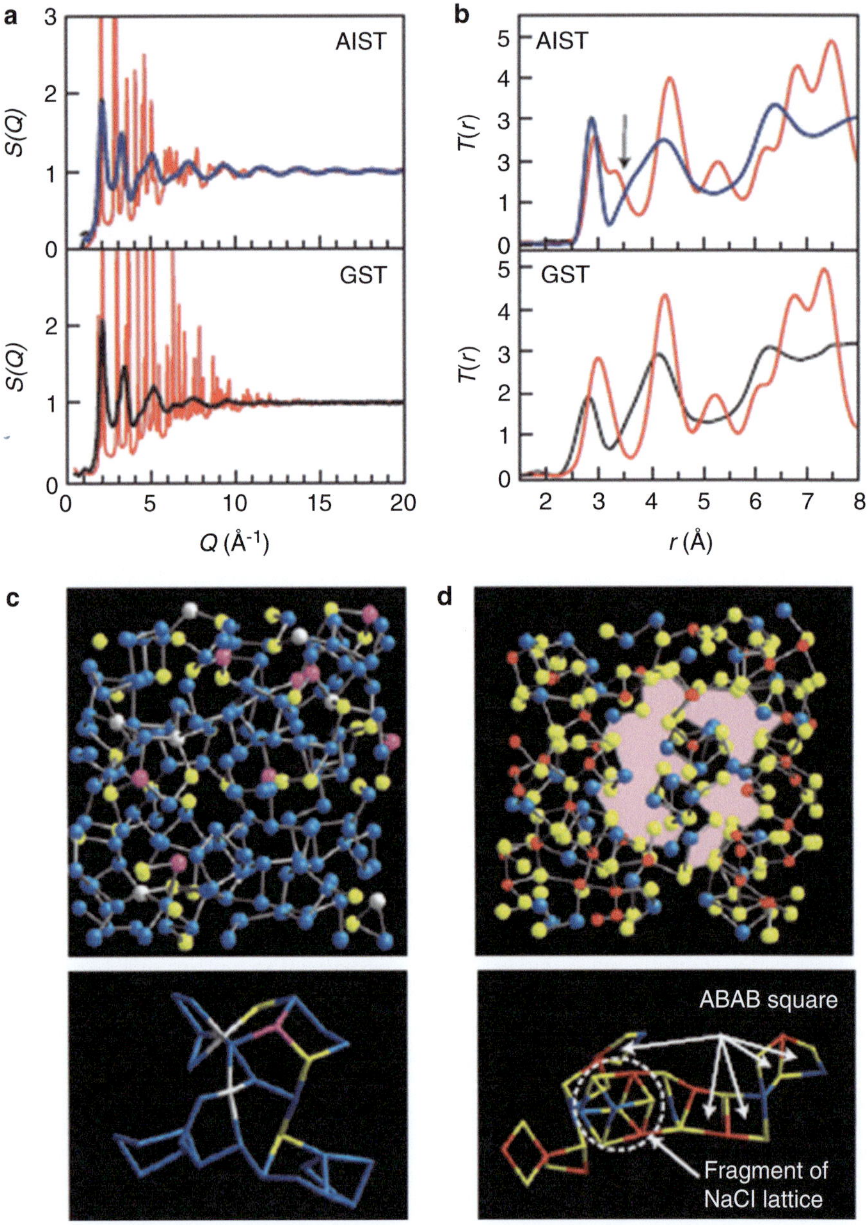

Fig. 9.7 HXRD data for AIST and GST, and atomic configurations of a-AIST and a-GST. (**a**) and (**b**): Structure factors $S(Q)$ and total correlation functions $T(r)$ of AIST and GST [16]. *Red line*, experimental data of crystalline phase; *black line*, experimental data of amorphous phase; *blue line*, DF-RMC model of a-AIST. The DF-RMC and experimental results are practically indistinguishable. (**c**) Section of 640-atom DF-MD model of a-AIST (24 Å × 24 Å × 12 Å). Ag—*silver*; In—*magenta*; Sb—*blue*; Te—*yellow*. (**d**) Section of 460-atom DF-MD model of a-GST (24 Å × 24 Å × 12 Å). Ge—*red*; Sb—*blue*; Te—*yellow*; large cavity, *pink*. Reprinted by permission from Macmillan Publishers Ltd: Nature Materials [16], copyright 2011

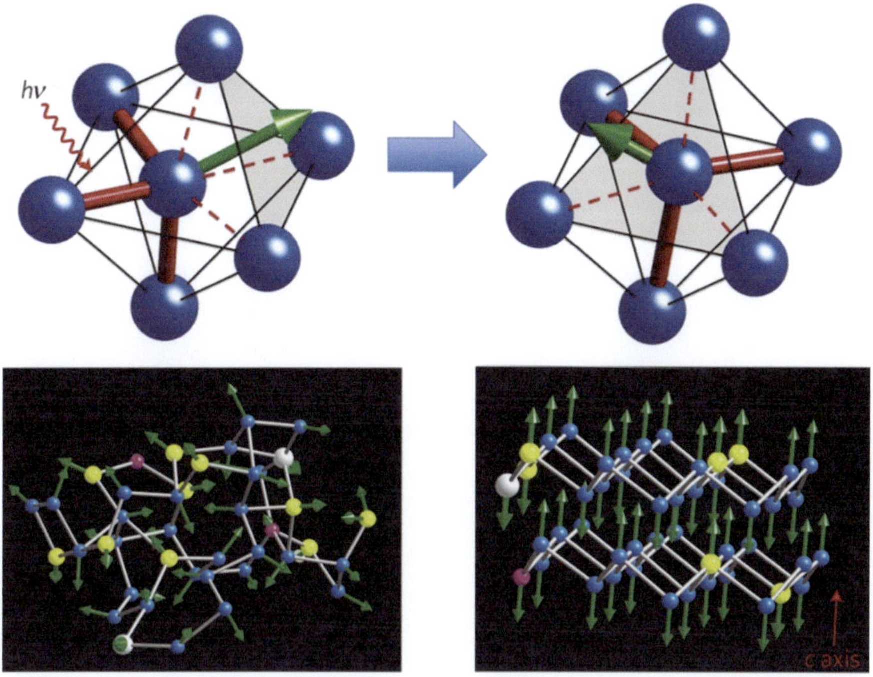

Fig. 9.8 The bonding electrons are excited by laser light, causing the atoms in the amorphous phase to move. Finally, the central atom with three short (*red*) and three long (*dashed*) bonds crosses the center of the distorted octahedron, interchanging a short and a long bond. *Green*: resultant vector of short bonds. Resonant bonding between periodic short and long bonds leads to the crystalline A7 network. The *grey sticks* (*lower right*) correspond to the *red* bonds (*upper right*). Atom colors as in Fig. 9.7c. Reprinted by permission from Macmillan Publishers Ltd: [Nature Materials] [16], copyright (2011)

The structure of amorphous (a-) AIST is found to consist of a range of atomic ring sizes, whereas a-GST shows mainly small rings and cavities. The local environment of Sb atoms resembles a distorted 3C3 octahedron in both crystalline and amorphous phases, and the bond lengths in a-AIST are slightly shorter (stronger) than in c-AIST as seen in Fig. 9.7, with enhanced insulating or semiconducting features. Crystalline AIST has a Jahn Teller distorted octahedral structure 16 and more metallic features, as reported. The similarity of the DOS is consistent with small atomic displacements during the phase change (PC). The atomic motion and/or diffusion accompanying the PC are larger in GST than in AIST, where the PC is accompanied by small changes in bond lengths.

The structures suggest a bond-interchange model, where a sequence of small displacements of Sb atoms accompanied by interchanges of short and long bonds is the origin of the rapid crystallization of a-AIST. It differs profoundly from crystallization in a-GST.

Reverse Monte Carlo (RMC)-refined density functional (DF) calculations were performed [16] in order to investigate the local environment of Sb and are shown in Fig. 9.8. The structure (bond lengths and angles) and chemical coordination numbers were calculated. Both a-AIST (left panel) and c-AIST (right panel) resemble a (distorted) 3C3 octahedron. Each vector shown is the sum of (typically three) short bonds and has been computed for each atom by summing bond vectors and applying a smooth Fermi function (cut off at 3.2 Å) for individual weights. The norm of a vector is largest for three short bonds (orthogonal directions, distorted octahedron). The vectors are reminiscent of a "pseudospin" in phase transition models and are randomly oriented in a-AIST. Laser irradiation or electric heating can lead to a sequence of small atomic shifts that align the vectors along the *c* axes of crystalline (A7, hexagonal) cells in the rim, so the entire amorphous mark takes on the A7 structure. It is suggested that the PC is simply a small displacement of the central atom as it interchanges one short (red) and one long (dashed) bond Figure 9.9 shows the difference between the crystallization processes for a-AIST and a-GST. This includes the distribution of closed loops (ring statistics) as see in Fig. 9.9a. In a-AIST a sequence of ring reconstructions by way of bond interchanges results in six fold rings with short Sb bonds (Fig. 9.9b) accompanied by small changes in bond lengths. Growth dominated crystallization dominates in these samples. On the other hand for a-GST many four and six fold rings act as nuclei for crystallization (Fig. 9.9c) and require larger atomic displacements than in a-AIST. This leads to predominance of nucleation driven crystallization in a- GST.

9.2.6 Other (Non-chalcogenide) Materials for Data Storage Devices

Photochromic materials such as diarylethenes, spyropyrans, spirooxazines, bacteriorhodopsin, and fulgides have been suggested as promising candidates for future optical data storage with high density and fast data processing rate. These materials have the ability to change between two distinct states; each state can represent "0" or "1" of a digital mode, via irradiation at different wavelengths. In this all photon mode recording, the light characteristics, such as wavelength, polarization, and phase, can be multiplexed to enable data storage and thus have the potential to increase dramatically the achievable memory density [27].

9.3 Optical Data Storage

A number of problems arise when high-capacity memory devices are being designed, particularly in obtaining a high density of information storage. One way to solve this problem consists in the application of optical methods of storage, whose possibilities were considerably extended after the invention of the laser.

The optical methods of information storage that use lasers can be divided into two classes: (1) those based on the principles of photography (point-by-point storage), and (2) those based on the principles of holography.

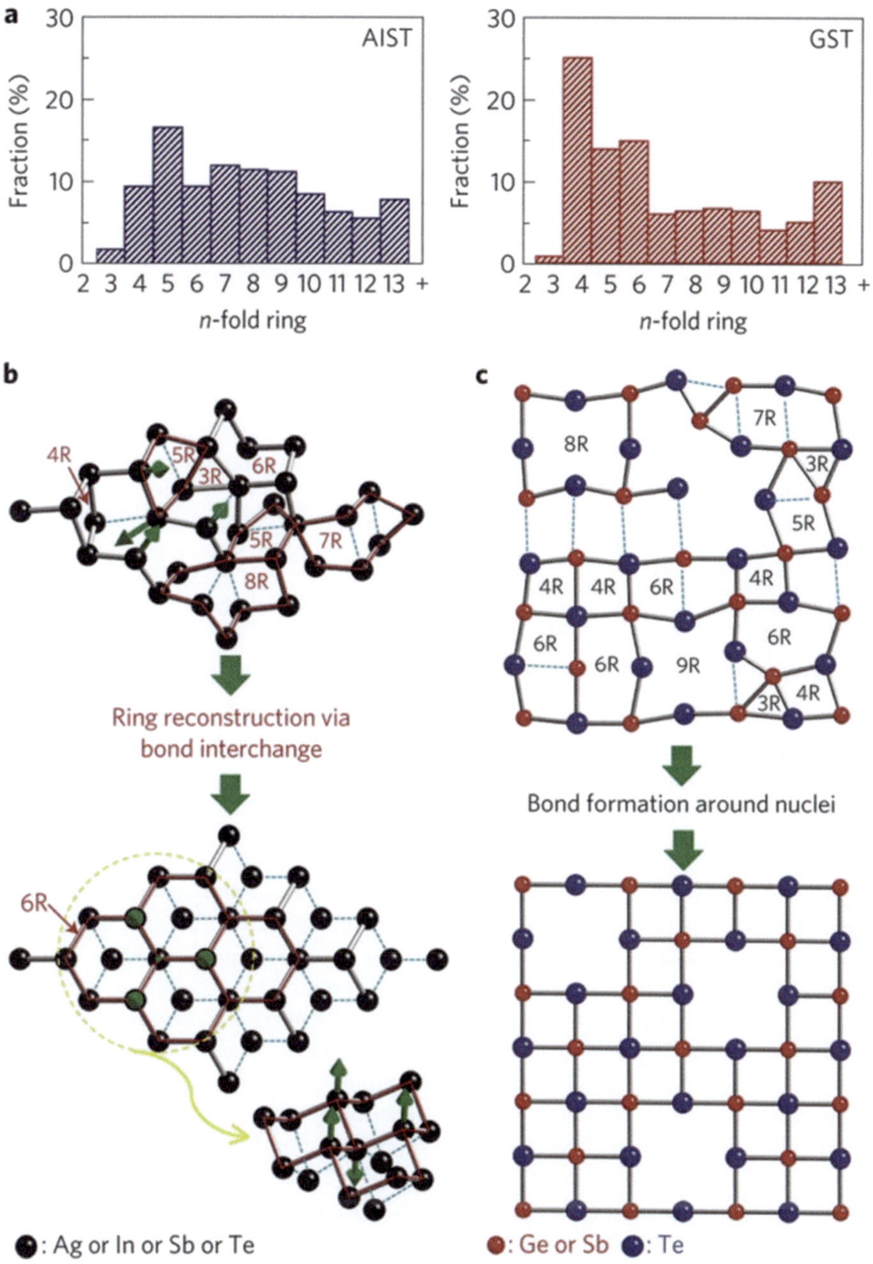

Fig. 9.9 (**a**) Ring statistics (bond cutoff 3.2 Å) reflect differences in $T(r)$ near 3.0–3.5 Å. (**b**, **c**) PC scheme in a-AIST and a-GST, respectively. *Dashed lines* show bonds corresponding to the peak in $T(r)$ near 3.5 Å (**b**). Reprinted by permission from Macmillan Publishers Ltd: Nature Materials [16] (2011)

In the point-by-point storage, use is made of the capability of the laser beam to be focused in a spot of a few microns in diameter. Due to the high degree of energy concentration in this spot, it is possible to use for the information storage not only ordinary photosensitive materials (photoemulsions, photochromic materials, and others), but also the materials in which the storage is accomplished by the heating effect of the laser beam. These include polyether film [28], europium oxide ferromagnetic film [29, 31] and ferroelectric ceramics [30].

Despite the high information density of the point-by-point storage technique, the method has a number of drawbacks: the decrease of information storage reliability with the increase of storage density caused by carrier micro defects and the presence of dust particles; the requirement of very thin photosensitive layers because of the light defocusing effect in the layer; very precise alignment of the reading devices (of the order of a few microns), since otherwise, inadmissibly large displacement of the projected image would occur. All these drawbacks of the optical method are over come to a considerable degree in the holographic method of the information storage.

9.3.1 Challenges Faced by Industry

Currently, the data storage technology faces four major challenges:

1. An appropriate optical source
2. Material properties
3. Speed of writing
4. Efficient use of the real-estate (conventional terminology used for effective surface area)

The source has been moving to shorter wavelengths, but reducing the wavelength much further may not be simple, without changing the material properties dramatically, such as improving transparency for multilayer operation. However, single layer writing at a wavelength of 200 nm may certainly be possible, quadrupling storage capacity or increasing by eightfold by simply moving to dual layer writing, with a maximum of 50 gigabytes (GB) capacity. Thus, the transparency of material properties might remain a challenge for truly multilayer storage. The third challenge is the speed of writing TB data on disks, which require large amounts of energy for ultrarapid writing at GB/s rates. For large capacity data storage to become useable, writing rates must improve to 10's GB/s or beyond. Today's technology is limited to 250 Mb/s, which takes over >8 h/TB to write. With 10 GB/s, this can be reduced to ~20 min. However, the rotation speed of the recording media is limited to ~100 RPS, and is unlikely to increase. Techniques must be found to solve this issue. Finally, the fourth challenge is the efficient use of the real-estate. This has been difficult to reduce, and is dependent on the speed of rotation of the storage medium and the duration of the light pulse in current systems.

9.3.2 Three Generations of Current Optical Disks

Since the discovery of reversible light induced crystallization of amorphous semiconductors films in the 1970s, phase change optical recording has evolved to a mature technology that is applied in rewritable (RW) versions of optical data storage systems like compact disk (CD) and digital versatile disk (DVD). The recording of information is based on writing and erasing amorphous marks in crystalline layer of a phase change material. Due to a difference in the optical properties of the crystalline and amorphous state, information can be read out as a change in reflectance. Writing of amorphous marks occurs by a pulsed focused laser beam, which heats the phase change layer locally above the melting point. After the pulse, the molten area is quenched to the amorphous state. The recorded information can be recrystallized (erased) by heating the material with the same focused laser beam above the crystallization temperature but below the melting temperature. Due to the increased mobility of the atoms at elevated temperatures, the amorphous state rapidly converts to the crystalline state.

The basic components of a digital optical recording media are:

- Drive Unit + Storage Medium in Rotating Disk form.
- Disks are preformatted using grooves and lands (tracks) → enables the positioning of an optical pick-up and recording head to access the information on the disk.
- Information is recorded as a change in materials characteristics → tiny pits in the tracks correspond to the zeroes and ones of the digital information. Stored data is retrieved by detecting the reflection difference of the reflected laser beam.

An optical disk is characterized to have a large capacity, and it is used widely as a medium for circulating and keeping images, music, or information on computers. Optical disks can be classified into "read only memory (ROM) type" which only reproduces information recorded in advance, "write-once read-many type" to which information can be written only once, and "rewritable type" to which information can be rewritten as many times as desired.

The capacity of the optical disk is determined according to the size of recording marks. The smaller the mark is, the more the capacity can be increased. The size of the recording mark basically depends on a condensed light spot size of a laser beam used for reproducing information. That is, with a smaller spot size, information of still higher density can be reproduced with a finer reproduction quality.

The spot formed by condensing laser beams by an objective lens does not converge into one point even at the condensed point and has a limited size due to a diffraction effect of light. This is generally called a diffraction limit, and it is a limit of mark length reproduced by $\lambda/(4\text{NA})$ where the wavelength of the laser beam is λ and the numerical aperture of the objective lens is NA.

For example, with an optical system where $\lambda = 405$ nm and NA = 0.85, 119 nm is the reproduction limit of the mark length, and a mark with length shorter than that cannot be read out precisely. In order to increase the capacity of the optical disk, there is a method of shortening the wavelength of the laser beam or a method of increasing NA of the objective lens. However, there is a limit in shortening the wavelength of the laser beam and increasing NA of the objective lens in terms of manufacture of optical components.

9.3.2.1 Compact Disk (CD)-Brief Review

Compact disk (CD) is a digital optical data storage format brought to the market by excellent team work of two major industries, Philips and Sony. The format was originally developed to store and play only sound recordings but was later adapted for storage of data (CD-ROM). Several other formats were further derived from these, including write-once audio and data storage (CD-R), rewritable media (CD-RW), Video Compact Disk (VCD), Super Video Compact Disk (SVCD), Photo CD Picture CD, CD-i, and Enhanced music CD. Audio CDs and audio CDs have been commercially available since October 1982. A complete listing of the various formats is given in Fig. 9.10 [32].

With the increasing popularity of personal computers (PCs), user-friendly and cost-effective ways of distribution of software and data became of crucial importance. This could initially be done with only magnetic media (floppy) disks. In the early 1990s popular software releases encompassed a series of floppies. But then the software releases grew bigger: Windows 95, for example, was released on 13 floppy disks [32]. This made the alternative offering of an even more complete CD release of the same program very attractive. CD became a crucial enabler for the evolution of the PC industry in that period of time. In addition the CD also enabled the distribution of games. The extra storage space and the random access to content enabled games with increasing sophistication and with more appealing and realistic graphics.

The seemingly simple disk represents a lot more than meets the eye. People may be intrigued by the rainbow interference colors from a compact disk. On a microscopic scale each CD contains millions of bits, coded and stored in a globally standardized form, and reproduced with unprecedented precision in millions of low-cost copies. The interference colors are a macroscopic manifestation of this. Standard CDs (single sided CD-ROM) have a diameter of 120 mm (4.7 in.) and can hold up to about 80 min of uncompressed audio or about 700 MiB of data. The recording takes the form of a helical track of etched pits commencing at the center of the disk having track to track distance of 1.66 μm. Figure 9.11 shows the typical layered structure of a CD disk.

The information layer is covered with a metallic reflective coating to enable information to be extracted by means of reflected light. This is achieved by focusing the light from an Aluminum Gallium Arsenide (AlGaAs) laser onto the track. This diode is a light source of considerable less power than that used for writing the master disk.

The laser light concentrated into a spot of 1.87 μm in diameter, follows the track thereby striking pits and non pits alternately. Due to this, light will be lost because it

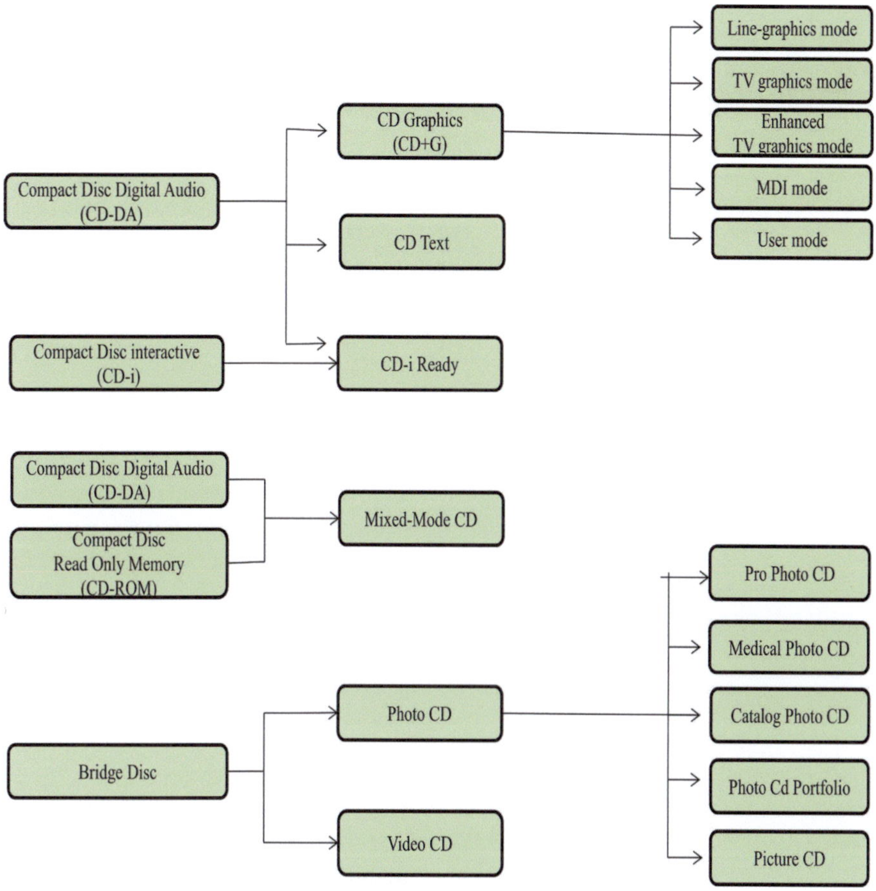

Fig. 9.10 Compact disk formats emerging from the same standard or from a combination of already established standards [32]. Permission from Springer

is diffracted over angles larger than the lens is capable of accepting. Thus the intensity of the reflected light is modulated by the physical structure of the disk and this is detected by a photodiode which, in turn, produces a modulated electrical signal.

The divergent light beam emitted by the laser is converted into a parallel beam by means of a lens. Optical pick-up unit of the prototype CD system is shown in Fig. 9.12. The parallel beam is directed toward the objective lens. It is here that the beam is focused onto the information track. The reflected, modulated light is directed at the detector diode by a prism, which serves as an output coupling mirror. A wedge is situated between this half mirror and the photodiode to split up the reflected beam into two parts, forming spots on different parts of the photo diode. The output currents of the diode parts contain the desired information signal as well as the error signals for radial tracking and focusing. The optical pick up unit is only

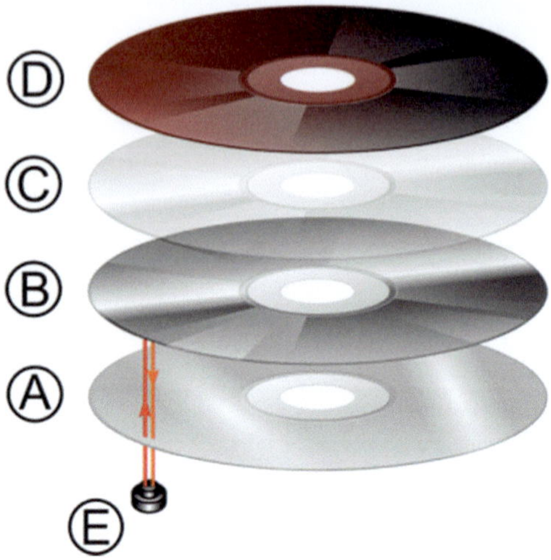

Fig. 9.11 Diagram of the basic CD layers [33]. (*A*) A polycarbonate disk layer has the data encoded by using bumps. (*B*) A shiny layer reflects the laser. (*C*) A layer of lacquer protects the shiny layer. (*D*) Artwork is screen-printed on the *top* of the disk. (*E*) A laser beam reads the CD and is reflected back to a sensor, which converts it into electronic data

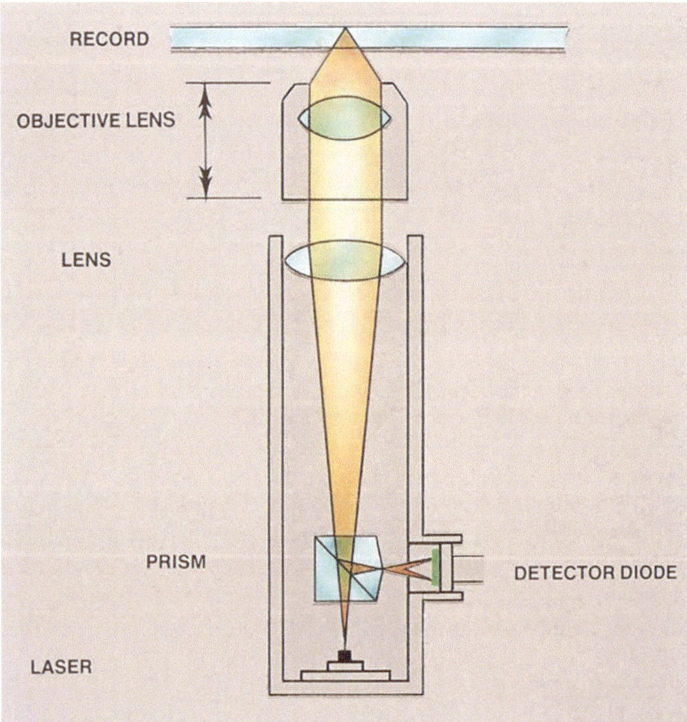

Fig. 9.12 Optical pick-up unit of the prototype CD system [32]. Permission from Springer

Fig. 9.13 An all-in-one optical pick unit for CD, DVD and BD developed by Philips. Courtesy-Philips Research Laboratories, Netherlands

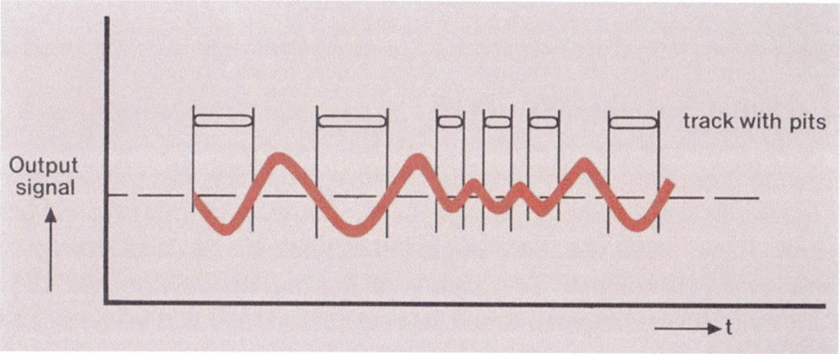

Fig. 9.14 Modulated output signal of the photo detector diode [32]. Permission from Springer

45 mm in length, 12 mm in diameter and weighs 14 g. It is mounted at the end of a moveable arm enabling it to follow the track in radial direction. The objective lens is mounted above the light-pen and with the help of a drive system of the principle of that of a loudspeaker it is possible to keep the spot focused on the information layer.

An all-in-one optical pick unit for CD, DVD, and BD was designed by Philips in 2004. The doublet lens configuration that is in the schematic diagram was only used at the very beginning, in experimental set ups. The second lens was at that time a lens of only 1 mm. Later Philips integrated it into a single lens format for BD. And in July 2004 they announced that they had made an all-in-one optical pick unit for CD, DVD, and BD as on the picture (Fig. 9.13).

The modulated output signal of the photo detector diode in relation to the pits and non pits on the CD-ROM disk is shown in Fig. 9.14.

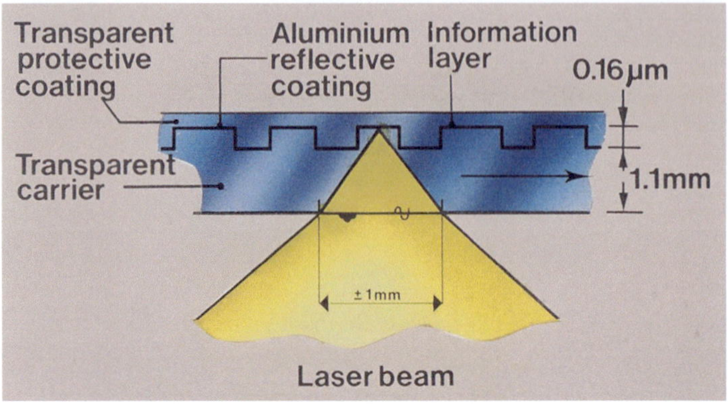

Fig. 9.15 Information readout through a transparent coating [32]. Permission from Springer.

Figure 9.15 shows that the point of information is found at a depth of 1.1 mm through the transparent disk material. The diameter of the light beam at the place it enters the disk surface is 1 mm, so 1000 μm. Dust particles and small scratches will be out of focus and intercept relatively little of the beam.

9.3.2.2 Digital Versatile Disk (DVD)

The growing consumer demand for higher storage density, faster devices and the ability to rewrite information on storage media motivated researchers and industrialists across the globe to search for improved technologies for data storage devices with improved performance. The history of the digital versatile disk (DVD) is relatively short when compared to the 20-year time period that was needed for so many CD standards to become established [32]. Digital optical disk storage format was invented and developed by Philips, Sony, Toshiba, and Panasonic in 1995. The Digital Versatile disk Read Only Memory was announced during the last months of 1995 by the DVD forum [32]. DVDs offer higher storage capacity than CDs while having the same dimensions. Prerecorded DVDs are mass-produced using molding machines that physically stamp data onto the DVD. Such disks are known as DVD-ROM, because data can only be read and not written or erased. Blank recordable DVD disks (DVD-R and DVD + R) can be recorded once using a DVD recorder and then function as a DVD-ROM. Rewritable DVDs (DVD-RW, DVD + RW, and DVD-RAM) can be recorded and erased many times.

Subsequent developments led to commercial semiconductor lasers that became available in the early 1990s [34]. It is worth mentioning here several specific technical problems that had to be solved, the stability of the laser output, quality of the readout spot, and the generation of sufficient light power.

Red and even green laser light was used during many experiments with read only optical disks, but worldwide research was also aiming at recordable and rewritable systems. In particular, the high-density media based on phase-change materials

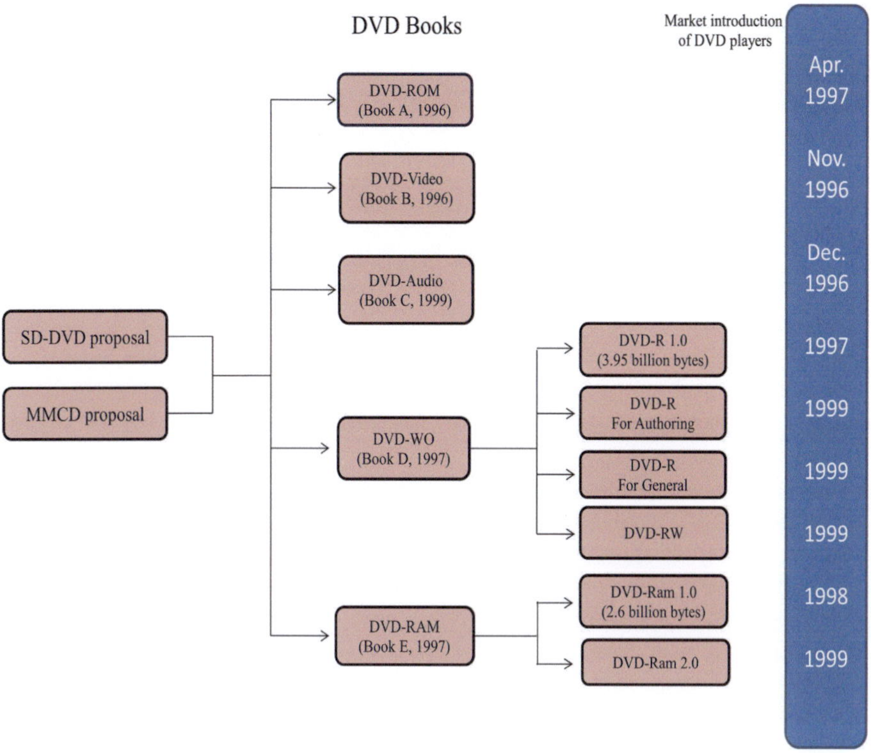

Fig. 9.16 Current classification of digital versatile disk formats endorsed by DVD forum [32] (Permission from Springer)

were considered very good candidates for use in future data storage systems [35] because no magnetic field was needed during recording and the required optical heads lacked the complexity of their magneto-optical counterparts. At about the same time, several companies pursued the miniaturization of optical and magneto-optical drives that would equally lead to small-size devices suitable for portable computers and to manufacturing expertise needed for mass-producing the more demanding components of a CD successor. Current classification of DVDs is given in Fig. 9.16.

DVD-RW, which emerged as a spin-off from the already-approved write-once specifications, must consequently be regarded as rerecordable. Looking back in time, the feasibility of a medium resembling many characteristics of the digital versatile disk and using the same recording material as CD-RW was proven already in 1996 [36]. Since, the real battle dealt with increasing the storage capacity on such media up to that of read-only DVDs. Some results were already achieved 1 year later [37], but the desired rerecordable DVD format was not yet in sight. An intermediate and quite reliable rewritable disk [38] with a storage capacity of 3.95 billion bytes was reported in 1997, although it was never standardized neither

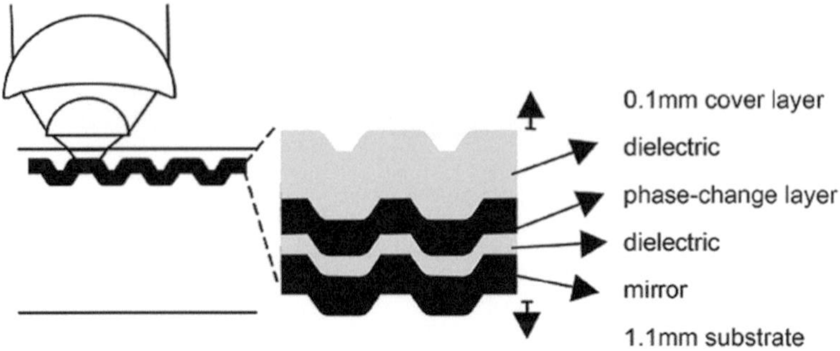

Fig. 9.17 Disk structure of DVR. Courtesy Philips Research Laboratories

commercialized. The development efforts finally led to a DVD-RW format [39, 40] that could store 4.7 billion bytes (4.38 GB) and was compatible with the version 2.0 of the DVD-R. Phase-change materials were used to achieve about 1000 direct overwrite (DOW) cycles. Note, also that, as a consequence of its spin-off from DVD-R, the new DVD-RW disk could be read out, written and erased with both 635- and 650-nm lasers. The recording speed was initially standardized to 1×, but an optional specification issued later under various revisions [41–43] allowed the media and drive manufacturers to write and erase up to 6×. The DVD-RW version 1.1 was also endorsed by European Computer Manufacturers Association (EMCA*) in an open standard [44] released in December 2002.

9.3.2.3 Disk Structure-Digital Video Recording (DVR)

A rigid 1.1-mm-thick poly-carbonate substrate is covered with a phase change stack, deposited in reversed order compared to the standard CD-RW or DVD + RW phase-change stacks (Fig. 9.17). On top of this stack a 0.1-mm-thick cover layer is applied by spin coating or foil lamination [5]. This thickness of 0.1 mm allows for sufficient tilt margin is approximately equal to that of DVD (650 nm, NA = 0.60, 0.6 mm substrate). The cover layer can be made with a thickness variation well within ±3 μm. With this thickness uniformity, there is no need for dynamical spherical-aberration correction, so a rigid dual-lens objective can be used. The substrate serves as a stiff and rigid carrier, containing the mastered information (embossed data and grooves).

9.3.2.4 Blu-ray

Blu-ray also known as Blu-ray Disk (BD), is a new optical disk format jointly developed by the Blu-ray Disk Association (BDA), a group of the world's leading consumer electronics [45], personal computer and media manufacturers (including Apple, Dell, Hitachi, HP, JVC, LG, Mitsubishi, Panasonic, Pioneer, Philips,

Samsung, Sharp, Sony, TDK, and Thomson). The format was developed to enable recording, rewriting, and playback of high-definition video (HD), as well as storing large amounts of data. The format offers more than five times the storage capacity of traditional DVDs and can hold up to 25 GB on a single-layer disk and 50 GB on a dual-layer disk. This extra capacity combined with the use of advanced video and audio codecs offer consumers an unprecedented HD experience.

- February 2002—The Blu-ray Disk specification was officially announced.
- 2003—Blu-ray Disk recorders were first launched in Japan.
- 2006—Blu-ray Disk players appeared globally.

Blu-ray or Blu-ray Disk (BD) was designed to supersede the DVD format, in that it is capable of storing high-definition video resolution (1080p). The plastic disk is 120 mm in diameter and 1.2 mm thick, the same size as DVDs and CDs. Conventional (pre-BD-XL) Blu-ray disks contain 25 GB per layer, with dual layer disks (50 GB) being the industry standard for feature-length video disks. Triple layer disks (100 GB) and quadruple layers (128 GB) are available for BD-XL re-writer drives. Recent developments by Pioneer [45] have pushed the storage capacity to 500 GB on a single disk by using 20 layers. Due to the fact that the data layer on a Blu-ray disk is placed much "closer" to the laser lens than in DVD (or even the HD-DVD proposal), there is less distortion resulting in significantly improved tolerances. Hence, more precision and ultrahigh storage densities are made possible.

The name "Blu-ray" refers to the blue laser (specifically, a violet laser) used to read the disk, which allows information to be stored at a greater density than is possible with the longer-wavelength red laser used for DVDs. Blu-ray products can easily be made backwards compatible with CDs and DVDs through the use of a BD/DVD/CD compatible optical pickup unit. The main application of Blu-ray is as a medium for video material such as feature films and physical distribution of video games for the PlayStation 3, PlayStation 4 and Xbox One. Besides the hardware specifications, Blu-ray is associated with a set of multimedia.

Two major technological breakthroughs have been achieved, which together allow a large increase in optical disk capacity. First, high numerical aperture (NA) objective lenses have become feasible by using two element lenses: NA = 0.85 lenses can be applied with sufficient system margin when readout is performed through a thin transparent cover layer of 0.1 mm thickness, instead of reading out through a 0.6-mm-thick substrate as is done for the digital versatile disk (DVD). Second, tremendous progress in the field of blue-violet diode lasers has been made diode laser samples with a wavelength around, $\lambda = 400$–410 nm and sufficient lifetime and output power have recently been realized. These combined breakthroughs allow a reduction of the focused spot size (which is proportional to (1/NA)) [2], by a factor of about 5 when compared with DVD, thus allowing a capacity of 22 GB on one layer of a 12-cm-diameter single-sided disk.

A comparison of the basic parameters for the three generations of optical disks is shown in Fig. 9.18 and the parameters are given in Table 9.4.

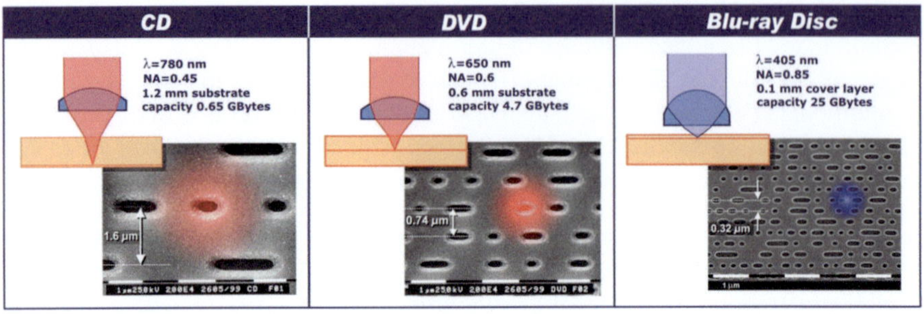

Fig. 9.18 Three generations of current optical disks. Courtesy Phillips Research Laboratories, Netherlands

Table 9.4 Parameters for the three generations of optical disks

Parameters	CD	DVD (single layer (SL))	Blu-ray (SL)
Storage capacity (GB)	0.65	4.7	25
Data transfer rate (Mbps)	1.41	11.08	36
Recording track pitch (mm)	1.6	0.74	0.32
Channel bit length (mm)	300	133	
Laser wavelength (nm)	780 (1.59 eV)	650 (1.91 eV)	405 (3.06 eV)
Numerical aperture	0.45	0.6	0.85

9.4 Multilayered Disks

Multilayered disk structures have been proposed in order to increase the disk capacity. The structure is sequentially laminated on the transparent substrate starting with a first dielectric layer; a phase control layer, the phase control layer having two areas are defined in a laser spot in which the irradiation with a reproduction beam causes a phase difference that alters the optical path of the reflected reproduction beam passing through the areas; a second dielectric layer; the phase change recording layer, which reversibly converts between a crystal phase and an amorphous phase by irradiation with a recording beam; a third dielectric layer; a reflective layer; and a protective layer. An alternative embodiment includes a fourth dielectric layer disposed between the reflective layer and the protective layer. The information of a recording mark can be accurately reproduced due to the phase difference between adjacent areas on the phase control layer, thereby reducing the size of the effective spot of the reproduction beam. Accordingly, the mixing of the signals between adjacent marks or adjacent tracks is decreased, thereby realizing a high density optical disk and improving the resolution power of a reproduction signal.

9.5 Alternative Technologies for Data Storage

9.5.1 *Holographic Data Storage*

The holographic memory is based on the concept that some reference beam can be used for recording holograms and for reconstructing the information to achieve the reconstructed images of various holograms to be placed in exactly the same position. Therefore, the virtual images of the holograms are used for reconstruction. Unlike conventional technologies, which record and recover digital data bit by bit, holography has the ability to read and write millions of bits of data with a single pulse of light, enabling data-transfer rates of billions of bits per second. The physical mechanism is photochromism, which is defined as a reversible transformation of a single chemical species between two states that have different absorption spectra and refractive indices. This allows for multiplexed holographic recording and reading, such as wavelength [46], angular, shift [48], and phase encoding.

Holographic memory offers the possibility of storing 1 terabyte (TB) of data in a sugar-cube-sized crystal [49]. A terabyte of data equals 1000 GB, one million megabytes or one trillion bytes. Data from more than 1000 CDs could fit on a holographic memory system. Most computer hard drives only hold 10–40 GB of data, a small fraction of what a holographic memory system might hold.

The general principles of holography were developed by Gabor in 1948 [50], however, the concepts were applied to data storage much later in the work of van Heerden in 1963 [31]. Predictions of storage densities on the order of $1/\lambda^3$ and the many unique properties of holography combined to serve as motivating factors in

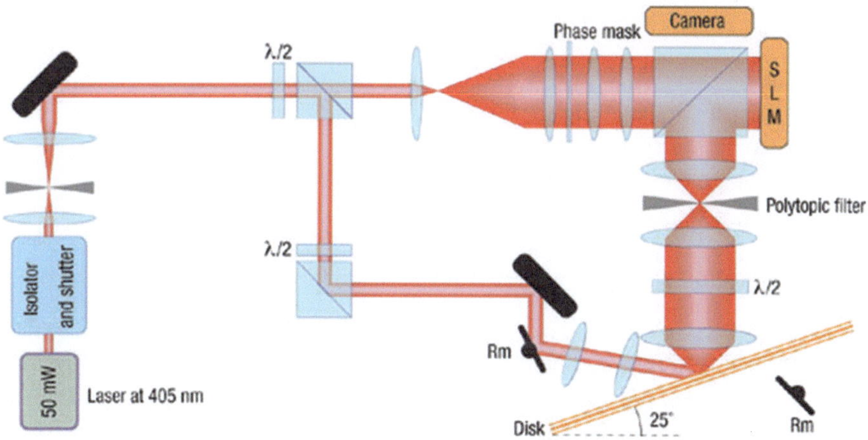

Fig. 9.19 The data is recorded using interference between the two paths. It is read using just the lower path, terminating at the camera instead of the spatial light modulator (SLM). $\lambda/2$ is a half-wave plate and Rm is a rotating mirror. Reprinted with permission from Macmillan Publishers Ltd., Nature Materials [51], Copyright 2008

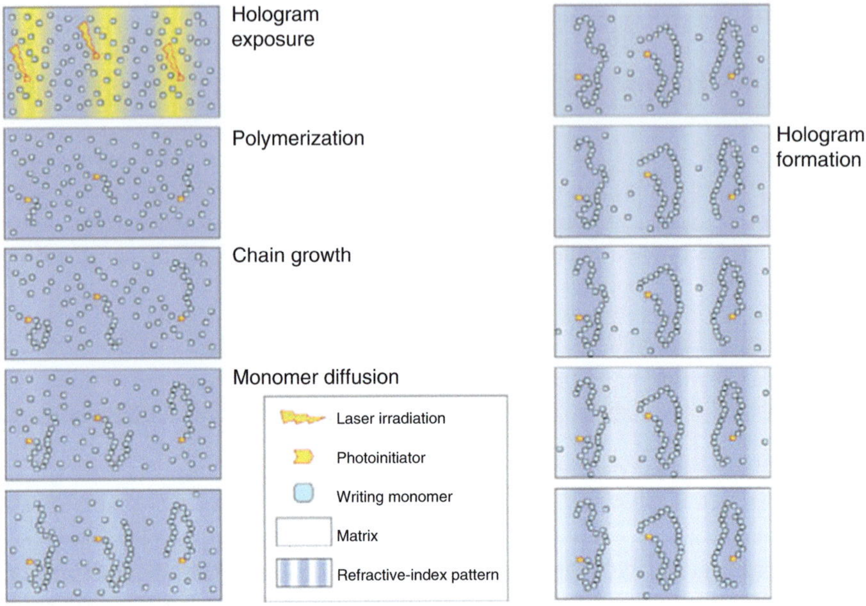

Fig. 9.20 Formation of a refractive-index pattern in a photopolymer following exposure to a hologram. Reprinted with permission from Macmillan Publishers Ltd: Nature Materials [51], Copyright 2008

the quest for the adaptation of this technology as a storage subsystem for the storage and retrieval of information.

A commercial holographic storage system requires a tunable, coherent laser source [51]. A typical setup is shown in Fig. 9.19.

Because holographic storage records throughout the volume of the medium and this volume can change anisotropically with temperature changes, a strategy to compensate for thermal effects is necessary. By changing the angle and wavelength of the reference beam used to readout the hologram, the effects of thermal changes can be compensated. Using a coated gallium nitride laser diode, a small, simple, stable, relatively high-power, tunable external-cavity laser has been developed, which uses a transmissive diffractive grating as the output coupler of the laser cavity. By using a grating, the wavelength of the diode is stabilized for holographic recording and can be changed by rotating the grating for readout. However, since the grating is transmissive (rather than the typically used reflective grating) the direction of the laser does not change. This makes the laser simple, compact and robust.

Formation of a refractive-index pattern in a photopolymer following exposure to a hologram is shown in Fig. 9.20 [52]. The unexposed recording medium consists of a cross linked polymer matrix with a low index of refraction into which a monomer with a high index of refraction is dissolved. During the holographic exposure, the coherent interference pattern of the signal and reference waves initiates a photopolymerization only in the bright (exposed) areas. This photopolymerization consumes the monomer in the bright areas, which leads to a diffusion of monomer molecules from the dark to the bright areas, further feeding the photopolymerization there. This process, leads to the desired local modulation of the index of refraction throughout the thickness of the recording medium, which forms a physical image of the interference pattern of the data page.

9.5.2 *Fabrication of Patterned Media by Nanoimprint Lithography (NIL)*

An additional challenge to the use of materials for device application is the investigation of electrical behavior where the material-metal junction for device integration and addressability offers several disadvantages. Patterned media using techniques such as photolithography, e-beam lithography, X-ray and interferometric lithography offer significant control over achieving high areal density structures. For example, a 50-nm-period patterned medium corresponds to a density of 250 Gbit/in^2, and a 25 nm-period one corresponds to a density of around 1 Tbit/in^2. Therefore a patterned medium is a promising candidate for the future recording system with a capacity beyond 1 Tbit/in^2 [53]. Nanoimprint lithography provides a way to fabricate sub 10 nm structures.

9.6 Multiphoton Technology

Multiphoton bit-by-bit memory [54] is another promising 3D optical data storage system, which gives rise to a higher storage density than single-photon excitation. It is based on the confinement of multiphoton absorption to a very small volume because of its nonlinear dependence on excitation irradiance. This characteristic provides a tool for activating chemical or physical processes with high spatial resolution in three dimensions. As a result there is less cross talk between neighboring data layers. Another advantage of multiphoton excitation is the use of infrared illumination, which results in the reduction of scattering and permits the recording of layers at a deep depth in a thick material. Two-photon 3D bit recording in photopolymerizable [55], photo bleaching [56, 57], and void creation in transparent materials [58, 59] has been demonstrated with a femtosecond laser.

9.6.1 Two Photon Technology

The optical data-storage technologies, such as CD, DVD, and Blu-ray, use one-photon-absorption processes for writing and reading information. In particular, a bit of information is recorded when photons are absorbed and induce a change in the local reflectivity of the medium, often as phase or structural changes, or burning.

In contrast, a two-photon absorption process occurs when two photons are absorbed simultaneously at the same spatial location. As two-photon processes are nonlinear and require high-optical intensities to occur, they typically only take place at the focal spot of a light beam. This characteristic offers the ability to store data within the volume of a disk, thus achieving three-dimensional (3D) storage by simply changing the focusing position of the writing beam within the medium. Using two-photon processes it is possible to store hundreds of layers within the volume of a 1-mm-thick DVD-type disk.

The idea of two-photon absorption was first predicted theoretically in 1931, but had to wait 30 years before experimental proof was possible (1961), and it was not until 1988 that the first description of a commercialized product was proposed. The reason it has taken so long for the idea of two-photon processes to become a reality is that the process requires high-power, monochromatic light sources, such as pulsed lasers. Three-dimensional data storage, as proposed in the 1980s, also needed the appropriate two-photon storage media. The special sources are needed because the absorption cross-section for two-photon processes is extremely low compared with the one-photon processes that drive natural processes, such as photosynthesis and vision.

In two-photon 3D optical data storage, writing is performed by short laser pulses at a wavelength of 532 nm (green), for example. One 532-nm photon alone will

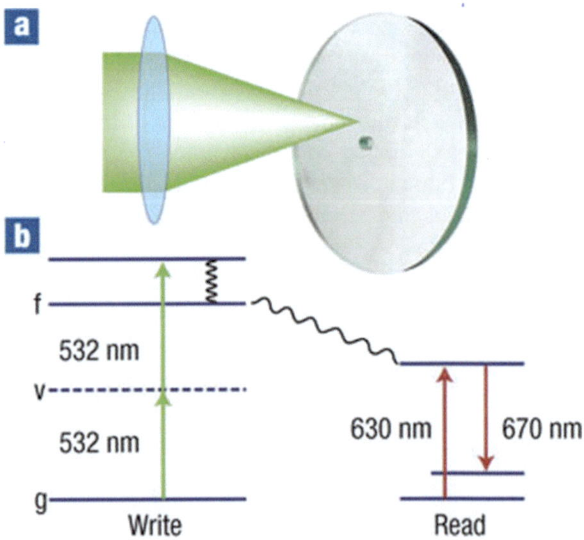

Fig. 9.21 (**a**) Writing within the volume of a disk. (**b**) Two-photon-absorption energy levels of writing and reading. *g* ground state, *v* virtual state, *f* final state. Reprinted with permission from Macmillan Publishers Ltd: Nature Materials [52] Copyright 2008

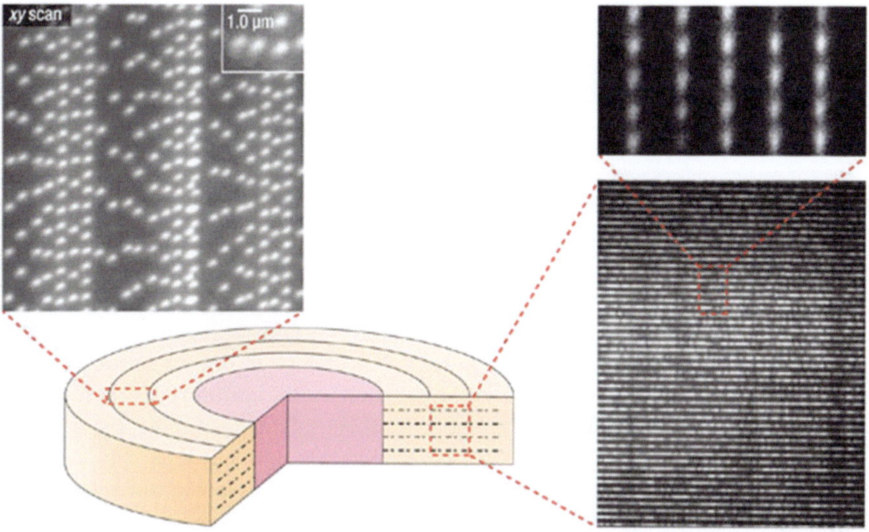

Fig. 9.22 Layers recorded in a 3D disk. Reprinted with permission from Macmillan Publishers Ltd: Nature Materials [52], Copyright 2008

either be absorbed at the surface (if its energy is enough to excite the medium) or it will be transmitted through the volume of the medium if it is of less energy.

However, when two photons overlap simultaneously where the laser beam is focused, two-photon absorption can occur imparting energy (with an equivalent wavelength of 266 nm) of twice that of the individual photon energy to the medium (Fig. 9.21). This energy can be used to locally change the medium's properties (such as making it fluorescent) and thus write a data bit of information. To read this data, the written bits are illuminated with a red laser diode that induces fluorescence from the written bits, which is collected by a photomultiplier tube or a photodiode array.

Figure 9.22 shows an image of a number of layers stored in a two-photon WORM memory. Experimental data show that the stored bits can be read over 10^6 times and that the write and read forms are thermally stable. The 3D memory disks are fabricated by the uniform dispersion of the medium's components in polymethylmethacrylate (PMMA), to produce 120-mm-diameter, 1.5-mm-thick disks, with parallel surfaces polished to optical quality.

The storage capacity can be increased even further by using blue light. They used Nichia 405-nm and 540-nm laser diodes to write and read, respectively. A standard laser-diode driver was used to modulate the laser diode, which is controlled by a standard interface. Using this system enables Blu-ray rather than DVD layer densities and effectively increases the capacity of the disk to 5 Tbyte.

It is has been proposed [52] two-photon 3D storage has the potential to be the next ultrahigh-density storage technology because of its compatibility with Blu-ray and DVD technology. It is expected that within a year a commercial prototype that is capable of storing at least 5 Tbyte on a DVD-size disk at a rate of 250 Mbits^{-1} will become available.

Although recording densities could reach terabits per cubic centimeter, these processes suffer from several drawbacks. The index modulation associated with high bit density limits the real data storage volume due to light scattering. The fluorescence can limit the data transfer rate and the lifetime of the device. Squier and Müller anticipated third-harmonic generation (THG) for 3D data reading [59].

9.6.2 *Third Harmonic Generation*

THG on 3D bits in a transparent material were first reported by Lionel et al. [60]. The contrast mechanism is neither a change in refractive index nor a change in absorption, but a change in the third-order susceptibility $\chi^{(3)}$ induced by femtosecond laser irradiation in a zinc phosphate silver containing glass. They demonstrated the possibility of 3D optical data storage inside a specific zinc phosphate glass containing Ag by using THG. The data are stored inside the glass by femtosecond laser irradiation below the refractive index modification threshold. By the accumulation effect, stable Ag clusters are created, giving rise to a formation of a nonlinear resonant interface. THG readout thus becomes possible.

Silver doped zinc phosphate glasses having composition $40P_2O_5$-$5Ag_2O$-55ZnO (mol.%) were prepared for 3D data storage by a standard melting technique in a platinum crucible [60]. The glass possesses an absorption cutoff wavelength at 280 nm (due to the Ag ions associated absorption at 260 nm) and emits fluorescence mainly at 380 nm when excited at 260 nm. The intrinsic fluorescence is due to $Ag+$ isolated in the glass. Glasses samples having thickness of 1 mm were prepared, and polished to obtain optical quality on both sides. These samples were irradiated using a femtosecond laser source emitting 440 fs, 9 MHz repetition rate pulses at 1030 nm. With a laser mode TEM00, and the output polarization TM. The maximum output average power is close to 6 W, which results in a maximum energy per pulse of 600 nJ.

Transmission, fluorescence, and THG readout images of bits recorded with different irradiance levels and laser shots is shown in Fig. 9.23c, d. The main advantages of this technique compared to usual 3D data storage are no photo bleaching, no change in linear refractive index, and therefore no scattering. Moreover, the THG signal is coherent and gives rise to a rather intense, directional, and less divergent beam with a high signal-to-noise ratio. Due to the fast THG response, the reading speed is limited only by the pulse duration. Thus a storage capacity in the range of Tbits cm^{-3} is anticipated in the glass. THG images present high signal-to-noise ratio and no cross talk.

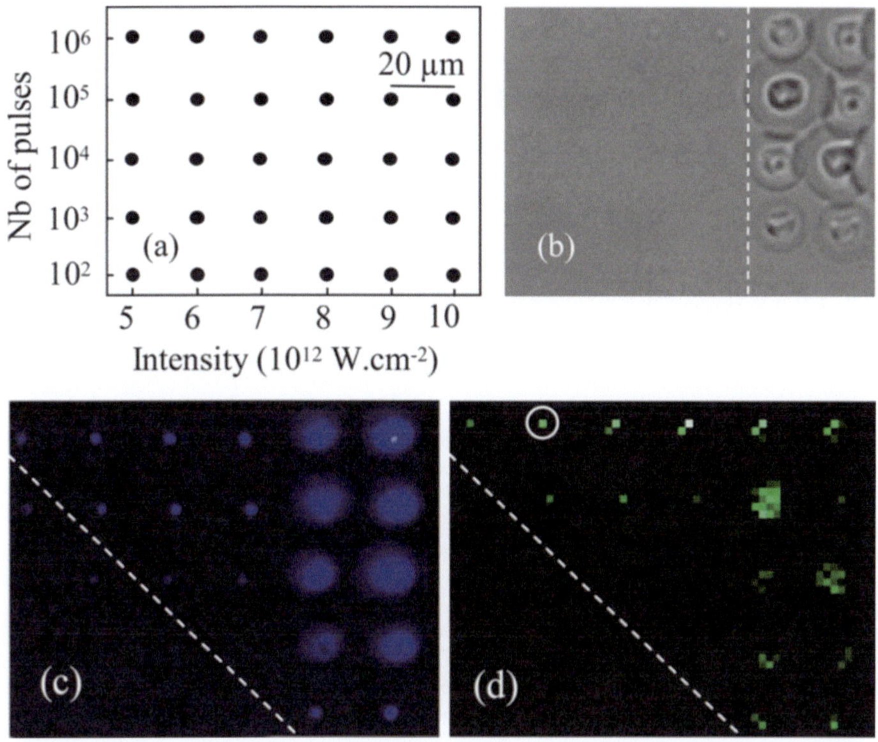

Fig. 9.23 Microscopy imaging of laser-induced species following the experimental map sketch (**a**) (*x* axis, laser intensity; *y* axis, number of laser pulses, 6_5 bits pattern; spacing, 20_m). (**b**) Epiwhite light microscopy image reveals linear refractive index modifications; *vertical dashed line*, damage threshold. Epifluorescence microscopy image (excitation wavelength, 365 nm; emission filter, 610 ± 40 nm); (**c**) *slanted dashed line*, threshold corresponding to induced fluorescent species. (**d**) THG image (excitation wavelength, 1030 nm; emission filter, 350 ± 50 nm); *slanted dashed line*, THG threshold; *encircled area*, data storage irradiation conditions ($I = 6 \times 10^{12}$ W cm^{-2}, $N = 10^6$) [60]. Reproduced with Permission from Optical Society under US Copyright Law

9.7 Near Field Phase Change Optical Recording and Super-Resolution Near Field Optical Disk—Super-RENS

In view of continuing efforts to increase the data density by reducing the spot size and improvements in optics, Blu-ray faces materials as well as technological constraints. On the technological side, optics lies at the heart of integrated circuit fabrication processes and is central to the operation of commercial optical disks. Conventional optics will have diminishing importance as the problem of interest push further into the nanometric regime, because the limit of its resolution, as originally determined by E. Abbe [61] is only ~$\lambda/2$ or ~200 nm for visible light (λ is

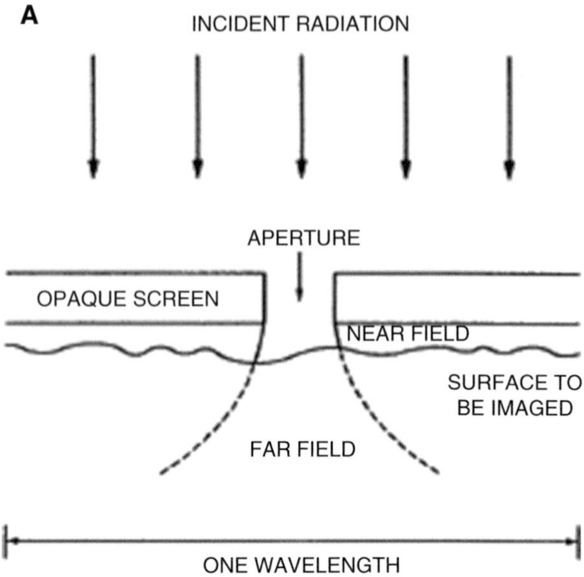

Fig. 9.24 Near field scanning optic microscopy as originally conceived. [From Near-Field Optics: Microscopy, Spectroscopy, and Surface Modification Beyond the Diffraction Limit, Eric Betzig and Jay K. Trautman, Science, New Series, Vol. 257, No. 5067 (Jul. 10, 1992), pp. 189–195 [62] With Permission from AAAS]

the wavelength of light). Under these circumstances, superior resolution could be produced in two ways (1) use of shorter wavelength radiation such as in electron or X-ray microscopy, and (2) various forms of scanning probe microscopy such as scanning tunneling microscopy. Near field optical recording is a promising approach being tried by researchers across the globe. It was first proposed and demonstrated by Betzig et al. [62] that by means of near-field scanning optical microscope (NSOM) it is possible to achieve ultrahigh density data storage. Fourier optics can be used [63, 64] to demonstrate easily that the diffraction limit to resolution in optical microscopy is not fundamental but rather arises from the assumption that the detection element (that is, a lens) is typically many wavelengths away from the sample of interest. However, by laterally scanning a source or detector of light in close proximity to the sample, one can generate an image at a resolution functionally dependent on only the probe size and the probe-to-sample separation, each of which can, in principle, be made much smaller than the wavelength of light. The first such probe to be proposed [65, 66], consists of a sub wavelength-diameter aperture in an optically opaque screen (Fig. 9.24). Light incident upon one side of the screen is transmitted through the aperture and, if the sample is within the near field, illuminates only one small region at any one time. The validity of this concept was first demonstrated by E.A. Ash and G. Nicholls [67], who attained $X/60$ resolution by using $X = 3$ cm microwaves. As originally conceived [65], an illuminated aperture acts as a light source of sub wavelength dimensions that can be scanned close to an object to generate a super-resolution image.

Later Terris et al. [68] adopted a solid immersion lens (SIL) to achieve higher recording density and speed. SIL is based on the principle that the numerical

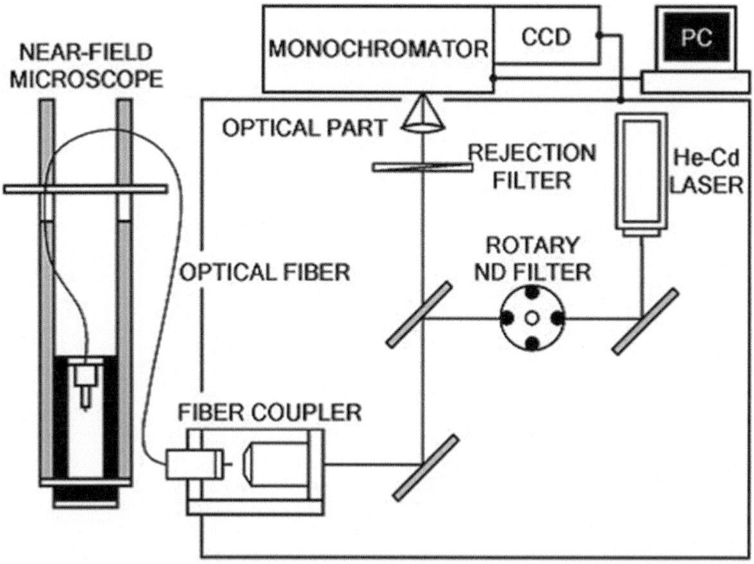

Fig. 9.25 Experimental setup for NSOM. Reprinted with permission from IOP [70]

aperture (NA) of the objective lens can be larger than 1 if the light is focused completely within the glass of the lens. Provided the air gap between the exit of the lens and the disk surface is very small, some photons of light can "jump" this small gap and reach the information layer on the disk. Shinoda et al., have investigated high-density near-field readout using SIL with a high refractive index [69]. By using a glass material with a high refractive index of 2.08, they developed an optical pick-up with the effective NA of 1.8. They have achieved 50 GB capacity disk in 120 mm diameter.

Kimura et al. [70] demonstrated an optical recoding on a CdSe nanocrystal thin film by using NSOM experimental setup shown in Fig. 9.25. Both positive and negative luminescent images were recorded on the CdSe NC thin films by utilizing photoluminescence (PL) enhancement and photobleaching respectively. Since the PL enhancement is reversible, CdSe NC thin films are exploitable as recording media (Fig. 9.26). Possibility of high density data storage was shown.

Tominaga et al. have proposed an approach of super-resolution near-field structure disk [71]. They named this structure as super-RENS. In this structure, a mask layer of SB is designed to be very close to the PC recording layer within the distance of near-field. During recording and readout process, a small aperture that can function as a local near-field SIL will be formed in the mask layer. Therefore this kind of optical disk is called aperture type super-RENS. Since the near-field distance is well controlled by the spacing layer between the mask layer and the recording layer, the difficulties of the feedback control of the probe can be overcome. Super-RENS is considered as a more feasible way of the near-field optical

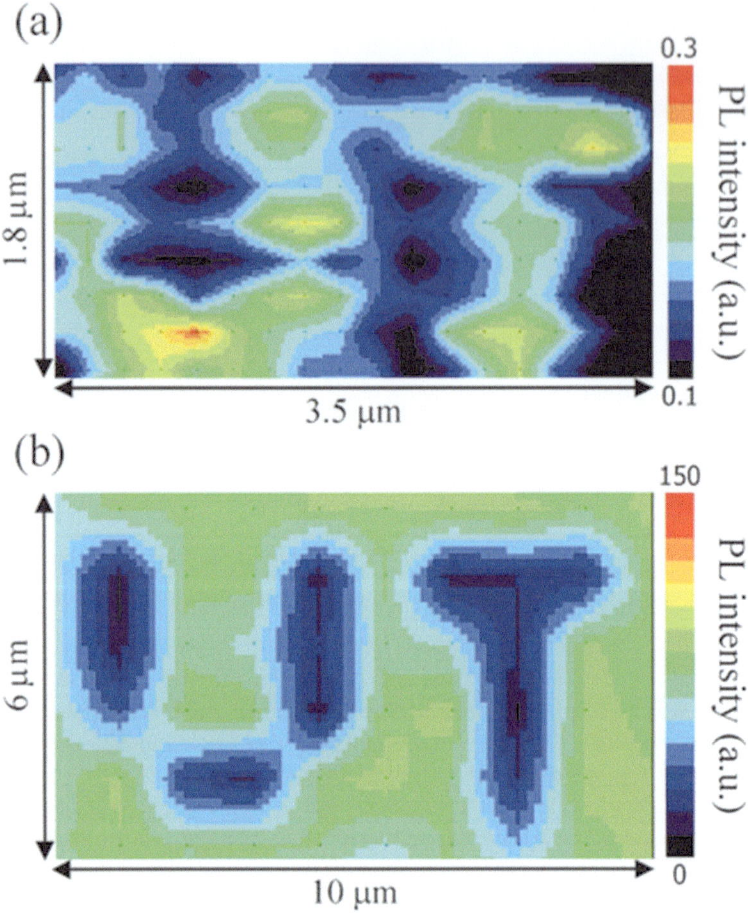

Fig. 9.26 (**a**) Positive luminescent image of the letters 'UT' recorded on the NC thin film using NSOM based on PL enhancement. The aperture diameter of the probe is 410 nm. The excitation intensities for writing and read-out processes were 2 and 0.5 nW, respectively. The image consists of 100 × 180 data points generated by an interpolation of 10 × 18 measured data points. (**b**) Negative image recorded on the NC thin film based on photobleaching. The aperture diameter is 500 nm. The excitation intensity was 200 nW for both writing and read-out processes. The image consists of 60 × 100 data points generated by an interpolation of 6 × 10 measured data points. Reprinted with permission from IOP [70]

recording with simpler recording head design, less mechanic damage, no contamination and higher recording speed.

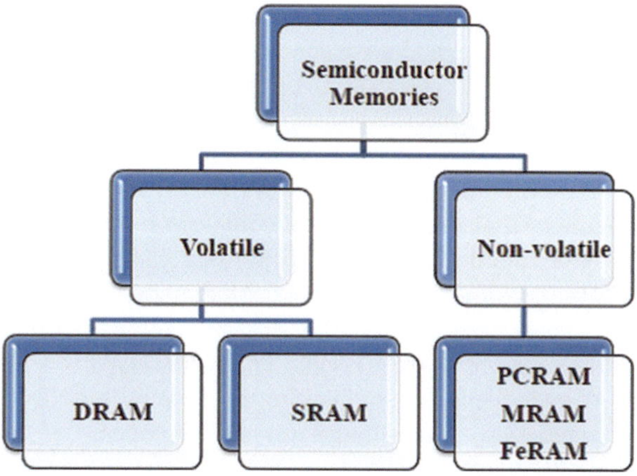

Fig. 9.27 Classification of semiconductor memories

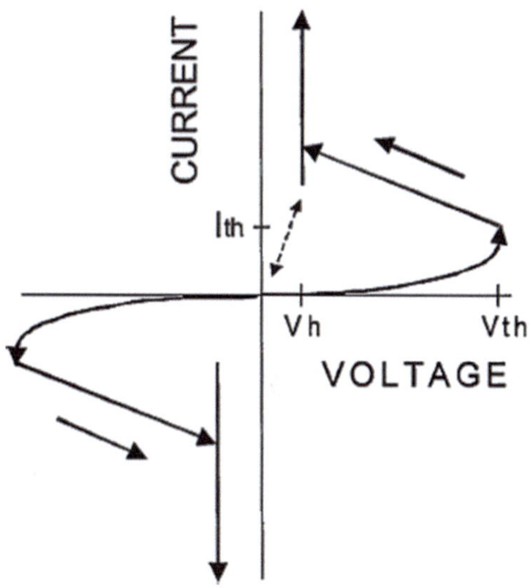

Fig. 9.28 Principle of threshold switching

9.8 Semiconductor-Electrical Memory

Solid-state memories have become the most attractive segment in the global semiconductor market. They have taken one third of the whole semiconductor market in revenue with the fastest growth. Advances in memory technology have come from the constant and almost predictable reduction in data bit structure size. Some of the advancements in memory technology have been structural and even procedural in nature. However, continuing to reducing the size of the transistor gate will reach the physical limit of around 10 nm and that may occur somewhere between 2010 and 2020.

Semiconductor memories can broadly be divided into two categories (refer Fig. 9.27):

1. Volatile, which would lose data in the absence of a power supply; this includes the dynamic random access memory (DRAM) and Static Random access memory (SRAM). The DRAMS are cost-effective and space-efficient as each cell consists of one transistor and one capacitor. SRAM is a faster memory with lower standby current compared to DRAM. However, a single memory cell of SRAM consists of four or six transistors, resulting in a lower chip density and makes it more expensive in comparison to DRAM.
2. Non-volatile memories (NVM) that can retain stored information even in the absence of a power supply. Each memory cell consists of a single transistor and enables high density data storage. This includes the flash memory which can save data for up to 10 years. The NAND type is widely used in camcorders, hand phones, laptops, etc. Despite the existing NAND's superior performance characteristics, it faces long-term scaling issues.

In memory switching, a bistable phase change material can retain its changed conduction state in the absence of current or voltage. As shown in Fig. 9.28, once in low resistance state, a sustaining current or voltage is not required.

This is because the resistance drop of memory switching is caused due to change in the atomic structure configuration of the material when subjected to electrical pulses.

OUM devices based on $Ge_2Sb_2Te_5$ can retain data for 10 years at 130 °C.

9.8.1 Phase Change Random Access Memories: PCRAMs

PCRAM is also known as Ovonic Unified Memory (OUM) and was originally developed by Energy Conversion Devices, Inc. (ECD), and licensed exclusively to Ovonyx, Inc. OUM uses a reversible structural phase-change—from the amorphous phase to crystalline phase—in a thin-film chalcogenide alloy material as the data storage mechanism. The small volume of active media in each memory cell acts as a fast programmable resistor, switching between high and low resistance with >40× dynamic range. The programming pulse drives the memory cell into a high or low resistance state, depending on current magnitude. Information stored in the cell is read out by measurement of the cells resistance.

GST is so far considered to be a desirable material for solid-state memories—Phase Change Random Access Memories: PCRAMs. Self-assembled nanowires-based phase change material memory devices offer an attractive solution owing to

Table 9.5 Binary, ternary, and quaternary alloys investigated for PCRAM applications

Binary	Ternary	Quaternary
Ga Sb	$Ge_2Sb_2Te_5$	Ag In Sb Te
In Sb	In Sb Te	(Ge Sn) Sb Te
In Se	Ga Se Te	Ge Sb (Se Te)
Sb_2 Te_3	Sn Sb_2 Te_4	$Te_{81}Ge_{15}Sb_2S_2$
Ge Te	In Sb Ge	

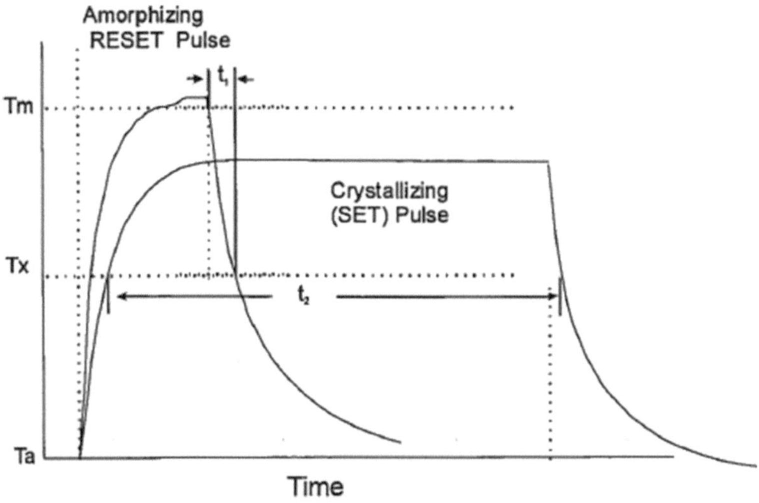

Fig. 9.29 SET and RESET states of a PCRAM device

their sub-lithographic sizes and unique geometry, coupled with the facile etch-free processes with which they can be fabricated.

Many phase-change alloys have been described in the technical literature. Some of these are listed in Table 9.5:

- OUM Devices use the GeSbTe alloy system.

Thermal and Mechanical Properties of $Ge_2Sb_2Te_5$ Alloys:

Thermal conductivity (W/cm K)	4.63E−3 (hcp)
Heat capacity (J/cm^3 K)	1.25
Density (g/cm^3)	6.20 (fcc)
Linear thermal expansion Coefficient (K^{-1}) [300–900 K]	23.5 ppm

For the Phase-change memory to be applied to computer uses, the required data writing cycle times must be at least 10^5 for replacing flash memory, and can be as high as 10^{15} to 10^{18} for replacing dynamic random access memory (DRAM). Crucial attributes of a universal non-volatile memory device for next generation memory chips are scalability, low power consumption, ultrafast switching, cyclability endurance, and compatibility with CMOS architectures.

Current Phase Change Memory (PCM) device technology has encountered technical limitations due to process-induced material damage at sub-100 nm sizes, thereby stimulating great interest in developing new materials and processing techniques that will allow continued scalability for ultrafast, low power, and high density memory devices, while maintaining data non-volatility. The conventional Flash memories face many reliability issues, such as possibility of failure of only a minor statistical "tail" of cells in an array.

Optimization of the active material may offer a way to improve the reliability and hence the scaling potential of PCM devices. Doping and alloying of the

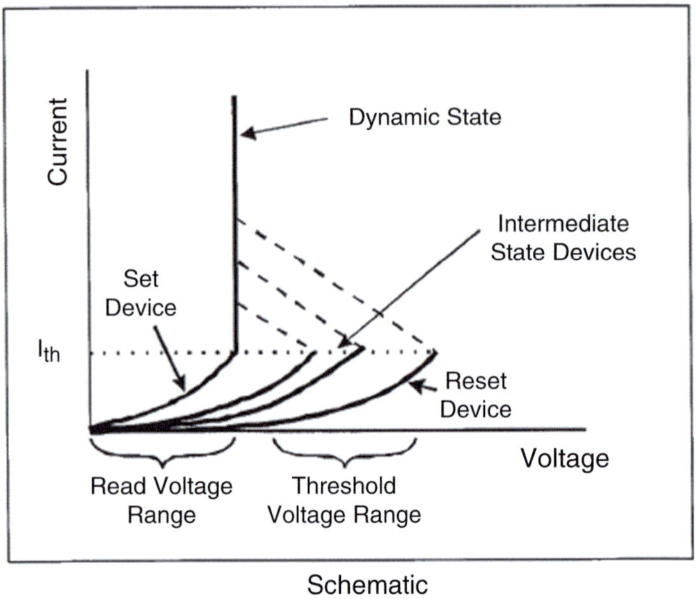

Fig. 9.30 Schematic of different states of PCRAM device

Fig. 9.31 Basic structure of an OUM device when programmed to the RESET or (high-resistance) state

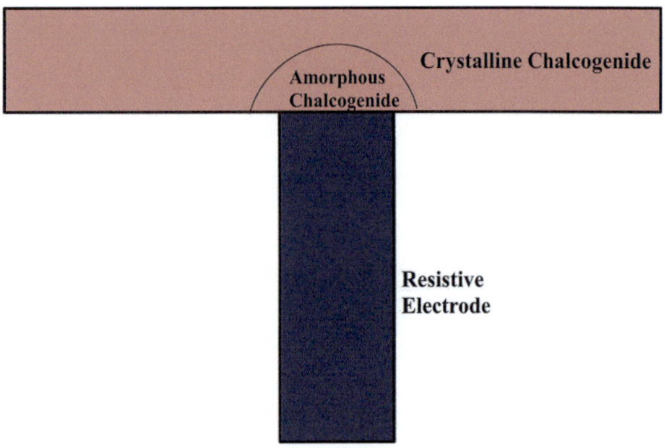

chalcogenide material with silicon [72] and Indium [73] have been shown to be beneficial in terms of data retention behavior, thus providing a solution to alleviate possible size dependent crystallization effects for ultra-scaled PCM cell. An efficient material engineering in the PCM device requires a detailed understanding and modeling of the relation between chemical composition, crystallographic and electronic structure, electrical/thermal properties and phase change parameters. A second challenge faced by the industry is the scaling down of materials to low dimensions in order to increase the data storage density. This can be best achieved by nano-structuring of the materials. The material $Ge_2Sb_2Te_5$ presently used for data storage applications has many limitations in terms of structural irregularities in its solid-state configuration. Instant data recording on these materials utilizes two

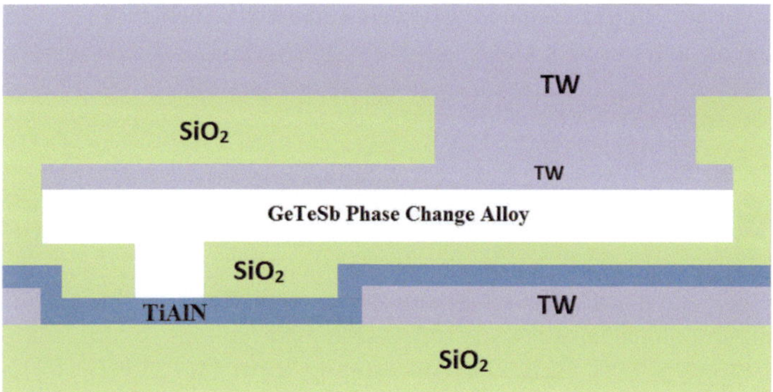

Fig. 9.32 Schematic of a typical OUM device

Fig. 9.33 Techniques being explored for scaling of data storage devices

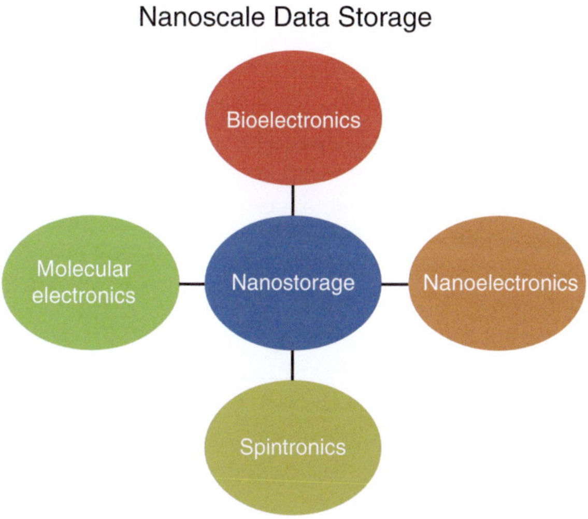

state process involving (1) melting/quenching (RESET) and (2) crystallization (SET) of the surface dots as shown in Fig. 9.29. Figure 9.30 shows the SET, RESET, and intermediate states during the operation of devices.

Magnitude and stability of electric current necessary for RESET and the time periods of both SET and RESET are the major bottlenecks in the existing technology. The required RESET current to melt the material and transform it back to the amorphous phase is rather high. The fastest SET/RESET pulses reported so far [74] using near-field optical lithography, varies from 90 ns/1.5 ns for 500 nm cells to 2.5 ns/ 0.4 ns for 19 nm cells. In order to provide enough RESET current, the size of the access device for the memory cell needs to be relatively large which directly increases the manufacturing cost. Furthermore, the performance of the chip will be degraded when the writing current of the unit cell is high since the overall writing speed will be reduced. In order to reduce the RESET current, scaling becomes a challenging issue for phase change memories; this can be achieved by confining the phase change material in a small volume. Figures 9.31 and 9.32 show the basic structure and cross-sectional view of an OUM device.

9.8.2 Scaling of PCM Devices

Chalcogenide films with reversible amorphous-crystalline phase transitions have been commercialized as optically rewritable data storage media [75, 76], and intensive effort is now focused on integrating them into electrically addressed non-volatile memory devices (Phase change random-access memory or PCRAM) [77, 78]. Since the energy required for phase transformation decreases with cell size, the write current scales with cell size and thus facilitates memory scaling. Major issues related to PCRAM devices are: (1) material optimization, (2) scaling of the memory cell by synthesizing materials at the nanoscale using bottom up approaches of chemical synthesis and vapor-transport method, and (3) investigation of phase change characteristics of nanomaterials and nanodevices. Figure 9.33 shows the various techniques being explored for scaling of data storage devices.

The challenge of bringing electronic phase change memory successfully to market has necessitated the application of several nonconventional methods for depositing thin films of phase change materials in order to reduce the RESET current. Some of these methods are as follows: (a) confining the phase change material (Ge-Sb-Te (GST) used presently) in a small volume [79]; (b) reactive ion etching (RIE) or trimming the blanket GST to scale down the volume becomes problematic due to the damage layer from RIE [80]; (c) filling GST into small vials by physical vapor deposition (PVD) is difficult due to the nature of the non-conformal PVD deposition profile; and (d) the chemical vapor deposited (CVD) films in general tend to be polycrystalline with large grains and rough surfaces and face the issue of inhomogeneous morphologies with incomplete lateral growth thus preventing its direct application for devices due to its non-reproducible nature.

Currently, phase change films are most commonly deposited by sputter deposition and patterned by conventional lithography [81]. However, as device features are pushed towards the deep sub-100-nm regime, the conventional scaling methods of the semiconductor industry face increasing technological and fundamental challenges [82–84]. For example, device size fluctuations may result in a large spread in device characteristics at the nanoscale, affecting key parameters such as the threshold voltage and on/off currents. The increasing costs associated with lithography equipment and operating facilities needed for traditional manufacturing creates an economic barrier to continued increases in the capabilities of conventional processor and memory chips. Scaling characteristics can be addressed by means of nanoscale bottom-up technology to assemble electronic/photonic devices at ultrasmall scales. Recently, the feasibility of Ge-Te and $Ge_2Sb_2Te_5$ nanowires memory has been reported [85]; however, critical issues related to memory switching efficiency, cyclability endurance and operation speed, in addition to size scaling effects on writing current and data retention properties are as yet undetermined.

Chemically synthesized nanomaterials [86] offer a chance to examine the scaling physics, and ultimate scaling limits, of phase change material properties. Phase change nanowires grown by vapor-liquid-solid mechanism are single crystals with diameters down to 30 nm or so. Devices fabricated from these wires have already begun to elucidate fundamental scaling properties of phase change memory at small dimensions. Nanopatterned thin films can illuminate the properties of phase change material confined to small volumes—a model for the memory devices of the future. Ultimately, sub 5 nm diameter nanoparticles of phase change materials are expected to reveal the physics of the amorphous-crystalline phase transformation in these materials in the small-volume limit and enable predicting of changing device behavior due to scaling. Coupling such predictions with the growing understanding of how materials characteristics control device properties will allow phase change memory to be successfully scaled to the extremely high densities required in decades to come.

9.9 Eternal 5D Data Storage

As per the recent press report by Professor Peter Kazansky [87], scientists at the University of Southampton have made a major step forward in the development of digital data storage that is capable of surviving for billions of years. Using nano-structured glass, scientists from the University's Optoelectronics Research Centre

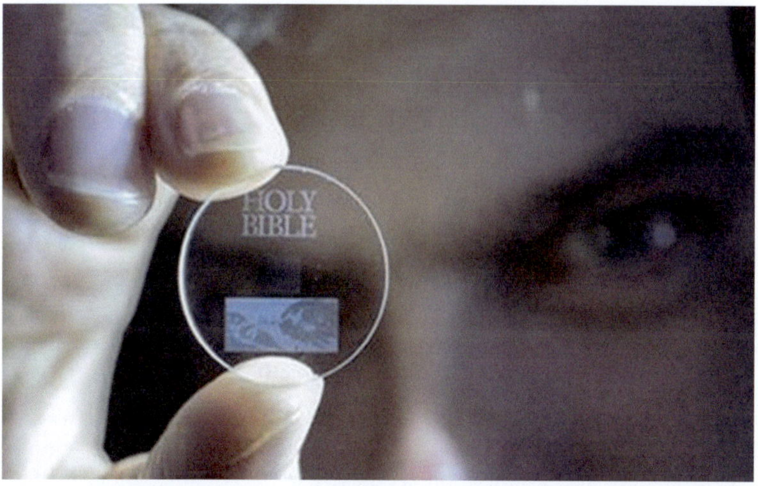

Fig. 9.34 Eternal 5D data storage, copyright of the Optoelectronics Research Centre, University of Southampton

(ORC) have developed the recording and retrieval processes of five-dimensional (5D) digital data by femtosecond laser writing under the guidance of Professor Kazansky (Fig. 9.34).

The storage allows unprecedented properties including 360 TB/disk data capacity, thermal stability up to 1000 °C and virtually unlimited lifetime at room temperature (13.8 billion years at 190 °C) opening a new era of eternal data archiving. As a very stable and safe form of portable memory, the technology could be highly useful for organizations with big archives, such as national archives, museums, and libraries, to preserve their information and records. The technology was first experimentally demonstrated in 2013 when a 300 kb digital copy of a text file was successfully recorded in 5D.

Now, major documents from human history such as Universal Declaration of Human Rights (UDHR), Newton's Opticks, Magna Carta, and Kings James Bible have been saved as digital copies that could survive the human race. A copy of the UDHR encoded to 5D data storage has been recently presented to UNESCO by the ORC at the International Year of Light (IYL) closing ceremony in Mexico.

The documents were recorded using ultrafast laser, producing extremely short and intense pulses of light. The file is written in three layers of nanostructured dots separated by five micrometers (one millionth of a meter). The self-assembled nanostructures change the way light travels through glass, modifying polarization of light that can then be read by combination of optical microscope and a polarizer, similar to that found in Polaroid sunglasses. Coined as the "Superman memory crystal," as the glass memory has been compared to the "memory crystals" used in the Superman films, the data is recorded via self-assembled nanostructures created in fused quartz. The information encoding is realized in five dimensions: the size and orientation in addition to the three-dimensional position of these nanostructures.

9.10 Future Outlook and Summary

The issue of addressing the need for improved data storage for next-generation optical memories has been led by Blu-ray. Blu-ray has addressed this issue by reducing the spot size for recording data by moving to 405 nm from 630 nm and has doubled the capacity. Nevertheless, it is difficult to envisage how shorter UV wavelengths can be used to further enhance the density of data storage, as few materials have the transparency in that region. Thus several proposals have been made to overcome the limits facing the industry, none being more promising than 3D storage using holography [88]. This technology is poised to offer truly gigantic increases in the capacity to Terabytes (TB) in disks of the size of current DVDs, but significant challenges exist in solving ghosting and higher order diffraction issues. Reading speeds are also a limitation, as these tend to be slow. Other technologies have also been suggested, such as second and third harmonic generation, which may allow multilayer (>10) writing to become reality, increasing storage capacity manyfold [51]. One serious problem that makes this task difficult is the deterioration in the signal as one tries to reach deeper layers. To solve this problem, third-harmonic has been offered as a solution by ensuring that only the material's third order nonlinearity (TON) is altered, without affecting its linear optical properties [60]. This approach works on the principle of locally enhancing the TON by precipitating tiny silver nano-crystals in silver zinc phosphate glass. However, even in this technique, the bit packing capacity is limited by the wavelength of the laser beam to approximately 3 μm separation. Another approach is based on two-photon absorption in materials, which can be used to change material properties locally, e.g., to make it fluorescent. A third different wavelength beam then does the reading. Multilayers are thus possible, as the writing process is nonlinear and therefore optical intensity dependent, localized to the focused region. These different technologies offer significant improvements over the current Blu-ray system; however, there are still many challenges that may impact the final outcome. It appears that no one technology will be the industry standard, as cost will drive the application, for either home or industrial use.

References

1. S. Moller, C. Perlov, W. Jackson, C. Taussig, S.R. Forrest, A polymer/semiconductor write-once read-many-times memory. Nature **426**, 161–169 (2003)
2. T. Matsunaga, N. Yamada, Crystallographic studies on high-speed phase-change materials used for rewritable optical recording disks. Jpn. J. Appl. Phys. **43**, 4704–4712 (2004)
3. a) Meinders, E. R., Mijiritskii, A. V., van Pieterson, L., Wuttig, M. Optical Data Storage, Vol. 4 (Philips Research Book Series, Springer, 2006). b) Volker L. Deringer, Richard Dronskowski, M. Wuttig, Microscopic complexity in phase change materials and its role for applications, Adv. Funct. Mater. **25**, 6343–6359, 2015, doi:10.1002/adfm.201500826
4. S.R. Ovshinsky, Phys. Rev. Lett. **21**, 1450 (1968)
5. S.R. Ovshinsky, Optical cognitive information processing—a new field. Jpn. J. Appl. Phys. **43** (7B), 4695–4699 (2004). doi:10.1143/JJAP.43.4695
6. T. Ohta, K. Nishiuchi, K. Narumi, Y. Kitaoka, H. Ishibashi, N. Yamada, T. Kozaki, Jpn. J. Appl. Phys. **39**, 770 (2000)
7. T. Ohta, S.R. Ovshinsky, in *Photo-induced Metastability in Amorphous Semiconductors*, ed. by A.V. Kolobov (Wiley-VCH, Berlin, 2003) pp. 310–326
8. A.V. Kolobov, P. Fons, I.F. Anatoly, L.A. Alexei, J. Tominaga, T. Uruga, Nat. Mater. **3**, 703 (2004)
9. S.R. Elliott, in *Materials Science and Technology*, vol. 9, ed. by J. Zarzycki (VCH, Weinheim, 1991), p. 375
10. N. Yamada, E. Ohno, K. Nishiuchi, N. Akahira, M. Takao, Rapid Phase transitions of GeTe-Sb2Te3 pseudobinary amorphous thin films for an optical disc memory. J. Appl. Phys. **69**, 2849–2856 (1991)
11. H. Iwasaki, Y. Ide, M. Harigaya, Y. Kageyama, I. Fujimura, Jpn. J. Appl. Phys. **31**, 461–465 (1992)
12. K. Shimakawa, A. Kolobov, S.R. Elliott, Photoinduced metastability in amorphous semiconductors and insulators. Adv. Phys. **44**(6), 475–588 (1995)
13. H. Nishihar, M. Haruna, T. Suhara, in *Optical & Electro-Optical Engineering Series*, vol. 215, ed. by R.E. Fisher, W.E. Smith (McGraw Hill, New York, 1989), p. 151

14. C. Bichara, J.-Y. Raty, J.-P. Gaspard, Structure and bonding in liquid selenium. Phys. Rev. B **53**, 206–211 (1996)
15. H. Iwasaki, Y. Ide, M. Harigaya, Y. Kageyama, I. Fujimura, Jpn. J. Appl. Phys. **25**, 1992 (1991)
16. T. Matsunaga, J. Akola, S. Kohara, T. Honma, K. Kobayashi, E. Ikenaga, R.O. Jones, N. Yamada, M. Takata, R. Kojima, From local structure to nanosecond recrystallization dynamics in AgInSbTe phase-change materials. Nat. Mater. **10**, 129–134 (2011)
17. M.H.R. Lankhorst, L. van Pieterson, M. van Schijndel, B.A.J. Jacobs, J.C.N. Rijpers, Jpn. J. Appl. Phys. 1 **42**, 863 (2003)
18. B. Liu, F.-X. Gan, Appl. Phys. A: Mater. Sci. Process. **77**, 905 (2003)
19. L. van Pieterson, M.H.R. Lankhorst, M. van Schijndel, A.E.T. Kuiper, J.H.J. Roosen, Phase-change recording materials with a growth-dominated crystallization mechanism: "A materials overview". J. Appl. Phys. **97**, 083520 (2005)
20. M.H.R. Lankhorst, J. Non-Cryst. Solids **297**, 210 (2002)
21. J.H. Coombs, A.P.J.M. Jongelis, W. van Es-Spiekman, B.A.J. Jacobs, J. Appl. Phys. **78**, 4906 (1995)
22. C. Trappe, B. Béchevet, B. Hyot, O. Winkler, S. Facsko, H. Kurtz, Jpn. J. Appl. Phys. 1 **39**, 766 (2000)
23. T. Matsunaga, R. Kojima, N. Yamada, K. Kifune, Y. Kubota, Y. Tabata, M. Takata, Inorg. Chem. **45**, 2235 (2006)
24. I.I. Petrov, R.M. Imamov, Z.G. Pinsker, Electronographic determination of the structures of Ge_2Sb_2Te5 and $GeSb_4Te_7$. Sov. Phys. Cryst. **13**, 339–344 (1968)
25. J. Gonzalez-Hernandez et al., Free carrier absorption in the Ge: Sb: Te system. Solid State Commun. **95**, 593–596 (1995)
26. J. Tominaga et al., Ferroelectric catastrophe: beyond nanometer-scale optical resolution. Nanotechnology **15**, 411–415 (2004)
27. S. Yanlin, D. Zhu (eds.), *High Density Data Storage-Principle, Technology, and Materials* (World Scientific, Singapore, 2009)
28. A 10^{13} bit mass memory reads and writes with laser. Comput. Des. **6**(3), 38–39 (1967)
29. Hot spot storage, Electronics **40**, 50 (1967)
30. More for your memories with ceramics, Des. News **22**(25), 10–11 (1967)
31. P.J. van Heerden, A new optical method of storing and retrieving information. Appl. Opt. **2**(4), 387–392 (1963)
32. H. Peek, J. Bergmans, J. van Haaren, F. Toolenaar, S. Stan, Origins and successors of the compact disc—contributions of Philips to optical storage, in *Philips Research Book Series*, ed. by F. Toolenaar, vol. 11 (Springer, The Netherlands, 2009)
33. Wikipedia. https://en.wikipedia.org/wiki/Compact_disc
34. G. Hakkatoshi, K. Nitta, K. Itaya, Y. Nishikawa, M. Ishikawa, M. Okajima, High power InGaAlP laser diodes for high density optical recording. Jpn. J. Appl. Phys. 1 **31**(2B), 501–507 (1992)
35. K. Uchino, K. Takada, T. Ohno, H. Yoshida, Y. Kobayashi, High-density pulse width modulation recording and rewritable capability in GeSbTe phase-change system using visible laser beam at low linear velocity. Jpn. J. Appl. Phys. 1 **32**(11B), 5354–5360 (1993)
36. M. Shinotsuka, T. Shibaguchi, M. Abe, Y. Ide, Potentiality of the Ag-In-Sb-Te phase change recording material for high density erasable optical disc. Jpn. J. Appl. Phys. 1 **36**(1B), 536–538 (1997)
37. K. Nagata, T. Saimi, S. Furukawa, K. Nishiuchi, N. Yamada, N. Akahira, 4.7 GB phase-change optical disk for an authoring system of digital versatile disc. Jpn. J. Appl. Phys. 1 **37**(4B), 2236–2240 (1998)
38. E. Muramatsu, A. Yamaguchi, K. Horikawa, M. Kato, S. Taniguchi, S. Jinno, M. Yamaguchi, H. Kudo, A. Inoue, The new rewritable disc system for digital versatile disc. Jpn. J. Appl. Phys. 1 **37**(4B), 2257–2258 (1998)

39. M. Yamaguchi, T. Togashi, S. Jinno, H. Kudo, E. Muramatsu, S. Taniguchi, A. Inoue, 4.7 GB phase change optical disc with in-groove recording. Jpn. J. Appl. Phys. 1 **38**(3B), 1806–1810 (1999)
40. M. Yoshida, Y. Shimoda, H. Ishii, A. Inoue, 4.7 Gbyte Re-writable disc system based on DVD-R system. IEEE Trans. Consum. Electron. **45**(4), 1270–1276 (1999)
41. DVD Forum. *DVD Specifications for Re-Recordable Disc (DVD-RW)*. Optional Specifications. 2X Speed DVD-RW. Revision 1.0, 2002
42. DVD Forum. *DVD Specifications for Re-Recordable Disc (DVD-RW)*. Optional Specifications. 4X Speed DVD-RW. Revision 2.0, 2003
43. DVD Forum. *DVD Specifications for Rewritable Disc. DVD-RAM (4.7 Gbyte). Part 1: Physical Specifications*. Version-up Information (2.1 to 2.2), 2004
44. ECMA-338. 80 nm (1, 46 Gbytes per side) and 120 nm (4, 70 Gbytes per side) DVD Re-recordable Disk (DVD-RW), 2002
45. Josh Dreuth, Pioneer Increases Disc Size to 500GB. http://www.blu-ray.com/news/?id=1616
46. D. Yu, J. Wu, G. Qian, H. Park, J. Am. Chem. Soc. **128**, 8148–8149 (2006)
47. N. Yamada, Development of Materials for Third Generation Optical Storage Media, in *Phase Change Materials*, eds. by S. Raoux, M. Wuttig. (Springer, New York, 2008), pp. 199–226
48. F.H. Mok, Opt. Lett. **18**, 915 (1993)
49. D. Psaltis, M. Levene, A. Pu, G. Barbastathis, Opt. Lett. **20**, 782 (1995)
50. D. Gabor, A new microscopic principle. Nature **161**, 777–778 (1948). doi:10.1038/161777a0
51. L. Dhar, K. Cortis, T. Facke, Nature Photonics, Technol. Focus **2**, 403–411 (2008)
52. E. Walker, P.M. Rentzepis, Two photon technology—a new dimension. Nature Photonics, Technol. Focus **2**, 406–408 (2008)
53. G. Della Valle, R. Osellame, P. Laporta, Micromachining of photonic devices by femtosecond laser pulses. J. Opt. A: Pure Appl. Opt. **11**, 049801 (2009)
54. X. Li, C. Bullen, J.W.M. Chon, R.A. Evans, M. Gu, Appl. Phys. Lett. **90**, 161116 (2007)
55. J.H. Strickler, W.W. Webb, Opt. Lett. **16**, 1780 (1991)
56. S. Pan, A. Shih, W. Liou, M. Park, J. Bhawalkar, J. Swiatkiewicz, J. Samarabandu, P.N. Prasad, P.C. Cheng, Scanning **19**, 156 (1997)
57. D. Day, M. Gu, Appl. Opt. **37**, 6299 (1998)
58. H. Jiu, H. Tang, J. Zhou, J. Xu, Q. Zhang, H. Xing, W. Huang, A. Xia, Opt. Lett. **30**, 774 (2005)
59. J.A. Squier, M. Müller, Appl. Opt. **38**, 5789 (1999)
60. L. Canioni, M. Bellec, A. Royon, B. Bousquet, T. Cardinal, Three-dimensional optical data storage using third harmonic generation in silver zinc phosphate glass. Opt. Lett. **33**(4), 360–362 (2008)
61. E. Abbe, Archiv. Mikros. Anat. **9**, 413 (1873)
62. E. Betzig, J.K. Trautman, Near-field optics: microscopy, spectroscopy, and surface modification beyond the diffraction limit. Science, New Series **257**(5067), 189–195 (1992)
63. J.W. Goodman, *Introduction to fourier optics* (McGraw-Hill, New York, 1968)
64. G.A. Massey, Appl. Opt. **23**, 658 (1984)
65. E.H. Synge, Philos. Mag. **6**, 356 (1928)
66. D. McMullan, The near-field concept has been independently formulated by several researchers, but E. H. Synge's work was first brought to the attention of the near field community. Proc. R. Microsc. Soc. **25**, 127 (1990)
67. E.A. Ash, G. Nicholls, Nature **237**, 510 (1972)
68. B.D. Terris, H.J. Marnin, G.S. Kino, Appl. Phys. Lett. **65**, 388 (1994)
69. M. Shinoda, K. Saito, T. Kondo, T. Ishimoto, A. Nakaoki, Jpn. J. Appl. Phys. **42**, 1101 (2003)
70. J. Kimura, S. Maenosono, Y. Near-field optical recording on a CdSe nanocrystal thin film. Yamaguchi, Nanotechnology **14**, 69 (2003). doi: 10.1088/0957-4484/14/1/316
71. T. Tominaga, T. Nakano, N. Atoda, Appl. Phys. Lett. **73**, 2078 (1998)
72. B. Qiao, J. Feng, Y. Lai, Y. Cai, Y. Lin, T. Tang, B. Cai, B. Chen, Semicond. Sci. Technol. **21**, 1073–1076 (2006)

73. T. Morikawa, K. Kurotsuchi, M. Kinoshita, N. Matsuzaki, Y. Matsui, Y. Fujisaki, S. Hanzawa, A. Kotabe, M. Terao, H. Moriya, T. Iwasaki, M. Matsuoka, F. Nitta, M. Moniwa, T. Koga, N. Takaura, Doped In-Ge-Te phase change memory featuring stable operation and good data retention. IEDM Tech. Dig. **12.3**, 307–310 (2007)
76. Q.F. Wang, L. Shi, S.M. Huang, X.S. Miao, K.P. Wong, T.C. Chong, Dynamics of ultrafast crystallization in As-deposited Ge2Sb2Te5 films. Jpn. J. Appl. Phys. **43**, 5006–5008 (2004)
77. M. Mansuripur, Rewritable optical disk technologies. Proc. SPIE **4109**, 162–176 (2000)
78. N. Yamada, Erasable phase-change optical materials. Mater. Res. Soc. Bull. **21**, 48–50 (1996)
79. M.H.R. Lankhorst, B.W.S.M.M. Ketelaars, R.A.M. Wolters, Low cost and nanoscale non-volatile memory concept for future silicon chips. Nat. Mater. **4**, 347–352 (2005)
80. S. Hudgens, B. Johnson, Overview of phase change chalcogenide non-volatile memory technology. Mater. Res. Soc. Bull. **29**, 829–832 (2004)
81. S.L. Cho, J.H. Yi, Y.H. Ha, B.J. Kuh, C.M. Lee, J.H. Park, S.D. Nam, H. Horii, B.K. Cho, K.C. Ryoo, S.O. Park, H.S. Kim, U.-I. Chung, J.T. Moon, B.I. Ryu, Symp. VLSI Tech. Dig. **2005**, 96–97 (2005)
82. E.A. Joseph, T.D. Happ, S.H. Chen, S. Raoux, C.F. Chen, M. Breitwisch, A.G. Schrott, S. Zaidi, R. Dasaka, B.Yee, Y. Zhu, R. Bergmann, H.L. Lung, and C. Lam, in *International Symposium on VLSI Technology, Systems and Applications, VLSI-TSA* (2008), pp. 142–143
83. D.J. Milliron, D.B. Mitzi, M. Copel, C.E. Murray, Solution processed metal chalcogenide films for p-type transistors. Chem. Mater. **18**, 587–590 (2006)
84. A. Allan, *International Technology Roadmap for Semiconductors* 2005 ed., http://public.itrs.net/
85. D.J. Frank et al., Device scaling limits of Si MOSFETs and their application dependencies. Proc. IEEE **89**, 259–288 (2001)
86. K.K. Likharev, Electronics below 10 mm, in *Nano and Giga Challenges in Microelectronics*, ed. by J. Greer, A. Korkin, J. Labanowski (Elsevier, Amsterdam, 2003), pp. 27–68
87. Professor Peter Kazansky, Eternal 5D data storage could record the history of human kind, 18 February 2016, http://www.southampton.ac.uk/news/2016/02/5d-data-storage-update.page
88. G.A. Rakuljic, V. Leyva, A. Yariv, Optical data storage by using orthogonal wavelength-multiplexed volume holograms. Opt. Lett. **17**, 1471 (1992)

Chapter 10
Photovoltaics

P.K. Shishodia and Gurinder Kaur Ahluwalia

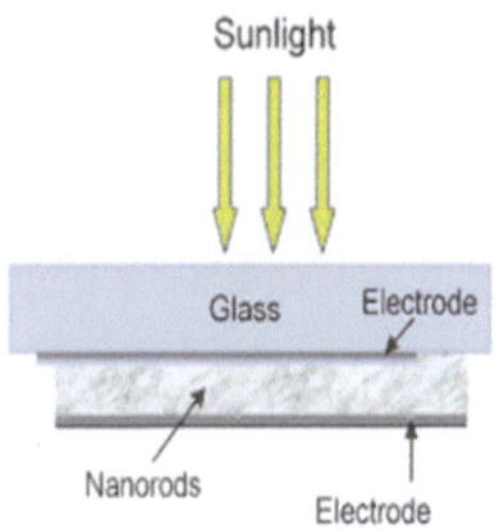

P.K. Shishodia, Ph.D. (✉)
Department of Physics & Electronics, Zakir Husain Delhi College, University of Delhi, Jawaharlal Nehru Marg, Delhi, 110002, India
e-mail: Shishodiaprem@gmail.com

G.K. Ahluwalia, Ph.D
Materials and Nanotechnology Research Laboratory, Department of Physics College of The North Atlantic, Labrador West Campus, 1600 Nichols Adam Highway, Labrador City, A2V 0B8, NL, Canada
e-mail: gurinder.kaur@cna.nl.ca

G.K. Ahluwalia (ed.), *Applications of Chalcogenides: S, Se, and Te*,
DOI 10.1007/978-3-319-41190-3_10

10.1 Introduction

10.1.1 Energy Crisis and Green Energy Solutions for Future

With the increasing energy demand, depletion of fossil fuels, climate changes associated with global warming; there is an urgent need for environmentally sustainable energy technologies. Renewable energy technologies, including photovoltaic, solar thermal, wind turbines, hydropower, wave and tidal power, biomass-derived liquid fuels and biomass-fired electricity generation supply only 14 % of the energy consumed worldwide. Among all the technologies allowing the production of renewable energy, photovoltaic technology is a promising alternative as a clean energy source.

Renewable energy resources such as solar, wind, and hydrothermal energy, by their nature will never be exhausted. Among alternative sources of energy, solar energy is considered as environmentally sustainable and non-polluting source. It is the most readily obtainable source of energy; it is cost-effective and helps in lessening the greenhouse effect. Simply through the photovoltaic cells, solar radiation gets converted into electricity and the power can be used or stored in the backup battery to be used at night. Solar energy has been used in different forms since the ancient times. After a rise in oil prices in 1970s, many countries began to conduct extensive research and development programs to harness solar energy. The price of photovoltaic cells is also going down. As a result, the usage of solar energy has increased. The benefits of using solar power are becoming evident every day. As per the "PV Roadmap Toward 2030 (PV2030)" the target production cost for PV module is expected to be 50 yen/W [1]. It is expected that the PV power generation can supply approximately 50 % of residential electricity consumption (approximately 10 % of total electric consumption) in 2030. But it would be very difficult to reach this goal only by the conventional technologies. One important concept to reduce the solar cell cost and to increase the conversion efficiency is to use nanotechnology, i.e., nanostructured material in solar cell. China, Japan, and USA are expected to be the largest markets, whereas the largest contributors in terms of absolute growth will be China, the USA, and India.

Photovoltaic power generation employs solar panels composed of a number of solar cells containing a PV material. Cells require protection from the environment and are usually packaged tightly behind a glass sheet. When more power is required than a single cell can deliver, cells are electrically connected together to form PV modules or solar panels. A single module is enough to power an emergency telephone, but for a house or a power plant the modules must be arranged in multiple arrays. A photovoltaic module is a packaged, connected assembly of typically 6×10 solar cells. Each module is rated based on the application level. The functional requirements of solar photovoltaics are efficiency, durability, easy installation, light weight, and disposability. For large-scale installations, prices below $1.00/W are now common.

10.1.2 Four Generations of Solar Cells

On the application level, the functional requirements of solar photovoltaics are efficiency, durability, easy installation, light weight, and disposability. Nanotechnology-based solutions may be used to improve the functional requirements of solar photovoltaics and to speed competitive PV-applications market entry. It may also increase the efficiency of semiconductor materials and enable novel applications of solar PV technology, such as integration with building materials or even clothing.

As for photovoltaics, half of the sun's power lies in the visible spectral and, and thus much effort in both conventional and solution processed solar cells has been devoted to harvesting within the visible regime. The sun's distribution of spectral power across the visible, near-infrared (NIR), and short wave infrared (SWIR) mandates that if a solar cell is to be made from a single junction, the bandgap of the light absorbing medium is best chosen to lie at 1.1 μm, at the beginning of the SWIR. The sun's spectrum spans the visible, the NIR, and a major portion of the SWIR. Indeed half the sun's power lies in the IR beyond 700 nm. Even if two junctions (a tandem solar cell) or three junctions are employed, all of the constituent materials are optimally chosen to have bandgap in the near or short-wave infrared. Colloidal quantum dots provide an attractive materials system in which to implement these functions, including in the infrared wavelengths. The most familiar chalcogenide-based colloidal quantum dot materials employing Cadmium as cation do not generally extend well into the infrared. In contrast, PbS and PbSe, with their bulk bandgap in the vicinity of 0.4 eV, provide remarkably wide size-effect tuning. The absorption onset of PbS can be tuned to lie anywhere from the beginning of the SWIR at 1000 nm all the way to 2000 nm. In view of the development of practical PV devices, four stages (generations) under investigations are as follows:

First generation solar cells are mainly based on silicon wafers and typically demonstrate a performance about 15–20 % [2]. These types of solar cells dominate the market and are mainly those seen on rooftops. The benefits of this solar cell technology lie in their good performance, as well as their high stability. However, they are rigid and require a lot of energy in production.

The second generation solar cells are based on amorphous silicon, CIGS and CdTe, where the typical performance is 10–15 % [3]. Since the second generation solar cells avoid use of silicon wafers and have a lower material consumption, it has been possible to reduce production costs of these types of solar cells compared to the first generation. The second generation solar cells can also be produced so they are flexible to some degree. However, as the production of second generation solar cells still include vacuum processes and high temperature treatments, there is still a large energy consumption associated with the production of these solar cells. Further, the second generation solar cells are based on scarce elements and this is a limiting factor in the price.

Third generation solar cells uses organic materials such as small molecules or polymers. Thus, polymer solar cells are a sub category of organic solar cells. The third generation also covers expensive high performance experimental

Table 10.1 Comparison of highest efficiency levels achieved by solar PV technologies

Technology	Efficiency (%)	References
Concentrator PV	46.0	[6]
Crystalline silicon	25.6	[7]
Thin film—CIGS	21	[8]
Thin film—CdTe	21.5	[9]
Thin film—a-Si	10.2	[10]
Dye sensitized	11.9	[11]
Organic PV	11.0	[12]
Perovskite thin film	15.0	[13]
Thin film—CZTS	12.6	[14]

multi-junction solar cells which hold the world record in solar cell performance. This type has only to some extent a commercial application because of the very high production price. A new class of thin film solar cells currently under investigation are perovskite solar cells and show huge potential with record efficiencies beyond 20 % on very small area. Polymer solar cells or plastic solar cells, on the other hand, offer several advantages such as a simple, quick and inexpensive large-scale production and use of materials that are readily available and potentially inexpensive. Polymer solar cells can be fabricated with well-known industrial roll-to-roll (R2R) technologies that can be compared to the printing of newspapers [4]. Although the performance and stability of third generation solar cells is still limited compared to first and second generation solar cells, they have great potential and are already commercialized.

Fourth Generation solar cells defined by R. Silva [5] are a hybrid that combine the low cost and flexibility of conducting polymer films (organic materials) with the lifetime stability of novel nanostructures (inorganic materials). This "inorganics-in-organics" technology improves the harvesting of solar energy and its conversion into electricity, offering better efficiency than the current 3G solar cells while maintaining their low cost base (Table 10.1).

10.2 Basic Principles of p–n Junction Solar Cell

Trapping of sunlight energy has been made easy with the introduction of solar cells. Energy conversion in solar cell consists of generation of electron–hole pairs in p–n junction by the absorption of light and separation of electrons and holes by an internal electric field. Charge carriers collected by two electrodes give rise to a photocurrent when the two terminals are connected externally. When a resistance load is connected to the two terminals, the separation of the charge carriers sets up a potential difference. The emergence of an electric potential between two electrodes attached to a solid or liquid system, upon light irradiation was discovered by Becquerel in 1839 [15] which is known as photovoltaic effect. The first solid state silicon p–n junction solar cell was developed in 1954 by Chapin et al. [16].

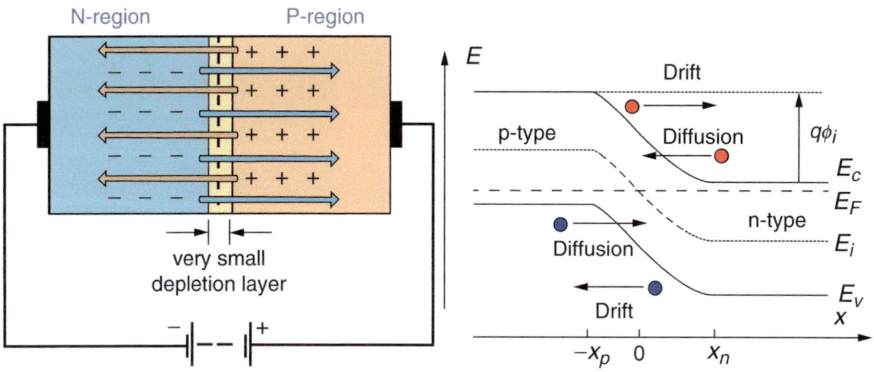

Fig. 10.1 Schematic diagram and energy band diagram of p–n junction at equilibrium [Source from internet: http://ecee.colorado.edu/~bart/book/book/chapter4/ch4_2.htm]

A simple solar cell made from silicon or gallium arsenide, consists of n-type semiconductor in contact with p-type semiconductor. The excess majority carriers will diffuse across the junction as shown in Fig. 10.1. Diffused electrons leave behind positively charged donor atoms and diffused holes leave behind negatively charged acceptor atoms. These exposed atoms create an electric field which opposes the force created by the concentration gradient. At equilibrium, the diffusion force matches the force due to the induced electric field and a depletion region is formed at the p–n junction.

In a p–n junction solar cell, a semiconducting material takes over the function of a light absorber and the charge transport material. Interband excitation by absorption of a photon of $h\nu > E_g$ promotes an electron from the valence band to the conduction band. Electrons and holes are generated in pairs called exciton. An exciton is a bound state of electron and hole pair due to a Columbic attraction between the two oppositely charged particles. When an exciton is generated inside or close to the depletion region, so they can diffuse into the depletion region before recombining, the internal built-in field will split the electron–hole pair and drive the electron towards the n-type and the hole towards the p-type material. The n-type semiconductor already has an excess of electrons and the p-type semiconductor has an excess of holes. If an alternate current path is provided, connecting the n-type and p-type semiconductors together, the exciton electrons will flow back to the p-type semiconductor and exciton holes will flow back to the n-type semiconductor. This results in a current and the depletion region electric field delivers a driving voltage which in turn provides power.

10.2.1 Solar Cell Performance Parameters

The solar cell is generally characterized using the equivalent circuit (single diode model) of a solar cell as shown in Fig. 10.2. The relation between the current *I* and the voltage *V* is given by:

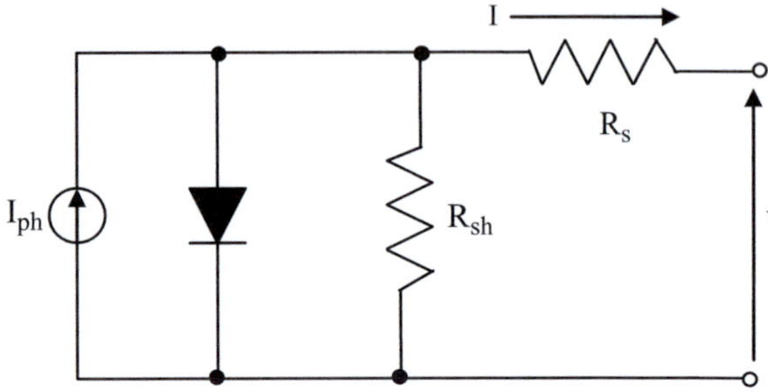

Fig. 10.2 Equivalent circuit (single diode model) of a solar cell

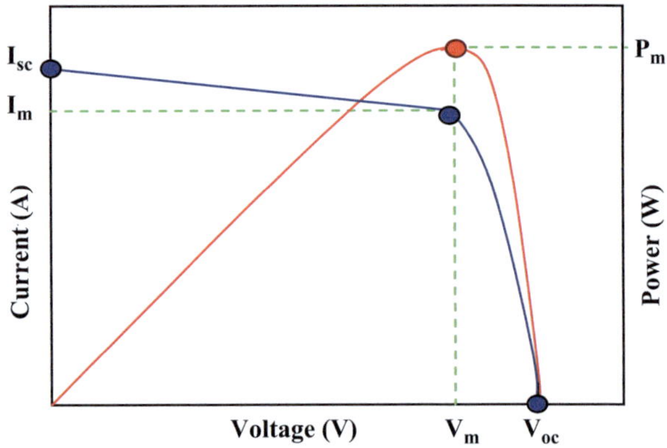

Fig. 10.3 Representation of *I*–*V* and the power curves

$$I = I_{ph} - I_o\left[\exp\left\{\frac{q(V + R_s I)}{nk_B T}\right\} - 1\right] - \frac{V + R_s I}{R_{sh}}$$

where I_{ph}, I_0, $R_{s,}$ R_{sh} q, n, k_B, and T are the photocurrent, the saturation current of the diode, the series resistance, the shunt resistance, the electron charge, the ideality factor, the Boltzmann constant, and the temperature, respectively. An ideal solar cell can be modeled by a current source in parallel with a diode; in practice no solar cell is ideal, so a shunt resistance and a series resistance component are added to the model as shown in Fig. 10.2. For a given particular operating voltage V, the equation may be solved to determine the operating current I at that voltage. However, because the equation involves I on both sides in a transcendental function the equation has no general analytical solution.

Typical current–voltage (*I*–*V*) characteristics of a solar cell are given in Fig. 10.3. Analysis of the *I*–*V* curves includes the determination of the following parameters.

10.2.1.1 Short-Circuit Current (I_{sc})

When the cell is operated at short circuit, i.e., $V = 0$, the current I that flows through the terminals is defined as the short-circuit current. It can be shown that for a high-quality solar cell (low R_s and I_o, and high R_{sh}), the short-circuit current (I_{sc}) is:

$$I_{sc} = I_{ph}$$

The short-circuit current I_{sc} depends on the number of absorbed photons and shows a linear dependence on the incident light intensity as long as no saturation effects occur within the active layer. I_{sc} can be enhanced by enlarging the absorption spectrum of the photoactive layer to harvest more photons within the solar spectrum.

10.2.1.2 Open-Circuit Voltage (V_{oc})

When the cell is operated at open circuit, i.e., $I = 0$, the voltage across the output terminals is defined as the open-circuit voltage [17]. Cell potential is measured when current in the cell is zero, corresponding to almost flat valence and conduction bands. V_{oc} is obtained from equation by setting the current to zero:

$$V_{oc} = \frac{n \cdot k \cdot T}{q} \ln\left(\frac{I_{ph}}{I_0} + 1 - \frac{V_{oc}}{I_0 \cdot R_p}\right) \approx \frac{n \cdot k \cdot T}{q} \ln\left(\frac{I_{ph}}{I_0} + 1\right)$$

The experimentally determined open-circuit voltage is generally somewhat lower, owing to the recombination of free charge carriers. At open-circuit conditions, all charge carriers recombine within the photoactive layer. Thus, if recombination can be minimized, the open circuit voltage V_{oc} can more closely approach the theoretical limit. However, based on thermodynamic considerations of the balance between photogeneration and recombination of charge carriers, it has been found that charge recombination cannot be completely avoided, resulting in a lower maximum open-circuit voltage.

10.2.1.3 Maximum Power Output (P_m)

A solar cell operates over a wide range of voltage (V) and current (I). By increasing the resistive load on an illuminated cell continuously from zero (a short circuit) to a very high value (an open circuit) one can determine the maximum power point, that maximizes the product of voltage and current ($V \times I$), i.e., the load for which the cell can deliver maximum electrical power at that level of illumination. The output power is zero in both the short circuit and open circuit extremes. The maximum power output of the cell is

$$P_{\mathrm{m}}(V) = I_{\mathrm{m}} + V_{\mathrm{m}}$$

I_m and V_m are the coordinates of the maximum in the $P(V)$ curve.

10.2.1.4 Fill Factor (FF)

Another defining term in the overall behavior of a solar cell is the fill factor (FF). This is the ratio of the maximum power point (P_m) divided by the open circuit voltage (V_{oc}) and the short circuit current (I_{sc}):

$$\mathrm{FF} = \frac{V_{\mathrm{m}} \cdot I_{\mathrm{m}}}{V_{\mathrm{oc}} \cdot I_{\mathrm{sc}}}$$

This parameter describes the quality of the solar cell. It indicates the deviation of the current–voltage characteristic from a square like curve. The fill factor is directly affected by the values of the cell's series and shunt resistance. Increasing the shunt resistance (R_{sh}) and decreasing the series resistance (R_s) will lead to higher fill factor, thus resulting in greater efficiency.

10.2.1.5 Efficiency (η)

A solar cell energy conversion efficiency (η), is the percentage of power converted (from absorbed light to electrical energy) and collected, when a solar cell is connected to an electrical circuit. This term is calculated using the ratio of the maximum power point, P_m divided by the input light irradiance (P_{in}, in W/m^2) under standard test conditions (STC) and the surface area of the solar cell (A_c in m^2).

$$\eta = \frac{P_{\max}}{P_{\mathrm{in}}} = \frac{V_{\mathrm{oc}} \cdot I_{\mathrm{sc}} \cdot \mathrm{FF}}{P_{\mathrm{in}} \cdot A_{\mathrm{c}}}$$

The efficiency is a function of the V_{oc}, I_{sc} and FF of the cell. Accordingly, improvement of the photovoltaic yield is the result of the optimization of these three parameters.

10.2.1.6 Incident Photon to Current Density Conversion Efficiency (IPCE)

In order to study the photovoltaic action of a given device, it is quite useful to measure the wavelength dependence of its photocurrent. The profile of a photocurrent to the wavelength of the incident light is called the action spectrum of photocurrent. The incident photon to current density conversion efficiency (IPCE) gives the ratio between the number of generated charge carriers contributing to the photocurrent density and the number of incident photons, as given by

$$\mathrm{IPCE}(\%) = \frac{1240 \times j_{sc}(\mathrm{mA/cm^2})}{\lambda(\mathrm{nm}) \times P_{in}(\mathrm{mW/cm^2})}$$

where j_{sc} is the short circuit current density, P_{in} is the light power density and λ is the wavelength in nanometer. j_{sc} of the cells is determined from the *J–V* characteristics of the cells obtained at different wavelength of light.

(a) Internal Quantum Efficiency

Internal Quantum Efficiency (η_{IQE}) is the ratio of the number of charge carriers collected by the solar cell to the number of photons of a given energy that shine on the solar cell from outside *and* are absorbed by the cell. A low η_{IQE} indicates that the active layer of the solar cell is unable to make good use of the photons. The internal quantum efficiency, is the product of four efficiencies [18] each corresponding to a step in the charge generation process:

$$\eta_{IQE} = \eta_A \eta_{ED} \eta_{CT} \eta_{CC}$$

where, η_A is the absorption efficiency of light within the active region of the solar cell; η_{ED} is the exciton diffusion efficiency to a dissociation site; η_{CT} is the charge transfer efficiency, which is the efficiency for dissociation of an exciton into a free electron and hole pair at that site; and η_{CC} is the charge collection efficiency.

For the case of organic photovoltaic cells, there is an inherent tradeoff between the absorption and the exciton diffusion efficiencies. That is, the exciton diffusion length, L_D, is typically much less than the optical absorption length, $1/\alpha$. For an absorbing organic layer of thickness *d*,

$$\eta_A = \left(1 - e^{-\alpha d}\right)$$

whereas, assuming that the arrival of excitons at a dissociation site is independent of electric fields or other extrinsic conditions, the diffusion efficiency is given by

$$\eta_{ED} = e_D^{-d/L}.$$

In general, it has been found that the charge collection and charge transfer efficiencies at organic donor/acceptor (DA) interfaces commonly used in thin-film molecular organic semiconductor PV cells approach 100 %, in which case the internal quantum efficiency is determined by the product $\eta_A \eta_{ED}$.

(b) External Quantum Efficiency

The External Quantum Efficiency (EQE) is defined as the number of electrons flowing in the external circuit per photon incident on the PV cell. For a device based on exciton dissociation by charge transfer at a donor–acceptor (DA) interface, the EQE is the product of the efficiencies of four sequential steps:

1. Photon absorption leading to the generation of an exciton
2. Diffusion of the exciton to the DA interface
3. Exciton dissociation by charge transfer at DA interface
4. Collection of the free charge carriers at the electrodes.

Taking into consideration the optical losses that occur on coupling light in the device active region, the external quantum efficiency is given by [19]:

$$\eta_{\mathrm{EQE}} = (1 - R)\,\eta_{\mathrm{IQE}}$$

where, R is the reflectivity of the substrate–air interface. The η_{IQE} is always larger than the η_{EQE}.

10.2.1.7 Series and Shunt Resistance

As series resistance increases, the voltage drop between the junction voltage and the terminal voltage becomes greater for the same flow of current. The result is that the current controlled portion of the I–V curve begins to sag toward the origin, producing a significant decrease in the terminal voltage V and a slight reduction in I_{sc} as shown in Fig. 10.4. Very high values of R_s will also produce a significant

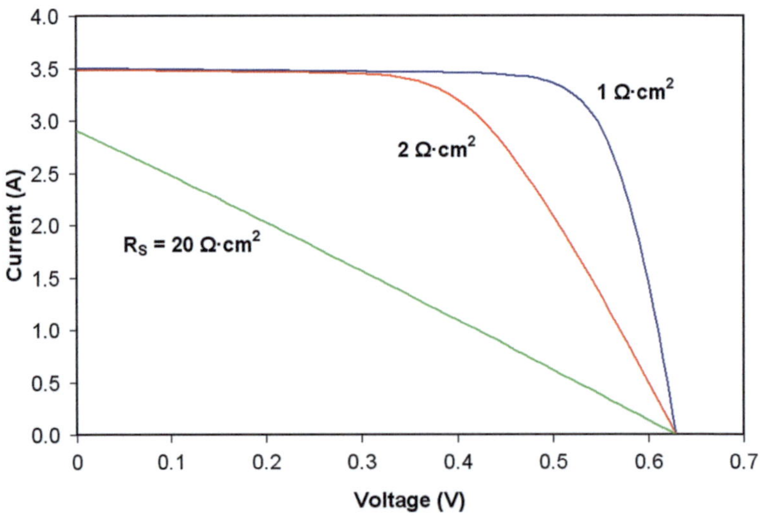

Fig. 10.4 Effect of series resistance on the current–voltage characteristics of a solar cell

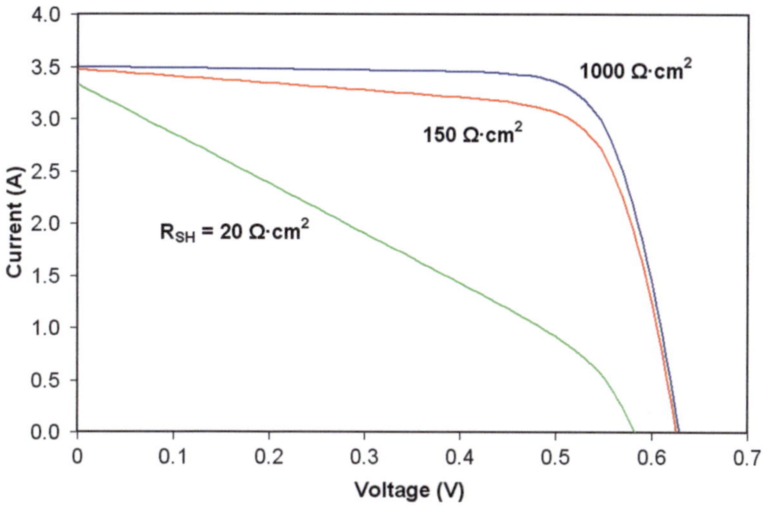

Fig. 10.5 Effect of shunt resistance on the current–voltage characteristics of a solar cell

reduction in I_{sc}; in these regimes, series resistance dominates and the behavior of the solar cell resembles that of a resistor. These effects are shown for crystalline silicon solar cells in the *I–V* curves displayed in Fig. 10.4.

As shunt resistance decreases, the flow of current diverted through the shunt resistor increases for a given level of junction voltage. The result is that the voltage controlled portion of the *I–V* curve begins to sag toward the origin, producing a significant decrease in the terminal current *I* and a slight reduction in V_{oc}. Very low values of R_{sh} will produce a significant reduction in V_{oc}. These effects are shown for crystalline silicon solar cells in the *I–V* curves displayed in Fig. 10.5.

10.3 Materials for PV Devices

First generation solar cells based on single crystalline p–n junction silicon cells are the most common PV converters having a current market share of approximately 85 %. However, the problems associated with this material are:

1. Small spectral mismatch between absorption of the semiconductor and the sunlight spectrum.
2. Silicon is an indirect semiconductor, that means valence band maximum and conduction band minimum are not opposite to each other in k-space. Light absorption is much weaker in an indirect gap semiconductor than in a direct semiconductor.
3. The conventional photovoltaics, based on the junction devices, such as crystalline or amorphous silicon, have exceptional solar energy conversion to

electricity efficiencies ~20 %. However, the fabrication of these photovoltaics using energy intensive high temperature and high vacuum processes is expensive.

In view of these restrictions, a lot of work has been devoted into the search for new materials. The ideal solar cell material possesses a band gap between 1 and 2 eV and a direct bandgap structure. Moreover the material should be available in abundance, nontoxic, and reproducible with easy deposition techniques to provide large area production at low cost.

For making a single semiconductor junction, all photons more energetic than the light-absorbing semiconductor's bandgap are substantially absorbed and harvested, while those lying below the bandgap remain unabsorbed, then simple analysis of the sun's power spectrum reaching the earth leads to an optimal choice of the bandgap. While further decrease of the bandgap beyond this optimum would increase the fraction of the sun's power absorbed, it would also lower the open circuit voltage of the device. For unconcentrated rays (1 sun), the optimal bandgap of such a single-junction cell is 1.13 eV, corresponding to a wavelength of 1.1 μm. Silicon, with its 1.11 eV bandgap, is an excellent choice from the standpoint of bandgap; however, since direct transitions only become possible in silicon above 3.0 eV, its absorption is relatively weak throughout much of the visible and all of silicon's infrared-absorbing region. This entails the need for a large thickness of material if light is to be substantially absorbed so the large silicon thickness is required to achieve the high external quantum efficiency (EQE) and thus power conversion efficiency.

Higher overall solar power conversion efficiencies can be achieved when different-bandgap semiconductors are stacked atop one another. A first cell absorbs higher-energy photons only, and provides a large open circuit voltage; the next cells absorb the lower energy photons, and provide additive contributions to the open circuit voltage. Assuming that each layer substantially absorbs its assigned spectral band, one can derive optimum bandgap choices for each member of the multijunction stack. If a rigorously series connected multijunction architecture is employed, current matching throughout the stack adds the further constraint that the absorbed photon flux must match in each layer.

In contrast to traditional semiconductors, conjugated polymers provide ease of processing, low cost, physical flexibility, and large area coverage [20]. These active optoelectronic materials produce and harvest light efficiently in the visible spectrum. The same functions are required in the infrared for telecommunications (1300–1600 nm), thermal imaging (1500 nm and beyond), biological imaging (transparent tissue windows at 800 nm and 1100 nm), thermal photovoltaics (>1900 nm), and solar cells (800–2000 nm). Photoconductive polymer devices have yet to demonstrate sensitivity beyond ~800 nm [21, 22]. Sensitizing conjugated polymers with infrared-active nanocrystal quantum dots provides a spectrally tunable means of accessing the infrared while maintaining the advantageous properties of polymers.

10.4 Chalcogen Based Thin Film Solar Cells

Thin film solar cells based on sulfide, selenide, or telluride semiconductors are known as chalcogenide solar cells. Thin films are prepared by depositing a very thin layer of semiconductor material on a substrate such as glass, plastic or metal. Thin-films of chalcogen based materials absorb light more easily than c-Si, further reducing the amount of material required and thus lead to lower manufacturing cost. Since twenty-first century, these solar cells have been of interest for production and their manufacturing volume is growing swiftly. Many companies have entered in this field to produce solar modules in PV market. Each company is working on efficiency and cost advantages from their specific cell design or production process. At present, the leading thin film materials used in the manufacture of solar cells include: cadmium telluride (CdTe), copper indium selenide ($CuInSe_2$ (CIS)), and copper indium gallium selenide ($CuInGaSe_2$ (CIGS)) due to their ease and cost effective manufacturing. The grain boundaries and interfaces in these materials are well passivated and have a low recombination velocity for excess carriers. Lower recombination activity at the grain boundaries permits high solar cell efficiencies even when the material is polycrystalline with grain sizes in the order of only a few 1–100 μm. Moreover, due to polycrystalline nature, different types of substrates like glass and steel/Al foil can be used to grow thin films over them which makes them compatible with low-cost temperature deposition techniques, as there is no need for epitaxial growth or high temperatures to obtain large grain sizes. The most important property of these materials is the bandgap tailoring, for instance, replacing Se by S in $CuInSe_2$ structure results in a higher bandgap and different stoichiometric ratios of the elements in compound can be used for tuning the bandgap. This property allows one to build bandgap gradients assisting the collection of excess carriers. At the same time increasing the number of components in ternaries and even quaternaries enables large number of possible material combinations.

10.4.1 CdTe Based Solar Cells

CdTe is very well matched for use as absorber material in thin-film solar cells. It has direct bandgap energy gap of 1.45 eV and well suitable to the spectrum of solar radiation. Due to the strong light absorption coefficient of about 10^5 cm^{-1} for light below wavelength of 800 nm, a film thickness of a few micrometers 3–7 μm is sufficient for complete light absorption. CdTe has been found to grow as highly stoichiometric material, with p-type semiconductor film and can form a p–n heterojunction with CdS. The first thin film solar cell of the type n-CdTe/p-Cu_2Te was introduced in 1963 by a GE Research Laboratory [23]. Then Bonnet and Rabenhorst have shown a p-CdTe/n-CdS heterojunction solar cell which had 6 % efficiency in 1972 [24]. The CdTe layer was deposited by vapor transport

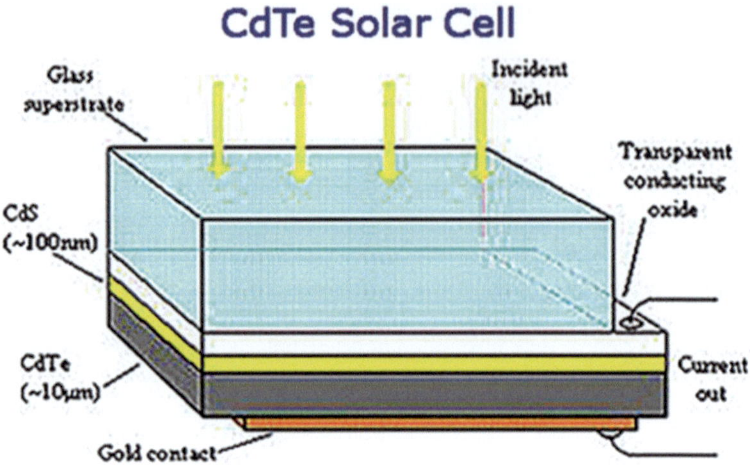

Fig. 10.6 Schematic view of the typical superstrate structure of CdTe solar cells [Source from internet: http://solarcellcentral.com/companies_page.html]

deposition. The cells were of the substrate type, having a Mo back contact and a CdTe thickness of more than 10 μm. Different from the previous structures, a superstrate thin film solar cell was developed by Tyan et al. [25] using closed-space sublimation method with more than 10 % efficiency. After this achievement, several companies started in-house developments.

A typical structure of a p–n junction cell is shown in Fig. 10.6. The first layer is transparent conducting oxide (TCO) film such as indium tin oxide (ITO) deposited onto glass substrate. Then an n-CdS film is deposited, followed by the active p-conducting CdTe film. At last a low-resistance aluminum or gold contact is deposited onto the CdTe. When light enters from the transparent side of the cell through the glass-substrate, photons transverse the TCO and CdS layers as they have a large bandgap. The CdTe film is the active absorber layer of the solar cell. Therefore maximum photons are absorbed and electron–hole pairs are generated near the junction. The electrons are drifted by the built-in voltage to the interface into the active film. The holes remain in the CdTe which promotes the p-type conduction and finally leave the cell via the back contact. Electric power is drawn by metallic contacts attached to the TCO film and the back contact. Best device efficiency of 21.5 % was reported for the superstrate structure and a substrate configuration resulted in a 13.6 % efficiency for CdS/CdTe structures [9, 26]. Superstrate configuration is preferable for CdTe devices because the CdTe surface is exposed for making contact. This feature of CdS diffusion during the fabrication reduces the lattice mismatch between CdTe and CdS. CdTe has also been deposited on thin metallic foils such as stainless steel, Mo, Ni and Cu, out of which Mo is preferred due to better thermal matching [27]. Being a binary compound, the occurrence of native defects renders a precise control of the doping density difficult [28].

At present, almost all efficient CdTe solar cells have a CdS buffer layer. The buffer layer may increase the efficiency by reducing reflection losses but may also decrease the efficiency due to light absorption and carrier recombination. Some of the leading manufacturing companies producing CdTe based photovoltaic modules commercially are: First Solar in the US, GE global research and National renewable energy laboratory (NREL). Current record efficiency of 21.5 % for a small-area 0.35-cm^2 CdTe cell has been fabricated by First Solar, USA [9]. CdTe cells, as well as cells based on several other materials, continue to move closer to the theoretical efficiency limit. Besides the development of CdTe solar cell on glass substrates, flexible substrates have been used to develop light weighted solar cell modules. In superstrate configuration, researchers at Swiss Federal Laboratories for Materials Science and Technology have reported an efficiency of 13.8 % using polyimide substrate based CdTe cell [29] whereas 14.0 % efficiency have been reported by NREL using flexible glass substrate [30]. Few companies have also commercialized CdTe solar modules like Advance Solar Power, Antec Solar with the efficiency above 10 % in the superstrate configuration on glass substrates [31, 32]. Exceptionally a company Xunlight 26 solar has start to develop CdTe solar modules on flexible glass substrates and as well as on polyimide substrates [33]. The recent improvement in CdTe record cells has been obtained by improving the J_{sc}. A comparison of several recent record cells is shown below in Fig. 10.7a current density–voltage and (b) quantum efficiency (QE) curves for CdTe cells, including record cells.

Figure 10.7 shows 13 % conversion efficiency of cell fabricated at Colorado State University (CSU) with 100 nm thin film of CdS, with an alternative Cd(S,O) window layer [34], the NREL cell which held the CdTe efficiency record for a decade, and two recent First Solar records, The impact of thinning or replacing CdS is clear in the short wavelength performance of these cells. In the case of CSU cells, the replacement of CdS with Cd(S, O) increased the J_{sc} by 5.5 mA/cm^2.

10.4.2 $CuInSe_2$/$CuInGaSe_2$ (CIS/CIGS) Based Solar Cells

Elemental Cadmium is a highly toxic material which can cause environmental issues resulting in limitations for industrial applications. Thus researchers look towards chalcopyrite structures, the most advanced absorber materials are based on $CuInSe_2$ (CIS). CIS has narrow bandgap of 1.0 eV which can be increased by doping gallium atom, to 1.15 eV approximately to obtain Cu(In,Ga)Se_2 compound. The most commonly used methods for the deposition of CIS and CIGS are vacuum based techniques such as co-evaporation, thermal evaporation and sputtering processes. Evaporation provides a direct control during film formation, including deliberate depth profiles for bandgap engineering as well as making optimum use of the fluxing activity of Cu(S, Se) phases. This may be the reason of yielding the highest cell performance from evaporation process. The device properties are dependent on the device structure and in present scenario, another configuration

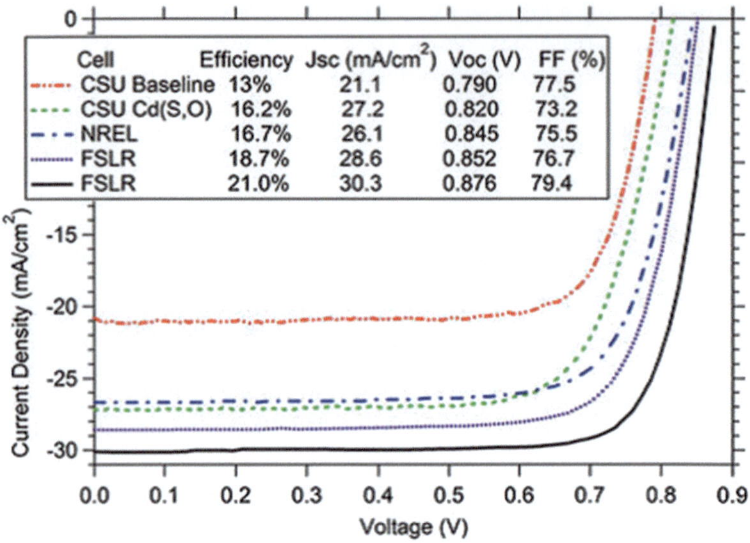

(a)

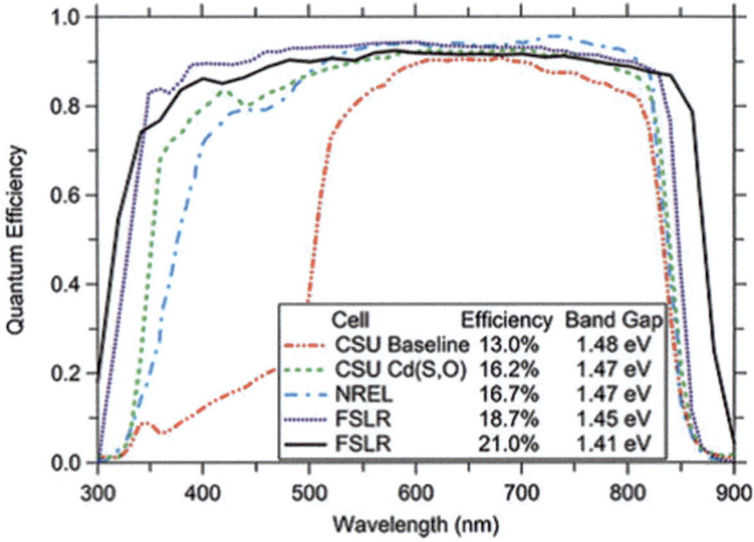

(b)

Fig. 10.7 (**a**) Current density–voltage and (**b**) QE curves for CdTe cells shows the highest efficient CdTe based solar cell (*J–V*) and QE curves. Reprinted with permission from IEEE journal of photovoltaics, 5 (4) 2015 1217 © IEEE

such as substrate device structures are being used for CIGS thin film solar cell fabrication as shown in Fig. 10.8a. These CIGS solar cells, based on the superstrate structure, are inferior to the substrate structure, because of the inter-diffusion of CdS during high temperature CIGS film growth. CIGS absorber layer have been grown successfully by both vacuum and non-vacuum methods.

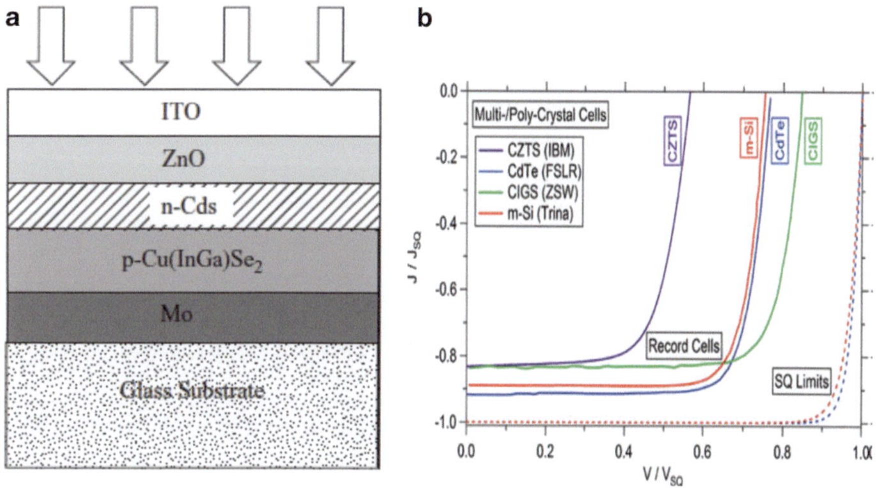

Fig. 10.8 (**a**) Schematic view of the typical substrate structure of CIGS solar cells [28]. (**b**) Record (*full lines*) and ideal (*dashed lines*) *J–V* curves for multi- and polycrystalline cells. Reprinted with permission from IEEE journal of photovoltaics, 5 (4) 2015 1217 © IEEE

Figure 10.8b shows current density–voltage curves for the best solar cell efficiency acquired till date at different laboratory scale. Annealing in H_2S environment has helped to overcome the problem of aggregation of metals to a certain degree but it also adds to the overall complexity of the processes and the device [35]. In evaporated films the standard bandgap (of 1.15 eV) can be obtained simply by incorporating sufficient amounts of gallium.

Some research groups have been able to achieve efficiencies of 20 % or above, using CIGSe absorber layers grown by co-evaporation on Mo coated substrate with CdS buffer layer by chemical bath deposition and i-ZnO/ZnO: Al window layer by sputtering [36]. The researchers at the (ZSW, Germany) and NREL (USA) have used a high-temperature (>550 °C) 3-stage co-evaporation process for deposition on glass substrates. Scientists at the Swiss Federal Laboratories for Materials Science and Technology (Empa, Switzerland) have achieved 20.4 % efficiency using a low-temperature (<450 °C) multi-stage co-evaporation process on flexible polyimide foil [37].

Various groups like Institute for Advanced Industrial Science and Technology (AIST, Japan) [38], the Helmholtz-Zentrum Berlin (HZB, Germany) [39], Aoyama Gakuin University (AGU, Japan) [34], and the University of Uppsala (Sweden) [40] have also reported the efficiencies above 18 % by using co-evaporation processes. The typical absorber layers energy bandgap are in the range of 1.1–1.25 eV in all the high efficiency cells. A wider bandgap absorber material of 1.4 eV is desirable to have the efficiencies near the theoretical optimum for a single junction solar cell for various reasons including lower temperature coefficient of power loss under solar illumination. To increase the bandgap the Institute of Energy Conversion at Delaware University (IEC, USA) added Ag or Al and achieved an efficiency of 17.9 % on a polyimide substrate [41].

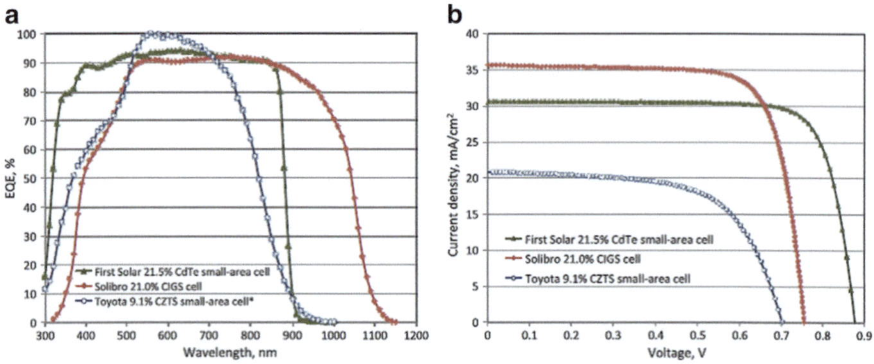

Fig. 10.9 (**a**) External quantum efficiency (EQE) for the new CdTe, CIGS, and CZTS results in this issue and (**b**) corresponding current density–voltage curves. Reprinted with permission from [43]. Copyright (2015) Wiley online library

The Solar Frontier has reported the highest efficiency of 19.7 % using CIGSSe layer grown by sputtering method [42] in collaboration with Japan's New Energy and Industrial Technology Development Organization. They have used chemical bath deposited Zn (O, S, OH) buffer layer, ZnO:B transparent conducting front contact. Figure 10.9 shows the best reported efficiency of CdTe, CIGS and CZTS based solar cell efficiency at laboratory scale [43].

Over the past decade, non-vacuum deposition methods have been extensively used for low cost production. These include (1) solution based deposition such as electro deposition (ED), electroplating and chemical bath deposition (CBD). (2) -Molecular-based precursor solutions printing and deposition from particulate-based solutions. In Electro deposition method, the deposition of multicomponent semiconductors like CIGSe is generally a difficult process as the different elements have different redox potentials and reduction kinetics. Among the approaches mentioned, the particulate- and precursor-based approaches have recently received a great deal of attention due to their potential to simplify the fabrication processing for large-scale manufacturing [44]. Various chemical modifiers and sacrificial ligands have been used to control the reactions rate and their kinetics toward the solubility of precursor materials. Thermal treatment has to be provided to convert the precursors into the desired phase and to remove the residual organic species. The important factors are the concentration and viscosity of the precursor solution that may highly impact the film formation and in its drying kinetics. Besides, the quality of the final metal chalcogenide film can be strongly influenced by the bulkiness and volatility of the sacrificial ligands. Further thermal treatment at high temperatures in a sulfur or selenium atmosphere is necessary in order to complete the phase formation.

Mitzi et al. [45] have reported the highest device efficiencies among all solution-processed solar cells using a hydrazine solution with a PCE of 15.2 %. Hydrazine is an exceptional reducing agent for dissolving many different metal chalcogenides. It tends to strongly coordinate and has a small size and high vapor pressure that allows formation of the desired product with minimum impurities at relatively low temperatures. The hydrazine is highly toxic and explosive solvent in nature and the solution of this problem can be obtained by substitutes of hydrazine solvent. Zhao et al. have reported the efficiencies up to 10.1 % by utilizing a molecular precursor from butyldithiocarbamic acid compounds [46]. These toxic reagents have been substituted with benign solvents or added binder materials to stabilize the metals but no success was achieved in improving device efficiencies [47]. The improvement in the efficiencies up to 8.01 % was obtained by substituting the nitrates precursors with acetates and actylacetonates and in a mixture of H_2 and N_2 gasses atmosphere to avoid the formation of the residual carbon [48] and Park et al. [49] have reported 8.28 % efficient cells by using wide band gap CIGS absorber layers from a precursor route that methanol solutions of metal nitrates was used with polyvinyl acetate binder material. This binder based was found to decompose during the thermal treatment with diluted H_2S environment to left a minimum of carbon residue and low degree of porosity. In early 1980s, electro-deposition (ED) technique has emerged as a promising low-cost industrial alternative to phase vapor deposition method (PVD) which later become the most established solution-based method for CIGS absorbers thin films [50]. ED can deposit directly the compound semiconductors $CuInSe_2$ or $CuInS_2$ [51, 52] or to electroplate metallic Cu-In or Cu-In-Ga precursors shown in Fig. 10.10a, b.

From metallic precursor route thermal treatment in the presence of chalcogens (S, Se) atmosphere at temperatures between 500 °C and 600 °C has been done to form the semiconductor pure phase. In ED technique all the precursor elements (Cu-In-Se-Ga-S) are initially deposited on substrates and then subjected for annealing. An effective annealing was performed in closed reactors in order to maintain a high partial pressure of chalcogen which is favorable for the p-type doping of CIGS [53]. An exhaustive research had led to efficiency improvements up to 11.3 % on ED-based processes with subsequent annealing in 2004 [54].

Figure 10.11a–d shows the photovoltaic parameters of the CIGS based solar cell. Air-stable CIGS thin film, conductive i-ZnO/AZO all functional layers except molybdenum are deposited from solution methods based on metallic precursor salts [55].

Mo deposited by magnetron sputtering to make a back contact, CIGSSe absorber layer deposited by ED from metal precursors solution followed by annealing in Se

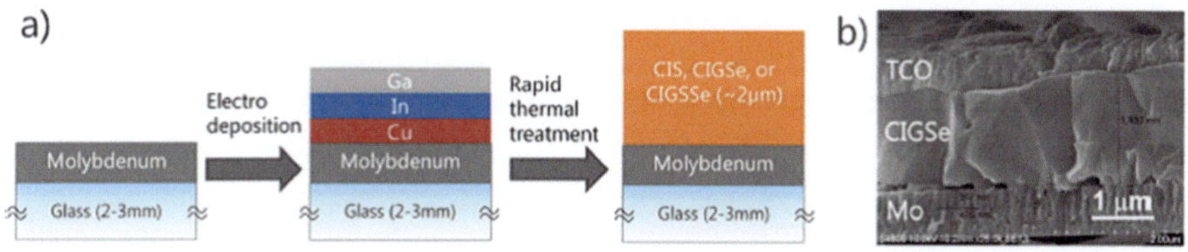

Fig. 10.10 (**a**) Schematic diagram for the deposition of CIGSe thin film by ED technique. (**b**) Cross sectional view of the films. Reprinted with permission from Advanced functional materials, 25, 2015, 12. © WILEY-VCH

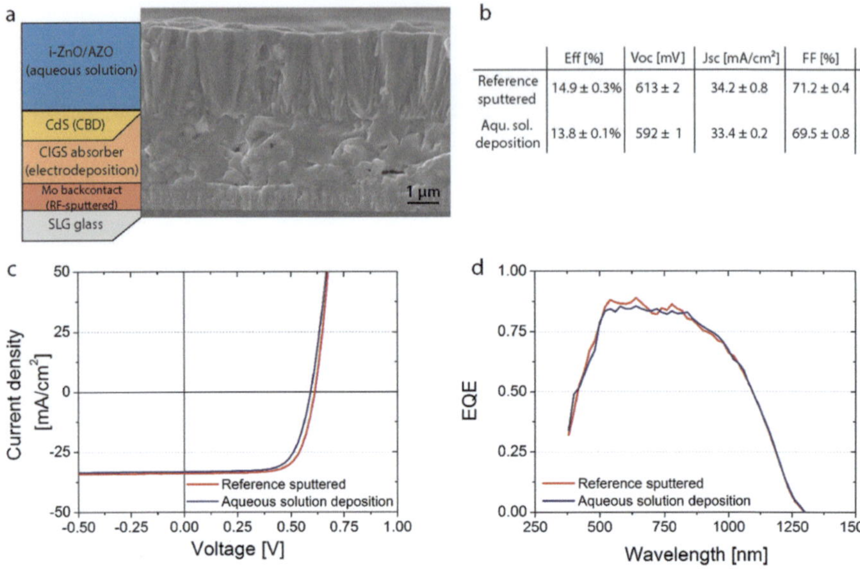

Fig. 10.11 (**a**) Cross section view of non-vacuum processed CIGS based solar cell, (**b**) best reported efficiency of CIGS solar cell, (**c**) *J–V* curves, (**d**) EQE of the cells. Reprinted with permission from [55]. Copyright 2014 WILEY-VCH

atmosphere, CdS buffer deposited by chemical bath deposition method (CBD), i-ZnO layer deposited by spin-coating method and finally, transparent conductive contact AZO is again deposited by CBD. The cross-sectional view of the cell is presented in Fig. 10.11a. The efficiency of 13.8 % for the solution-processed CIGS solar cells was obtained for a cell area of 0.09 cm^2 which is in proximity to the sputtered CIGS based solar cell as listed in Fig. 10.11b. The *J–V* and EQE curves of respective cells are also presented in Fig. 10.11c, d.

10.4.3 Cu_2ZnSnS_4 (CZTS) Based Thin Film Solar Cell

Researchers have found a low cost alternative to CIGS as Cu_2ZnSnS_4 (CZTS), which consists of elements available in abundance on earth as shown in Fig. 10.12. It has optimum photovoltaic absorber properties similar to CIGS, direct bandgap of 1.5 eV and high absorption coefficient. Current record highest efficiency of 12.6 % has been reported on solution processed $CZT(S, Se)_2/CdS$ structure by Mitzi et al. [14]. Many research groups have acquired two step deposition approach first by depositing precursor solution at low temperature followed by annealing in chalcogen-containing atmosphere [56]. Sputtered precursor stack films followed by sulfurization in H_2S atmosphere results in highest efficiency achieved of 10.8 % [57]. ED method can also be used to deposit the metal precursor by sequential

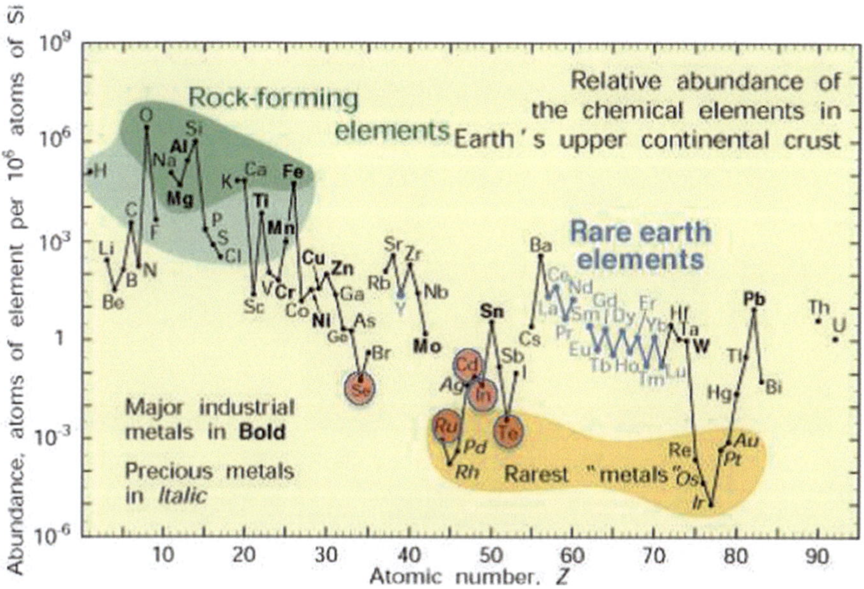

Fig. 10.12 Earth abundant materials available on earth (*Source*: Abundance of chemical elements source: Wikipedia)

electro deposition of Cu, Zn, and Sn metals [58] or by Cu-Sn-Zn layers [59] that gives near around 7 % and 5.8 % devices efficiency, respectively. The highest efficient kesterite solar cell with the energy conversion efficiency was achieved by a hydrazine solution based CZTSSe. Hydrazine is first choice among the solvent since it has a strong reducing action and decomposed into carbon free residues, that provide high-quality CZTSSe absorbers but its toxicity and explosive nature is the major drawback. Researchers have work on other than hydrazine base solvents which are very promising and less toxic like dimethyl sulfoxide (DMSO) as the solvent for metal salts and thiourea as sulfur source [60]. The highest reported efficiency has been achieved by using DMSO solution is 10.0 % [61]. The DMSO based precursor solution is typically spin-coated onto Mo-coated glass with and baked at 200–320 °C with repetitive cycles that results in formation of fine-grained Cu-Zn-Sn-S containing precursor film. The influence of difference annealing atmosphere have been shown in Fig. 10.13. The morphology of the grown CZTSSe grains was strongly depends on the annealing condition in a N_2 or mixed $N_2 + Se$ atmosphere in a tube flow furnace as shown in Fig. 10.13.

In an inert atmosphere, large-grain CZTS material was observed when excess amount of Sn is present whereas in Se atmosphere annealing results in a smaller grain growth due to significant loss of Sn [62]. If sodium (Na) dopant ions are present in the precursor solution, solar cell based on such absorber layers yields power conversion efficiency of 8.3 % [63].

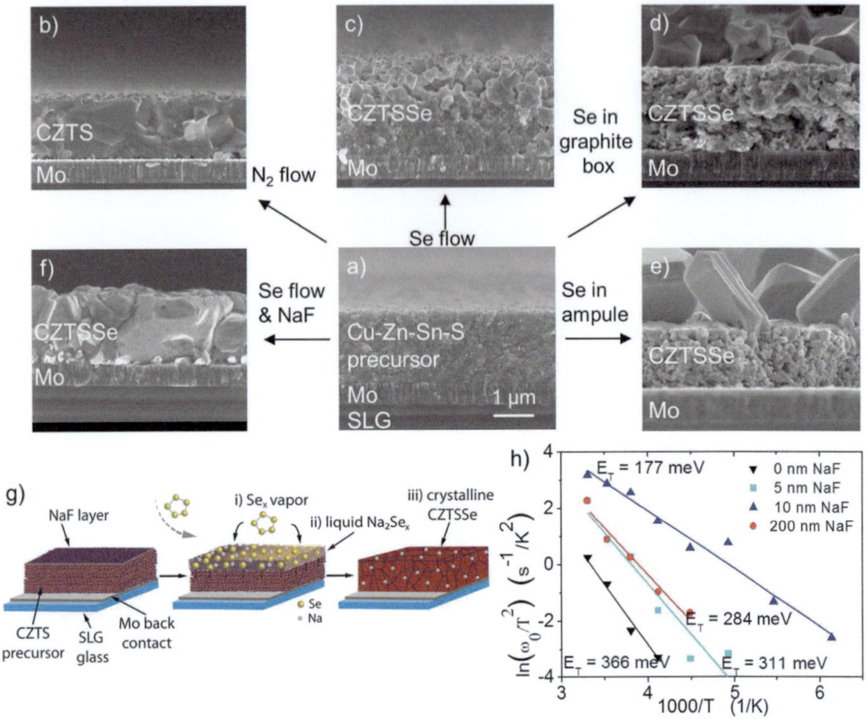

Fig. 10.13 (**a**) SEM cross-sectional images of the Cu-Zn-Sn-S precursor that is converted into a crystallized layer upon annealing in N_2 flow in a tube furnace (**b**), in selenium flow with N_2 carrier gas in a tube furnace (**c**), in selenium atmosphere in a closed graphite box (**d**), with selenium overpressure in a sealed quartz ampule (**e**), and with additional 100 nm NaF top layer during selenization in a tube furnace (**f**). The increased grain size in (**f**) as compared to (**c**) can be explained by the liquid Na_2S ex-assisted crystallization mechanism (**g**). The presence of the NaF top layer does not only influence the layer morphology but also changes electrical properties of the semiconductor layer, which can be seen from the variation of the defect activation energy derived from temperature-dependent capacitance-voltage measurements (**h**). Reprinted with permission from [55]. Copyright 2014, WILEY-VCH

10.5 Chalcogen Nanoparticles Based Solar Cells

Application of nanotechnology in photovoltaics offers flexible and versatile solutions for conventional photovoltaic systems that are bulky and fragile. The substrate used in the manufacturing these cells can range from glass to plastics to other polymers, and hence offer non-rigid and versatile systems of use. Use of nanostructures has been suggested to build new highly efficient solar cells considering the following key points:

1. Hot carriers generated by absorbing photons of energy more than the energy bandgap can be employed to generate additional current and or voltage.

2. Energy states in the energy band gap can be used to trap carriers generated from photons of energy lower than the energy band gap.

Companies using nanotechnology for solar cells include US based Konarka, Nanomat, Inc., Nanosys, Inc., Nanosolar, and Solaris Nanosciences.

10.5.1 Challenges Faced by Nano-PV Technologies

Single junction PV cells are typically limited by four fundamental loss mechanisms:

- Incomplete solar spectrum absorption.
- Thermalization of hot carriers (or excitons) in the form of excess heat.
- Chemical potential (thermodynamic) losses dictating that the photovoltage must be less than the band gap (E_g) for relaxed carriers.
- Radiative recombination.

The presence of other non-radiative losses (e.g., Auger recombination, internal or phonon conversion, and surface recombination) can further reduce cell performance.

Considering only radiative losses, Shockley and Queisser (SQ) predicted a limiting power efficiency of 30 % under simulated black body radiation that has been subsequently recalculated as 33 % under AM1.5 solar irradiance.

10.6 Hybrid Polymer Solar Cell

Third generation solar cells include low cost technology as found in tandem solar cell [64] dye-sensitized solar cells [65] and polymer solar cell (combination of small organic molecules and conjugated polymer) [66]. The fourth generation of solar cells known as bulk heterojunction (BHJ) solar cells provides the combined advantages of both organic and inorganic semiconductors [67]. Towards photovoltaic applications based on chalcogenide compounds nanoparticles (NPs) or Quantum dots (QDs) have to be combined with a second material, which acts as a hole transporter, such as conjugated polymers and an electron transporter material, such as metal oxides or fullerenes, or a metal electrode. In such type of devices, the hole transport material or electron donor and the electron acceptor or electron transport material are the conducting polymer and inorganic nanocrystals (NC) such as CdX or PbX ($X = S$, Se, or Te). NCs have attracted great attention nanoparticle based solar cells due to their tunability energy bandgap with respect to their size and the most beneficiary phenomenon of multiple exciton generation (MEG) [68]. The performance of the BHJ devices based on Pb and Cd chalcogenides NCs is strongly depends on their size and interface structure in the polymer matrix. The major problem arises in BHJ solar cells during the processing of mixture based film due to absence of high solubility of NCs in typical solvents [69].

BHJ solar cells originates from organic solar cell structure where small organic molecules is replaced by inorganic semiconductor NCs as electron acceptors. It is well known that by tuning the size of the NCs, their band gap as well as their energy level can be varied based on the quantum size effect. Furthermore, quantum confinement effect leads to an enhancement of the absorption coefficient compared to that of the bulk materials. Therefore, in the BHJ solar cell, both NCs and polymer have the ability to absorb incident photons. BHJ cells include conjugated polymers that absorb light as the donor and transport holes. The first hybrid polymer solar cell was came into existence by Alivisatos et al. [70] they have proposed a way to make out the use of semiconducting NPs to provide which strongly contributes increase in J_{sc}. Typically chalcogenide semiconductor NPs have a very good absorption values. The high absorption values for NPs can extend to near-IR wavelengths, which is an additional advantage for hybrid solar cell. Semiconducting NPs have been synthesized by using solution based routes (i.e., solvothermal methods, hydrothermal, and hot injection method). Solution processed techniques allows to control over NP size and thus one can control the energy bandgap values [71]. Since fullerenes have a only of type of geometry but NP geometries can be designed or modified into nanorods (NRs) and other geometries. Whilst NCs (Fig. 10.14a) and NRs (Fig. 10.14b) are other common morphologies have been used in hybrid polymer SCs. CdSe is one of the most versatile compounds in generating different nanostructures and can be prepared as NCs, NRs, tetrapods (Fig. 10.14c) and hyper branched forms (Fig. 10.14d) [72–75]. Semiconducting NPs usually aggregate when they dispersed in the organic solvents to prepare hybrid polymer SCs. By coating the NPs with ligands that may have binding groups or hydrophobic tails, semiconducting NPs can be dispersible. However, the ligands are insulating in nature, which interrupts the charge transport mechanism in the NPs and active polymer films. Therefore, these ligands must be removed in order to allow effective charge separation and transport during operation. After removing ligands from NPs surfaces, their dispersion ability decreases therefore an approach has been used to removing bulky ligands used ligand exchange with pyridine for CdSe NPs. After subsequently heating NP, Pyridine removed from the CdSe surface. Achieving reproducible and controlled morphologies are especially important in the context of future translation of hybrid SCs to large scale deployment.

The structure of BHJ solar cell devices consist of a photo-active layer sandwiched between two electrodes of different work functions (Fig. 10.15). Transparent indium tin oxide (ITO) substrate covered with the conducting polymer mixture PEDOT:PSS is used as the anode. The PEDOT:PSS not only acts as a hole transporter and electron blocker but it is also used as an anode buffer material for smoothening the ITO surface. Thus, the ITO surface enhances the adhesion of the upper light absorbing layer, sealing the active layer from oxygen, and keeping the anode material from diffusing into the active layer.

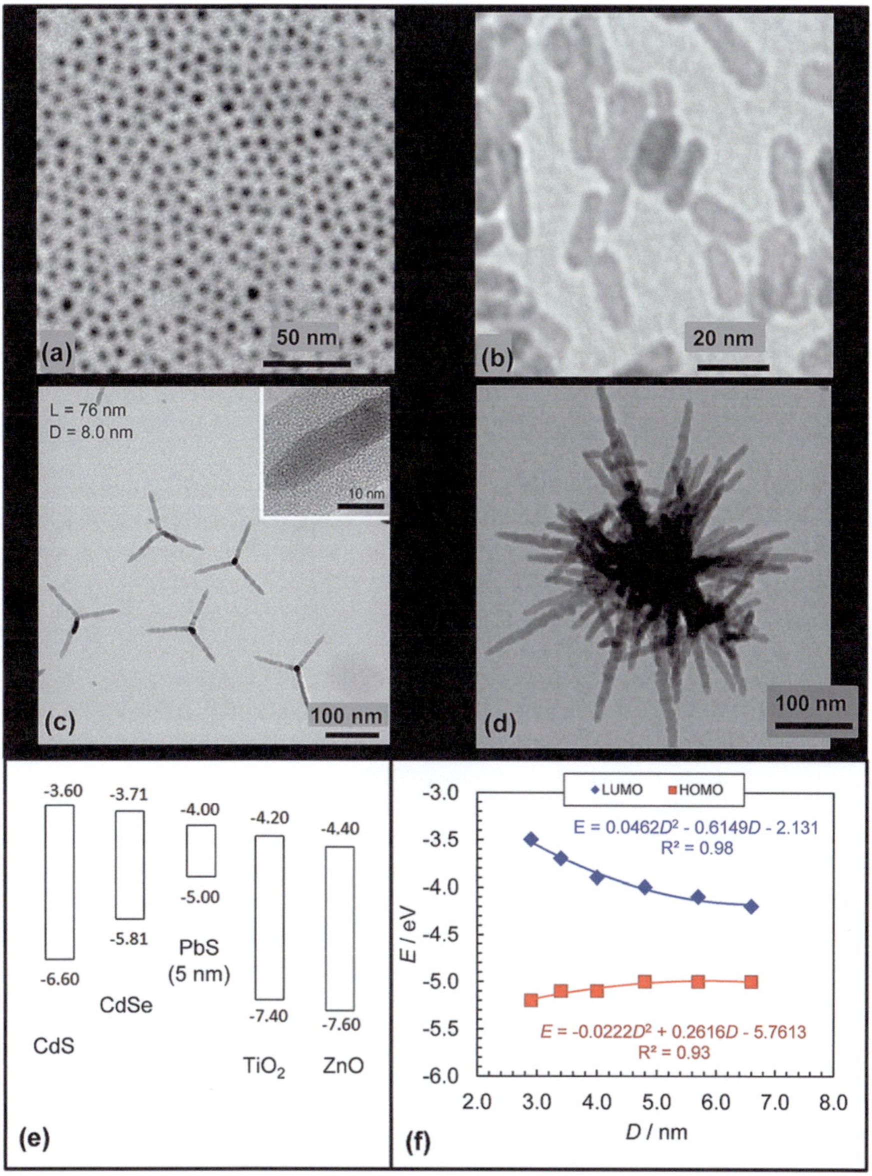

Fig. 10.14 Architectures and energy levels for selected inorganic acceptors. TEM images of PbS NCs [72] (**a**), ZnO NRs [73] (**b**), CdSe tetrapods [74] (**c**) and hyperbranched CdSe [75] (**d**). (**e**) Energy levels for various NCs. (**f**) Size dependence of PbS NC energy levels. Reprinted with permission from [72]. Copyright 2014 RSC advances

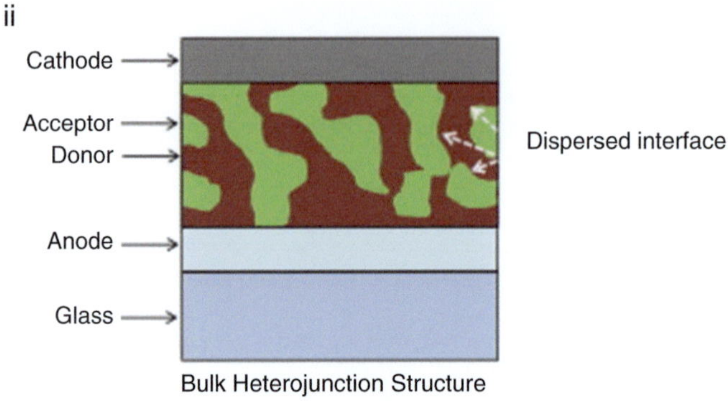

Fig. 10.15 Schematic diagram of typical BHJ solar cell device. Reprinted with permission from Solar energy materials and solar cells 107, 87 (2012) ©Elsevier

On the top of the PEDOT: PSS, the photoactive layer is usually deposited by spin-coating NCs/conjugated polymer nano composite solution to form a thin-film of around 100–200 nm. The cathode, typically made of Al, Ca, Ag, and Au is vacuum deposited on top of the photo active layer.

The photo excitation of organic semiconductors generally results in a strongly bound electron–hole pair, called an exciton. These excitons are only effectively separated at an interface between a p-type (electron donating) and n-type (electron accepting) materials. Hence, in BHJ solar cell devices, the second step after absorption of a photon is a photo induced electron transfer between donor and acceptor type semiconducting materials, yielding a charge-separated state.

This photo induced electron transfer between donor and acceptor boosts the photo generation of free charge carriers compared to the individual, pure materials. In general, the exciton life time and diffusion length in inorganic semiconductors are limited by radiative and non-radiative decay. In conjugated polymers the exciton diffusion length is about 10–20 nm [76]. This implies that for effective charge separation, the photogeneration of excitons should occur within a few nanometers from the p–n junction. These short exciton diffusion lengths impose an important condition for efficient charge generation. Anywhere in the active layer, the distance to the interface should be of the order of the exciton diffusion length. The photocurrent generation in hybrid polymer solar cell occurs in various steps as shown in Fig. 10.16.

In the first step, incident photons ($h\nu$) excites the electrons from the highest occupied molecular orbital (HOMO) to the lowest unoccupied molecular orbital (LUMO) level by the donor/acceptor (D/A) nano composite and creating excitons with a certain binding energy. Excitons diffuse at DA interface. In order to dissociate exciton and generate separated positive and negative charges, the exciton diffusion length of conjugated polymers (around 10–20 nm) helps in the charge transfer and excitons dissociate into free charge carriers before recombination takes place. Excitons dissociate at the DA interface: the HOMO level of the polymer is higher than that of the NCs and the LUMO level of the NCs is lower than that of the

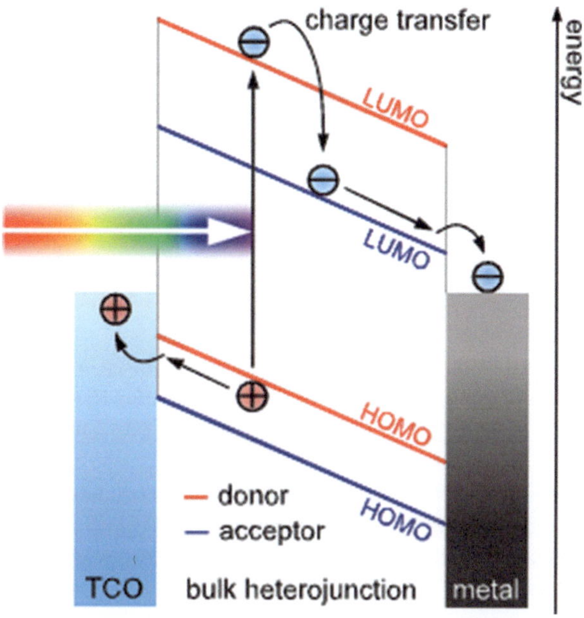

Fig. 10.16 Schematic diagram of charge transport mechanism in hybrid polymer solar cell. Reprinted from Internet source: http://www.physik.uni-wuerzburg.de/EP6/research-oe.html

polymer. After charge transfer at the DA interface, separated electrons and holes are distributed within the acceptor and donor phases, respectively. Further, an internal electric field (deriving from the Fermi level difference of the electrodes) contributes in transportation of holes and electrons towards the respective electrodes [77].

Based on polymer/fullerene mixtures organic solar cell, Greenham et al. [70] used a blend of poly (1-methoxy-4-(2-ethylhexyloxy-2,5-phenylenevinylene)) (MEH-PPV) with CdS or CdSe nanoparticles. The efficient photoconductivity achieved with high nanocrystal concentrations, when electrons and holes could be transported, respectively, through the inorganic and polymer components of the composite material. In addition with the size-energy gap tunable properties, chalcogenide NPs has also the energy levels or bandgap tuning ability by alloying.

The CdSe spherical nanoparticles and CdSe nanorods have also been combined with poly (3-hexylthiophene) (P3HT) to form the BHJ based device. The device provides maximum external quantum efficiency (EQE) of 20 % with CdSe nanospheres and 55 % with nanorods [78]. The charge transport is also dependent on the aspect ratio of the nanorods. In case of small size nanoparticles, electronic transport mechanism dominates through hopping process while in bigger sized nanoparticles conduction charge transport followed through a continuous bands. There are several other reports on different nanostructures, such as tetrapods [79, 80] and hyperbranched nanocrystals [75] based on CdSe, CdTe, and $CdSe_xTe_{1-x}$, used with P3HT or poly[2-methoxy-5-(3,7-dimethyloctyloxy)]-1,4-phenylenevinylene (MDMO-PPV), resulting in devices with efficiencies around 1–2 %. A comparison between the CdSe nanorods or tetrapods structures based BHJ device have been reported [79]. Tetrapod structures results into more efficient devices, due to their extended branched parts which helps in electric conduction. CdSe branch nanostructure enhanced charge separation process in contact with P3HT phase due the longer armed tetrapods, the three-dimensional connectivity increases which decreases the charge hopping events, and contributes into improved charge carrier transport. The increase in aspect ratio from dots to rods to tetrapods helps in conducting to move electrons from the recombination sites. Thus, such type of structures have been used to generate the long-lived charge carriers.

PbS QDs have also been comprehended in BHJ device in combination with MEH-PPV. The device efficiency of 1.1 % was obtained by using single wavelength irradiation [81]. With the inclusion of Se atoms, the bandgap of PbS decreases and helps in harvesting the infrared solar energy. The PbSe QDs of size 6 nm diameter capped with oleate molecules to stabilize the blending solution with P3HT. The device delivered a short circuit current J_{sc} of 1.08 mA/cm^2, V_{oc} of 0.35 V, FF of 0.37, and PCE of 0.14 % [82]. It is important to note that, the efficiencies obtained for devices based on lead chalcogenides/polymer and traditional conjugated polymers and P3HT, MEHPPV, and MDMO-PPV were inferior to cadmium chalcogenides/polymer based devices [83–85].

Sargent et al. [86, 87] have used MEH-PPV and PbS NCs composites and their optical absorption was tuned by controlling nanoparticle size. An ITO as front electrode and Mg as back contact was used to build the PV device. The structure delivered an extremely small EQE of 0.0008 %. Later they have performed thermal annealing investigations on PbS NCs: MEH-PPV nanocomposites. Nanocomposites were treated above 160 °C, which slightly improves the changes in J_{SC} and a very small rise in EQE of about 0.002 %. Wang et al. [88], have synthesized PbS NPs and PbS nanorods using MDMO-PPV as a capping material and again dispersed those NPs and nanorods into MDPO-PPV. In both cases, the efficiencies were extremely low. Jiang et al. [89] were the first to report a PV effect in nanocomposite of P3HT and PbSe NCs via direct blending approach. The thin film devices showed a good diode behavior with an efficiency of 0.04 %, providing evidence of electron transfer between the polymer and PbSe NCs. Yang et al. [90] have reported an alternative aqueous processing technique to prepare nanocomposite of water soluble polymer matrix and CdTe NCs. The PV device based on the nanocomposite of water soluble polymer and CdTe NCs had shown PCE of 0.86 % (Fig. 10.17).

10.6.1 Chalcogenide Quantum Dots Solar Cells

10.6.1.1 Quantum Dots Sensitized Solar Cells (QDSSC)

Among the third-generation solar cells based on quantum dots, the quantum dot-sensitized solar cells (QDSSCs) have received substantial attention probably because they are a straightforward extension of the very popular dye-sensitized

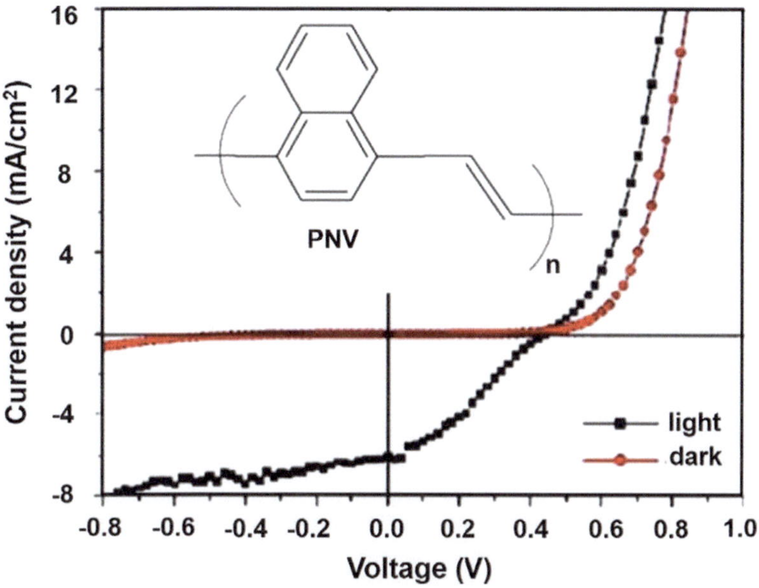

Fig. 10.17 *J–V* curves of CdTe NCs QDSSC in dark and light. Reprinted with permission from [90], copyrights American Chemical Society

solar cells (DSSC). Brian O'Regan and Michael Gratzel first developed dye sensitized PV cells in 1991 [91]. In contrast to the semiconductor p–n junction solar cells where light absorption and charge transport occurs in the same material, the DSSC separates these functions. In DSSC, photons are absorbed by the dye molecules and transport of charges is carried out in a wide bandgap semiconductor electrode and electrolyte. The charge separation in the semiconductor p–n junction cells is induced by the electric field across the junction.

10.6.1.2 Dye Sensitized Solar Cell (DSSC)

Sensitization of wide band gap semiconductors such as ZnO and TiO_2 using dyes has a long history, dating back to early days of photography of the nineteenth century. Systematic studies have been reported for dye sensitization in DSSC on ZnO [92–95] and SnO_2 [96, 97] electrodes with various dyes and electrolytes. The breakthrough of dye sensitized devices as solar cell was based on the exchange of the bulk semiconductor by mesoporous structures of high surface area, which allowed a drastic enhancement of the amount of adsorbed dye. The development of stable metal organic sensitizers bearing attachment groups which allow the chemical bonding of the sensitizer to the surface of the TiO_2 lead to a drastic improvement in charge injection efficiency. The use of ruthenium (II) polypyridyl complexes, substituted with carboxylic groups which specifically bind to the TiO_2 surface. *cis*-Ru(dcbpy)$_2$(NCS)$_2$ represents the most efficient sensitizer known so far, combining high charge injection efficiency with good light harvesting over the total visible range of the spectrum, down to the near infrared.

The fundamental advantage of DSSCs arises from the fact that bulk recombination is suppressed, as only majority charge carriers are present in each of the two phases, confining recombination to the interface. Despite the very promising performance of these cells and contrary to earlier predictions, the series production for commercial applications faced unexpected problems arising from the corrosive and volatile nature of the liquid iodine/iodide electrolyte used in such cells. While sealing of solid state electronic components like computer chips or silicon solar cells towards water and oxygen is a standard technique, sealing of photoelectrochemical solar cells turned out to be a much more difficult task. Additional prerequisites towards the sealant like non permeability for the electrolyte solvent and inertness towards the corrosive nature of iodine have to be guaranteed on the long term. A second major problem is that, conventional current collector grids cannot be applied to these photoelectrochemical solar cells, as even silver or gold grids do not resist the corrosive nature of the electrolyte. Practical advantages may therefore be gained by the replacement of the liquid electrolyte with a solid charge transport material. Inorganic p-type semiconductors have been tested in this regard, but photon to electron conversion efficiencies remained low.

The basic operating principle of a DSSC is shown in Fig. 10.18. The cell consists of a porous nanocrystalline wide bandgap semiconductor such as TiO_2 or ZnO in order to increase the internal surface area of the electrode to allow large amount of dye to be adsorbed at the TiO_2 electrode. The TiO_2 layer is coated on the ITO coated glass to form the front electrode (cathode). The working principle of the

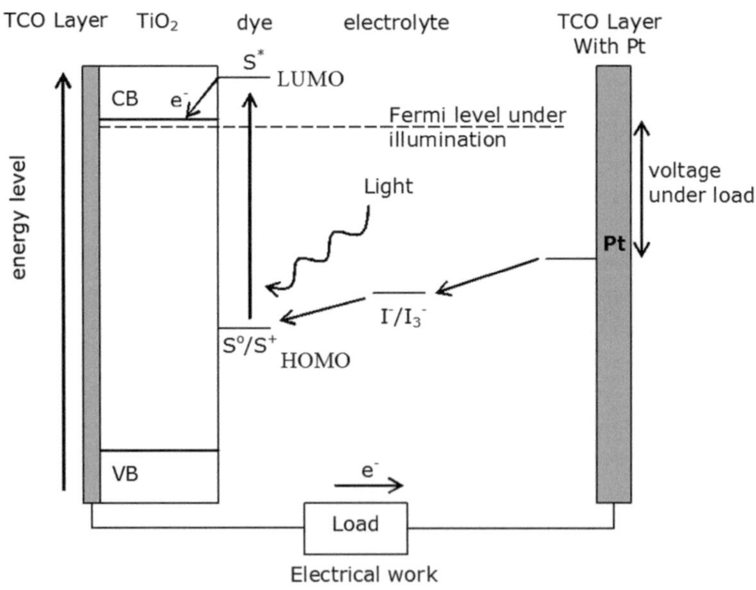

Fig. 10.18 Working principle of DSSC

DSSC is based on excitation of the dye followed by fast electron injection into the conduction band of the TiO_2, semiconductor leaving an oxidized dye molecule on the TiO_2 surface. Injected electrons percolate through the TiO_2 and are fed into the external circuit. At the counter-electrode (anode), triiodide is reduced to iodide by metallic platinum under uptake of electrons from the external circuit.

$$I_3^- + 2e^- \rightarrow 3I^-$$

Iodide is transported through the electrolyte towards the photoelectrode, where it reduces the oxidized dye. The dye molecule is then ready for the next excitation–oxidation–reduction cycle.

Quantum dot-sensitized solar cells (QDSSCs) are renowned energy devices known for their distinct qualities, including (1) the ability to harvest sun light that generates multiple electron–hole pairs, (2) simplicity in fabrication, and (3) low cost. Various research groups have studied the ability of QDs as sensitizers in QDSSCs. The QD properties mentioned in the preceding sections are not only limited to Cd chalcogenides but applicable to other QD materials as well. The main motivation of using QDs as sensitizers in solar cell is due to their tunable energy bandgaps, which can control their absorption range and multiple exciton generation. There are several reports in the literature showing that CdS and CdSe with tunable band gaps property are capable of converting visible light to electric energy. The power conversion efficiency (η) rates of many QDSSCs are lower than those of dye-sensitized solar cells, reaching a maximum of 12 % as a result of narrow absorption ranges and of the charge recombination occurring at the QD– and TiO_2–electrolyte interfaces. New types of sensitizers are necessary to be developed to further increase the efficiency of QDSSCs. Various QD materials have been investigated in QDSSCs because of their particular properties for photosensitization. These materials include CdS, CdSe, PbS, ZnSe, CdHgTe, CdTe, InP, and Ag_2S. Among these QD materials, CdSe and CdS are frequently used because of their ease of fabrication and characterization.

The QDSSC working principle is very similar to DSSC. Here Dye molecules are replaced by QDs of (PbX or CdX, X = S, Se, or Te). Upon illumination, electron–hole pairs generates through QDs. The electrons transfer into the conduction band (CB) of the QD and the hole remains in the valence band (VB). The excited electrons moves further into the CB of the wide band gap semiconductor (e.g., TiO_2) and become itself oxidized state while the hole remains in the valence band. The injected electron further travel via porous TiO_2 network and reached to the conducting electrode and completes the circuit by entering back through the counter-electrode. The whole transport processes of charge carriers in QDSSC can be illustrated in Fig. 10.19.

The major advantage of using QDs as sensitizers in QDSSCs are the tunable energy gap with varying size due to the quantum confinement effect. The quantum confinement effect is manifested when the QD solution exhibits a different color that corresponds to the change in the particle size. This change in size affects a different light absorption band as shown in Fig. 10.20.

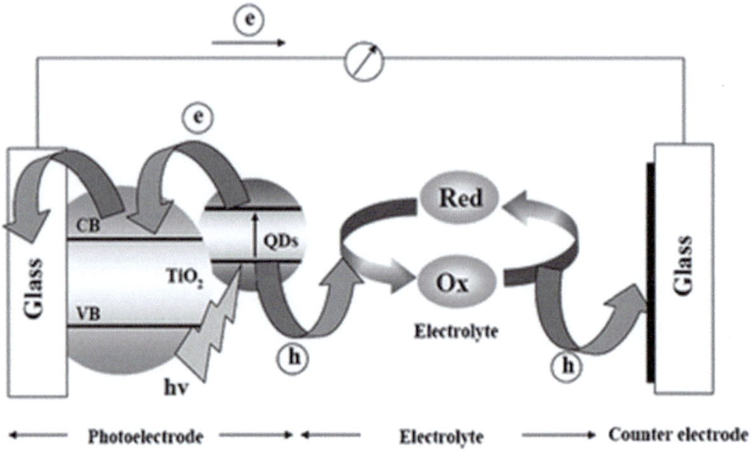

Fig. 10.19 Operating principle of typical QDSSC. Reprinted with permission from Chem. Commun., 2011, 47, 9561. Copyrights Royal Society of Chemistry

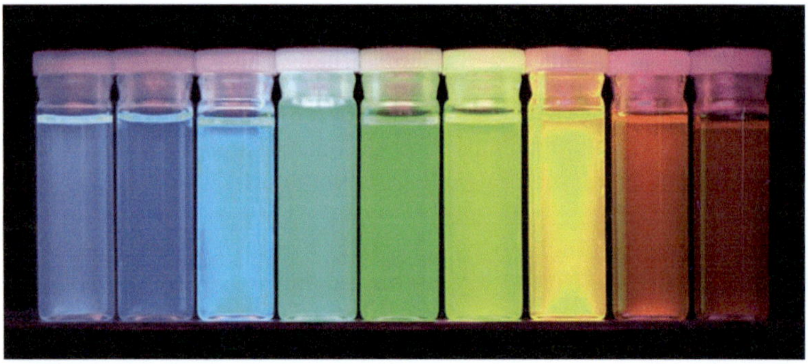

Fig. 10.20 Luminescence emission from CdSe(ZnS) quantum dot solutions: the size of the nanoparticles increases from 2 to 6 nm (*left* to the *right*). The emission spans across the visible part of the electromagnetic spectrum from the *blue* (2 nm nanocrystals) to the *red* (6 nm nanocrystals). Reprinted with permission from Progress in Polymer Science, 34(5), 393–430 (2009) © Elsevier

When the size of the QDs is decreases, the bandgap of QDs increases. Subsequently, optical absorptions and emissions in relation to excitations across the bandgap shift toward higher energies. The blue shift of the optical spectra of CdSe nanoparticles has been obtained as the size decreases. QDSSC based on different sizes of CdSe, i.e., 2.2, 2.5, and 3.3 nm QDs yield different IPCE values with an efficiency of 1.26 % [98]. Mixed combination of CdSe QDs of size 2.5 and 3.5 nm used with TiO_2 nanoparticles exhibits longer lifetime than that of CdSe QD of size 3.5 nm alone.

Yu et al. [99] have reported an increase in efficiency by 4.21 % using CdSe/CdS core shell structure than single CdSe QD sensitizer solar cell.

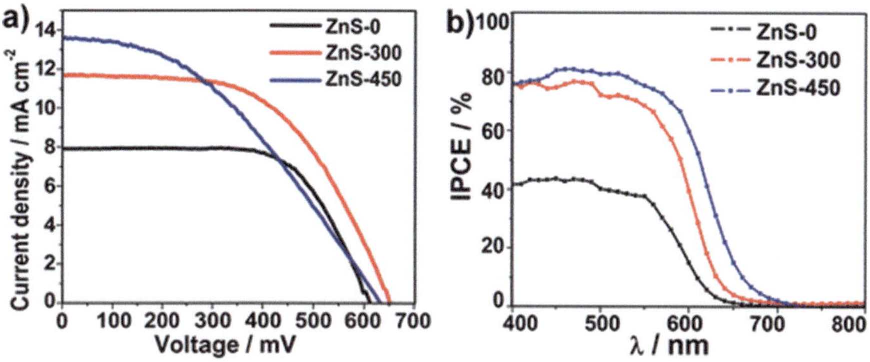

Fig. 10.21 Schematic diagram and (**a**) *J–V* and (**b**) *IPCE spectra* of QDSSCs based on ZnS-capped photoelectrodes (TiO_2–QDs–ZnS) annealed at 400 °C for different time periods. Reprinted with permission from [99]. Copyright 2012 Royal Society of Chemistry

Another CdSe/CdS QD employed in QDSSC which prevents charge recombination inside QDs by reducing the surface states. Whereas single QDs are incapable of increasing energy conversion efficiency of QDSSCs. Jung et al. [100] have developed another approach to improve the performance of QDSSC. They have used ZnS passivation layer in CdS QDs which reduced the recombination rate of excitons that results in increase in efficiency. Similar results have also been demonstrated in PbS QD SSCs with a CdS coating layer by Si-Xin et al. [101]. Multilayered QDs fabricated using the chemical bath deposition (CBD) method can likewise improve QDSSC performance. Hu et al. [102] reports the increase in efficiency up to 1.47 % for CdS/ $CuInS_2$ multilayer QDs approaches. Whereas in single QD structure with CdS and $CuInS_2$ QD SSCs the efficiencies were only 0.34 % and 0.38 % respectively. The influence of different annealing temperatures (100–250 °C) on the performance of ZnS capped QDSSC reported by Yu et al. [99] as shown in Fig. 10.21. the remarkable increase in IPCE from 0.46 to 2.83 % was achieved with thermal treatment of CdSe/CdS QDs. Sung-Huwan et al. [103] have report PbS based QDSSC with TiO2 wide bandgap semiconductor with Pt counter-electrode. They have reports the efficiency of 0.59 % using TiO_2 and 0.80 % single walled carbon nanotubes and TiO_2 with PbS as sensitizer. Mali et al. [104] reported the effect of using nanocoral TiO_2 structured oxide semiconductor with PbS QDs. The device exhibits efficiency of 1.23 %.

10.7 Concluding Remarks and Future Outlook

The chapter reviews the possible approaches and strategies to use the chalcogenide materials for the photovoltaic devices. Though Cd and Pb based chalcogenide solar cells have shown the better efficiency but their toxic nature limits their large production or further usage. Cu based chalcogenide solar cells have been found to possess an efficiency of more than 12 %. They can be further studied in details to prove green photovoltaic technology with great efforts to address all kinds of important issues and pave a way toward the ultimate successful realization of low-cost and high efficiency solar cells. In this path, active communication between engineers, chemists, physicists, and materials scientists and their concerted efforts are highly desired to make breakthroughs in near future.

Recently, inorganic–organic halide perovskites represent a major revolution in the growth of photovoltaic materials. These materials are composed of Pb, an element which is of great environmental concern and the halide salts which tend to dissociate in the presence of moisture, and causes stability issues for long-term usage. Therefore chalcogenide perovskites based on S and Se can be realized towards photovoltaic applications. Till date there is no accurate knowledge of the band gaps and, in particular, optical properties, available or the chalcogenide perovskite materials, based on which their potential for photovoltaic applications may be evaluated. The chalcogenide perovskites are based on the Ca, Sr, Ba, Ti, Zr, and Hf elements, which are more environment friendly as opposed to the Pb-based halide perovskites, it is of great interest to study these materials for photovoltaic applications.

References

1. Overview of "PV Roadmap Toward 2030" (PV 2030), New energy and Industrial Technology Development Organization (NEDO), New Energy Technology Development Department, June 2004
2. Y. Lee, C. Park, N. Balaji, Y. Lee, V. Dao, Israel J. Chem. **55**, 1050–1063 (2015)
3. W.A. Badaw, J. Adv. Res. **6**, 123–132 (2015)
4. J. Willmann, D. Stocker, E. Dörsam, Org. Electron. **15**, 1631–1640 (2014)
5. K.D.G.I. Jayawardena, L.J. Rozanski, C.A. Mills, M.J. Beliatis, N.A. Nismy, S. Ravi, P. Silva, Nanoscale **5**, 8411–8427 (2013)
6. Simon P. Philipps, Andreas W. Bett, Current status of concentrator photovoltaic (CPV) technology. Fraunhofer ISE NREL CPV Report 1.1, September 2015 TP-6A20-63916, 2015
7. K. Masuko, M. Shigematsu, T. Hashiguchi, D. Fujishima, M. Kai, N. Yoshimura, T. Yamaguchi, Y. Ichihashi, T. Yamanishi, T. Takahama, M. Taguchi, E. Maruyama, S. Okamoto, IEEE J. Photovolt. **4**, 1433–1435 (2014)
8. H. Sai, T. Matsui, K. Matsubara, M. Kondo, I. Yoshida, IEEE J. Photovolt. **4**, 1349–1353 (2014)
9. First Solar Press Release, First Solar builds the highest efficiency thin film PV cell on record (2014)
10. T. Matsui, H. Sai, T. Suezaki, M. Matsumoto, K. Saito, I. Yoshida, M. Kondo, in *28th European Photovoltaic Solar Energy Conference*, 2213–2217, 2013
11. R. Komiya, A. Fukui, N. Murofushi, N. Koide, R. Yamanaka, H. Katayama, in *21st International Photovoltaic Science and Engineering Conference*, Fukuoka, November, 2C-5O-08 2011.
12. M. Hosoya, H. Ooka, H. Nakao, T. Gotanda, S. Mori, N. Shida, R. Hayase, Y. Nakano, M. Saito, in *Proceedings of the 93rd Annual Meeting of the Chemical Society of Japan*, 21–37, 2013

13. K. Kaneko, NIMS achieves 15% efficiency with perovskite PV cell. Nikkei Business Publications, Inc., 12 May 2015. http://techon.nikkeibp.co.jp/english/NEWS_EN/20150512/417866/?ST=msbe. Accessed 22 May 2015
14. W. Wang, M.T. Winkler, O. Gunawan, T. Gokmen, T.K. Todorov, Y. Zhu, D.B. Mitzi, Adv. Energy Mater. **4**, 1301465 (2014)
15. E. Becquerel, C. R. Acad. Sci. Paris **9**, 561 (1839)
16. D.M. Chapin, C.S. Fuller, G.L. Pearson, J. Appl. Phys. **25**, 676 (1954)
17. I.D. Parker, J. Appl. Phys. **75**, 1656 (1994)
18. S.R. Forest, MRS Bull. **30**, 28–32 (2005)
19. P. Peumans, A. Yakimov, S.R. Forrest, J. Appl. Phys. **93**, 3693 (2003)
20. Y. Xia, Y.J. Xiong, B. Lim, S.E. Skrabalak, Angew. Chem. Int. Ed. **48**, 60–103 (2009)
21. C.B. Murray, C.R. Kagan, M.G. Bawendi, Annu. Rev. Mater. Sci. **30**, 545–610 (2000)
22. U. Jeong, X. Teng, Y. Wang, H. Yang, Y. Xia, Adv. Mater. **19**, 33–60 (2007)
23. D.A. Cusano, Solid State Electron. **6**, 217 (1963)
24. D. Bonnet, H. Rabenhorst, in *Proceedings of the 9th IEEE Photovoltaic Specialist Conference*, 129, 1972.
25. Y.S. Tyan, E.A. Perez-Albuerne, in *Proceedings of the 16th IEEE Photovoltaic Specialist Conference, IEEE, San Diego*, 794, 1982
26. L. Kranz et al., Nat. Commun. **4**, 2306 (2013)
27. I. Matulionis et al., Mater. Res. Soc. Symp. Proc. **668**, H8.23.1 (2001)
28. J. Poortmans, V. Arkhipov, *Thin Film Solar Cells Fabrication, Characterization and Applications* (Wiley, England, 2006), p. 278
29. http://www.empa.ch/plugin/template/*/110807
30. W. Rance, J.M. Burst, M.O. Reese, in *39th IEEE Photovoltaic Specialist Conferecne*, Tampa, USA, 2013
31. http://www.advsolarpower.com/index.php/en/producten/productsilien
32. M. Bauer, http://www.calyxo.com/en/news-event/news/117-calyxo-cdte-thin -film-producer-reaches-efficiency-of-13.4 percent and secure financing for the year 2012 html
33. http://www.xunlight.com/26/technology.html
34. J. Kephart, R. Geisthardt, W.S. Sampath, Prog. Photovolt. **23**, 1484–1492 (2015). doi:10.1002/pip.2578
35. Y. Akaki, K. Nomoto, S. Nakamura, T. Yoshitake, K. Yoshino, J. Phys. Conf. Ser. **100**, 082022 (2008)
36. M. Powalla, P. Jackson, W. Witte, D. Hariskos, S. Paetel, C. Tschamber, W. Wischmann, Sol. Energ. Mater. Sol. Cells **119**, 51–58 (2013)
37. P. Reinhard, S. Buecheler, A.N. Tiwari, Technological status of Cu(In, Ga)(Se, S)2-based photovoltaics. Sol. Energ. Mater. Sol. Cells **119**, 287–290 (2013)
38. S. Ishizuka, A. Yamada, P. Fons, S. Niki, Prog. Photovolt. Res. Appl. **21**, 544–553 (2013)
39. J. Haarstrich, H. Metzner, M. Oertel, C. Ronning, T. Rissom, C.A. Kaufmann, T. Unold, H.W. Schock, J. Windeln, W. Mannstadt, E. Rudigier-Voigt, Sol. Energ. Mater. Sol. Cells **95**, 1028–1030 (2011)
40. J. Lindahl, U. Zimmermann, P. Szaniawski, T. Törndahl, A. Hultqvist, P. Salomé, C. Platzer-Björkman, M. Edoff, J. Photovolt. **3**, 1100–1105 (2013)
41. G.M. Hanket, C.P. Thompson, J.K. Larsen, E. Eser, W.N. Shafarman, in *Proceedings of the 38th IEEE Photovolt. Specialists Conference*, Austin, USA, 662–667, 2012
42. M. Nakamura, Y. Kouji, Y. Chiba, H. Hakuma, T. Kobayashi, T. Nakada, in *Proceedings of the 39th IEEE Photovolt. Specialists Conference*, Tampa, USA, 2013
43. M.A. Green, K. Emery, Y. Hishikawa, W. Warta, E.D. Dunlop, Prog. Photovolt. Res. Appl. **23**, 805–812 (2015)
44. H. Azimi, Y. Hou, C. Brabec, J. Christroph, Energy Environ. Sci. **7**, 1829 (2014)
45. T.K. Todorov, O. Gunawan, T. Gokmen, D.B. Mitzi, Prog. Photovolt. Res. Appl. **21**, 82 (2013)
46. W. Zhao, Y. Cui, D. Pan, Energy Technol. **1**, 131 (2013)

47. M. Kaelin, D. Rudmann, F. Kurdesau, H. Zogg, T. Meyer, A.N. Tiwari, Thin Solid Films **480–481**, 486 (2005)
48. W. Wang, S.-Y. Han, S.-J. Sung, D.-H. Kim, C.-H. Chang, Phys. Chem. Chem. Phys. **14**, 11154 (2012)
49. S.J. Park, J.W. Cho, J.K. Lee, K. Shin, J.-H. Kim, B.K. Min, Prog. Photovolt. Res. Appl. **22**, 122 (2014)
50. V.S. Saji, I.H. Choi, C.W. Lee, Sol. Energy **85**, 2666 (2011)
51. R.N. Bhattacharya, J. Electrochem. Soc. **130**, 2040 (1983)
52. R.N. Bhattacharya, K. Rajeshwar, Sol. Cells **16**, 237 (1986)
53. J.F. Guillemoles, P. Cowache, S. Massaccesi, L. Thouin, S. Sanchez, D. Lincot, J. Vedel, Adv. Mater. **6**, 379 (1994)
54. D. Lincot, J.F. Guillemoles, S. Taunier, D. Guimard, J. Sicx-Kurdi, A. Chaumont, O. Roussel, O. Ramdani, C. Hubert, J.P. Fauvarque, N. Bodereau, L. Parissi, P. Panheleux, P. Fanouillere, N. Naghavi, P.P. Grand, M. Benfarah, P. Mogensen, O. Kerrec, Sol. Energy **77**, 725 (2004)
55. Y.E. Romanyuk, H. Hagendorfer, P. Stücheli, P. Fuchs, A.R. Uhl, C.M. Sutter-Fella, M. Werner, S. Haass, J. Stückelberger, C. Broussillou, P.-P. Grand, V. Bermudez, A.N. Tiwari, Adv. Funct. Mater. **25**, 12–27 (2015)
56. C. Guillén, J. Herrero, J. Electrochem. Soc. **143**, 493 (1996)
57. R.N. Bhattacharya, W. Batchelor, J.F. Hiltner, J.R. Sites, Appl. Phys. Lett. **75**, 1431 (1999)
58. L. Guo, Y. Zhu, O. Gunawan, T. Gokmen, V.R. Deline, S. Ahmed, L.T. Romankiw, H. Deligianni, Prog. Photovolt. Res. Appl. **22**, 58 (2014)
59. C. Gougaud, D. Rai, S. Delbos, E. Chassaing, D. Lincot, J. Electrochem. Soc. **160**, 485 (2013)
60. W. Ki, H.W. Hillhouse, Adv. Energy Mater. **1**, 732 (2011)
61. T. Schnabel, T. Abzieher, E. Ahlswede, in *40th IEEE Photovoltaics Specialists Conference* (*PVSC 40*), Denver, CO, USA, 2014
62. C.M. Sutter-Fella, A.R. Uhl, Y.E. Romanyuk, A.N. Tiwari, Phys. Status Solidi A **22**, 121 (2014)
63. M. Werner, C.M. Fella, Y.E. Romanyuk, A.N. Tiwari, unpublished
64. Q. Cheng et al., Funct. Mater. Lett. **8**, 1550054 (2015)
65. A.R. Marshall, M.C. Beard, J.M. Luther, in *Photovoltaic Specialist Conference* (*PVSC*), *2014 I.E. 40th*, 8–13 June 2014, 1077–1079, Denver, CO
66. C. Chen, W. Chang, K. Yoshimura, K. Ohya, J. You, J. Gao, Z. Hong, Y. Yang, Adv. Mater. **26**, 5670–5677 (2014)
67. A.J. Heeger, J. Phys. Chem. B **105**(36), 8475–8491 (2001)
68. H.W. Hillhouse, M.C. Beard, Curr. Opin. Colloid Interface Sci. **14**(4), 245–259 (2009)
69. J. Seo, W.J. Kim, S.J. Kim, K.S. Lee, A.N. Cartwright, P.N. Prasad, Appl. Phys. Lett. **94**, 133302 (2009)
70. N.C. Greenham, Z. Peng, A.P. Alivisatos, Phys. Rev. B: Condens. Matter Mater. Phys. **54**, 17628–17637 (1996)
71. W.W. Yu, L. Qu, W. Guo, X. Peng, Chem. Mater. **15**, 2854–2860 (2003)
72. J. Yan, B.R. Saunders, RSC Adv. **4**, 43286 (2014)
73. K.F. Jeltsch, M. Schädel, J.-B. Bonekamp, P. Niyamakom, F. Rauscher, H.W.A. Lademann, I. Dumsch, S. Allard, U. Scherf, K. Meerholz, Adv. Funct. Mater. **22**, 397–404 (2012)
74. R. Rhodes, M. Horie, H. Chen, Z. Wang, M.L. Turner, B.R. Saunders, J. Colloid Interface Sci. **344**, 261–271 (2010)
75. I. Gur, N.A. Fromer, C.-P. Chen, A.G. Kanaras, A.P. Alivisatos, Nano Lett. **7**, 409–414 (2007)
76. D.E. Markov, E. Amsterdam, P.W.M. Blom, A.B. Sieval, J.C. Hummelen, J. Phys. Chem. A **109**, 5266–5274 (2005)
77. J. Patela, F. Mighrian, A. Ajjib, T.K. Chaudhuri, J. Nananoen. **5**, 36 (2014)
78. W.U. Huynh, J.J. Dittmer, A.P. Alivisatos, Science **295**, 2425 (2002)
79. Y. Zhou, Y.C. Li, H.Z. Zhong, J.H. Hou, Y.Q. Ding, C.H. Yang, Y.F. Li, Nanotechnology **17**, 4041 (2006)

80. B.Q. Sun, E. Marx, N.C. Greenham, Nano Lett. **3**, 961 (2003)
81. R. Watt, D. Blake, J.H. Warner, E.A. Thomsen, E.L. Tavenner, H. Rubinsztein-Dunlop, P. Meredith, J. Phys. D: Appl. Phys. **38**, 2006 (2005)
82. D. Cui, J. Xu, T. Zhu, G. Paradee, S. Ashok, M. Gerhold, Appl. Phys. Lett. **88**, 138111 (2006)
83. S. Gunes, K.P. Fritz, H. Neugebauer, N.S. Saricici, S. Kumar, G.D. Scholes, Sol. Energ. Mater. Sol. Cells **91**, 420 (2007)
84. K.P. Fritz, S. Guenes, J. Luther, S. Kumar, N.S. Sacricici, G.D. Scholes, J. Photochem. Photobiol. A **195**(39) (2008)
85. Z. Tan, T. Zhu, M. Thein, S. Gao, A. Cheng, F. Zhang, C. Zhang, H. Su, J. Wang, R. Henderson, J.-I. Hahm, Y. Yang, J. Xu, Appl. Phys. Lett. **95**, 063510 (2009)
86. S.A. McDonald, P.W. Cyr, L. Levina, E.H. Sargent, Appl. Phys. Lett. **85**, 2089–2091 (2004)
87. S.A. McDonald, G. Konstantatos, S. Zhang, P.W. Cyr, E.J.D. Klem, L. Levina, Nat. Mater. **4**, 138–142 (2005)
88. Z. Wang, S. Qu, X. Zeng, J. Liu, C. Zhang, M. Shi, F. Tan, Z. Wang, Curr. Appl. Phys. **9**, 1175 (2009)
89. X. Jiang, S.B. Lee, I.B. Altfeder, A.A. Zakhidov, R.D. Schaller, J.M. Pietryga, V.I. Klimov, Proc SPIE **5938**, 59381F-1–59381F-9 (2005)
90. Z. Fan, H. Zhang, W. Yu, Z. Xing, H. Wei, Q. Dong, W. Tian, B. Yang, ACS Appl. Mater. Interfaces **3**, 2919–2923 (2011)
91. B. O'Regan, M. Graztel, Nature **353**, 737 (1991)
92. H. Gerischer, H. Tributsch, Berich. Buns. Gesell. **72**, 437 (1968)
93. S. Rani, P. Suri, P.K. Shishodia, R.M. Mehra, Sol. Energ. Mater. Sol. Cells **92**(12), 1639–1645 (2008)
94. S. Rani, P.K. Shishodia, R.M. Mehra, J. Renew. Sust. Energy **2**(4), 043103 (2010)
95. H. Tributsch, H. Gerischer, Berich. Buns. Gesell. **73**, 251 (1969)
96. R. Memming, F. Schroppel, Chem. Phys. Lett. **62**, 207 (1979)
97. R. Memming, F. Schroppel, U. Bringmann, J. Electroanal. Chem. **100**, 307 (1979)
98. J. Chen, W. Lei, W.Q. Deng, Nanoscale **3**, 674 (2011)
99. X.-Y. Yu, B.-X. Lei, D.-B. Kuang, C.-Y. Su, J. Mater. Chem. **22**, 12058–63 (2012)
100. J. Sung Woo, K. Jae-Hong, K. Hyunsoo, C. Chel-Jong, A. Kwang-Soon, J. Curr. Appl. Phys. **12**(6), 1459 (2012)
101. J. Jie, Z. Zheng-Ji, Z. Wen-Hui, W. Si-Xin, Mater. Sci. Semicond. Process. **16**(2), 435 (2013)
102. H. Xing, Z. Quanxin, H. Xiaoming, L. Dongmei, L. Yanhong, M. Qingbo, J. Mater. Chem. **21**, 15903 (2011)
103. W. Lee, J. Lee, K.M. Sun, T. Park, W. Yi, S. Han, Mater. Sci. Eng. **156**, 48 (2009)
104. S. Mali, S.K. Desai, S. Kalagi, C.A. Betty, P.N. Bhosale, R.S. Devan et al., Dalton Trans. **41**, 6130 (2012)

Chapter 11
Infrared Detectors

Gurinder Kaur Ahluwalia and Ranjan Patro

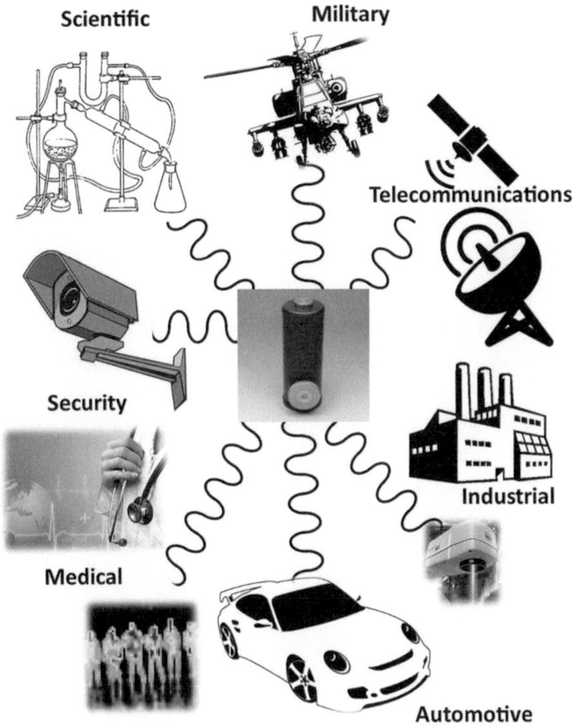

Infrared detector applications

G.K. Ahluwalia, Ph.D. (✉)
Materials and Nanotechnology Research Laboratory, Department of Physics College of The North Atlantic, Labrador West Campus, 1600 Nichols Adam Highway, Labrador City, A2V 0B8, NL, Canada
e-mail: gurinder.kaur@cna.nl.ca

R. Patro, Ph.D.
Physics Department, College of the North Atlantic, 219 Hamilton River Road, P.O. Box 1720 Station B, Happy Valley Goose Bay, A0P1E0, NL, Canada

G.K. Ahluwalia (ed.), *Applications of Chalcogenides: S, Se, and Te*,
DOI 10.1007/978-3-319-41190-3_11

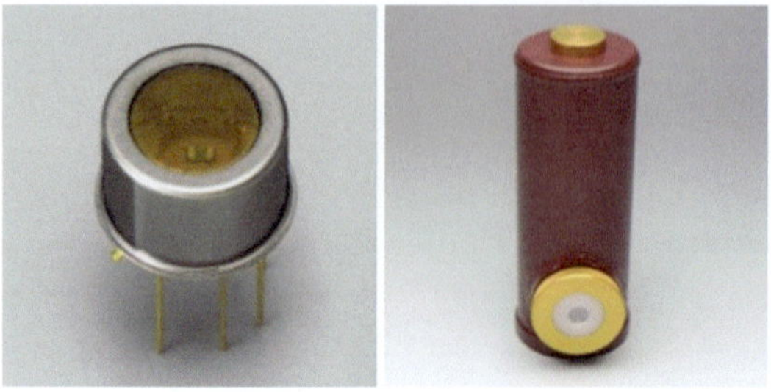

Left: Non-cooled type and TE-cooled type photoconductive MCT detector, continuous operation cutoff wavelength: 5.4 μm, Photoconductive element that decreases its resistance by input of infrared light: *Right*: 10 μm band infrared detector with high sensitivity and high-speed response. Courtesy: Hamamatsu Corporation

11.1 Introduction

Infrared photodetectors (IRPD) are devices that respond to incident photons with wavelengths in the infrared spectrum (~1–1000 μm). Applications of infrared detectors include a wide spectrum [1] such as military and security, scientific, earth resources, medical, industrial, automotive, and energy conservation. Traditionally, IR technologies were used for controlling functions and night vision problems by forming IR images from temperature and emissive differences (systems for recognition and surveillance, tank sight systems, anti-tank missiles, air–air missiles, etc.). Common peaceful applications using infrared detectors include rail safety, space operations, gas leak detection, flame detection, alcohol level testing, anesthesiology testing, petroleum exploration, temperature sensing, water and steel analysis. Infrared motion detectors are used in home alarm systems. Thermal night vision goggles collect infrared radiation particles by sensing the emitted heat and using these particles in order to create a picture. IR detectors assist night driving, night flying and night surveillance, wildlife observation, and search and rescue missions. Law enforcement officials can also use IR detectors to help them find criminals that may be hiding in dark areas.

In general, all objects emit an IR spectrum based on their temperature. This emission spectrum can be approximated by wavelength, λ, as blackbody radiation, which can be characterized according to the blackbody's temperature, T, by Eq. (11.1) [2, 3]:

$$e_{\mathrm{B}}(\lambda, T)\,d\lambda = \frac{2\pi hc^2 d\lambda}{\lambda^5\left[e^{hc/\lambda kT} - 1\right]} \tag{11.1}$$

It implies that an object at room temperature will emit IR radiation with a peak intensity of around 9.5 μm, with detectable IR emissions for wavelengths microns away from this center value. While not every portion of the IR spectrum is ordinarily present due to natural blackbody radiation (a blackbody would need to be at a temperature of over 4000 K to have peak light emissions at the edge of the visible spectrum), these sources do emit across large portions of the IR spectrum.

According to Wien's law, hotter objects emit more of their energy at shorter wavelengths. A blackbody source at 300 K has a peak exitance (power per unit area leaving a surface) at a wavelength of about 9.5 μm. For a source at 1000 K, the maximum exitance occurs at 2.9 μm. Therefore, detectors operating in the LWIR band are well suited to image room temperature objects (people, buildings, etc.), while MWIR (mid-wave infrared) band imagers are good for viewing objects at higher temperatures (hot engines and exhaust gases). In general, LWIR and MWIR bands will produce thermal images if small temperature changes or varying emissivities exist within a scene. However, while the LWIR band imagery may exhibit a higher sensitivity for room temperature objects, the MWIR band imagery presents a better resolution.

Recently, there has been considerable progress towards new materials development and device design innovations for IR detection. Wavelength 1 μm is a sensitivity limit of popular Si detectors. Similarly wavelength 3 μm is a long-wavelength sensitivity of PbS and InGaAs detectors; wavelength 6 μm is a sensitivity limit of InSb, PbSe, PtSi detectors, and HgCdTe detectors optimized for 3–5 μm atmospheric window, and finally wavelength 15 μm is a long-wavelength sensitivity limit of HgCdTe detectors optimized for an 8–14 μm atmospheric window [1]. In particular, significant advances have been made during the last two decades in the area of band-gap engineering of various compound semiconductors that has led to new detector architectures. New emerging strategies include barrier structures such as nBn detector, low-dimensional solids (especially T2SLs) with lower generation–recombination leakage mechanisms, photon trapping detectors, and multistage/cascade infrared devices.

At present, HgCdTe is the most widely used variable gap semiconductor for IR photodetectors. The specific advantages of HgCdTe are the direct energy gap, ability to obtain both low and high carrier concentrations, high mobility of electrons, and low dielectric constant. The extremely small change of lattice constant with composition makes growth of high quality layers and heterostructures possible. HgCdTe can be used for detectors operating under various modalities (photoconductors, photodiodes, and metal–insulator–semiconductor (MIS) detectors), and can be optimized for operation within an extremely wide range of the IR spectrum (from visible region to 30 μm) and at temperatures ranging from that of liquid helium to room temperature.

From technical point of view, the main limitation of LWIR HgCdTe-based detectors is that they need cooling to temperatures near that of liquid nitrogen (77 K), to reduce noise due to thermally excited current carriers (see cooled infrared camera). MWIR HgCdTe cameras can be operated at temperatures accessible to thermoelectric coolers with a small performance penalty. Hence, HgCdTe detectors are relatively heavy compared to bolometers and require maintenance. On the other side, HgCdTe enjoys much higher speed of detection (frame rate) and is significantly more sensitive than some of its more economical competitors.

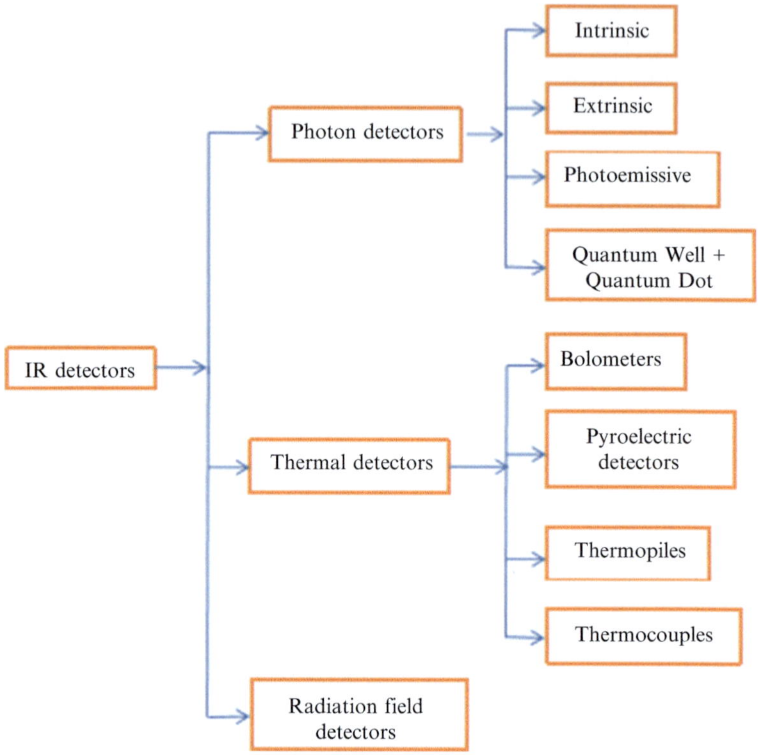

Fig. 11.1 Modes of operation of IR detectors

11.2 Development and Classification of IR Detectors

Ever since William Herschel discovered the presence of infrared radiation in sunlight in the early nineteenth century [4], people have tried various means to detect and analyze the spectrum of invisible light by developing infrared detectors. The earliest detectors were thermopiles developed by Marcedonio Melloni in the mid-nineteenth century [5].

Important milestones in the history of development of IR detectors include:

- Seebeck (1921) discovered the thermoelectric effect and soon thereafter demonstrated the first thermocouple.
- Nobili (1829) constructed the first thermopile by connecting a number of thermocouples in series.
- Melloni (1933) modified thermocouple design and used bismuth and antimony for it.

The first modern photodetector using the quantum effects in materials such as lead sulfide was developed by Case in 1917 [6]. Figure 11.1 gives a classification of existing IR detectors based on the mechanism of detection.

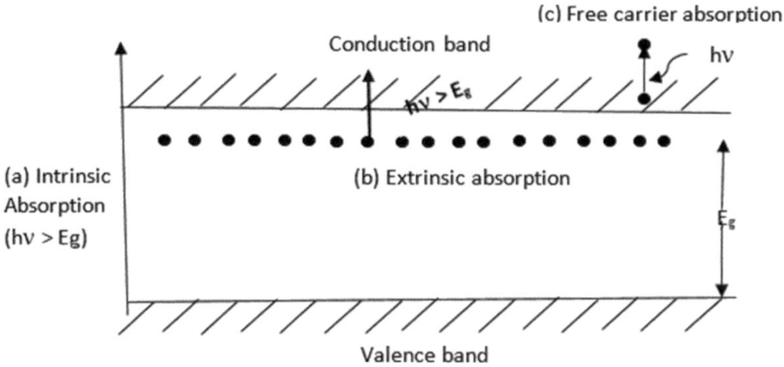

Fig. 11.2 Fundamental optical excitation processes in semiconductors: (**a**) intrinsic absorption, (**b**) extrinsic absorption, (**c**) free carrier absorption

In photoconductive detectors the radiation is absorbed within the material by interaction with electrons, either bound to the lattice atoms, or to impurity atoms or with free electrons as shown in Fig. 11.2. The observed electrical output signal results from the changed electronic energy distribution.

The photon detectors show a selective wavelength dependence of the response per unit incident radiation power. They exhibit both perfect signal-to-noise performance and a very fast response. But to achieve this, the photon detectors require cryogenic cooling. Cooling requirements are the main obstacle to the more widespread use of IR systems based on semiconductor photodetectors making them bulky, heavy, expensive and inconvenient to use. Photon detectors can be subdivided into several categories including intrinsic detectors, extrinsic detectors, photoemissive detectors, quantum wells, and quantum dots. Intrinsic detectors consist of direct bandgap semiconductors of III–V (InGaAs, InAs, InSb, InAsSb), IV–VI (PbS, PbSe, PbSnTe), and II–VI (HgCdTe) material systems where photon-induced optical transitions primarily occur between the valence and conduction bands. Extrinsic detectors consist of indirect bandgap semiconductors such as Si doped with Ga or As (Si: Ga, Si:As) or Ge doped with Cu or Hg(Ge: Cu, Ge: Hg) to facilitate photon absorption from the impurity band within the forbidden gap to the conduction band. Photoemissive detectors are typically metal silicide Schottky barrier devices made of PtSi, Pt_2Si, or IrSi. Quantum well and quantum dot IR photodetectors (QWIP, QDIP) consist of superlattice structures and heterojunction configurations of type-I (GaAs/AlGaAs, InGaAs/AlGaAs) or type-II (InAs/InGaSb, InAs/InAsSb) materials. Depending on the design, materials, and application, these devices can be operated in photoconductive (PC), photovoltaic (PV), photoelectromagnetic, or photoemissive modes.

IR thermal detectors are employed in various forms [7–9], most of which are commonly identified as bolometers, pyroelectric detectors, and thermopiles and thermocouples. When subjected to incident photons, the detector material is heated by the IR radiation resulting in a change in temperature. A temperature-dependent property such as resistance, pyroelectric voltage, or thermoelectric voltage is then used to generate an electrical signal. The detector response time is limited by the heating

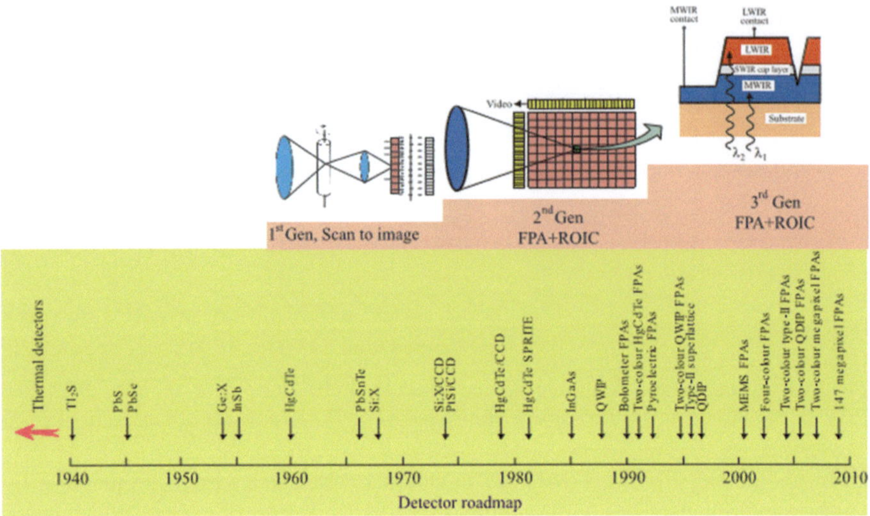

Fig. 11.3 History of the development of infrared detectors and systems. Three generation systems can be considered for principal military and civilian applications: first Gen (scanning systems), second Gen (staring systems electronically scanned) and third Gen (multicolor functionality and other on-chip functions). Reprinted with permission from [10]

and cooling time of the detector element. The thermal detector spectral response is broad and generally wavelength independent, relying on the incident power rather than the energy of the incoming photons. Thermal detectors are usually operated at room temperature and typically found in low-cost, low-sensitivity applications.

11.3 Four Generations of IR Detector Devices

HgCdTe detector architectures have developed from the initial simple photoconductor structure to standard p–n junction photodiode structure and, more recently, to more complex detector focal plane arrays (FPAs). With the increased requirements for high quality imaging, large format HgCdTe two-dimensional (2D) FPAs, viz., assemblage of individual detector picture elements (pixels) located at the focal plane of an imaging system, were needed, which could not be met with HgCdTe photoconductors due to their high power dissipation. In contrast, photovoltaic detectors are p–n junction diodes operating under reverse bias conditions, and thus have high impedance and very low power dissipation, which provides an excellent device structure for fabricating high performance large format 2D FPAs. Apart from very low power dissipation, photodiode detectors also have advantages of negligible 1/f noise and easy multiplexing to a silicon readout integrated circuit (ROIC). Furthermore, the response of photodiodes remains linear over a significantly higher photon flux range compared to that of photoconductors. As a result, photodiodes have become the dominant device structure for IR imaging applications. The four generations of development of IR detectors as identified by A. Rogalsky [10] are shown in Fig. 11.3.

First Generation: These were developed during the late 1960s and early 1970s based on photoconductor linear arrays. These systems relied on scanning optics and did not include multiplexing functions in the focal plane.

Second Generation: Starting mid-1970s, the second generation detectors were developed based on photodiodes for passive IR imaging applications. In contrast to photoconductors, photodiodes with their very low power dissipation, inherently high impedance, negligible $1/f$ noise, and easy multiplexing on focal plane silicon chip can be assembled in two-dimensional (2-D) arrays containing more than a megapixel of elements, limited only by existing processing technologies. The main attention turned towards HgCdTe photodiodes with the idea of an all-solid-state electronically scanned 2-D IR detector array instigated with the invention of charge coupled devices (CCDs) by Boyle and Smith [11]. By the end of 1970s, the emphasis was directed towards large photovoltaic HgCdTe arrays in the MW and LW spectral bands for thermal imaging. Recent efforts have been extended to short

Table 11.1 Comparison of IR detectors

Detector type	Advantages	Disadvantages
Thermal	Light, rugged, reliable, and low cost	Low detectivity at high frequency
	Room temperature operation	Slow response (ms order)
Photon		
Intrinsic		
IV–VI	Available low-gap materials	Poor mechanical
	Well studied	Large permittivity
II–VI	Easy band-gap tailoring	Nonuniformity over large area
	Well-developed theory and exp.	High cost in growth and processing
	Multicolor detectors	
III–V	Good material and dopants	Heteroepitaxy with large lattice mismatch
	Advanced technology	
	Possible monolithic integration	
Extrinsic	Very long-wavelength operation	Extremely low temperature operation
	Relatively simple technology	
Free carriers	Low-cost, high yields	Low quantum efficiency
	Large and close packed 2D arrays	Low temperature operation
Quantum wells		
Type I	Matured material growth	Low quantum efficiency
	Good uniformity over large area	Complicated design and growth
	Multicolor detectors	
Type II	Low Auger recombination rate	Complicated design and growth
	Easy wavelength control	Sensitive to the interfaces

Reprinted with permission from [12]

wavelengths, e.g., for starlight imaging in the SW range, as well as to VLWIR space borne remote sensing beyond 15 μm.

Third Generation: The third generation HgCdTe and type-II superlattice (T2SL) systems continue to be currently developed, and at the same time concept development towards the so-called fourth generation systems has recently been initiated.

Fourth Generation: The definition of fourth generation systems is not well established. These systems provide enhanced capabilities in terms of greater number of pixels, higher frame rates, better thermal resolution, as well as multicolor functionality and other on-chip functions. Multicolor capabilities are highly desirable for advanced IR systems. Collection of data in distinct IR spectral bands can discriminate for both the absolute temperature and the unique signature of objects within the scene. Comparison of different kinds of IR detectors is shown in Table 11.1. By providing this new dimension of contrast, multiband detection also offers advanced color processing algorithms to further improve sensitivity compared to that of single-color devices. It is expected that the functionalities of fourth generation systems could manifest themselves as spectral, polarization, phase, or dynamic range signatures that could extract more information from a given scene.

11.4 Mercury Cadmium Telluride (HgCdTe): Material and Device Characteristics

Since the initial reports of its crystal growth in 1959 and its sensitive infrared quantum detector properties in 1962, the pseudo-binary alloy semiconductor mercury-cadmium telluride (HgCdTe or Hg; Cd Te) has emerged as the most widely applicable infrared detector material [12]. HgCdTe or mercury cadmium telluride (MCT, MerCad Telluride, MerCadTel, MerCaT, or CMT) is a pseudo-binary alloy semiconductor that crystallizes in the zinc blende structure. It is a commonly used in photodetectors of Fourier transform infrared spectrometers, military applications, remote sensing, and infrared astronomy research. In particular, military technology has depended on HgCdTe for night vision, the US Air Force makes extensive use of HgCdTe on all aircraft, and to equip airborne smart bombs. A variety of heat-seeking missiles are also equipped with HgCdTe detectors [13]. HgCdTe detector arrays can also be found at most of the world's major research telescopes including several satellites. Many HgCdTe detectors (such as Hawaii and NICMOS detectors) are named after the astronomical observatories or instruments for which they were originally developed.

HgCdTe was first synthesized in 1958 by Lawson at the Royal Radar Establishment in England. The important features which make it well suited for application in IR detectors and are a consequence of its zinc blende structure are:

- Tailorable energy band gap over the 1–30 μm range,

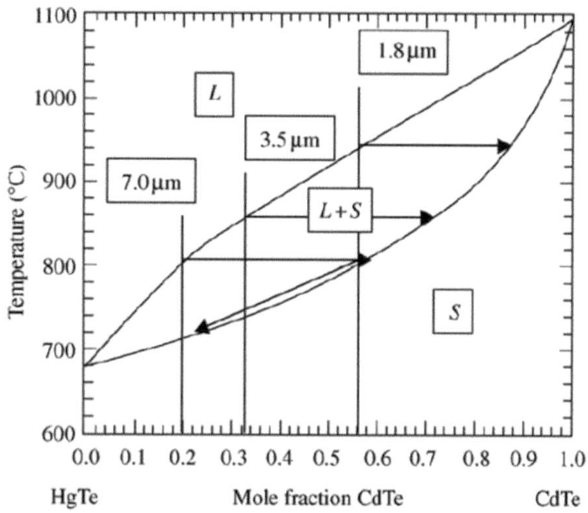

Fig. 11.4 Pseudobinary phase diagram for HgTe-CdTe system (*L* liquid, *S* solid). Reprinted with permission from [13]

- Large optical coefficients that enable high quantum efficiency and
- Favorable inherent recombination mechanisms that lead to high operating temperature (HOT)
- Ability to obtain both high and low carrier concentrations
- High mobility of electrons and
- Low dielectric constant
- Small change in lattice constant with composition make it possible to grow high quality layered and graded gap structures
- As a result HgCdTe can be used for photodetectors operated in various modes such as photoconductor, photodiode or metal-insulator (MIS) semiconductor detector

The detector is composed of a thin layer (10–20 μm) of HgCdTe with metalized contact pads defining active area. It exhibits different wavelengths spanning from shortwave infrared to the very long-wave infrared regions proportional to the alloy composition. The amount of cadmium (Cd) in the alloy (the alloy composition) can be chosen so as to tune the optical absorption of the material to the desired infrared wavelength. The wavelength of peak response depends on the material's band-gap energy. Figure 11.4 gives the pseudobinary phase diagram for HgTe-CdTe system.

11.4.1 Material and Device Characteristics of HgCdTe Detectors

Three fundamental properties determine the degree to which a semiconductor will make a successful infrared detector.

Table 11.2 Summary of the material properties of $Hg_{1-x}Cd_xTe$ ternary alloy, listed for the binary components HgTe and CdTe, and for several technologically important alloy compositions

Property	HgTe	$Hg_{1-x}Cd_xTe$						CdTe
x	0	0.194	0.205	0.225	0.31	0.44	0.62	1.0
a (Å)	6.461	6.464	6.464	6.464	6.465	6.468	6.472	6.481
T	77 K	77 K	77 K	77 K	140 K	200 K	250 K	300 K
E_g (eV)	−0.261	0.073	0.091	0.123	0.272	0.474	0.749	1.490
λ_c (μm)	–	16.9	13.6	10.1	4.6	2.6	1.7	0.8
n_i (cm^{-3})	–	1.9×10^{14}	5.8×10^{13}	6.3×10^{12}	3.7×10^{12}	7.1×10^{11}	3.1×10^{10}	4.1×10^{5}
m_c/m_0	–	0.006	0.007	0.010	0.021	0.035	0.53	0.102
g_c	–	−150	−118	−84	−33	−15	−7	−1.2
$\varepsilon_s/\varepsilon_0$	20.0	18.2	18.1	17.9	17.1	15.9	14.2	10.6
$\varepsilon_\infty/\varepsilon_0$	14.4	12.8	12.7	12.5	11.9	10.8	9.3	6.2
n_r	3.79	3.58	3.57	3.54	3.44	3.29	3.06	2.50
μ_c (cm^2 V^{-1} s^{-1})	–	4.5×10^5	3.0×10^5	1.0×10^5	–	–	–	–
μ_{hh} (cm^2 V^{-1} s^{-1})	–	450	450	450	–	–	–	–
$b = \mu_e/\mu_\eta$	–	1000	667	222	–	–	–	–
τ_R (μs)	–	16.5	13.9	10.4	11.3	11.2	10.6	2
τ_{A1} (μs)	–	0.45	0.85	1.8	39.6	453	4.75×10^3	
$\tau_{typical}$ (μs)	–	0.4	0.8	1	7	–	–	–
E_p (eV)				19				
Δ (eV)				0.93				
m_{hh}/m_0				0.40–0.53				
ΔE_v (eV)				0.35–0.55				

Reprinted with permission from [14, 15]

τ_R and τ_{A1} calculated for n-type HgCdTe with $N_d = 1 \times 10^{15}$ cm^{-3}. The last four material properties are independent of or relatively insensitive to alloy composition

Table 11.3 Some physical properties of narrow gap semiconductors. Reprinted with permission from [16]

	E_g (eV)		n_i (cm^{-3})			μ_e (10^4 cm^2 V^{-1} s^{-1})		μ_h (10^4 cm^2 V^{-1} s^{-1})	
Material	77 K	300 K	77 K	300 K	ε	77 K	300 K	77 K	300 K
InAs	0.414	0.359	6.5×10^3	9.3×10^{14}	14.5	8	3	0.07	0.02
InSb	0.228	0.18	2.6×10^9	1.9×10^{16}	17.9	100	8	1	0.08
$In_{0.53}Ga_{0.47}AS$	0.66	0.75	–	5.4×10^{11}	14.6	7	1.38	–	0.05
PbS	0.31	0.42	3×10^7	1.0×10^{15}	172	1.5	0.05	1.5	0.06
PbSe	0.17	0.28	6×10^{11}	2.0×10^{16}	227	3	0.10	3	0.10
PbTe	0.22	0.31	1.5×10^{10}	1.5×10^{16}	428	3	0.17	2	0.08
$Pb_{1-x}Sn_xTe$	0.1	0.1	3.0×10^{13}	2.0×10^{16}	400	3	0.12	2	0.08
$Hg_{1-x}Cd_xTe$	0.1	0.1	3.2×10^{13}	2.3×10^{16}	18.0	20	1	0.044	0.01
$Hg_{1-x}Cd_xTe$	0.25	0.25	7.2×10^8	2.3×10^{15}	16.7	8	0.6	0.044	0.01

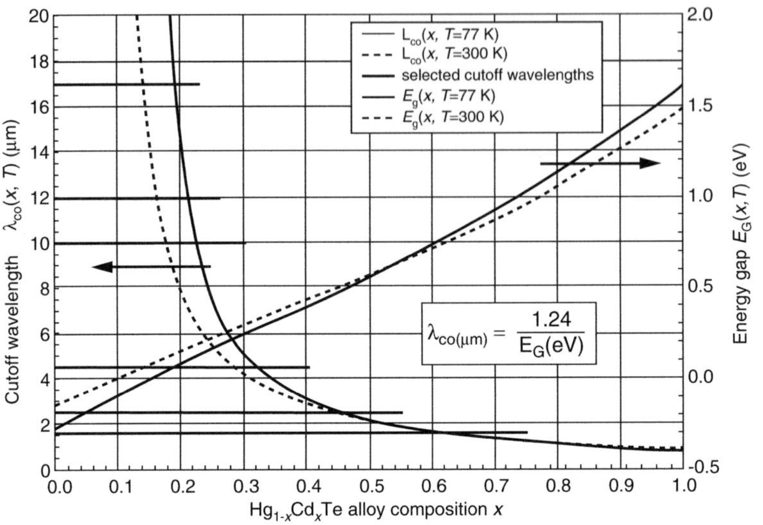

Fig. 11.5 Energy bandgap $E_G(x, T)$ and cutoff wavelength $\lambda_{CO}(x, T)$ for $Hg_{1-x}Cd_xTe$, plotted versus alloy composition parameter x, for temperature 77 K (*solid curves*) and 300 K (*dashed curve*). *Horizontal bars* indicate important cutoff wavelengths for $Hg_{1-x}Cd_xTe$ detector technology. Reprinted with permission from [16]

- The energy bandgap must be approximately equal to the smallest photon energy to be detected.
- The optical absorption coefficient must be large enough so that nearly all incident photons of interest are converted to electron hole pairs.
- The inherent recombination mechanisms must allow long carrier lifetimes to be realized so that thermal generation rates, and hence thermal noise, can be minimized.

Table 11.2 summarizes the important properties of $Hg_{1-x}Cd_xTe$ for different values of x. A comparison of parameters of HgCdTe and other narrow gap semiconductors is given in Table 11.3.

11.4.1.1 Energy Bandgap

For $Hg_{1-x}Cd_xTe$ the energy bandgap varies continuously, and nearly linearly, with alloy composition parameter x, ranging from 1.6 to −0.3 eV as shown in Fig. 11.5 [16]. CdTe is a semiconductor with a bandgap of approximately 1.6 at room temperature whereas HgTe is a semimetal, having a negative bandgap of −0.3 eV. Mixing these two substances allows one to obtain any bandgap between 0 and 1.6 eV for an alloy having the composition $Hg_{1-x}Cd_xTe$. The bandgap passes through zero at an alloy composition of $x = 0.16$ (for $T = 0$ K). This broad range of energy bandgap provides a continuum of materials whose absorption edge can be

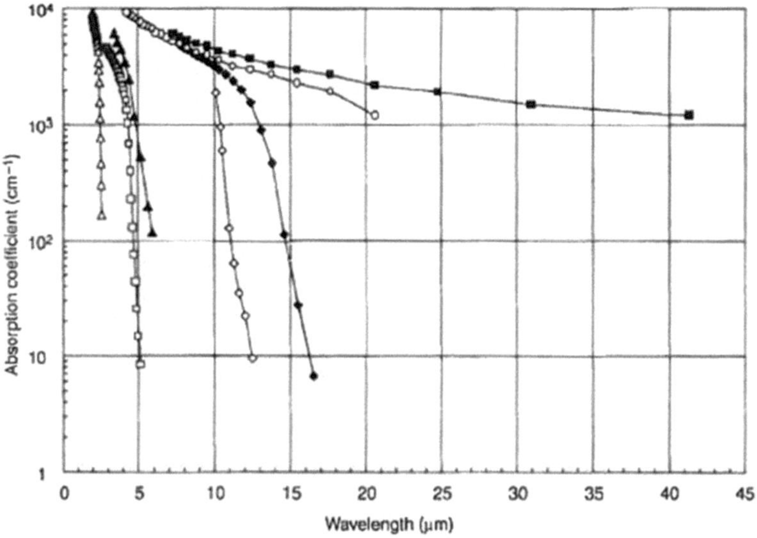

Fig. 11.6 Optical absorption coefficient data for several $Hg_{1-x}Cd_xTe$ alloy compositions, for photon energies near the fundamental absorption edge, plotted versus wavelength. Reprinted with permission from [16]

tailored to provide high performance quantum detectors for wavelengths over the 1–30 μm spectral range (Fig. 11.6).

As the Cd composition increases, the $Hg_{1-x}Cd_xTe$ energy bandgap gradually increases from a negative value for HgTe to a positive value for CdTe. The bandgap energy tunability enables the material to be used in IR detector applications spanning the short wavelength IR (SWIR: 1–3 μm), middle wavelength IR (MWIR: 3–5 μm), long-wavelength IR (LWIR: 8–14 μm), and very LWIR (VLWIR: 14–30 μm) ranges. Because of its band gap tunability with x, $Hg_{1-x}Cd_xTe$ has evolved to become the most important material for detector applications over the entire IR range [17]. Several expressions approximating the bandgap energy are available in literature [15]. The most widely used expression is due to Hansen et al. [18]:

$$E_g = -0.302 + 1.93x + 5.35 \times 10^{-4}(1 - 2x)T - 0.81x^2 + 0.832x^3 \quad (11.2)$$

where E_g is bandgap energy in electron volts, x is cadmium concentration, T is temperature in Kelvin.

11.4.1.2 Intrinsic Carrier Concentration

The most widely used expression for intrinsic carrier concentration (n_i) is that of Hansen and Schmit [19], who used their own $E_g(x, T)$ relationship given by Eq. (11.2), the $k \cdot p$

method and a value of 0.443 m_0 (where m_0 is the free electron mass) for heavy hole effective mass ratio and given by

$$n_i = (5.585 - 3.82x + 0.001\,753T - 0.001\,364xT) \times 10^{14} E_g^{3/4} T^{3/2} \exp\left(-\frac{E_g}{2kT}\right) \tag{11.3}$$

11.4.1.3 Carrier Lifetimes

Two fundament mechanisms that determine the minority carrier lifetime in defect free HgCdTe are radiative recombination and Auger recombination (Auger-1 in n-type and Auger-7 in p-type). Since these involve only interactions among electrons and holes in the valence and conduction bands, and do not require the intermediary of defect states, these mechanism are regarded as inherent to the material itself, depending only on the band structure, the donor or accept concentration, and the temperature. The important advantage of HgCdTe as a high-performance detector material is that these fundamental mechanism provide, for practically achievable carrier concentrations, lifetimes that are long enough (and hence thermal generation rates that are low enough) and the background-limited sensitivity is achieved at temperatures substantially higher than for other classes of infrared detectors, such as extrinsic detectors (As-doped Si, Cu-doped Ge) and quantum-well infrared photodetectors (QWIPs).

$$G_A - R_A = \frac{n_i^2 - np}{2n_i^2}\left[\frac{n}{(1+an)\tau_{A1}^i} + \frac{P}{\tau_{A7}^i}\right]$$

11.4.1.4 Effective Mass

The electron and light hole effective masses m_e^* and m_{lh}^* respectively in the narrow gap mercury compounds are close and they can be established according to Weiler's [20] expression

$$\frac{m_0}{m_e^*} = 1 + 2F + \frac{E_p}{3}\left(\frac{2}{E_g} + \frac{1}{E_g + \Delta}\right),$$

where $E_p = 19$ eV, $\Delta = 1$ eV, and $F = -0.8$. This relationship can be approximated by $m_e^*/m = 0.071E_g$, where E_g is in electron volts. The effective mass of heavy hole m_{hh}^* is high; the measured values range between 0.3 and $0.7m_0$. The value of $m_{hh}^* = 0.55m_0$ is frequently used in modeling of IR detectors.

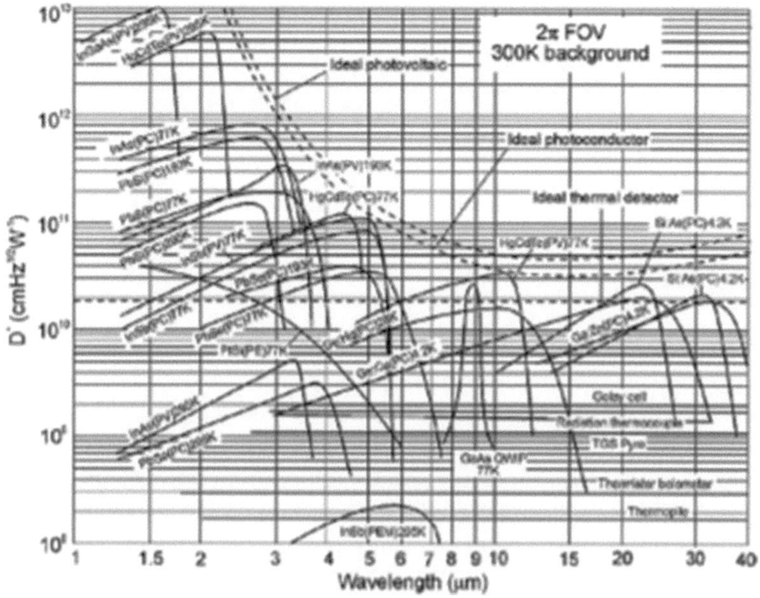

Fig. 11.7 $D*$ values as a function of incident wavelength for commercially available infrared detectors operating at indicated temperature. Theoretical curves for the BLIP $D*$ values represented by *dashed lines* for ideal photovoltaic, photoconductor, and thermal detectors. Reprinted with permission from [16]

11.4.1.5 Optical Absorption Coefficient

Optical absorption coefficient data for several alloy compositions of $Hg_{1-x}Cd_xTe$ are plotted versus wavelength (λ) in Fig. 11.7. The absorption strength generally decreases as the gap becomes smaller due to both the decrease in the conduction band effective mass and to the $\lambda^{-1/2}$ dependence of the absorption coefficient. There is an exponential tail at energies just below the bandgap energy [16].

11.4.1.6 Mobility

The electron mobility in $Hg_{1-x}Cd_xTe$ (expressed in cm^2 V^{-1} s^{-1}), in composition range $0.2 \le x \le 0.6$ and temperature range $T > 50$ K, can be approximated as [21]:

$$\mu_e = \frac{9 \times 10^8 s}{T^{2r}}, \quad \text{where } r = \left(\frac{0.2}{x}\right)^{0.6} \quad \text{and} \quad s = \left(\frac{0.2}{x}\right)^{7.5}$$

Higgins et al. [22] give an empirical formula (valid for $0.18 \le x \le 0.25$) for the variation of μ_e with x at 300 K for the very high quality melt grown samples that they studied:

$$\mu_e = 10^4(8.754x - 1.044)^{-1} \quad \text{in}\, \text{cm}^2(\text{V s})^{-1}.$$

The hole mobilities at room temperature were found to be in the range from 40 to 80 $cm^2(V\ s)^{-1}$, and the temperature dependence is relatively weak. The hole mobility at 77 K is by one order of magnitude higher. The hole mobility [23], measured at 77 K falls as the acceptor concentration is increased and in the composition range 0.20–0.30 yields the following empirical expression:

$$\mu_h = \mu_0 \left[1 + \left(\frac{p}{1.8 \times 10^{17}} \right)^2 \right]^{-1/4}$$

where $\mu_0 = 440\ cm^2\ (V\ s)^{-1}$. For modeling IR photodetectors, the hole mobility is usually calculated assuming that the electron-to-hole mobility ratio $b = \mu_e/\mu_h$ is constant and equal to 100.

Direct allowed valence to conduction band transitions are responsible for large values of the optical adsorption coefficient, which enable quantum efficiencies approaching 100 % to be achieved in practical device designs. Long minority carrier lifetimes result in low thermal noise, and permit high performance to be achieved at operating temperature that are the highest reported among infrared detectors of comparable cutoff wavelengths. HgCdTe is well established today as the most broadly applicable high-performance infrared detector material. Typically, detectors are manufactured in a square or rectangular configuration to maintain a uniform bias current distribution throughout the active region.

11.4.1.7 Responsivity

The spectral current responsivity of photon detectors is given by [7]

$$R_i = \frac{\lambda\eta}{hc} qg,$$

where λ is the wavelength, h is the Planck's constant, c is the velocity of light, q is the electron charge, and g is the photoelectric current gain. The current that flows through the contacts of the device is noisy due to the statistical nature of the generation and recombination processes, fluctuation of optical generation, thermal generation, and radiative and nonradiative recombination rates. Assuming that the current gain for the photocurrent and the noise current are the same, the noise current is

$$I_n^2 = 2q^2 g^2 (G_{op} + G_{th} + R)\Delta f,$$

where G_{op} is the optical generation rate, G_{th} is the thermal generation rate, R is the resulting recombination rate, and Δf is the frequency band.

11.4.1.8 Detectivity

Detectivity, D^*, is the main parameter to characterize normalized signal-to-noise performance of photoconductive detectors and can be defined as [7]

$$D^* = \frac{R_i (A_d \Delta f)^{1/2}}{I_n}$$

At equilibrium, the generation and recombination rates are equal. In this case

$$D^* = \frac{\lambda \eta}{2hc(Gt)^{1/2}}$$

Background radiation frequently is the main source of noise in an IR detector. Assuming no contribution due to recombination, I_n is defined as

$$I_n^2 = 2\Phi_B A_d \eta q^2 g^2 \Delta f,$$

where Φ_B is the background photon flux density. Therefore, at the background limited performance conditions (BLIP performance)

$$D^*_{BLIP} = \frac{\lambda}{hc}\left(\frac{\eta}{\Phi_B}\right)^{1/2}$$

Once background-limited performance is reached, quantum efficiency η is the only detector parameter that can influence a detector's performance.

In general, the detectivity (D^*) of an infrared photodetector is limited by generation and recombination rates G and R in the active region of the device. It can be expressed as

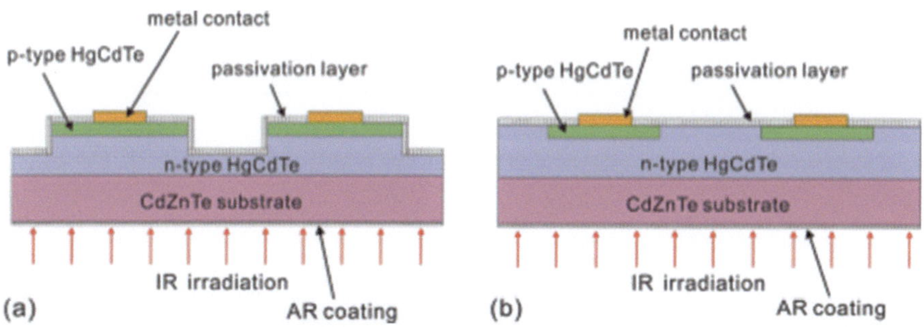

Fig. 11.8 Schematic diagram of (**a**) mesa structure and (**b**) planar structure of p-on-n HgCdTe photodiodes. Reproduced with permission from [24]

$$D^* = \frac{\lambda}{2^{1/2}hc(G+R)^{1/2}}\left(\frac{A_o}{A_e}\right)^{1/2}\frac{\eta}{t^{1/2}},$$

where λ is the wavelength, h is the Planck's constant, c is the velocity of light, and η is the quantum efficiency. Here, the detector is considered as a slab of homogeneous semiconductor with actual "electrical" area, A_e, thickness, t, and volume, $A_{e}t$. Usually, the optical (A_o) and electrical areas of the device are the same or similar. However, the use of some kind of optical coupler/concentrator can increase the A_o/A_e ratio by a large factor.

For a given wavelength and operating temperature, the highest performance can be obtained by maximizing the ratio of the quantum efficiency to the square root of the sum of the sheet thermal generation and recombination rates $\eta/[(G+R)t]^{1/2}$. This means that high quantum efficiency must be obtained with a thin device.

The spectral detectivities for commercially available detectors are shown in Fig. 11.8. Most commercial IR detectors achieve $D*$ values around 10^{10} cm $Hz^{0.5}$/W or higher in all spectral ranges spanning SWIR to VLWIR [16]. It can be seen that ideal PV detectors outperform both PC and thermal detectors. More notably, cryogenically cooled II–VI compound mercury cadmium telluride (HgCdTe, MCT) PV detectors achieve $D*$ values near the theoretical limit in both MWIR and LWIR spectral ranges, surpassing numerous detectors based on other materials (GaAs, InAs, InGaAs, InSb, Ge:Hg,Cu, PbS, PbSe, PtSi, Si:As). These performance characteristics strongly motivate the pursuit of next-generation HgCdTe photovoltaic IR devices, laying the foundation for this study.

11.4.1.9 Resolution of Thermal detectors

Thermal resolution of infrared thermal systems is usually characterized by the noise equivalent temperature difference (NETD) given by [1]

$$\mathrm{NETD} = \frac{4F^2\Delta f^{1/2}}{A_d^{1/2}t_{op}}\left[\int_{\lambda_a}^{\lambda_b}\frac{dM}{dT}D^*(\lambda)d\lambda\right]^{-1}$$

where F is the optics f-number, Δf is the frequency band, A_d is the detector area, t_{op} is the optics transmission, and M is the spectral emittance of the blackbody described by Planck's law.

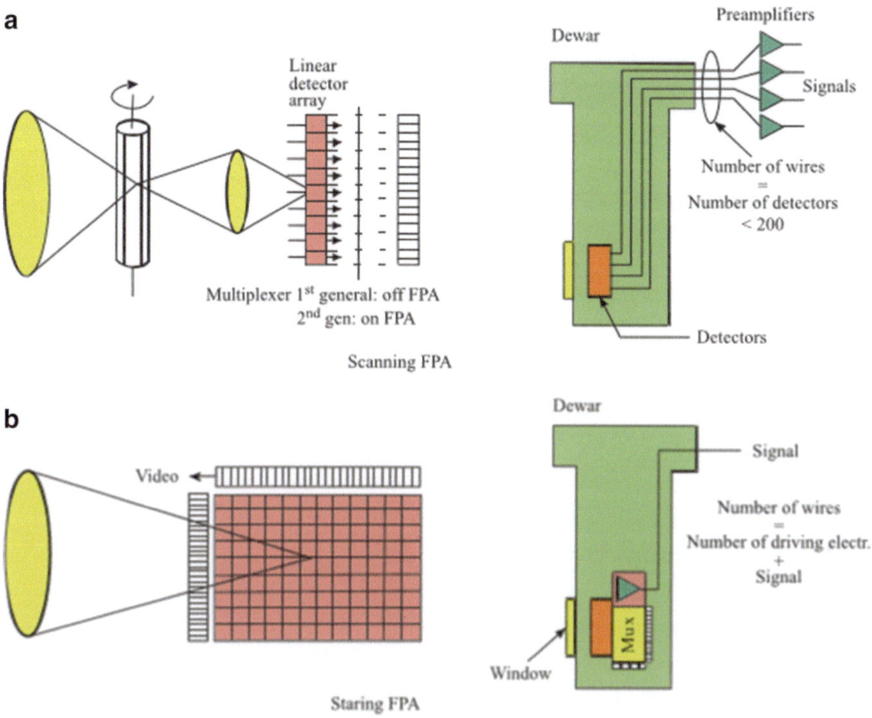

Fig. 11.9 (**a**) Scanning focal plane array (first generation) and (**b**) staring focal plane array (second generation). Reprinted with permission from) [16]

11.5 Photodiodes

For photodiode device structures, the critical region is the p–n junction, which can be formed using different approaches, including intentional extrinsic p and n type doping, ion implantation, Hg diffusion, impurity diffusion, and type conversion. For example, n-on-p homojunctions can be formed by implanting n-type dopants into p-type HgCdTe (intrinsic Hg vacancy doped p type) or by conversion of the p-type material into n-type using techniques such as plasma processing and ion milling. If the HgCdTe materials are grown using MBE or MOCVD, the p–n junction can be formed during the growth process via in-situ extrinsic p and n type doping.

Regarding the fabrication technology of p–n junction photodiodes, there are generally two classes of processing structure used: mesa structures and planar structures [24], shown in Fig. 11.9. The processing procedure of a planar device structure usually involves surface passivation, window etch, p–n junction formation, followed by metal contact deposition, whereas that of mesa device structures usually involves p–n junction formation, mesa isolation etching, followed by surface passivation and metal contact formation.

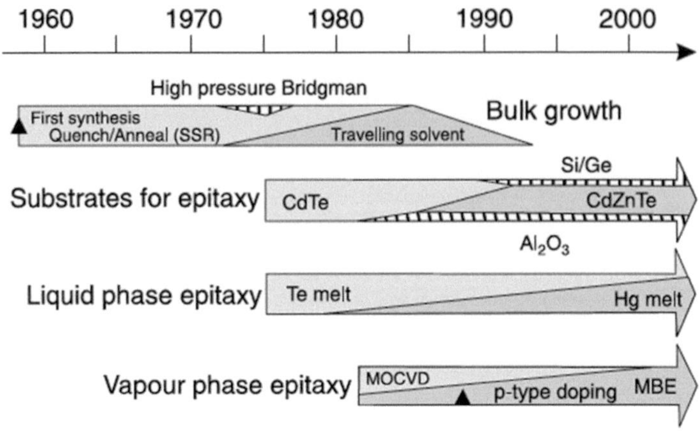

Fig. 11.10 Evolution of HgCdTe crystal growth technology from 1958 to present. Reprinted with permission from [25]

11.6 Focal Plane Arrays (FPAs)

Infrared focal plane arrays are critical components in many of the military and civilian applications for advanced imaging system. There is continuing requirement for extended detector capability to support applications for future generations of infrared sensor systems. In the first-generation linear arrays developed in the 1960s and 1970s, an electrical contact for each element of a multielement array is brought off the cryogenically cooled focal plane to the outside, where there is one electronic channel at ambient temperature for each detector element—see Fig. 11.7 of intrinsic photoconductive PbS, PbSe, HgCdTe detectors [24]. The first generation scanning system does not include multiplexing functions in the focal plane. These allowed LWIR FLIR systems to operate with a single-stage cryo-engine, making the systems much more compact, lighter, and requiring significantly less power consumption. The simplest scanning linear FPA consists of a row of detectors. An image is generated by scanning the scene across the strip using, a mechanical scanner. At standard video frame rates, at each pixel (detector) a short integration time has been applied and the total charges are accommodated. The US common module HgCdTe arrays employ 60, 120, or 180 photoconductive elements depending on the application. An example of 180-element common module FPA mounted on a Dewar stem is shown in Fig. 11.10.

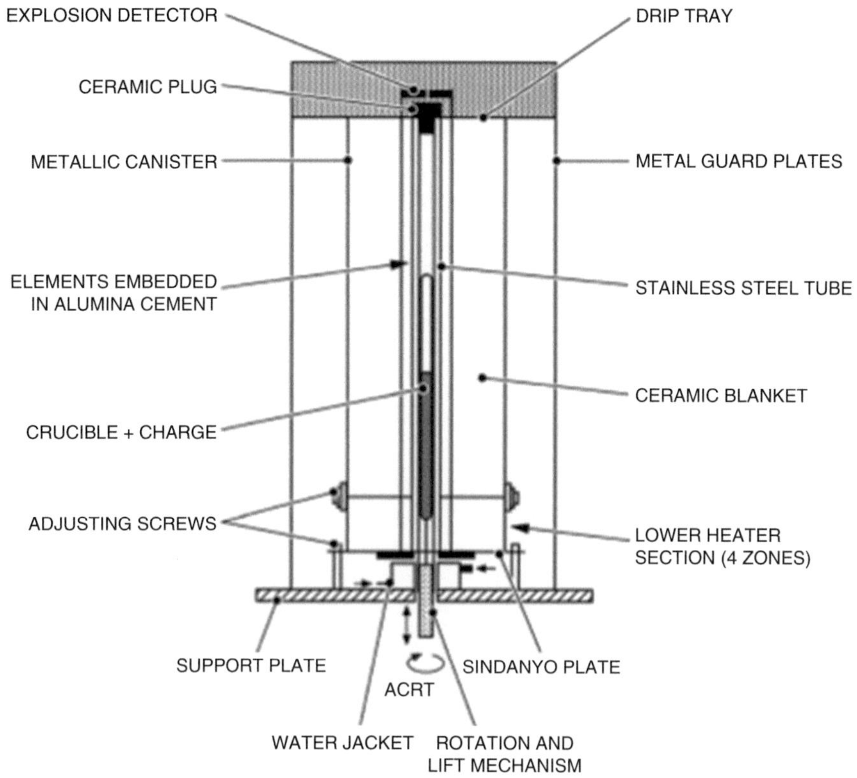

Fig. 11.11 Bridgman/ACRT growth apparatus. Reprinted with permission from [33]

11.7 Techniques for Synthesis of MCT Material

Main parameters that impact the performance of IR detectors, especially FPAs, are material quality, device architecture, device processing, and surface passivation. To achieve high performance HgCdTe detectors, it is critical to develop a reliable fabrication process and passivation technology.

Principle methods developed for the growth of HgCdTe fall under three main categories and are discussed below:

- Bulk Crystal growth
- Liquid Phase epitaxy
- Vapor Phase epitaxy

The time line for the evolution of these growth technologies is illustrated in Fig. 11.11.

11.7.1 Bulk Growth

Bulk growth process is typically used for large area single detectors for applications such as spectrometry. However, for photovoltaic arrays there are challenges associated with crystal grain boundaries, which are electrically active and contribute to line defects. Also there are limitations in the ingot diameter, which makes bulk growth suitable for only quad or single detector arrays. Several methods [13, 26–31] have been developed for growing MCT bulk crystals including:

- Solid-State Recrystallization (SSR)
- Traveling Heater Method (THM)
- Bridgman Growth method
- Slush Growth, and
- Zone Melting

The general challenge with melt grown MCT is to maintain a relatively high Hg vapor pressure during growth; otherwise, it is difficult to control the stoichiometry of the grown crystal. Also, the large separation between the liquidus and solidus compositions across a constant thermal tie line can result in a steady variation in the composition of a moving growth interface.

11.7.1.1 Solid State Recrystallization

The SSR technique is used to alleviate the compositional variation at the growth interface in ternary systems, such as MCT, where the solidus and liquidus lines are widely separated. In the basic technique, the three high purity elements of MCT are cleaned and loaded into a thick walled, small diameter quartz ampoule that is evacuated, sealed and placed into a furnace. The ampoule is heated to approximately 950 °C and the melt is mixed by constantly rocking the furnace. The MCT melt is rapidly quenched to produce a uniform composition. The rapid quenching produces a dendritic structure in the MCT that is reduced by an extended (several days) recrystallization step, at temperatures just below the melting point. Grain growth occurs during the re-crystallization step and remaining compositional inhomogeneities are removed. On the down side, the high melt temperatures and very high Hg vapor pressures used in SSR to produce MCT can cause the ampoule to explode. Also, very long annealing times are required and the resulting crystals are small. The typical diameter of the ingot obtained by this method is limited to about 2.0 cm in order to control impurity segregation in the crystal.

11.7.1.2 Traveling Heater Method

In the THM method, a solvent zone is created between a solid seed and the feedstock material. In the case of MCT, the crystal is grown by passing the solvent

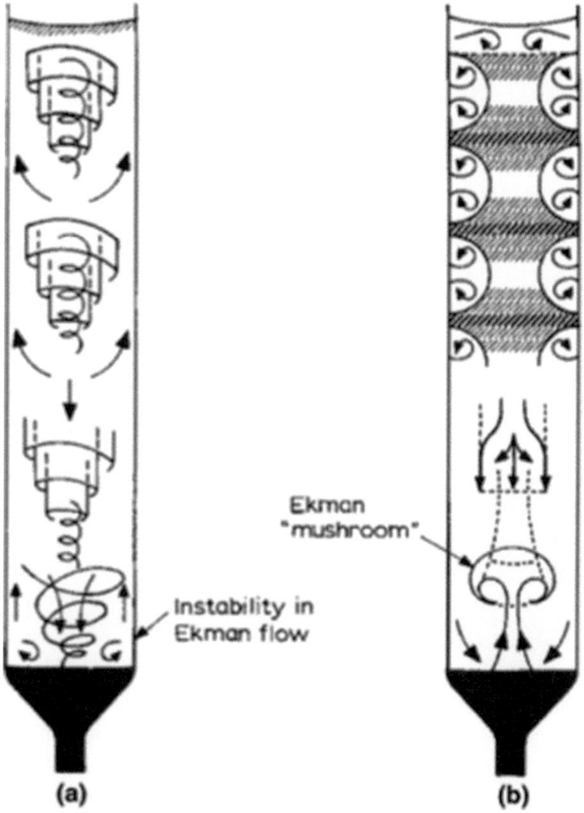

Fig. 11.12 Flows in a flat-based container during, (**a**) spin-up and (**b**) spin-down. Reprinted with permission from [40]

zone (e.g., Te rich) through a polycrystalline MCT rod having a composition that is to be replicated in grown crystal. The motion of the molten interface is produced by the slow movement of the heater along the charged crucible. Crystallization takes place at the advancing seed–solvent interface and dissolution of feed material occurs at the solvent–feed phase boundary. Through convection and diffusion, the solid material is dissolved at the high temperature interface and deposited at the low temperature interface of the zone. Crystal growth occurs in the 500–700 °C range, lower than the temperature used for Solid State Recrystallization growth method. The lower growth temperature used in THM reduces the incidence of antisite defects, resulting in crystals with more reproducible composition and higher resistivity. Also, the lower temperatures reduce contamination from the crucible walls and decrease the evaporation of the constituent species. One successful implementation of THM resulted in crystals up to 5 cm in diameter [32]. The perfect quality of crystals grown by this method is achieved at the cost of a low growth rate [33].

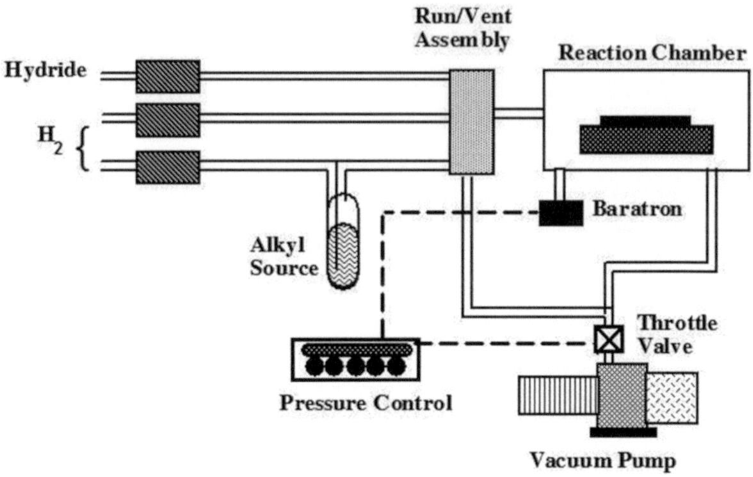

Fig. 11.13 MOCVD apparatus [48]

11.7.1.3 Bridgman Growth Method

Bridgman growth takes place in simple 2-zone furnaces with the pure elements contained in thick-walled high-purity silica ampoules, as shown in Fig. 11.12. The thick ampoule walls are needed to contain the high (up to *70 atm) mercury vapor pressures within the ampoules. Mercury and tellurium are purified on site before use in order to obtain the required electrical properties.

The marked segregation of the matrix elements in standard Bridgman growth is both a disadvantage and an advantage. The advantage is that all regions of interest can be produced in the same crystal, to some extent. A further advantage is that segregation of impurities also occurs and this leads to low background donor levels in Bridgman material. The main disadvantage is the low yield of material of specific composition for the most commonly required regions. The other disadvantages of the standard Bridgman technique are that the material obtained is nonuniform in composition in the radial direction, as well as in the growth direction, resulting in low yields, along with the presence of numerous grain and sub-grain boundaries.

The Accelerated crucible rotation technique (ACRT) was proposed by Scheel [34] and was utilized to increase the size of crystals, eliminate unwanted phases, and to increase stable growth rates. The first reported use of ACRT in MCT Bridgman growth was given in [35]. These effects were developed and discussed in more detail in later papers and book chapters, see [36–39] for reviews. In essence, bidirectional rotation is used with maximum rotation rates governed by the ampoule diameter (e.g., 25 rpm for 20-mm and *5 rpm for 50-mm diameter crystals, respectively) with short stop and run times, e.g., 1 and 8 s. The main stirring mechanism is Ekman flow near the flat solid–liquid interface; the schematic of simulations are shown in Fig. 11.13. This, in turn, improves the radial and axial compositional uniformity of the material, and reduces somewhat the grain and sub-grain densities. An additional marked advantage of ACRT is that the improved radial compositional uniformity enables larger-diameter material to be produced (up from 13 mm to as much as 70 mm in our case).

11.7.2 Liquid Phase Epitaxy (LPE)

LPE is a solution growth technique that involves the controlled precipitation of a solute dissolved in a solvent onto a single crystal substrate. For LPE growth of MCT, bulk grown CdZnTe single crystal substrates are suitable, since they are thermodynamically compatible and nearly lattice matched. The solvent can either be Hg [41] or Te-rich [42].

As compared to bulk growth techniques, the LPE method involves lower growth temperatures, shorter growth times, multilayered device structures, and better compositional homogeneity over large substrate areas. The versatility of LPE for growth of high performance device quality MCT epitaxial layers, with different Cd mole fractions and excellent compositional uniformity, is discussed in [43–47]. Today, detector arrays prepared from LPE based materials exhibit best performance, and majority of military IR applications use this technology. Both Te-solution growth (420–500 °C) and Hg-solution growth (360–500 °C) are used with equal success in a variety of configurations. The design of the Te-rich LPE system can be configured to allow the melt to contact the substrate by either sliding or tipping or dipping techniques. A sliding boat system uses a small melt volume and is adaptable for changing composition, thickness and doping. Tipping and dipping systems can be scaled up for large melts to provide thick, uniform layers. Both the tipping and dipping designs are being used for Te and Hg-rich solutions, while only the sliding technique is used for Te-rich solutions. The major difference

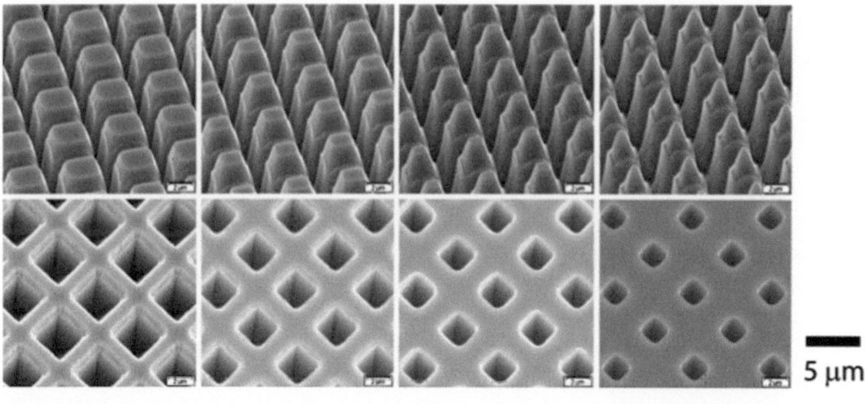

Fig. 11.14 Examples of photon trapping HgCdTe microstructures with test photonic crystal fields with varying fill factor for FTIR demonstration. Reprinted with permission from [62]

between the Hg and Te-rich solvents is that in the former case, the vapor pressure of Hg over the melt is much higher than in the latter case.

11.7.3 Metal-Organic Chemical Vapor Deposition (MOCVD)

In the metal organic chemical vapor deposition (MOCVD) technique, reactant gases are combined at elevated temperatures in the reactor to cause a chemical interaction, resulting in the deposition of materials on the substrate. The schematic of MOCVD is shown in Fig. 11.14.

MOCVD growth of MCT depends on transporting the elements Cd and Te (and dopants In, I and As) at room temperature as volatile organometallics [31]. These species react with Hg vapor in the hot gas stream above the substrate or catalytically on the substrate surface at about 400 °C. The pyrolytic nature of the reaction requires that only the substrate be heated to ensure efficient deposition. In practice, the growth of MCT by MOCVD using dimethyl cadmium (DMCd) and diethyl telluride (DETe) is accomplished by two processes: (a) CdTe synthesis from DMCd and DETe, and (b) formation of HgTe from DETe and Hg at the heated substrate. The challenge with this growth technique is to control the composition of the epitaxial layer and achieve uniformity over large surface areas. MOCVD-MCT composition is influenced by substrate temperature, DMCd and Hg partial pressures. Compositional control and layer uniformity are addressed using the inter-diffused multilayer process (IMP) technique in which very thin layers (0.1–0.2 μm) of HgTe and CdTe are deposited sequentially. These layers, with high diffusion coefficients, inter-diffuse during growth at about 400 °C to form a homogeneous ternary epilayer with a composition that is controlled by the thickness ratio of HgTe: CdTe layers [49]. The preferred precursor for Te is di-isopropyl telluride (DiPTe) that allows for a reduction in the MCT growth temperature from 400 °C to 350 °C. DiPTe in conjunction with DMCd can allow the deposition CdTe to occur at lower temperatures (300 °C). Doping for MOCVD MCT layers is done using Group III metals for p-type doping and Group VII halogens for n-type doping. The main morphological problem for MOCVD are macro defects called hillocks, which are caused by preferred (111) growth, nucleated from a particle or polishing defect. Hillocks can cause clusters of defects in focal plane arrays. Orientations 3–4 off (100) are used primarily to reduce both the size and density of hillocks. The MOCVD technique is used to manufacture high-quality, large-area infrared focal plane arrays for many applications [50]. Consequently, there is a renewed interest in using MOCVD because of its ability to: achieve low surface defect densities, deposit MCT films on large-area, low-cost substrates such as GaAs and control n- and p-type doping levels. In comparison to an MBE system, the overall maintenance and operational costs of an MOCVD system is lower.

11.7.4 Molecular Beam Epitaxy (MBE)

Molecular beam epitaxy is an important technique for large area epitaxial growth of thin films and involves slow growth rates ~3000 nm/h in high vacuum or ultrahigh vacuum (~10^{-8} Pa). The slow deposition rate requires proportionally better vacuum to achieve the same impurity levels as other deposition techniques. The absence of carrier gases as well as the ultrahigh vacuum environment results in the highest achievable purity of the grown films. Growth of MBE-MCT is carried out under an ultrahigh vacuum environment with Knudsen-type effusion source cells charged with Hg, Te_2, and CdTe [51, 52]. MBE-MCT deposition temperature plays a critical role in the introduction of extended defects. Typically, growth is carried out at 180–190 °C on (211) CdZnTe substrates. The low growth temperature and the ability to rapidly shutter the sources are key features that allow MBE to produce sharp interfaces for multilayered IR devices that operate in two or three different spectral bands. The ultrahigh vacuum growth chamber allows for in-situ analytical tools to monitor and control the MCT growth process and evaluate the properties of the grown layers [53, 54]. At the lower temperature range, a Hg-rich condition prevails at the substrate because the sticking coefficient of Hg increases as the temperature is reduced. The condition with excess Hg results in the formation micro-twins that are detrimental to the performance of the MCT IR focal plane array. Typical etch pit densities (EPD) of material grown under such Hg-rich conditions are high (10^6–10^7 cm^{-2}). If the growth temperature is raised to about 190 °C, then a deficiency of Hg leads to the formation of voids in the MCT layer. Hg is incorporated in the film only by reacting with free Te, thus the MCT composition is contingent on the Te to CdTe flux ratio. The structural perfection of the film depends strongly on the Hg to Te flux ratio and growth is usually restricted to a tight temperature range. By optimizing the Hg to Te flux ratio, the concentration of voids is about 100 cm^{-2} which may be attributable to dust particles or substrate related imperfections. The EPD values for epilayers grown under these conditions are in the low 10^5 cm^{-2} ranges. Indium is the most widely used n-type extrinsic dopant in MCT epitaxial layers and is well activated. At low Indium doping levels, Hg vacancies can compensate some of the n-type impurities and affect dopant control. p-Type dopants, such as Arsenic, are less conveniently incorporated into the epilayers. Significant efforts are being expended to improve the incorporation of As and Sb during the MBE process and to reduce the temperature required for activation. The metal saturation conditions cannot be reached at the temperatures required for high-quality MBE growth. The necessity to activate acceptor dopants at high temperatures diminishes the gains of low-temperature deposition. Nearly 100 % activation has been achieved for a 2×10^{18} cm^{-3} As concentration, with as low as 300 °C activation anneal, followed by a 250 °C stoichiometric anneal [55]. Because of its various advantages, MBE-MCT technology is becoming more attractive than the other epitaxial technologies and is required for the fabrication of IR detectors with advanced architectures. The MBE-MCT technology has developed to the point where MBE layers grown on bulk CZT substrates exhibit characteristics comparable to those prepared by LPE.

Table 11.4 Comparison of the various methods used to grow HgCdTe. Reprinted with permission from [25]

	Bulk			Liquid phase epitaxy		Vapor phase epitaxy	
		Traveling heater method					
	SSR	HCT melt	Te melt	Hg melt	Te melt	MOCVD	MBE
Temperature (°C)	950	950	500	350-550	400-550	275-400	160-200
Pressure (Torr)	150,000	150,000	760–8000	760–11,400	760–8000	300–760	10^{-3}–10^{-4}
Growth rate (μm/h)	250	250	80	30–60	5–60	2–10	1–5
Dimensions							
w (cm)	0.8–12 dia	0.8–1.2 dia	2.5 dia	5	5	7.5 dia	7.5 dia
l (cm)	–	–	–	6	5	4	4
t (cm)	15	15	15	0.0002–0.0030	0.0005–0.012	0.0005–0.001	0.0005–0.001
Dislocations (cm^{-2})	$<10^5$	–	$<10^5$	$<10^5$	$<10^5$–10^7	5×10^5–10^7	5×10^4–10^6
Purity (cm^{-3})	$<5 \times 10^{14}$	$<5 \times 10^{14}$	$<5 \times 10^{14}$	$<5 \times 10^{14}$	$<5 \times 10^{14}$	$<1 \times 10^{15}$	$<1 \times 10^{15}$
n-Type doping (cm^{-3})	N/A	N/A	N/A	1×10^{14}–1×10^{18}	1×10^{14}–1×10^{16}	5×10^{14}–5×10^{18}	5×10^{14}–1×10^{19}
p-Type doping (cm^{-3})	N/A	N/A	N/A	1×10^{15}–1×10^{18}	1×10^{15}–5×10^{16}	3×10^{15}–5×10^{17}	1×10^{16}–5×10^{18}
X-ray rocking curve (arcsec)	–	–	20–60	<20	<20	50–90	20–30
Compositional uniformity (Δx)	<0.002	<0.004	<0.005	<0.002	<0.002	±0.01–0.0005	±0.01–0.0006

MBE technology is now routinely used for multicolor detector arrays. The main challenge of MBE-MCT technology is to grow very high quality layers on low-cost, large-area substrates. The issues that complicate MBE growth on alternative large-area substrates are: lattice mismatch, nucleation phenomena, thermal mismatch, and contamination [56, 57]. Sapphire, Si, and GaAs are some of the low-cost, large-area materials that have been successfully employed as substrates for MCT epitaxial growth [58–61]. However, appropriate buffer layers of CdTe or CZT are required on the alternative substrates before MCT films are deposited. The best MBE-MCT layers grown on buffer/Si substrates achieved thus far exhibit defect densities of 2–5 × 10^6 cm^{-2}. Novel thermal cycle annealing schedules have been used to further reduce the defect density. More effort is necessary to reduce this defect density by at least an order of magnitude to make MBE based materials for many military applications. The ability to grow MCT on large diameter Si wafers will enable low-cost, large-format infrared focal plane arrays.

11.7.5 Comparison of Growth Techniques

The properties of HgCdTe grown by the variety of techniques discussed above are summarized in Table 11.4. The continuing evolution of HgCdTe materials and crystal growth technology is reviewed at a number of conferences each year as well as at other conferences held less frequently. A convenient reference is the proceedings of the U.S. Workshop on Physics and Chemistry of II–VI materials, published by the Journal of Electronic Materials.

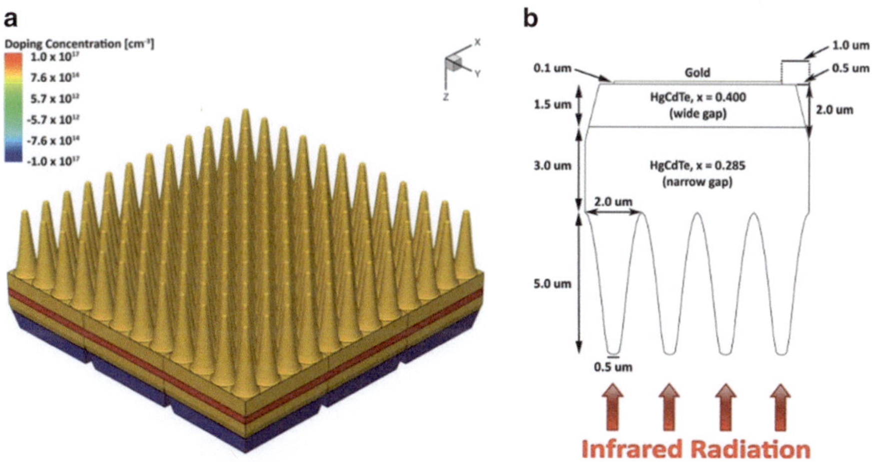

Fig. 11.15 (**a**) Three-dimensional view of a 3 × 3 pixel array with 8 μm pixels incorporating a PT structure and (**b**) Schematic representing the geometry of a single pixel of an array with 8 μm pixels incorporating a PT structure. Reprinted with permission from [16]

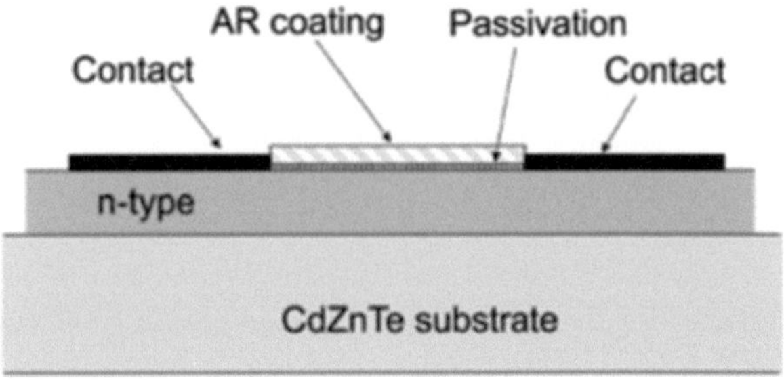

Fig. 11.16 Cross section of a basic HgCdTe photoconductor. Reprinted with permission from [16]

11.8 MCT Infrared Detectors

11.8.1 Photon Trapping (PT) Detectors

The performance of infrared detector can be improved by reducing volume of detector's active region as indicated earlier in Sect. 11.4.1.8. Reducing of the detector material volume can be done via a concept of photon trapping (PT) [10]. Reduction of the dark current should be achieved without degrading the quantum efficiency.

Photon trapping detectors have been demonstrated independently in II–VI [62–64] and III–V [64–67] based epitaxial materials. Semiconductor pillar arrays of sub-wavelength size have been designed and structured as an ensemble of 3D photonic structure units using either top-down or bottom-up process scheme to significantly increase absorption and quantum efficiency. The sub-element architecture of different shapes such as pyramidal, sinusoidal, or rectangular [65] can be obtained depending on the specific etching process. Figure 11.15 shows the photon trap structures with pillars and holes of varying volume fill factors. The samples were fabricated from HgCdTe layers on Si grown by MBE with a cutoff wavelength of 5 μm at 300 K.

HgCdTe-based FPAs are normally backside illuminated where the incident light is absorbed through the transparent cadmium zinc telluride (CdZnTe) substrate on which the detector array is grown. Such conventional detectors suffer from significant reflection losses at the air–HgCdTe or air–CdZnTe interface. Consequently an anti-reflection (AR) coating is incorporated to reduce these losses. However, it is very difficult to engineer an AR coating that works effectively over a wide spectral range. Additionally, an AR coating can introduce stress into a detector limiting its inter-operability or lifespan. An alternative approach is to instead remove the CdZnTe substrate and etch a micro-structured photon trapping (PT) structure directly into the HgCdTe detector absorber layer. The three-dimensional view of PT structure is shown in Fig. 11.16a. When properly engineered, the PT structure will significantly reduce reflection losses, negating the need for an AR coating [68]. Furthermore, by removing device volume, the detector dark current is potentially lowered, allowing higher operating temperatures to be achieved. It has been shown [69] that PT structures offer superior device performance in terms of cross

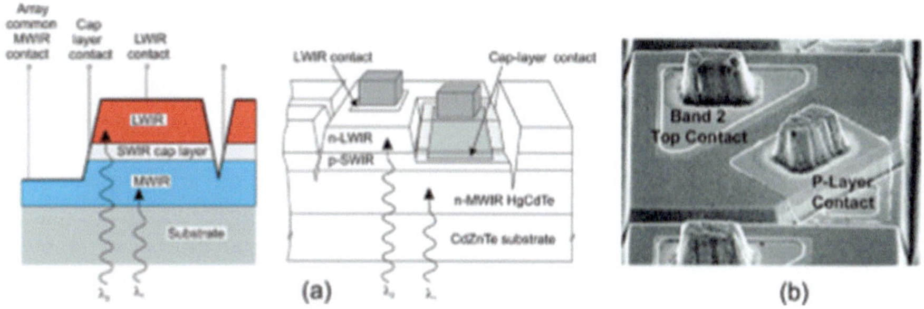

Fig. 11.17 Two-color n–p–n detector structure for simultaneous operating mode: (**a**) cross section schematic and (**b**) an SEM image. Reprinted with permission from [16, 70]

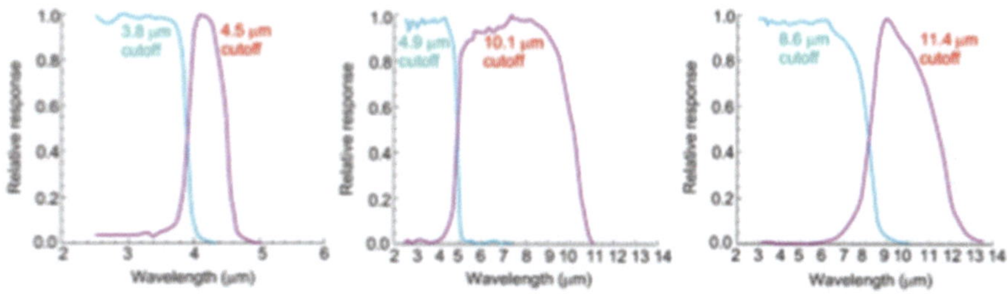

Fig. 11.18 Spectral response curves for two-color HgCdTe detectors in various dual-band combinations of MWIR and LWIR spectral bands. Reprinted with permission from [16]

talk compared to non-PT structures. The optical and electrical characteristics have been investigated with an emphasis on cross talk in mercury-cadmium telluride ($Hg_{1-x}Cd_xTe$) pixel arrays that incorporate a periodic micro-structured surface to enhance broadband absorption in the mid-wave infrared (MWIR) spectral region (0.5–5.0 μm) without use of an AR coating.

Utilization of InAsSb absorber on GaAs substrates instead of HgCdTe absorber enables fabrication of MWIR low-cost, large-format HOT FPAs. Souza and coworkers [65–67] have described research efforts in developing visible to mid-wave (0.5–5.0λm) broadband PT InAsSb-based detectors operating at high temperature (150–200 K) with low dark current and high quantum efficiency. Schematic representation of the geometry of a single pixel of an array incorporating a PT structure is shown Fig. 11.16b.

11.8.2 Barrier Infrared Detectors

The basic HgCdTe photoconductor device is illustrated in Fig. 11.17. The pure n-type material is sandwiched between two metal electrodes. Typical photoconductors are passivated with anodic oxide and anti-reflection coated with zinc sulfide.

The simplest two-color HgCdTe detector, and the first to be demonstrated [16], is the bias-selectable n–p–n triple-layer heterojunction (TLHJ), back-to-back photodiode shown in Fig. 11.18 (capital letter means wider band gap structure). The n-type base absorbing regions are deliberately doped with indium at a level of about $(1–3) \times 10^{15}$ cm^{-3}. A critical step in device formation is connected with the in situ p-type, as-doped layer with good structural and electrical properties to prevent internal gain from generating spectral cross talk. The band gap engineering effort consists of increasing the CdTe mole fraction and the effective thickness of the p-type layer to suppress out-of-band carriers from being collected at the terminal. The sequential-mode detector has a single indium bump per unit cell that permits sequential bias selectivity of the spectral bands associated with operating back-to-back photodiodes. When the polarity of the bias voltage applied to the bump contact is positive, the top (LW) photodiode is reverse biased and the bottom (SW) photodiode is forward biased. The SW photocurrent is shunted by the low impedance of the forward-biased SW photodiode, the only photocurrent to emerge in the external circuit is the LW photocurrent. When the bias voltage polarity is reversed, the situation reverses; only SW photocurrent is available. Switching times within the detector can be relatively short, in the order of microseconds, so detection of slowly changing targets or images can be done by switching rapidly between the MW and LW modes. One bump contact per unit cell, as for single-color hybrid FPAs, is the big advantage of the bias-selectable detector. It is compatible with existing silicon readout chips. The major drawbacks of bias selectable device are as follows: (1) its construction does not allow independent selection of the optimum bias voltage for each photodiode and (2) there can be substantial MW cross talk in the LW detector.

Many applications require true simultaneity of detection in the two spectral bands. Simultaneous dual-band detector architectures require an additional electrical contact to an underlying layer in the multijunction structure of both the SW and LW photodiodes. An implementation of the simultaneous mode using a second indium bump in the unit cell is shown in Fig. 11.18. The most important distinction is the requirement of a second readout circuit in each unit cell. Figure 11.18 shows the Spectral response curves for two-color HgCdTe detectors in various dual-band combinations of MWIR and LWIR spectral bands [16].

Photodetectors are sub-divided into different types based on how electric and magentic fields are developed [71]. Table 11.5 also shows the schematics of detectors, their advantages, disadvantages and performance at different temperatures. Advantages of photoconductive detectors are that they offer low-cost technology and have high responsivity; however, they require bias. The stacked multi-heterojunction photodiode configuration has high dynamic resistance, good quantum efficiency, and fast response. The response time for photoconductor and Dember detector is ≤ 1 ns whereas PEM detector has ≤ 10 ns at 330 K.

Table 11.5 HgCdTe uncooled photodetectors

Mode of operation	Schematic of detector	Advantages	Disadvantages	Performance (optically immersed, $\lambda = 10.6$ μm)
Photoconductor	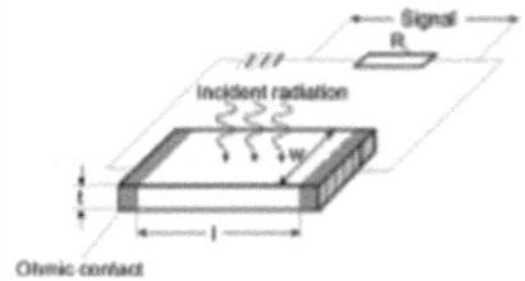	• Low-cost technology • High responsivity	• Bias required	At 300 K: • D^* up to 2×10^8 cm $Hz^{1/2}$ W^{-1} • Response time ≤1 as
				At 200 K: • D^* up to 3×10^9 cm $Hz^{1/2}$ W^{-1} • Response time ≤10 as
PEM detector	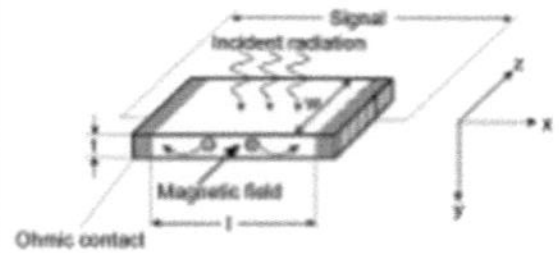	• No bias required • No flicker noise • Very short response time	• Bulky • Low resistance	At 300 K: • D^* up to 1×10^8 cm $Hz^{1/2}$ W^{-1} • Response time ≤1 as
Dember detector	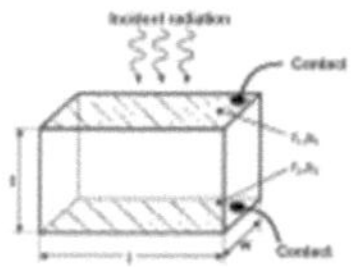	• No bias required • No flicker noise	• Low resistance • Require low resistance contacts	At 300 K: • D^* up to 1.5×10^7 cm $Hz^{1/2}$ W^{-1} • Response time ≤1 as
Photodiode	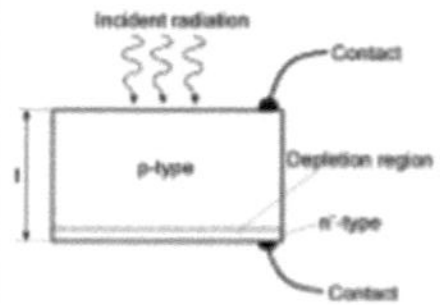	• Fast response • No flicker noise	• Very low dynamic resistance • Low quantum efficiency • Influence of series resistance	At 220 K (multielement array): • D^* up 2×10^8 cm $Hz^{1/2}$ W^{-1} • Response time ≤3 as
Stacked multiheterojunctions photodiode	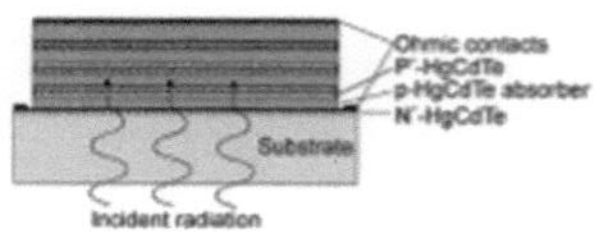	• No flicker noise • Extremely fast response • Can be use from IX to very high frequencies	• High dynamic resistance • Good quantum efficiency	• Generation–recombination limit of performance can be achieved in practice at any wavelength • Practical implementation requires well established epitaxial technology

Reprinted with permission from [71]

11.9 Future Outlook and Summary

The present development of IR photodetectors has been dominated by complex band-gap heterostructures [72–77]. Among various variable bandgap semiconductor alloys, HgCdTe is the only one material covering the whole IR spectral range having nearly the same lattice parameter. The difference of lattice parameter between CdTe ($E_g = 1.5$ eV) and $Hg_{0.8}Cd_{0.2}Te$ ($E_g = 0{:}1$ eV) is $\approx 0.2\,\%$. Replacing small fraction of Cd with Zn or Te with Se can compensate the residual lattice mismatch. The independence of lattice parameter on composition is a major advantage of HgCdTe over any other competing materials for IR detector applications.

Heterojunctions are helpful in achieving high performance in practice. For example, the narrow gap HgCdTe that absorbs IR radiation can be buried encapsulated in wider gap HgCdTe preventing instabilities due to the weak Hg–Te bonds. When background-photon noise is the dominant noise mechanism, the detector is operating in an ideal mode, and is said to exhibit background limited performance (BLIP). BLIP temperature is defined that the device is operating at a temperature at which the dark current equals the background photocurrent, given a field of view (FOV), and a background temperature. The operating temperature of "bulk" intrinsic IR detectors (HgCdTe and PbSnTe) is higher than for other types of photon detectors. Photodiode technology is being vigorously extended to wavelengths beyond 12 μm. In the next few years, photodiodes are expected to replace photoconductors at wavelengths out to about 20 μm. Single element photoconductors used in spectrometers to about 25 μm at liquid nitrogen temperature will continue to occupy that niche application.

HgCdTe technology will continue in future to expand the envelop of its capability because of its excellent properties. Despite serious competition from alternate technologies and slower progress than expected, MCT is unlikely to be seriously challenged for high-performance applications, applications requiring multispectral capabilities and fast response. A vision for future of IR detector technology is photodetectors integrated with powerful hardware performing smart algorithm. We also may see arrays with each pixel sensing the whole IR spectrum.

References

1. A. Rogalski, *Infrared Detectors*, 2nd edn. (CRC Press, Boca Raton, FL, 2011)
2. D.L. Chubb, *Fundamentals of Thermophotovoltaic Energy Conversion* (Elsevier, Amsterdam, The Netherlands, 2007), p. 515
3. C. Downs, T.E. Vandervelde, Progress in infrared photodetectors since 2000. Sensors **13**, 5054–5098 (2013). doi:10.3390/s130405054
4. W. Herschel, Experiments on the refrangibility of the invisible rays of the sun. Phil. Trans. Royal Soc. Lond. **90**, 284–292 (1800)
5. E.S. Barr, The infrared pioneers—II. Macedonio Melloni. Infrared Phys. **2**, 67–73 (1962)

6. T.W. Case, Notes on the change of resistance of certain substrates in light. Phys. Rev. **9**, 305–310 (1917)
7. A. Rogalski, History of infrared detectors. Opto-Electron. Rev. **20**(3), 279–308 (2013). doi:10.2478/s11772-012-0037-7
8. A. Rogalski, K. Adamiec, J. Rutkowski, *Narrow-Gap Semiconductor Photodiodes* (SPIE-The International Society for Optical Engineering, Bellingham, WA, 2000)
9. J. Piotrowski, A. Rogalski, *High Operating Temperature Infrared Photodetectors* (SPIE, Bellingham, WA, 2007)
10. P. Martyniuk, J. Antoszewski, M. Martyniuk, L. Faraone, A. Rogalski, New concepts in infrared photodetector designs. Appl. Phys. Rev. **1**, 041102 (2014)
11. W.S. Boyle, G.E. Smith, Charge-coupled semiconductor devices. Bell. Syst. Tech. J. **49**, 587–593 (1970)
12. M. Razeghi, Current status and future trends of infrared detectors. Opto-Electron. Rev. **6**, 55–194 (1998)
13. P. Capper, C. Maxey, C. Butler, M. Grist, J. Price, Bulk growth of near-IR cadmium mercury telluride (CMT). J. Mater. Sci. Mater. Electron. **15**, 721–725 (2004)
14. M.B. Reine, Fundamental properties of mercury cadmium telluride, in *Encyclopedia of Modern Optics* (Academic, London, 2004)
15. M.B. Reine, *Encyclopedia of Modern Optics 2005* (Academic, London, 2005), pp. 392–402
16. A. Rogalski, HgCdTe infrared detector material: history, status and outlook. Rep. Prog. Phys. **68**, 2267–2336 (2005). doi:10.1088/0034-4885/68/10/R01
17. A. Rogalski, Infrared detectors: status and trends. Prog. Quantum Electron. **27**, 59–210 (2003)
18. G.L. Hansen, J.L. Schmit, T.N. Casselman, Energy gap versus alloy composition and temperature in $Hg_{1-x}Cd_xTe$. J. Appl. Phys. **53**, 7099–7101 (1982)
19. G.L. Hansen, J.L. Schmit, J. Appl. Phys. **54**, 1639 (1983)
20. M.H. Weiler, Magnetooptical properties of $Hg_{1-x}Cd_xTe$ alloys, in *Semiconductors and Semimetals*, ed. by R.K. Willardson, A.C. Beer, vol. 16 (Academic, New York, 1981), pp. 119–191
21. J.P. Rosbeck, R.E. Star, S.L. Price, K.J. Riley, Background and temperature dependent current–voltage characteristics of $Hg_{1-x}Cd_xTe$ photodiodes. J. Appl. Phys. **53**, 6430–6440 (1982)
22. W.M. Higgins, G.N. Seiler, R.G. Roy, R.A. Lancaster, Standard relationships in the properties of $Hg_{1-x}Cd_xTe$. J. Vac. Sci. Technol. A **7**, 271–275 (1989)
23. P.N.J. Dennis, C.T. Elliott, C.L. Jones, A method for routine characterization of the hole concentration in p-type cadmium mercury telluride. Infrared Phys. **22**, 167–169 (1982)
24. W. Lei, J. Antoszewski, L. Faraone, Progress, challenges, and opportunities for HgCdTe materials and detectors. Appl. Phys. Revs. **2**(041303) (2015)
25. P. Norton, Opto-Electron. Rev. **10**, 159 (2002)
26. W.F.H. Micklethwaite, The crystal growth of mercury cadmium telluride, in *Semiconductors and Semimetals*, ed. by R.K. Willardson, A.C. Beer (Academic, New York, 1981), pp. 48–119
27. J.H. Tregilgas, Developments in recrystallized bulk HgCdTe. Progress in Crystal Growth and Characterization of Materials **28**, 57–83 (1994)
28. R. Triboulet, The travelling heater method (THM) for $Hg_{1-x}Cd_xTe$ and related materials. Prog. Cryst. Growth Charact. Mater. **28**, 85–114 (1994)
29. P. Capper, Bulk crystal growth-methods and materials, in *Springer Handbook of Electronic and Photonic Materials*, ed. by S. Kasap, P. Capper (Springer Science, New York, 2006)
30. P. Capper, Bridgman growth of $Cd_xHg_{1-x}Te$: a review. Prog. Cryst. Growth Charact. Mater. **19**, 259–293 (1989)
31. N.K. Dhar, R. Dat, A.K. Sood, Advances in infrared detector array technology, Ch 7 (open access), in *Optoelectronics—Advanced Materials and Devices*, ed. by S.L. Pyshkin, J.M. Ballato (Croatia, InTech, 2013). d (© 2013 Dhar et al., licensee InTech. This is an open access article distributed under the terms of the Creative Commons Attribution License (http://creativecommons.org/licenses/by/3.0), which permits unrestricted use, distribution, and reproduction in any medium, provided the original work is properly cited
32. L. Colombo, R.R. Chang, C.J. Chang, B.A. Baird, Growth of Hg-based alloys by the travelling heater method. J. Vac. Sci. Technol. A **6**, 2795–2799 (1988)

33. P. Capper et al., J. Cryst. Growth **46**, 575 (1979)
34. H.J. Scheel, J. Cryst. Growth **13/14**, 560 (1972)
35. P. Capper, J.J.G. Gosney, U.K. Patent 8115911, 1981
36. P. Capper, Prog. Cryst. Growth Charact. **28**, 1 (1994)
37. P. Capper (ed.), *Bulk Crystal Growth of Electronic, Optical and Optoelectronic Materials* (Wiley, Chichester, 2005)
38. P. Capper, J. Garland (eds.), *Mercury Cadmium Telluride: Growth, Properties and Applications* (Wiley, Chichester, 2011)
39. P. Capper, E. Zharikov, *Handbook on Crystal Growth*, 2nd edn. (Elsevier, Amsterdam, 2015)
40. P. Capper et al., Copyright (1988) reproduced with permission from The Minerals, Metals and Materials Society.
41. P.E. Herning, Experimental determination of the mercury-rich corner of the Hg-Cd-Te phase diagram. J. Electron. Mater. **13**, 1–14 (1984)
42. T.C. Harman, Liquidus isotherms, solidus lines and LPE growth in the TeRich corner of the Hg-Cd-Te system. J. Electron. Mater. **9**, 945–961 (1980)
43. M. Lanir, K.J. Riley, Performance of PV HgCdTe Arrays for 1-14-μm Applications. IEEE Trans. Electron. Dev. ED **29**, 274–279 (1982)
44. C.C. Wang, Mercury cadmium telluride junctions grown by liquid phase epitaxy. J. Vac. Sci. Technol. B **9**, 1740–1745 (1991)
45. G.N. Pultz, P.W. Norton, E.E. Krueger, M.B. Reine, Growth and characterization of p-on-n HgCdTe liquid-phase epitaxy heterojunction material for 11-18 μm applications. J. Vac. Sci. Technol. B **9**, 1724–1730 (1991)
46. T. Tung, M.H. Kalisher, A.P. Stevens, P.E. Herning, Liquid-phase epitaxy of $Hg_{1-x}Cd_xTe$ from Hg solution: a route to infrared detector structures. Mater. Res. Soc. Symp. Proc. **90**, 321–56 (1987)
47. T. Tung, L.V. De Armond, R.F. Herald, P.E. Herning, M.H. Kalisher, D.A. Olson, R.F. Risser, A.P. Stevens, S.J. Tighe, State of the Art of Hg-melt LPE HgCdTe at Santa Barbara Research Center. Proc. SPIE **1735**, 109–31 (1992)
48. https://en.wikipedia.org/wiki/Metalorganic_vapour_phase_epitaxy
49. P. Mitra, F.C. Case, M.B. Reine, Progress in MOVPE of HgCdTe for advanced infrared detectors. J. Electron. Mater. **27**(6), 510–520 (1998)
50. L.G. Hipwood, I.M. Baker, C.L. Jones, C. Maxey, H.W. Lau, J. Fitzmaurice, M. Wilson, P. Knowles, LW IRFPAs made from HgCdTe grown by MOVPE for use in multispectral imaging. Proc. SPIE **6940** (2008)
51. O.K. Wu, Status of HgCdTe MBE technology for IRFPA. *Proceedings of SPIE, 202179*. Optoelectronics - Advanced Materials and Devices, 186, 1993
52. J.P. Faurie, Molecular beam epitaxy of HgCdTe: growth and characterization. Prog. Crystal Growth Charact. **29**, 85–159 (1994)
53. R. Dat, F. Aqariden, W.M. Duncan, D. Chandra, H.D. Shih, In situ spectroscopic ellipsometry for real time composition control of HgCdTe grown by molecular beam epitaxy. Mater. Res. Soc. Symp. Proc. **484**, 613–618 (1997)
54. S.I.C. Irvine, J. Bajaj, Recent progress with in situ monitoring of MCTGrowth. Proc. SPIE **2274**, 24–36 (1994)
55. T.S. Lee, J. Garland, C.H. Grein, M. Sumstine, A. Jandeska, Y. Selamet, S. Sivananthan, Correlation of arsenic incorporation and its electrical activation in MBE HgCdTe. J. Electron. Mater. **29**, 869–872 (2000)
56. W.E. Tennant, C.A. Cockrum, J.B. Gilpin, M.A. Kinch, M.B. Reine, R.P. Ruth, Key issue in HgCdTe-based focal plane arrays: an industry perspective. J. Vac. Sci. Technol. B **10**, 1359–1369 (1992)
57. R. Triboulet, A. Tromson-Carli, D. Lorans, T. Nguyen Duy, Substrate issues for the growth of mercury cadmium telluride. J. Electron. Mater. **22**, 827–834 (1993)
58. E.R. Gertner, W.E. Tennant, J.D. Blackwell, J.P. Rode, HgCdTe on sapphire: a new approach to infrared detector arrays. J. Cryst. Growth **72**, 462–467 (1985)

59. D.D. Edwall, J.S. Chen, J. Bajaj, E.R. Gertner, MOCVD HgCdTe/GaAs for IR detectors. Semicond. Sci. Technol. **5**, 221–224 (1990)
60. N.K. Dhar, C.E.C. Wood, A. Gray, H.Y. Wei, L. Salamanca-Riba, J.H. Dinan, Heteroepitaxy of CdTe on {211} Si using crystallized amorphous ZnTe templates. J. Vac. Sci. Technol. B **14** (3), 2366–2370 (1996)
61. J.M. Peterson, J.A. Franklin, M. Readdy, S.M. Johnson, E. Smith, W.A. Radford, I. Kasai, High-quality large-area MBE HgCdTe/Si. J. Electron. Mater. **35**, 1283–1286 (2006)
62. J.G.A. Wehner, E.P.G. Smith, G.M. Venzor, K.D. Smith, A.M. Ramirez, B.P. Kolasa, K.R. Olsson, M.F. Vilela, HgCdTe photon trapping structure for broadband mid-wavelength infrared absorption. J. Electron. Mater. **40**, 1840–1846 (2011)
63. K.D. Smith, J.G.A. Wehner, R.W. Graham, J.E. Randolph, A.M. Ramirez, G.M. Venzor, K. Olsson, M.F. Vilela, E.P.G. Smith, High operating temperature mid-wavelength infrared HgCdTe photon trapping focal plane arrays. Proc. SPIE **8353**, 83532R (2012)
64. N.K. Dhar, R. Dat, Advanced imaging research and development at DARPA. Proc. SPIE **8353**, 835302 (2012)
65. A.I. D'Souza, E. Robinson, A.C. Ionescu, D. Okerlund, T.J. de Lyon, R.D. Rajavel, H. Sharifi, D. Yap, N. Dhar, P.S. Wijewarnasuriya, C. Grein, MWIR InAs1xSbx nCBn detectors data and analysis. Proc. SPIE **8353**, 835333 (2012)
66. H. Sharifi, M. Roebuck, T. De Lyon, H. Nguyen, M. Cline, D. Chang, D. Yap, S. Mehta, R. Rajavel, A. Ionescu, A. D'Souza, E. Robinson, D. Okerlund, N. Dhar, Fabrication of high operating temperature (HOT), visible to MWIR, nCBn photon-trap detector arrays. Proc. SPIE **8704**, 87041U (2013)
67. A.I. D'Souza, E. Robinson, A.C. Ionescu, D. Okerlund, T.J. de Lyon, R.D. Rajavel, H. Sharifi, N.K. Dhar, P.S. Wijewarnasuriya, C. Grein, MWIR InAsSb barrier detector data and analysis. Proc. SPIE **8704**, 87041U (2013)
68. C.A. Keasler, E. Bellotti, A numerical study of broadband absorbers for visible to infrared detectors. Appl. Phys. Lett. **99**, 091109 (2011)
69. J. Schuster, E. Bellotti, Analysis of optical and electrical crosstalk in small pitch photon trapping HgCdTe pixel arrays. Appl. Phys. Lett. **101**, 261118 (2012)
70. P.R. Norton, Infrared detectors in the next millennium. Proc. SPIE **3698**, 652–665 (1999)
71. J. Piotrowski, A. Rogalski, Uncooled long wavelength infrared photon detectors. Infrared Phys. Technol. **46**, 115–131 (2004)
72. A. Turner et al., Producibility of VIPTM scanning focal plane arrays. Proc. SPIE **2228**, 237–248 (1994)
73. I.M. Baker, Photovoltaic IR detectors, in *Narrow-gap II–VI Compounds for Optoelectronic and Electromagnetic Applications*, ed. by P. Capper (Chapman and Hall, London, 1997), pp. 450–473
74. P. Tribolet, J.P. Chatard, P. Costa, A. Manissadjian, Progress in HgCdTe homojunction infrared detectors. J. Cryst. Growth **184/185**, 1262–1271 (1998)
75. J. Bajaj, State-of-the-art HgCdTe materials and devices for infrared imaging, in *Physics of Semiconductor Devices*, ed. by V. Kumar, S.K. Agarwal (Narosa Publishing House, New Delhi, 1998), pp. 1297–1309
76. V.S. Varavin, V.V. Vasiliev, S.A. Dvoretsky, N.N. Mikhailov, V.N. Ovsyuk, G. Sidorov Yu, A.O. Suslyakov, M.V. Yakushev, A.L. Aseev, HgCdTe epilayers on GaAs: growth and devices. Opto-Electron. Rev. **11**, 99–111 (2003)
77. T.J. DeLyon, J.E. Jensen, M.D. Gorwitz, C.A. Cockrum, S.M. Johnson, G.M. Venzor, MBE growth of HgCdTe on silicon substrates for large-area infrared focal plane arrays: a review of recent progress. J. Electron. Mater. **28**, 705–711 (1999)

ERRATUM TO

Part II
Sulfur

G.K. Ahluwalia (ed.), *Applications of Chalcogenides: S, Se, and Te*,
DOI 10.1007/978-3-319-41190-3

DOI 10.1007/978-3-319-41190-3_12

In the original publication Part II opener, (page 159), Illustrations and text were mistakenly left out of the print and online versions of this book. Please see the next page for the missing content.

The online version of the original book can be found at
DOI 10.1007/978-3-319-41190-3

G.K. Ahluwalia (ed.), *Applications of Chalcogenides: S, Se, and Te*,
DOI 10.1007/978-3-319-41190-3_12

Part II
Sulfur

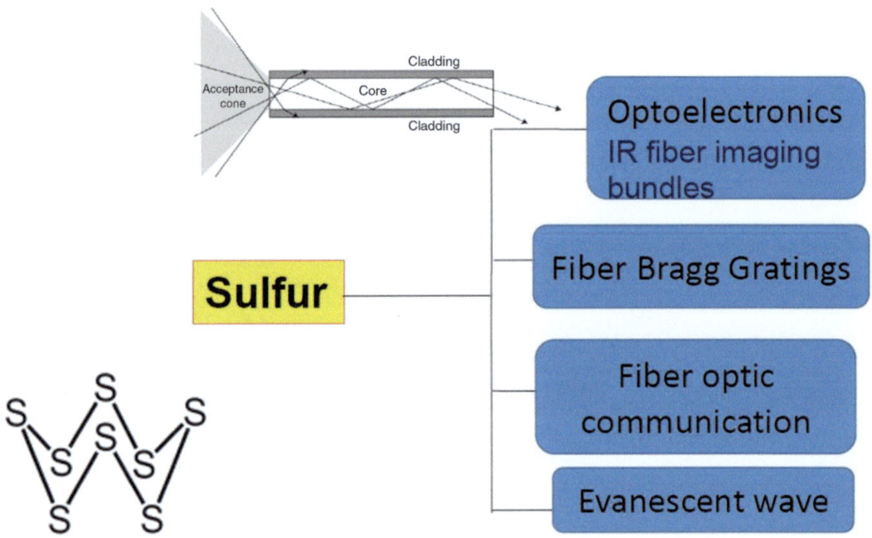

Sulfur- The Element

S - 16 - Ne $3s^2$ $3p^4$

Introduction

Sulfur is a polyatomic non-metallic chemical element with symbol S and atomic number 16. The term sulfur/sulphur originates from Sanskrit sulvere and Latin sulphurium. It has an atomic weight 32.059; 32.076, melts at. 115.21°C; and has a boiling point at 444.61°C. Specific gravity (rhombic) 2.07, (monoclinic) 1.957 (20°C). Sulfur is referred to in bible as brimstone (burn stone) in English. It is widely distributed in nature as pyrites (iron sulfide; FeS), galena (lead sulfide; PbS), spharelite (zinc sulfide; (Zn, Fe)S), cinnabar (mercury sulfide; HgS)), stibnite (antimony sulfide, Sb_2S_3), gypsum (calcium sulphate dihydrate; $CaSO_4{\cdot}2H_2O$), epsom salts (heptahydrate magnesium sulphate; $MgSO_4{\cdot}7H_2O$), celestite (strontium sulphate; ($SrSO_4$), barite (barium sulphate; $BaSO_4$), alunite (potassium aluminium sulfate; $KAl_3(SO_4)_2(OH)_6$) etc.

Sulfur is a pale yellow, odorless, brittle solid, which is insoluble in water but soluble in carbon disulfide. In every state, whether gas, liquid or solid, elemental sulfur occurs in more than one allotropic form or modification. Amorphous or "plastic" sulfur is obtained by fast cooling of the crystalline form. X-ray studies indicate that amorphous sulfur may have a helical structure with eight sulfur atoms per spiral. Crystalline sulfur seems to be made of rings, each containing eight sulfur atoms, amorphous sulfur may have a helical which fit together to give a normal X-ray pattern. Twenty-one isotopes of sulfur are now recognized. Four occur in natural sulfur, none of which is radioactive. Sulfur is essential to life. It is a minor component of fats, body fluids, and skeletal minerals. Sulfur is essential to all living things. It is taken up as sulfate from the soil (or seawater) by plants and algae. It is used to make two of the essential amino acids needed to make proteins. It is also needed in some co-enzymes. The average human contains 140 grams and takes in about 1 gram a day, mainly in proteins.

- Elemental sulfur is produced as a byproduct of removing sulfur-containing contaminants from natural gas and petroleum fertilizers, and in the manufacture of sulfuric acid.

Fig. 1 Sulfur recovered from hydrocarbons in Alberta, stockpiled for shipment in North Vancouver, B.C., Canada. [1]

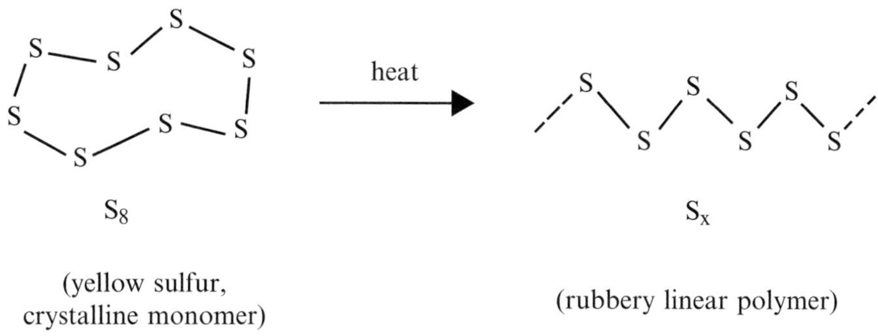

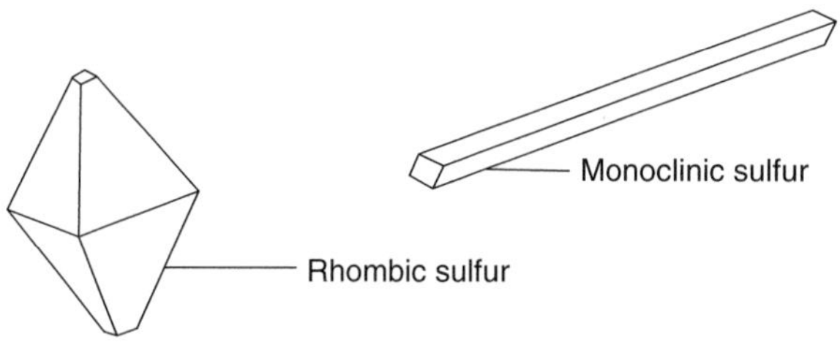

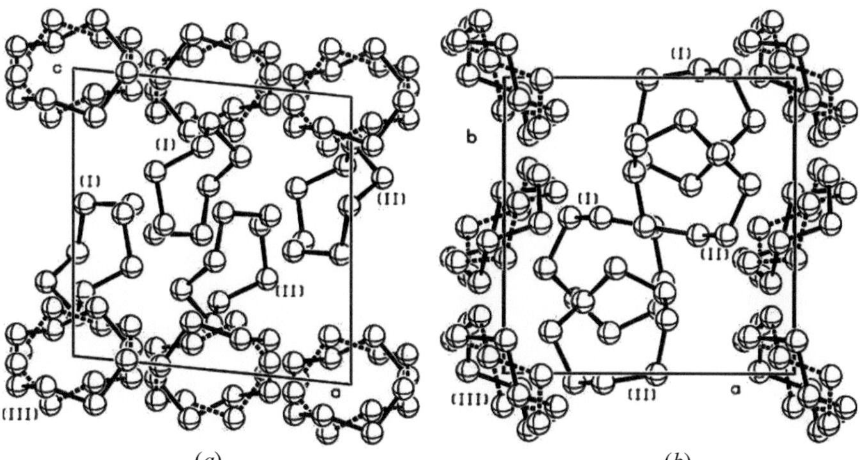

Fig. 2 Illustration of the crystal structure of monoclinic sulfur, viewed (a) along the b axis and (b) along the c axis. The three S8 molecules in the asymmetric unit are denoted as (I) (atoms S1-S8), (II) (atoms S9-S16) and the disordered molecule (III) (atoms S17-S24, S17' – S24') Copyright permission obtained [2].

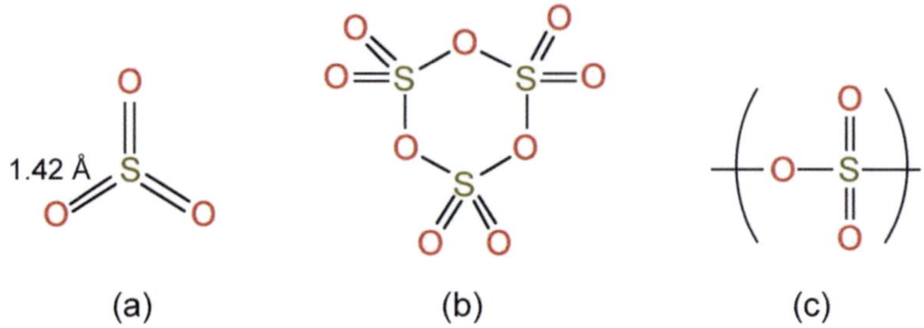

S_2, disulfur

S_2 is the predominant species in sulfur vapour above 720°C. at low pressure (1mm Hg) at 530°C it comprises 99% of vapour species. It is a singlet diradical like dioxygen with an S-S bond length of 188.7 pm. The blue colour of burning sulfur is due to emission of light by the S_2 molecule produced in the flame.

S_3, trisulfur

S_3 is found in sulfur vapour, comprising 10% of vapour species at 440°C and 10mm Hg. It is cherry red in colour. Its structure is bent, similar to ozone, O_3.

S_4, tetrasulfur

This has been detected in the vapour phase but has not been fully characterised. Various forms, (e.g chains, branched chains and rings) have been proposed. The latest view, based on theoretical calculationsis that it has a ring structure.

Solid cyclo-sulfur allotropes

i) cyclo-S_5, pentasulfur - This has not been isolated, but has been detected in molten sulfur.
ii) S_6, cyclo- hexasulfur - This was first prepared by M.R. Engel in 1891 by reacting concentrated HCl with thiosulfate, $HS_2O_3^-$. Cyclo-S_6 is orange-red and forms a rhombohedral crystal. It is called ρ-sulfur, ε-sulfur, Engel's sulfur and Aten's sulfur.
iii) cyclo-S_6.cyclo-S_{10} adduct - This is produced from a solution of cyclo-S_6 and cyclo-S_{10} in CS_2. It has a density midway between cyclo-S_6 and cyclo-S_{10}. The crystal consists of alternate layers of cyclo-S_6 and cyclo-S_{10}. For all the elements this may be the only allotrope which contains molecules of different sizes.
iv) S_7, cyclo- heptasulfur - It is a bright yellow solid. Four (α-, β-, γ-, δ-) forms of cyclo-heptasulfur are known. Two forms (γ-, δ-) have been characterised. The cyclo-S_7 ring has an unusual range of bond lengths of 199.3pm - 218.1pm.
v) **Cyclo-S_8**

α- Sulfur - α-sulfur is the form most commonly found in nature. When pure it has a greenish-yellow colour (traces of cyclo-S_7 in commercially available samples make it appear yellower). It is practically insoluble in water and is a good electrical insulator with poor thermal conductivity. It is quite soluble in carbon disulfide 35.5g/100g solvent at 25°C. It has a rhombohedral crystal structure. This is the predominant form found in "flowers of sulfur", "roll sulfur" and "milk of sulfur". It contains S_8 puckered rings, alternatively called a crown shape. The S-S bond lengths are all 206pm and the S-S-S angles are 108° with a dihedral angle of 98°. At 95.3°C α-sulfur converts to β-sulfur.

β- Sulfur - This is a yellow solid with a monoclinic crystal form. Like the α- form it contains puckered S_8 rings and only differs from it in the way the rings are packed in the crystal. It is less dense than the α-form. It is unusual because it is only stable above 95.3°C, below this it converts to α-sulfur. It can be prepared by crystallising at 100°C and cooling rapidly to slow down formation of α-sulfur. It has a melting point variously quoted as 119.6°C and 119.8°C but as it decomposes to other forms at around this temperature the observed melting point can vary from this. The ca. 119°C melting point has been termed the "ideal melting point" and the typical lower value (114.5°) when decomposition occurs, the "natural melting point".

γ- Sulfur - This form was first prepared by F.W Muthmann in 1890, sometimes called "nacreous sulfur" or "mother of pearl sulfur" because of its appearance, it crystallises in pale yellow monoclinic needles. It contains puckered S_8 rings like α-sulfur and β-sulfur and only differs from them in the way that these rings are packed. It is the densest form of the three. It can be prepared by slowly cooling molten sulfur that has been heated above 150°C or by chilling solutions of sulfur in carbon disulfide, ethyl alcohol or hydrocarbons. It is found in nature as the mineral rosickyite.

cyclo- S_n, (n=9-15, 18, 20) - These allotropes have been synthesised by various methods for example, reacting titanocene pentasulfide and a dichlorosulfane of suitable sulfur chain length, $S_{n-5}Cl_2$. S_{12}, S_{18} and S_{20} can be prepared from S_8.

Unlike α-sulfur and cyclo-S_6 the rings contain S-S bond lengths and S-S-S bond angle that differ one from another.

Cyclo-S_{12} is the second most stable cyclo- allotrope after cyclo-S_8. Its structure can be visualised as having sulfur atoms in three parallel planes, 3 in the top, 6 in the middle and three in the bottom.

Two forms (α-, β-) of cyclo-S_9 are known one of which has been characterised. Two forms of cyclo-S_{18} are known where the conformation of the ring is different. To differentiate these structures rather than use the normal crystallographic convention of α-, β- etc. which in other cyclo-S_n compounds refer to different packings of essentially the same conformer, these two conformers have been termed endo- and exo-.

Solid catena sulfur allotropes - The production of pure forms of catena-sulfur has proved to be extremely difficult. Purity of the starting material and the thermal history of the sample posing major constraints.

(i) **ψ- Sulfur** - This form, also called (confusingly) fibrous sulfur or ω1-sulfur, has been well characterised. It has a density of 2.01g cm^{-3} (α-sulfur 2.069g cm^{-3}) and decomposes around its melting point of 104°C. It consists of parallel helical sulfur chains. These chains have both left and right handed "twists" and a 95pm. The S-S bond length is 206.6pm, the S-S-S bond angle is 106° and the dihedral angle is 85.3°. (Comparable figures for α-sulfur are 203.7pm, 107.8° and 98.3°).

(ii) **Lamina sulfur** - This has not been well characterised but is believed to consist of criss-crossed helices. It is also called χ-sulfur or ω2-sulfur.

Catena sulfur forms - The naming of the different forms is very confusing and care has to be taken to determine what is being described as the same names are used interchangeably.

Amorphous sulfur - This is the quenched product of molten sulfur (>160°C). Its form changes from an initial plastic form gradually to a glassy form, hence its other names of plastic, glassy or vitreous sulfur. It is also called χ-sulfur. It contains a complex mixture of catena-sulfur forms mixed with cyclo-forms.

Insoluble sulfur - This is obtained by washing quenched liquid sulfur with CS_2. It is sometimes called polymeric sulfur, μ-S or ω-S.

φ- Sulfur, fibrous sulfur - This is a mixture of the allotropic ψ- form and γ-cycloS_8.

ω- Sulfur - This is a commercially available product prepared from amorphous sulfur that has NOT been stretched prior to extraction of soluble forms with CS_2. It sometimes called "white sulfur of Das" or super sublimated sulfur.. It is a mixture of ψ-sulfur and lamina sulfur. The composition depends on the exact method of production and the samples history. One well known commercial form is "Crystex". ω- Sulfur is used in the vulcanization of rubber.

λ- Sulfur - This name is given to the molten sulfur immediately after melting. Cooling this gives predominantly β-sulfur.

μ- Sulfur - This name is applied to solid insoluble sulfur and the melt prior to quenching.

π- Sulfur - Dark coloured liquid formed when λ-sulfur is left to stay molten. Contains mixture of S_n rings.

S_∞ - This term is applied to biradical catena- chains in sulfur melts or the chains in the solid.

Applications of elemental Sulfur:
Sulfur is a component of black gun powder, and is used in the vulcanisation of natural rubber and a fungicide. It is also used extensively in making phosphatic fertilizers. A tremendous tonnage is used to produce sulfuric acid. It is used in making sulfite paper and other papers as a fumigant, and in the bleaching of dried fruits. The element is a good electrical insulator. Some important compounds of

sulfur include calcium sulfate, ammonium sulfate, carbon disulfide, sulfur dioxide and hydrogen sulfide. Of these carbon disulfide, hydrogen sulfide and sulfur dioxide particularly should be handled carefully.

References

1. https://en.wikipedia.org/wiki/Sulfur
2. David, Ibberson, Cox & Wood (2006). Acta Crystallographica B62, pp. 953–959.

Index

G.K. Ahluwalia (ed.), *Applications of Chalcogenides: S, Se, and Te*,
DOI 10.1007/978-3-319-41190-3

T

MIX
Papier aus verantwortungsvollen Quellen
Paper from responsible sources
FSC® C105338

If you have any concerns about our products,
you can contact us on
ProductSafety@springernature.com

In case Publisher is established outside the EU,
the EU authorized representative is:
Springer Nature Customer Service Center GmbH
Europaplatz 3, 69115 Heidelberg, Germany

Printed by Libri Plureos GmbH
in Hamburg, Germany